Particle Technology Forum

Core Programming Topic at the 2012 AIChE Annual Meeting

Pittsburgh, Pennsylvania, USA
28 October - 2 November 2012

ISBN: 978-1-62276-737-3

Printed from e-media with permission by:

Curran Associates, Inc.
57 Morehouse Lane
Red Hook, NY 12571

Some format issues inherent in the e-media version may also appear in this print version.

Printed by Curran Associates, Inc. (2013)

For permission requests, please contact AIChE
at the address below.

AIChE
3 Park Avenue
New York, NY 10016-5991

Phone: (203) 702-7660
Fax: (203) 775-5177

www.aiche.org

Additional copies of this publication are available from:

Curran Associates, Inc.
57 Morehouse Lane
Red Hook, NY 12571 USA
Phone: 845-758-0400
Fax: 845-758-2634
Email: curran@proceedings.com
Web: www.proceedings.com

TABLE OF CONTENTS

Hopper Design for Binary Mixtures 1
Ben Freireich, Karl Jacob, Madhusudhan Kodam, Defne Kayrak-Talay

Relationships Between Particle Properties of Fine Powders and Flow Properties Under Low Compaction Stresses 2
Horng Yuan Saw, Clive E. Davies, Guillaume Brisson, Anthony (Tony) H. J. Paterson, James Jones

Investigating the Effects of Particle and Bulk Powder Properties On Microwave Sensor Measurements 3
John S. Austin III, Sip C. Liew, Ryan McDonnell, Pei-Fang Sung, Michael T. Harris

Influence of Physical Factors and Powder Properties On the Accuracy and Precision of Multivariate Calibration Models Using NIR Spectroscopy 4
Leonel Quiñones, Luis Obregon

The Modified Drucker-Prager/Cap Properties of Pharmaceutical Powder Blends: The Influence of MgSt and Acetaminophen Concentrations 5
Shrikant Swaminathan, Ariel Muliadi, Petrus Geldenhuis, Andrew S. Yuk, Carl R. Wassgren, James D. Litster

Quantitative Correlation of the Effect of Process Conditions On the Capping Tendencies of Tablet Formulations 6
Ilgaz Akseli, Adrian Stecula, Xiaorong He, Nadia Ladyzhynsky

Online Particle Size Monitoring During Roller Compaction 7
Rhye Hamey, David Good, Dongsheng Bu, Ziyang Su, Chandra Vema-Varapu

Enantioselective Separation On Chiral Au Nanoparticles 8
Nisha Shukla, Andrew J. Gellman

Specific Targeting Stealth Nanoparticles Coated with Ultra-Low Fouling Peptide 9
Ann K. Nowinski, Lei Zhang, Wei Yang, Shaoyi Jiang

Development of Magnetic Field-Responsive Surfactants 10
Sun Hae Ra Shin, Rajiv Misra, Peter Schiffer, Kyle J. M. Bishop

Deposition of Coated Titanium Dioxide Nanoparticles with Controllable Wetting Properties 11
Mikko Aromaa, Janne Haapanen, Hannu Teisala, Mikko Tuominen, Jurkka Kuusipalo, Milena Stepien, Jarkko J. Saarinen, Martti Toivakka, Jyrki M. Mäkelä

Nanoparticles Surface Design Process for Aggregation and Dispersion Behavior Control in Liquid to Apply Functional Material 12
Hidehiro Kamiya, Motoyuki Iijima

Reduction of a Mechanistic Drying Model of Pharmaceutical Granules for Inclusion in a Population Balance Model with Continuous Growth Term 13
Séverine T. F. C. Mortier, Ingmar Nopens, Thomas De Beer, Krist V. Gernaey

A Systematic Approach for Quantifying the Multi-Particle Interactions in Dry Milling: Rational Function Type Effectiveness Factor 15
Maxx Capece, Rajesh Dave, Ecevit Bilgili

Theoretical Breakage Distribution Functions 16
Priscilla J. Hill

Optimization and Control of Growth Modifier Concentration Profile for Achieving Target Crystal Shape Distribution 17
Aniruddha Majumder, Zoltan K. Nagy

Self Starting and Globally Convergent Parameters Estimation Method for Aggregation-Breakage Population Balance Model Using Genetic Algorithm 18
Akinola Falola, Xue Z. Wang, Antonia Borissova, Jingjing Liu

Fast Monte Carlo Algorithm for Heterogeneous Population Balance Models: Effect of Reactor Dissipation Energy Distribution and Blending Time On Metal Nanoparticle Formation 26
Roberto Irizarry

Computation of Minimum Bubbling Velocity for Large Particles 27
Dimitri Gidaspow, Fang Li

Constitutive Model for the Fluid-Particle Drag Coefficient in Filtered Two-Fluid Models for Gas-Particle Flows 28
Chris Milioli, Fernando Milioli, William Holloway, Sankaran Sundaresan

Predicting Gas-Solid Bubbling Bed Flows Using Continuum Modeling 30
Jennifer Sinclair Curtis, Deepak Rangarajan

Simulation of Mono- and Bi-Disperse Gas-Particle Riser Flow with Quadrature-Based Moment Methods 31
Alberto Passalacqua, Rodney O. Fox
A Constitutive Model for Intermediate, Quasi-Static and Inertial Regimes of Dense Granular Flow 32
Shankar Subramaniam, V. Vidyapati
Minimum Fluidization and Bubbling Properties - the Effects of Fines Content and Pressure 33
Allan S. Issangya, Roy C. Hays, S. B. Reddy Karri, Ray A. Cocco, Ted M. Knowlton
The Primary Particle Diameter of Aerosol Agglomerates & Aggregates From Mass-Mobility Characterization 34
Max Eggersdorfer, Arto J. Gröhn, C. M. Sorensen, Peter H. McMurry, Sotiris E. Pratsinis
Combined On-Line Differential Mobility and Particle Mass Analysis for Determination of Size Resolved Particle Density and Microstructure Evolution 36
Qin Liu, Xiaofei Ma, Michael Zachariah
Roughness Effects On Particle Adhesion 37
Nyah V. Zarate, James D. Litster, Stephen P. Beaudoin
Using Inverse Gas Chromatography to Assess Passivation of High Surface Energy Sites of Milled Pharmaceutical Crystals Via Dry Coating of Nano-Silica 38
Xi Han, Laila Jai Jallo, Daniel To, Chinmay Ghoroi, Rajesh N. Dave
Electrically and Thermally Conductive Oxide Particles Functionalized by Metallic W Atomic Layer Deposition 39
Staci A. Van Norman, Joseph W. Tringe, John Falconer, Alan W. Weimer
Phase Separation and Reaction Mechanism During Agcu Particle Formation by Cosolvent Assisted Pyrolysis Technology 40
Kai Zhong, George Peabody, Elizabeth Blankenhorn, Howard Glicksman, Sheryl H. Ehrman
Synthesis of Visible Light-Sensitive TiO_2 Coated Silica Beads by Cu(II) or Fe(III) Grafting 41
Junichi Ida, Kazumasa Tokumoto, Sho Suzuki, Soshi Watanabe, Tatsushi Matsuyama, Hideo Yamamoto
Experimental Studies in the Characterization Occurring At the Catalytic Surfaces of Silver On Titania Adsorbents 42
Zenda D. Davis, Bruce J. Tatarchuk
Preparation and Assembly of Anisotropically Polymer-Grafted Silica Nanospheres Via Adsorption On Charged Surface 43
Dieu Huong Pham, Ayae Sugawara-Narutaki, Atsushi Shimojima, Tatsuya Okubo
Metal:Metal Oxide Complexes Via Organic Coating Techniques: Understanding Electro-Optical Properties 44
Diane M. Steeves, Jason W. Soares, Jisun Im, Jagdeep Singh, Yiwen Wang, James E. Whitten
Surface Modification of Carbon Hard Sphere Via Surface Initiated Atom Transfer Radical Polymerization 45
Lakhinder Kamboj
Adsorption of Acrylic/Sulfonic Acid Copolymer-Coated Iron Oxide Nanoclusters On Silica Microspheres At High Divalent Salinity 46
Hitesh G. Bagaria, Zheng Xue, Andrew Worthen, Kiyoul Yoon, Bethany Neilson, Victoria Cheng, Jaeho Lee, Christopher W. Bielawski, Keith P. Johnston
Role of Aeration with Standpipe Flow 47
Ray Cocco, Ted Knowlton, S. B. Reddy Karri
Agglomeration and De-Agglomeration of Wetted Solids: A Summary of Recent Findings 48
Christine M. Hrenya, Carly M. Donahue, Robert H. Davis
A Novel Mppic Methodology for Numerical Simulations of Gas-Solids Flows 49
Rahul Garg, Jean-Francois Dietiker, Mehrdad Shahnam
Implementation and Verification of Population Balance Model Using FCMOM in CFD Codes 50
Emadoddin Abbasi, Hamid Arastoopour
DNS of Particle-Laden Gas Flow with Four-Way Coupling 51
John B. McLaughlin, Goodarz Ahmadi, Hojjat Nasr
ECVT Imaging of 3-D Flow Structures of Irregular Geomteric Multi-Phase Flow Systems 52
Aining Wang, Qussai Marashdeh, L.-S. Fan
Multi-Scale Modeling and Validation of Twin Screw Granulation Processes 53
Dana Barrasso, Rohit Ramachandran
Application of Euler-Lagrange Approach to Predict the Multiphase Flow with Simultaneous Heat and Momentum Transfer, and Solidification Predicted by A Shrinkage Unsolidified Core Model 54
Guilherme A. N. Ricardo, Waldir P. Martignoni, Dirceu Noriler, Henry F. Meier
Prediction of Single Drop Granule Formation Mechanisms for Regime Separated Granulation 61
Heather N. Emady, Defne Kayrak-Talay, William C. Schwerin, James D. Litster

Optimization and Scale up of High Shear Wet Granulation Process 62
Saurabh Sarkar, Apurva More, Bodhisattwa Chaudhuri
The Effect of Intra-Granular Binder Distribution On Single Granule Strength 63
Steven Dale, James D. Litster, Carl Wassgren
Evaluation of the Effect of Blade Angle On a High Shear Wet Granulation 64
David A. Mota-Aguilar, Carlos Velázquez
Boron Composite Nanoparticles for Enhancement of Biofuel Combustion 65
Neng Wang, Jacob Hanberry, Srinibas Karmakar, Sumanta Acharya, Kerry M. Dooley
Morphology and Molecular Weight Control of Core-Shell Polymer Nanoparticles 66
Ibrahim A. El-Hedok, Jennifer M. O'Donnell
Thermogravimetric Analysis of the Kinetics of the Reaction of Alkoxysilane with Silica 67
Yuk Yu Law, Donald L. Feke, Ica Manas-Zloczower
Atom Transfer Radical Polymerization Synthesis to Yield Copper-Free Block Copolymers 68
Justin R. Adams, Surya K. Mallapragada
Engineered Quantum Dots for Light Selective Polymer Films with High Photoluminescence and Photostability 69
Abdul Mumin, Golam Moula, Jenna Allan, Paul A. Charpentier
Self-Cleaning and Anti-Microbial Polyurethane Coatings 70
Adria F. Lotus, Yixing Tang, Koosha Azhie, Paul A. Charpentier
Magnetoresistive Conductive Polyaniline - Barium Titanate Nanocomposites with Negative Permittivity 71
Xi Zhang, Suying Wei, Zhanhu Guo
COAL Direct Chemical Looping Process: Metallurigical Coke and PRB COAL Conversions 72
L.-S. Fan
Important Length Scales in Dense Gas-Particle Flows 73
Stefan Radl, Chris Milioli, Fernando Milioli, Sankaran Sundaresan
CFD Simulation of Pharmaceutical Particle Drying in Bubbling Fluidized Bed Dryers At Different Scales 74
Hamid Arastoopour, Jungkee Jang
Dynamics of Particle Cratering Using High Speed Gas Jets 75
Casey Lamarche, Jennifer Sinclair Curtis, Shibo Kuang, Aibing Yu
Particle Scale Modelling and Analysis of Fluid Flow and Heat Transfer in Fluidised Beds 76
Zongyan Zhou, Qinfu Hou, Aibing Yu
Measurement of Density Fluctuations in a Vertically Oscillated Granular Bed At the Onset of Vibrofluidization 80
Hugo Caram, James F. Gilchrist, Kenneth J. Ford
Quantitative Predictions of Instabilities in Solid-Particle Flows 81
Peter P. Mitrano, Christine M. Hrenya
Layer Melt Crystallization of Long-Chain Aldehydes From Hydroformylation Processes 82
Thorsten Beierling, Gabriele Sadowski
Hybrid Filtration-Drying-Dissolution Device for Compact Pharmaceutical Manufacturing System 83
Shin Yee Wong, Jie Chen, Allan S. Myerson
Design and Characterization of a Continuous Stirred Tank Crystallizer 84
Gary Morris, Guangyang Hou, Mark Barrett, Brian Glennon
X-Ray Compatible Microfluidic Platforms for On-Chip Structure Determination of Membrane Proteins 91
Sudipto Guha, Ashtamurthy Pawate, Sarah L. Perry, Paul J. A. Kenis
Investigation of a Counter-Current Crystallization Process for Separation of Systems Forming Solid Solutions 92
Erik Temmel, Sebastian Wloch, Uwe Müller, Detlef Grawe, Robert Eilers, Heike Lorenz, Andreas Seidel-Morgenstern
Designing a Fit-for-Purpose Agglomeration Method for Pharmaceutical Crystals 94
Chenchi Wang, Daniel Hsieh, Mario Hubert, Chiajen J. Lai
Utilization of Population Balance Models to Develop a Continuous Crystallization Process; Determination of Crystallization Kinetics 95
Christopher L. Burcham, Michael A. Lovette, Christopher S. Polster, Sean K. Bermingham, Hassan Mumtaz
Utilization of Population Balance Models to Develop a Continuous Crystallization Process; Process Design Using Optimization 96
Christopher L. Burcham, Michael A. Lovette, Christopher S. Polster, Sean K. Bermingham, Hassan Mumtaz
Improved Mixing in Dry Catalyst Impregnation Using a Double Cone Blender: An Experimental and Computational Approach 97
Yangyang Shen, M. Silvina Tomassone

DEM Validation of Tablet Motion in a Vibratory Coating Pan 98
Rahul Kumar, Carl Wassgren
Axial Dispersion of Non-Spherical Particles in Horizontal, Rotating Cylinders 99
Guang Lu, James Third, Christoph R. Müller
Application of Enhanced DEM Simulations for the Improvement of a Given Tablet Coating Process 100
Gregor Toschkoff, Sarah Just, Georg Scharrer, Klaus Knop, Peter Kleinebudde, Dejan Djuric, Andreas Altmeyer, Adrian Funke, Johannes G. Khinast
A Renewal Theory Approach to Understanding Inter-Particle Coating Variability 102
Ben Freireich, Jianfeng Li
Sphero-Cylindrical Particle Stresses in a Vertical Axis Mixer 103
Xia Hua, Jennifer Sinclair Curtis, Bruno C. Hancock, William R. Ketterhagen, Carl R. Wassgren
Experimental Validation of a 3-D Finite Element Method Roll-Compaction Model 104
Ariel R. Muliadi, Carl R. Wassgren, James D. Litster
A Favre-Averaged Turbulence Model for Gas-Particle Flows 105
Thomas J. O'Brien
L-Valve Operation Under High Temperature Conditions 107
Dawei Wang, Hyung Kim, Samuel Bayham, L.-S. Fan
Study of Distribution and Stability of Solid Particles in a Spouted Bed 108
Ghanim M. Alwan Sr., Muthanna H. Aldahhan, Shreekanta Aradhya, Stoyan Nedeltchev
A Novel Bubble-Emulsion Model for Modeling Gas-Solid Bubbling Fluidized Beds 113
Xi Gao, Li-Jun Wang, Cheng Wu, You-Wei Cheng, Xi Li
Particle Segregation in a Liquid Fluidized Bed Incorporating Inclined Channels Subjected to Centrifugal Forces 114
Kevin P. Galvin, James E. Dickinson
An Analysis Method for Fluidized Bed Flow Behavior Based On Equivalent Circuit Model 117
Masanori Ishizuka, Yuji Yoshie, Yasuki Kansha, Atsushi Tsutsumi
A Numerical Model for the Flow Including Dense Solids with Large Size Difference 118
Takuya Tsuji, Kyohei Higashida, Yoshitomo Okuyama, Toshitsugu Tanaka
Atomic Layer Deposition As a Catalyst Synthesis Technique for Nickel Nanoparticles 119
Troy D. Gould, Alia M. Lubers, John L. Falconer, J. Will Medlin, Alan W. Weimer
Microfluidic Synthesis of Electro-Catalytically Active Nanoparticles 120
Anand Kumar, Klavs F. Jensen
High-Precision Manufacturing of Core-Shell Nanoparticles for Catalysis 121
J. Ruud Van Ommen, Barbara M. Kampa, David Valdesueiro, Gabrie Meesters, Freek Kapteijn, Michiel T. Kreutzer
Effects of Synthetic Variations On the Dispersion of Pt Catalysts Supported On Aluminum-Modified TiO_2 122
Rebecca E. Olsen, Calvin H. Bartholomew, Brian F. Woodfield
Polymer and Surfactant-Templated Synthesis of Nanostructured ZnS Nano- and Micro-Spheres in a Spray Pyrolysis Reactor 123
Munish Sharma, Parham Rohani, Sha Liu, Mark Kaus, Mark T. Swihart
Synthesis and Phase Transfer of Oxidation Resistant Copper Nanoparticles to Various Organic Solvents 125
Manish Niranjan, Jayanta Chakraborty, Aabid Shaikh
Towards a Refined Theory for Liquid Transfer in Dense and Dilute Particle Beds 126
Stefan Radl, Matthew Girardi, Charles Radeke, Johannes G. Khinast, Sankaran Sundaresan
Responsive Discrete Element Model for Preparation of Advanced Materials by Mechanical Milling Using an Attritor Mill 127
Priya R. Santhanam, Edward L. Dreizin
Assessment of Collision-Less Vs. Collisional Simulation of Lunar Regolith Ejection During Spacecraft Landing 129
Kyle Berger, Philip Metzger, Christine M. Hrenya
Development of an Accurate Granular Normal Contact Force Model Based On a Non-Newtonian Liquid Filled Dashpot 130
Ebrahim Alizadeh, Francois Bertrand, Jamal Chaouki
Population Density Dependence of Mechanical Multi-Particle Interactions 131
Maxx Capece, Rajesh Davé
Multi-Grid Monte Carlo Algorithm for the Simulation of Multi-Component Population Balance Models 132
Roberto Irizarry

Pressure Signal Analysis and Bubble Behavior in a 3D Fluidized Bed. Comparison of Experimental Data and Computational Results 133
Alireza Abbasi, Mohammad Ashraful Islam, Paul E. Ege, Hugo H. De Lasa
The Comparison of Large Square Cyclone and Centrifugal Cyclone by Numerical Simulation 134
Zhihong Liu, Toshiyuki Suda
Filtered Drag Models for Euler-Lagrange Simulations of Gas-Solid Flow 143
Stefan Radl, Sankaran Sundaresan
Effects of Pressure and Fines Content On Bubble Diameter in a Fluidized Bed Studied Using Fast X-Ray Tomography 145
J. Ruud Van Ommen, Gerrit Brouwer, Evert Wagner, Robert F. Mudde
Heat Transfer From Hot Particles to Cold Particles in Downward-Flow Binary Solids Systems 146
Yuji Yoshie, Masanori Ishizuka, Guoqing Guan, Chihiro Fushimi, Atsushi Tsutsumi
Experimental and Numerical Investigations On the Electrostatics in the Downer Reactor 147
Yongpan Cheng, Darren Yan Jun Lau, Guoqing Guan, Chihiro Fushimi, Atsushi Tsutsumi, Chi-Hwa Wang
Predicting Minimum Fluidization and Bubbling Velocities for Geldart Type A Particles 148
Janine Galvin, Sofiane Benyahia
Hybrid Bilayer Hollow Submicron Particles with Carbon and Silica Nanoshells Fabricated Through a Template Free Process and Their Applications As Multifunctional Materials 149
Vijay T. John, Yingqing Wang, Jingjing Zhan, Bhanukiran Sunkara
Creating Conductive Copper-Silver Bimetallic Nanostructured Coatings Using a High Temperature Reducing Jet Aerosol Reactor 150
Munish Sharma, Raymond D. Buchner, William J. Scharmach, Vasilis Papavassiliou, Mark T. Swihart
Combustion-Driven One Step Synthesis of Palladium Hybrid Films in a High Temperature Reducing Jet Reactor for Hydrogen Storage and Purification 152
Munish Sharma, Di Qi, Raymond D. Buchner, William J. Scharmach, Mark T. Swihart
Pilot-Scale Flame Spray Aerosol Reactor for Manufacture of ZrO_2 Nanoparticles 154
Arto J. Gröhn, Sotiris E. Pratsinis, Karsten Wegner
Spray Granulation of Nanopowders 155
Arto J. Gröhn, Karsten Wegner
Brownian Dynamics Simulation of Recognition Kinetics Between Lock and Key Colloidal Particles 156
Jun Liu, Ronald G. Larson
Measurement of Crystal Morphology: A New Method Based On Reflection of Light From Crystal Faces 157
Jayanta Chakraborty, Debasis Sarkar, Abhishek Singh
Metal Nanoparticle Growth by Molecular Dynamics 158
Beat Buesser, Sotiris E. Pratsinis
Interfacial Tension in Multiphase, Highly Supersaturated Aqueous Solutions for $CaCO_3$ Polymorphs 161
Ben Schroeder, Sean T. Smith, Derek D. Harris
Fabrication of Highly Ordered and Close Packed Colloidal Crystals From Large Microparticles 162
Melissa H. Lash, Kaladhar Kamalssanan, Shu Li, Joseph McCarthy, Steven R. Little
Small and Wide-Angle X-Ray Scattering for in Situ Monitoring of Particle Size Evolution During Platinum Nanoparticle Synthesis 163
Samuel St. John, Naiping Hu, Dale W. Schaefer, Anastasios Angelopoulos
A Numerical Study of Granular Shear Flows of Rod-Like Particles Using the Discrete Element Method 164
Yu Guo, Carl R. Wassgren, William R. Ketterhagen, Bruno C. Hancock, Benjamin D. James, Jennifer Sinclair Curtis
Experimental Validation of a Fast, 3D Physics-Based Cellular Automata Model of Granular Shear Flows 165
Martin C. Marinack Jr., C. Fred Higgs III
Slurry Flow Modeling Using Eulerian/Lagrangian CFD 166
Kuochen Tsai
Density and Velocity Distribution in Pneumatic Conveying System 167
Ankita Deshmukh, Manaswita Bose
Electrostatic Charging in Granular Flow 169
Daniel J. Lacks, R. Mohan Sankaran
Quantitative Validation for Frictional Granular Flows 170
Adel F. Alenzi, Jiaming Cheng, Martin C. Marinack Jr., C. F. Higgs III, Joseph McCarthy
Extending and Filling the Gaps in Solids Flow Theory 171
Kerry D. Johanson
Onset of Bubbling in Geldart Group A Powders 172
Clive E. Davies, Horng Yuan Saw

Effects of Internals and Solid Loading On the Bubble Dynamics in a Slurry Bubble Column ... 173
Moses Kagumba, Yasser Abdulaziz, Muthanna Al-Dahhan
Minimum Fluidization Velocity in a 3D Fluidized Bed Modified with an Acoustic Field ... 184
David R. Escudero, Theodore J. Heindel
Phosphorescent Tracing to Study Solids Mixing in Fluidized Beds ... 185
Mike Wormsbecker, Craig McKnight, Andrew Mezo, Darwin Kiel, Xiaotao Bi, John R. Grace
Magnetic Resonance Imaging (MRI) of Jets in Gas-Solid Systems ... 195
Maximilian Koehl, Guang Lu, James Third, Klaas Pruessmann, Christoph R. Müller
Magnetic Particle Tracking in Spouting and Bubbling Fluidized Beds ... 197
Jack Halow, Stuart C. Daw
Effect of Fluidized Bed Scale and Bed Height On Flow Regimes Transition Using Optical Probe and Pressure Transducer Measurements ... 199
Faraj M. Zaid, Muthanna H. Al Dahhan
Coupling of Crystallizers for Efficient Enantioseparation - Comparison of Two Different Process Strategies ... 200
Matthias Johannes Eicke, Guillaume Levilain, Martin Peter Elsner, Andreas Seidel-Morgenstern
PBE-Aided Design of Slurry Milling Process to Reduce Particle Size of Intermediate and Active Pharmaceutical Ingredients ... 202
Carla Luciani, Kevin D. Seibert, Daniel Jarmer, Eric Moher
Development of a High Shear Crystallization Process for a Pharmaceutical Intermediate ... 203
Brenda Remy, Amit Joshi, Junying Fan, Jason Sweeney, Kenneth Natalie, Aghogho Pedro, Shawn Pack
Developing a Selective Crystallization Process for a Complex Reaction System ... 204
Chenchi Wang, Truc Vu, Lucius Rossano
Crystal Nucleation Control Using Microscopic Online Imaging ... 205
Tod Canty
Precipitation Kinetics of Aluminum Solids Formation During the Caustic Side Solvent Extraction (CSSX) Process ... 206
Rebecca K. Toghiani, Punith P. Naik, Jeffrey S. Lindner, Laura T. Smith, Larry Pearson
Percolation Segregation in a Sheared Bed of Particles (Poster) ... 207
Kevin J. Brennan, Carl R. Wassgren
Oxidation Mechanism of Metal Particles Derived From Thermal Analysis ... 208
Shasha Zhang, Edward L. Dreizin
Lattice Boltzmann Method (LBM) Simulations of Gas-Solid Flows Containing Non-Spherical Particles ... 209
Yong Chen, James Third, Christoph R. Müller
Photocatalytic Activity of the Liquid Flame Spray Generated Titanium Dioxide Nanoparticle Coating On a Lacquered Metal Surface ... 210
Miia Luhtalampi, Mikko Aromaa, Kimmo Lahtonen, Oskari Kangasniemi, Janne Haapanen, Mika Valden, Jyrki M. Mäkelä
Fundamental Aspects of the Carbonation and Calcination Reactions and the Synthesis of Stable, Calcium-Based CO_2 Sorbents ... 211
Marcin Broda, Agnieszka Kierzkowska, Christoph R. Müller
Effect of Ignition Energy On Dust Explosion Behavior: Phenomena of Under-Driving and Over-Driving ... 213
Niansheng Kuai, Weixing Huang, Bing Du, Jingjie Yuan, Zongshan Li, Yuan Gan, Jingyi Tan
Effect of the Flow Field of An Internally Spout-Fluid Bed On Dispersion of Glass Fibre for Thermoplastic Composites Manufacture ... 221
Yuan Zong, Xiaogang Yang, Gance Dai, Ling Zhao
Increasing the Clay Content and Size of Iron Ore Pellets Reduces Their 'Dustiness' ... 224
Joe Halt
Selective Collection of Fine Particles by Water Drops ... 225
Kathika R. Liyanaarachchi, Grant B. Webber, Kevin P. Galvin
Deposition of Non-Spherical Particles in a Bifurcating Airway ... 230
Poom Bunchatheeravate, Jennifer Sinclair Curtis
Biomass Size Reduction of Microalgae ... 231
Rajasekhar Anumolu, Leonard F. Pease III
Experimental Investigation On the Influence of the Moisture Content On Coal Dust Explosion ... 232
Jingjie Yuan, Bing Du, Weixing Huang, Niansheng Kuai, Jingyi Tan
Impact of Operating Conditions On Physical Properties in an Agitated Filter Dryer ... 240
Ekneet Sahni, Bodhisattwa Chaudhuri
A New Type of Unimolecular Polymeric Nanoparticle: Janus Double-Brush Copolymers ... 241
Yukun Li, Efrosyni Themistou, Jiong Zou, Biswa P. Das, Marina Tsianou, Chong Cheng

Three-Dimensional Simulation of a Solid-Liquid Flow by a Lagrangian-Lagrangian Approach 242
Xiaosong Sun, Mikio Sakai, Yoshinori Yamada
Computational Investigation of Nanoparticle Synthesis by Flame Spray Pyrolysis 250
Arto J. Gröhn, Sotiris E. Pratsinis, Karsten Wegner
Numerical Parametric Investigation of Breakage in a Ball Mill 251
Matthew J. Metzger, Benjamin J. Glasser
Liquid Distribution in Wet Fluidized Beds 252
Matthew Girardi, Stefan Radl, Sankaran Sundaresan
Growth of Metal Nanoparticles by Molecular Dynamics 254
Beat Buesser, Sotiris E. Pratsinis
Sphero-Cylindrical Particle Dynamics and Stresses in a Vertical Axis Mixer 256
Xia Hua, Jennifer Sinclair Curtis, Bruno C. Hancock, William R. Ketterhagen, Carl Wassgren
Thermite Reaction Based On Nano-Al and Iodine Pentoxide: Ignition and Combustion Characterization 257
Guoqiang Jian, Snehaunshu Chowdhury, Jingyu Feng, Michael Zachariah
Development of a Boundary Model for Complex Systems in the Discrete Element Method 258
Yusuke Shigeto, Mikio Sakai
Biosensing with Plasmonic Silver Nanoparticles 267
Christoph O. Blattmann, Georgios A. Sotiriou, Sotiris E. Pratsinis
The Polydispersity of Primary Particles in Aggregates Undergoing Sintering 273
Max Eggersdorfer, Sotiris E. Pratsinis
Simulation of Flow Behavior of Particles in a Magnetic Liquid-Solid Fluidized Bed with Uniform Magnetic Field 274
Shuyan Wang, Lixin Wei, Jinsen Gao, Xingying Lan
Template-Directed Synthesis of Porous $CoFe_2O_4/Al_2O_3$ Reactive Structure with Atomic Layer Deposition for Solar Thermochemical Fuel Production 275
Darwin Arifin, Victoria J. Aston, Xinhua Liang, Anthony H. McDaniel, Alan W. Weimer
Gas Phase Synthesis of Nickel Catalyst Nanoparticles Via Atomic Layer Deposition 276
Troy D. Gould, Alia M. Lubers, J. Will Medlin, Alan W. Weimer, John L. Falconer
Dissipative Particle Dynamics Simulation of Sonication-Mediated Particle Interactions 277
Shu Li, Melissa H. Lash, Steven R. Little, Joseph McCarthy
Mixing of Solid Particles in Coarse Particle Fluidized Beds 278
Haihong Fan, Yanxin Chen, Hui Li, Jinhe Fan, Yong Min
Recovery of Iron From Gold Mine Tailings by Decentralized Magnetization Roasting-Magnetic Separation 279
Yanxin Chen, Yanfei Yao, Bo Zhao, Yong Min
Theoretical and Experimental Study On the Improvement of Powder Bulk Density 280
Yanxin Chen, Feng Wu, Yaru Hu, Haihong Fan, Yong Min
Thin Film, Big Difference – Atomic Layer Deposition Functionalized Oxide and Polymer Particles 281
Staci A. Van Norman, Joseph W. Tringe, John L. Falconer, Alan W. Weimer
Stabilization of Pt Catalyst Via Deposition of Ultrathin Porous Films 282
Alan Izar, Sandro Hutter, John L. Falconer, Alan W. Weimer
Continuous Pharmaceutical Hot-Melt Extrusion and Hot-Die Face Pelletizing 283
Daniel Treffer, Andreas Eitzlmayr, Clemens-Michael Smola, Gerold Koscher, Eva Roblegg, Johannes G. Khinast
Predictive Modeling of Hot Melt Extruders 284
Andreas Eitzlmayr, Daniel Treffer, Gudrun Hoerl, Sarah Windhab, Gerold Koscher, Johannes G. Khinast
Foam Granulation Extrusion: A Novel Method to Continuous Wet Granulation of Powder Drug Formulations 285
Michael R. Thompson, Sharleen Weatherley, Rohan Pukadyil, Paul Sheskey
Powder Feeder Pairing with Downsteam Continuous Mixer 287
William E. Engisch Jr., Fernando J. Muzzio
Powder Lubrication in a Continuous Mixing Process 288
Daniel O. Blackwood, Bruce C. Macdonald, Joseph Kushner Iv, Bruno C. Hancock, Yang A. Liu
Effect of Paddle Shape On Flow of Powder Through Feed Frames On Rotary Tablet Presses 289
G. Singh, J. Givand, P. Rajniak, R. Chern, S. D. Reynolds, I. C. Sinka
Supply Chain Analysis of Continuous Pharmaceutical Manufacturing 291
Arul Sundaramoorthy, James M. B. Evans, P. I. Barton
A Smart Oil Spill Dispersant Formulation for Reduced Environmental Impact and Consumption 292
Courtney A. Ober, Ram B. Gupta
A New Approach to Prepare Polystyrene(Pst)/$CaCO_3$ Core-Shell Composite Nanoparticles 293
Yang Hu, Guanzhi Cheng, Wenjiao Ji, Yiming Jiang, Chengyou Kan

Fabrication and Characterization of Porous Alumina Nanolayers Via Molecular Layer Deposition 294
Alan Izar, Sandro Hutter, John L. Falconer, Alan W. Weimer

Inkjet Printing and Spraying of Composite Calcium Alginate Particles with Solid Inner Inclusions 295
Jiri Dohnal, F. Štìpánek

Investigation of Specific Particle Adhesion Towards Model and Biological Substrates 297
Viola Tokarova, Frantisek Stepanek

Dynamic Encapsulation of Hydrophilic Nanomaterials in Hydrophobic Polymer-Based Particles by Simple Emulsion Technique 305
Shaowen Ji, Devesh Srivastava, Nathan J. Parker, Ankush A. Gokhale, Ilsoon Lee

Influence of [Alpha]-Zrp Particle Size and Crystalline Level On Dust Explosion Suppression Effectiveness 306
Diana Castellanos, Andres F. Mejia, Agustin Diaz, Chad Mashuga, Victor H. Carreto, Zhengdong Cheng, M. Sam Mannan

Efficient Parallelization of Large Scale Fluidized Bed Flow Computations 307
Amit Amritkar, Surya Deb, Danesh Tafti

Efficient Euler-Lagrangian Models for Large Scale Fluidized Beds 308
Shitanshu Ratilal Gohel, Shailesh Ozarkar, Ulrich Becker, Markus Braun, L. Srinivasa Mohan

Erosion of Granular Beds - Particle-Resolved Simulations 309
Jos Derksen

Multiphase Simulation of a Commercial Fluidized Catalytic Cracking Regenerator 310
Samuel M. Clark, Dale Snider, Ray Fletcher

Computational Analysis of the Air Injection Effect On the Regeneration Process in FCC 319
Bumjoon Cha, Hyungtae Cho, Sung Real Son, Suk-Joon Kim, Il Moon

Industrial Fluidized Bed Polyethylene Reactor Temperature Modeling in Condensed Mode Operation 321
Yefeng Zhou, Lijin Zhou, Zhengliang Huang, Jingdai Wang, Binbo Jiang, Yongrong Yang

Effect of Particle Size On Hydrogen Production by Reaction of Liquid Water and Aluminum Powders 322
Hongqi Nie, Mirko Schoenitz, Edward L. Dreizin

Aqueous Solubility, Octanol-Water Partition Coefficients and Carbon Adsorption of 1, 3, 3-Trinitroazetidine (TNAZ) 323
Veera Boddu, Carmen Costales-Nieves, Rayan Zilner, Reddy Damavarapu

Combustion Dynamics of Aluminum in Turbulent Flows 324
Amy Corcoran, Vern K. Hoffmann, Edward L. Dreizin

Evaluation of Predictive Capabilities for Pyrotechnics Using Different Thermodynamic Software Packages(INVITED) 325
Jan Puszynski, Paul Redner, Karl Oyler, Christopher Csernica, Jesse Sabatini, Jay Poret, Anthony Shaw, Neha Mehta, Gartung Cheng, Deepak Kapoor

Direct Calculation of 1-Octanol–Water Partition Coefficients From Adaptive Biasing Force Molecular Dynamics Simulations 326
Navendu Bhatnagar, Ganesh Kamath, Isaac Chelst, Jeffrey J. Potoff

Techno-Economic Evaluation of Integrated Renewable Acetic Acid, Glycolaldehyde, and Acetol Production From Biomass-Derived Pyrolysis Oil 327
André B. De Haan, Caecilia R. Vitasari, G. Wytze Meindersma

Clean Pyrolytic Sugars Solution 328
Marjorie Rover, Patrick Johnston, Laura Jarboe, Robert C. Brown

5-HMF Production in Piperylene Sulfone: A Bifunctional Solvent for Reaction and Separation 329
Christopher J Butch, Jakob Crowe, Gregory Marus, Pamela Pollet, Charles L. Liotta, Charles A. Eckert

Enhanced Compost-Drying Process of Sewage Sludge with Carbonized Sludge 330
Nobusuke Kobayashi, Yoshinori Itaya, Akira Suami, Yanliang Li, Masakazu Sawai, Hisashi Hamabe

In -Situ Precipitation of $CaCO_3$ within Pulp Fibers Enhances Optical and Printing Properties of Paper 335
Pradeep Kumar, Yuvraj Singh Negi, Surendra Pal Singh

A Preliminary Study On Uncertainties in CFD Simulations of a Pilot-Scale Circulating Fluidized Bed Riser 347
Tingwen Li, Aytekin Gel, Mehrdad Shahnam, Madhava Syamlal

Filtered Model for Energy Transport in Gas-Solid Flows 348
Kapil Agrawal, William Holloway, Sankaran Sundaresan

A Numerical Method for Large-Scale Dense Particulate Flows Based On an Improved MP-PIC Method and Extended Lattice-Boltzmann Scheme 349
Radompon Sungkorn, Jos Derksen

Parametric Analysis of the Effects of Particle Properties On the Granular Flows of Cylindrical Particles ... 350
Yu Guo, Carl R. Wassgren, William R. Ketterhagen, Bruno C. Hancock, Benjamin D. James, Jennifer Sinclair Curtis

Coupled CFD and DEM Simulation of Fluid-Granular Systems ... 351
Dalibor Jajcevic, Charles Radeke, Georg Scharrer, Johannes G. Khinast

Numerical Study On Large Scale Discrete Element Modeling for Fine Particles in a Fluidized Bed ... 353
Mikio Sakai, Naoto Sekimura, Tatsuya Itoi

A Physics-Based Cellular Automata Approach for Modeling Industrial Granular Flow Problems ... 360
Martin C. Marinack Jr., C. Fred Higgs III

Dissipative Particle Dynamics Simulations of Nanopowder Dispersion for Applications in Solid Propellants ... 361
Naida Lacevic, David Reese, Lori Groven, Darren Naud, Steven F. Son

Numerical Modeling and Simulation Analysis of Casting of Energetic Material in a Projectile Controlled by the Active Cooling/Heating Solidification and Mechanical Vibration Technologies ... 362
Ruslan S. Mudryy, Laurentiu Nastac

Squeeze Flow Rheometry ... 372
Dilhan Kalyon, Halil Gevgilili, Hansong Tang

Propellant Dough Screening using Squeeze Flow Characterization ... 379
Michael Fair, Robin Crownover, Joseph Rokosz

Scale-up of Capsule Filling Operations Using a Nozzle-Dosator Machine ... 380
Marcos Llusá, Eva Faulhammer, Stefano Biserni, Vittorio Calzolari, Simon Lawrence, Georg Scharrer, Johannes G. Khinast

Determination of Residence Time Distribution in Rotary Calciner ... 383
Yijie Gao, Alberto Cuitino, Benjamin J. Glasser, Fernando J. Muzzio, Marianthi G. Ierapetritou, Jean Beeckman, Natalie A. Fassbender, William G. Borghard

Studying the Effects of Fluidized Bed Scale On the Solid Dynamics Using Sophisticated Optical Probe ... 385
Faraj M. Zaid, Muthanna H. Al Dahhan

Residence Time Distribution As Predicted by a Kinematic Model for Mass Flow Powder Discharge From a Conical Hopper ... 386
Luke Fullard, Clive E. Davies, Graeme Wake

Effect of Water and Magnesium Stearate On the Strength of Ceramic Compacts ... 387
Sarang Oka, Alisar Tuncer, Alberto Cuitino, Fernando Muzzio

Measurement of Particle Impact Velocities in Vibrationally Fluidized Granular Flows ... 388
Kamyar Hashemnia, Amirhossein Mohajerani, Jan. K. Spelt

Rate-Based Separation by Exploiting Cohesive Interactions ... 390
Diana M. Liévano, Joseph McCarthy

Design of Better Particle Coatings with Respect to Attrition Resistance ... 391
Gabrie Meesters, Giacomo Perfetti, P. Van Hee

Improved Recovery and Dissolution of Poorly Water-Soluble Drug Nanoparticles From Dried Nanocomposite Microparticles ... 392
Anagha Bhakay, Mohammad Azad, Rajesh N. Dave, Ecevit Bilgili

One-Step Fabrication of Agent-Loaded Biodegradable Microspheroids for Drug Delivery and Imaging Applications ... 393
Michael J. Heslinga, Gabriella Willis, Omolola Eniola-Adefeso

Particle Engineering Via Dry Coating of Micronized API Powders for Improved Dissolution of Directly Compacted Tablets with High Drug Loading ... 394
Xi Han, Chinmay Ghoroi, Rajesh N. Dave

Controlled Release From Self-Assembled Polymer Microparticles That Are Responsive to pH and Temperature ... 395
James M. Myrick, Sitaraman Krishnan, Frederick A. Sexton

Wet Coating of Geldart-C Type Particles in a Rotating Fluidized Bed in a Static Geometry ... 396
Philippe Eliaers, Juray De Wilde

Solvent-Free Beta-Carotene Nanoparticle Manufature ... 398
Phong Huynh, Paul Takhistov

Relationship Between Solution Metastable Limit and Solubility ... 399
Michael Tomasini, M. Silvina Tomassone

Measuring the Energetic Heterogeneity of Pharmaceutical Powders and the Effects On Powder Processing and Formulation ... 400
Daniel J. Burnett, Jiyi Khoo

Formation of Itraconazole-Succinic Acid Cocrystals by CO_2 Antisolvent Cocrystallization ... 401
Courtney A. Ober, Ram B. Gupta

Preparation of Co-Ground Drug-Superdisintegrant Nanosuspensions As a Precursor to the Production of Fast Dissolving Surfactant-Free Nanocomposite Microparticles 402
Mohammad Azad, Afolawemi Afolabi, Rajesh Davé, Ecevit Bilgili

Wet-Media Milling of Colloidal Drug Suspensions Stabilized by Means of Charged Nanoparticles 403
Michael Juhnke, Edgar John

Interactive Mixtures for Dissolution Rate Enhancement of Cefuroxime Axetil 405
Namita Dalal, Ira Shea Buckner

Formation of Stable Nanocarriers by Tuning Pharmaceutically Active Ingredient Properties Via an in Situ Salt Precipitation 407
Nathalie M. Pinkerton, Arnaud Grandeury, Andreas Fisch, Jörg Brozio, Bernd Riebesehl, Robert K. Prud'Homme

Application of a Kinetics Model to Simulate Biomass Gasification Using Oxygen and Steam 408
Sudhanya Banerjee, Jordan Tiarks, Song-Charng Kong, Terry Meyer, Robert C. Brown, Brian Anderson

Gasifier Performance Modeling Including Mass Transfer Effects 409
Ronald W. Breault, Phil Nicoletti

Effects of Steam and Oxygen On Syngas Composition Derived From Biomass Gasification 410
Cuong Van Huynh, Song-Charng Kong

Economical Study of Bioparticles Used for Denitrification in Fluidized Bed Bioreactors (FBBR) 411
Ahmed Khiray Eldyasti, George Nakhla, Jingxu Zhu

Modeling Effects of Drag Force On Biomass Particles in a Fast Pyrolysis Reactor 435
Qingang Xiong, Song-Charng Kong

Effects of Acid-Washing Treatment On Pyrolysis Products of Biomass in a Downer Reactor 436
Tongli Ding, Songgeng Li, Wenli Song, Weigang Lin

Historical Review of Pneumatic Conveying and Solids Processing World Wide 437
George Klinzing

Electrostatic Charging of Particles in a Straight Pipe 441
Poom Bunchatheeravate, Yusuke Fuji, Jennifer Sinclair Curtis, Shuji Matsusaka

Power Considerations of Pneumatic Convey Systems 442
Jonathan O. Thorn

Effect of Outlet Gas Velocity and Solids Loading On Erosion in 2nd Stage Cyclones 451
S. B. Reddy Karri, Ray Cocco, Ted M. Knowlton

An Experimental Study of Uniflow Cyclone Design Parameters 452
U. Muschelknautz, Paul Pattis, Michael Reinalter, Michael Kraxner, Tobias Kofler, Matthias Aigner, Gregor Eichwalder

Influence of the Inlet Vane Geometry On the Uniflow Cyclones Performance 453
M. Kraxner, U. Muschelknautz, S. Wechner, S. Ackermann, V. Greif, J. Bolda

Toward Design of Semiconductor Ternary Quantum Dots with Optimal Optoelectronic Function 454
Sumeet C. Pandey, Xu Han, Dimitrios Maroudas

Surface Mineralization and Characterization of Palladium Nanoparticles On Genetically Engineered Tobacco Mosaic Virus (TMV) Templates 455
Alexander Freer, Lucas Guarnaccio, Kristin Wafford, Johanna Smith, Jayne Steilberg, James Culver, Michael Harris

Infrared Active Photocatalysis Through Morphological Control of Semiconductor Nanoparticles/Metal Tip Heterostructures 456
Doh C. Lee, Chaewon Pak

Tunable Localized Surface Plasmon Resonances in Tungsten Oxide Nanocrystals 457
Karthish Manthiram, A. Paul Alivisatos

Laser- Drawn Features On Nanoparticle Films 458
Sanjeev Kumar Kandpal, Kody Allcroft, Michael D. Mason, Douglas W. Bousfield, David J. Neivandt

Controllable Synthesis of Cu_{2-X} E (E=S, Se) Nanocrystals with Localized Surface Plasmon Resonance and Investigation of Conductivity of NC Thin Films 466
Xin Liu, Xianliang Wang, Mark T. Swihart

Nanostructured Energetic Materials for Insensitive Munitions 467
Victor Stepanov, Hongwei Qiu, Ashok Surapaneni

Spraying Effect On Particle Size and Morphology in Recrystallization of RDX Using Supercritical Carbon Dioxide Anti-Solvent 468
Byung-Chul Lee, Youn-Woo Lee, Hyoun-Soo Kim

Synthesis and Characterization of Nanostructured Pyrophoric Materials 469
Lauren Armstrong, Chris Haines, Zac Doorenbos, Kendall Mills, Deepak Kapoor

Synthesis and Characterization of Energetic Nanomaterials for Propulsion Applications 470
William Lewis, Christopher Bunker, Elena Guliants, Shiral Fernando, Marcus Smith, Barbara Miller, Venroy Watson, Xifan Wang

Structure – Reactivity Relationship: Milling-Induced Exothermic Nano-Materials 471
Khachatur V. Manukyan, Ya-Cheng Lin, Alexander Mukasyan
Simulation of the Breakage of Bonded Agglomerates in a Ball Mill 472
Matthew J. Metzger, Benjamin J. Glasser
A Microhydrodynamic Analysis of Wet Stirred Media Milling for Production of Stable Drug Nanoparticle Suspensions 473
Afolawemi Afolabi, Ecevit Bilgili
Jamming in Granular Systems 474
Shrikant Swaminathan, Carl Wassgren, Jennifer Sinclair Curtis, Bruno C. Hancock, William R. Ketterhagen
Experimental and Computational Investigation of Breakage of Pharmaceutical Spherical Particles 475
Shivangi S. Naik, Bodhisattwa Chaudhuri, Ramesh B. Malla
Microwave Assisted Breakage Through Quartz Inversion in Low Grade Iron Ore 476
Gurulaxmi Srikakulapu, Anshuman Parida, Pradip Banerjee, Mark Denys, T. Venugopalan
High Shear Rotor-Stator Wet Milling for Drug Substance: Expanding Capability and Improving Scalability 477
Allison Harter, Luke Schenck, Ivan Lee, Luis Sierra, Vincent Capodanno, Aaron Cote
Impact of Multi-Particle Interactions On the Attainment of A Self-Similar Particle Size Distribution During Dense-Phase Dry Milling 478
Ecevit Bilgili, Maxx Capece, Rajesh N. Dave
Molecular Simulation Studies of Quantitative Correlation Between the Stability of Drug Nanocrystals Suspension System and Certain Properties of Both the Drug API and the Stabilizer 480
Wusheng Zhu, Frank Romanski, M. Silvina Tomassone
The Properties of Various Coating Polymers and Their Effects On Surface Frictional Behavior As Coatings On Pharmaceutical Tablets 481
Padma Narayan, Karen Balwinski, Debora Holbrook, Madhusudhan Kodam, Meaghan Blake, Karl Jacob
Surface Modification of Pollen Shape Carriers for Improved Dry Powder Inhalation Efficiency 482
Hung Loong Giam, Thi Quynh Ngoc Nguyen, Raymond Lau
Encapsulation of Laccase in Chitosan Micro-Carriers by Two-Fluid and Three-Fluid Nozzle 483
Ondrej Kaspar, Frantisek Stepanek
Engineered Biodegradable Janus Particles for Drug Delivery 490
Jennifer Winkler, Frank Romanski, Maria S. Tomassone
Computer Simulation of a Full Chemical Looping Combustion System 491
James M. Parker
Oxygen Uncoupling Enhanced Iron Oxides for Solid Fuel Gasification – Oxygen Carrier Development, Reactive Testing, and Process Analyses 492
Feng He, Nathan Galinsky, Yan Huang, Fanxing Li
Three Dimensional Simulation of Catalytic Cracking Reaction in an Industrial Scale Riser Using a 10-Lump Kinetic Model 493
Ariane Barbosa, Gabriela Cantarelli Lopes, Leonardo Machado Da Rosa, Milton Mori, Waldir Pedro Martignoni
Development of Rotary Kiln Furnace for Iron Reduction and Slag Separation 502
Hui Li, Wen-Bin Yang, Wei Zhao, Xiaodan Fan, Qiang Zhi, Xingyan Zhang, Yong Min
Gas-Solid Flow Hydrodynamics Characteristics in a Circulating Fluidized Bed Riser with External Heat Exchanger 503
Guoliang Song, Xiaozhe Mou, Yunkai Sun, Haigang Wang, Qinggang Lv
Investigation the Gas-Solids Distribution in a Large-Scale Circulating Fluidised Bed with Electrical Capacitance Tomography 504
Haigang Wang Sr., Guorong Zhao, Guizhi Qiu, Yunkai Sun Sr., Qinggang Lu, Wuqiang Yang
Circulating Fluid Bed Riser Computational Fluid Dynamics (CFD) Model Validation with Electrical Capacitance Volume Tomography (ECVT) 505
Dirk Van Essendelft, Justin Weber, Tingwen Li
Tunable Assembly of Metallic and Magnetic Nanoparticle Clusters in DMSO Solvent Systems 506
Jennifer N. Duggan, Pranav S. Vengsarkar, Christopher B. Roberts
Structure and Properties of Surface Modified Magnetic Hollow Silica Spheres 507
Pavel Kovacik, Zuzana Kremlackova, Marek Soltys, Frantisek Stepanek
Distinguishing Between Amine and Carboxylate Bonds to Iron Oxide Nanoparticles 515
Myles A. Ikenberry, Keith L. Hohn
Magnetic High Density Polyethylene Nanocomposites Reinfored with in-Situ Synthesized Fe@FeO Core-Shell Nanoparticles 516
Qingliang He, Jiahua Zhu, Suying Wei, Zhanhu Guo
Two-Step Gas Phase Synthesis of Janus-Like and Core-Shell Fe_xO_y-TiO_2 Composite Nanoparticles 517
Juha Harra, Juha-Pekka Nikkanen, Mikko Aromaa, Heikki Suhonen, Mari Honkanen, Saara Heinonen, Erkki Levänen, Jyrki M. Mäkelä

Blade Design Effect in Powder Phenomena and Residence Time Distribution Inside the Feed Frame 518
Daniel Mateo, Fernando J. Muzzio, Carlos Velázquez, Rafael Mendez
Study the Effect of the Low Shear Continuous Tumble Mixer Parameters On Powder Phenomena, Residence Time, and Blend Uniformity 519
Miguel Florian, Carlos Velázquez, Rafael Mendez
Numerical Study of Solid Particle Axial Mixing in a Fixed Cylindrical Drum with Rotating Paddles 520
Zafer Zeren, Hervé Neau, Pascal Fede, Olivier Simonin, B. Descales, W. Stephen
Percolation Segregation in a Sheared Bed of Particles 527
Kevin J. Brennan, Carl R. Wassgren
Characterization of in-Situ near-Infrared Imaging to Aid Pharmaceutical Powder Blend Formulation Development 528
Juan G. Osorio, Gina Stuessy, Gabor J. Kemeny, Fernando J. Muzzio
Segregation and Mixing of Granular Mixtures in Heap Formation 529
Yi Fan, Paul B. Umbanhowar, Julio M. Ottino, Richard Lueptow
Mixing of Silica Aerogel (Cabot Nanogel) Particles of Different Sizes 530
Robert Pfeffer, Ding Wang
Assessing Powder Segregation Potential by near Infrared (NIR) Spectroscopy and Correlating Bench Scale Segregation Tendency to Tabletting Performance 531
Xiaorong He, Xi Han, Nadia Ladyzhynsky, Rich Deanne
Understanding and Preventing Complex Segregation Behavior 532
Kerry D. Johanson
Lead-Free Nanosolders and Their Application for Nanowire Assembly and Joining 533
Fan Gao, Zhiyong Gu
Electromechanically Driven Nonlinear Dynamics of Voids in Metallic Thin Films Under Anisotropic Mechanical Stress 534
Dwaipayan Dasgupta, Georgios I. Sfyris, Dimitrios Maroudas
Atomic Layer Deposition Enabled Synthesis of Nanostructured Composite $BiFeO_3/CoFe_2O_4$ Thin Films for Multiferroic Applications 535
Calvin D. Pham, Jane P. Chang
Atomic Layer Deposition of AlN Thin Films As Gate Dielectrics for Wide Bandgap Semiconductors 536
Ya-Chuan Perng, Jane P. Chang
Observations, Formation Mechanisms and Reduction Methods of 3-Dimensional Cross-Slip Dislocations in Silicon Single Crystal 537
Do Won Song, Hyo Kim
Numerical Methods for Solving Mode-Related Problems in Photonic Devices 542
Meng-Mu Shih
Analyzing the Dynamics of the Horizontal Ribbon Growth Process for Solar Silicon 543
Parthiv Daggolu, Andrew Yeckel, Jeffrey J. Derby
Melt Growth of CZT with Convex Interfaces VIA A Bell-Curve Gradient Freeze Profile 544
Nan Zhang, Andrew Yeckel, Jeffrey J. Derby
Examining Correlations of Electrostatic Discharge Sensitivity and Electrical Conductivity of Composite Energetic Materials 545
Eric Collins, Chelsea Weir, Michelle Pantoya
Nanoscale Aluminum, Silicon, and Aluminum/Silicon Mixture Formation for Use in Pyrotechnic Applications 546
Mark Miller, Zac Doorenbos, Matt Puszynski, Lauren Armstrong, Deepak Kapoor, Jan Puszynski
On Mechanism of Thermal Initiation in Fully-Dense 2Al.3CuO Nanocomposite Powder 553
Rayon Williams, Mirko Schoenitz, Alexandre Ermoline, Edward L. Dreizin
Particle Design of HMX Using Supercritical Anti-Solvent Processes 554
Bumjoon Seo, Youn-Woo Lee, Byung-Chul Lee, Hyoun-Soo Kim
Nanostructured Aluminum-Fluorocarbon Particles for Energetic Applications 555
Travis R. Sippel, Steven F. Son, Lori J. Groven
Combustion of Nanoscale Silicon with Polytetrafluoroethylene 556
Brandon C. Terry, Lori J. Groven, Paul Redner, Youdong Lin, Alexander Mukasyan, Deepak Kapoor, Steven F. Son
DEM-CFD Simulation of the Drying of Heterogenous Particles in a Wall-Heated Rotated Drum 557
Deliang Shi, Joseph McCarthy
Drying of Biomass Pellets: Experimental Study and Comparison of the Performance of a Conventional Fluidized Bed and a Rotating Fluidized Bed in a Static Geometry 558
Philippe Eliaers, Juray De Wilde

Application of Global Sensitivity Analysis As Preparatory Step for Reduction of a Drying Model of Pharmaceutical Granules 560
Séverine T. F. C. Mortier, Ingmar Nopens, Thomas De Beer, Krist V. Gernaey
Investigating Contact Drying in An Agitated Filter Dryer: Experiments and Simulations 562
Ekneet Sahni, Bodhisattwa Chaudhuri
Modelling Convective Drying of Iron Ore Pellets 563
Anna-Lena Ljung, Staffan Lundstrom
Nano-Al and Nano-Al/CuO Thermite Under Rapid Heating: A Mechanistic Study 564
Guoqiang Jian, Nicholas Piekiel, Michael Zachariah
Nano-Structured Aluminum Powders with Modified Protective Surface Layers 565
Shasha Zhang, Mirko Schoenitz, Edward L. Dreizin
Silicon-Based Nanocomposites As Novel Illuminants (INVITED) 566
Paul Redner, Karl Oyler, Christopher Csernica, Jesse Sabatini, Jay Poret, Anthony Shaw, Neha Mehta, Gartung Cheng, Deepak Kapoor
Measuring Thermal Histories Using Fragile Glass Crystallization 567
Shashank Vummidi Lakshman, Mirko Schoenitz
Aluminum-Based Reactive Materials with Customized Combustion Characteristics 575
Yasmine Aly, Edward L. Dreizin
Water Processed Nanothermite Inks for Direct Deposition On Low Energy Initiators 576
Robert R. Nellums, Steven F. Son, Richard M. Kellett, Craig J. Boucher, Lori J. Groven
Author Index

(19a) Hopper Design for Binary Mixtures

Ben Freireich, Karl Jacob, Madhusudhan Kodam and Defne Kayrak-Talay, Engineering Sciences - Solids Processing, The Dow Chemical Company, Midland, MI

At the previous several year's conferences our group presented work (Jacob and Vasquez, 2009, "The Role of Fines in Controlling the Flow of Bulk Solids", AIChE Conference, Session 56, Talk a; Freireich, Jacob, and Kodam, 2010, "Flowability of Binary Mixtures: Effects of Fines Concentration", AIChE Conference, Session 544, Talk c; Freireich, Jacob, Kodam, and Kayrak-Talay, 2011, "Flowability of Binary Mixtures: Effects of Fines Concentration", AIChE Conference, Session 357, Talk a) elucidating the effects of fines on the flowability of bulk solids. Since those presentations, even further studies have been conducted scrutinizing these effects. Specifically, using a Schulze ring shear tester, key parameters such as unconfined yield strength, angle of internal friction, and effective angle of internal friction as function of consolidation stress were determined for narrow size distributions of coarse and fine materials. Experimental data and analysis will be presented. In particular, for this talk we discuss the consequences of these material effects on the resulting hopper design.

(19b) Relationships Between Particle Properties of Fine Powders and Flow Properties Under Low Compaction Stresses

Horng Yuan Saw[1], Clive E. Davies[1], Guillaume Brisson[2], Anthony (Tony) H.J. Paterson[1] and James Jones[1], (1)School of Engineering and Advanced Technology, Massey University, Palmerston North, New Zealand, (2)Riddet Institute, Massey University, Palmerston North, New Zealand

In the food and pharmaceutical industries, significant amounts of process material are in the form of fine powder. Fine powders in general do not flow well because of their cohesive nature, and various characterization methods have been used to predict powder flowability. We present here results for thirteen model lactose powders, in which Hausner ratio and powder flow function data, i.e. compaction stress/powder strength data, have been correlated with size descriptors, as well as with each other. Particle size distribution was measured by the laser diffraction method (Mastersizer 2000, Malvern Instruments Ltd, UK); the key size descriptors used were surface-volume diameter, d_{32}, and span, $(d_{90}-d_{10})/d_{50}$. Hausner ratio, ρ_{tap}/ρ_B, was determined with tapped density, ρ_{tap}, measured according to a GEA Niro standard for dairy powders at 1250 taps in accordance with the European Pharmacopoeia; loose bulk density, ρ_B, was based on New Zealand Standard 3111. An annular shear cell (Brookfield Engineering Laboratories, USA) operated at low compaction stresses, in the range 0.31–4.85 kPa, was used to obtain powder flow functions, and hence estimates of particle cohesion, C, and Jenike flow index, σ_c/σ_y; σ_c is the major consolidating stress and σ_y is the unconfined yield stress. Hausner ratio and C increased with decreasing d_{32} and increasing span, while σ_c/σ_y increased with increasing d_{32} and decreasing span. A plot of Hausner ratio versus σ_c/σ_y at 4–6 kPa suggested a power correlation, and the relationship between Hausner ratio and C appeared linear.

(19c) Investigating the Effects of Particle and Bulk Powder Properties On Microwave Sensor Measurements

John S. Austin III[1], Sip C Liew[2], Ryan McDonnell[2], Pei-Fang Sung[1] and Michael T. Harris[3], (1)Chemical Engineering, Purdue Universtiy, West Lafayette, IN, (2)Chemical Engineering, Purdue University, West Lafayette, IN, (3)School of Chemical Engineering, Purdue University, West Lafayette, IN

In solids manufacturing, it is desirable to know the properties of the processed materials in real-time for effective process and quality control. Though many laboratory techniques have been well developed to measure particulate properties, most are not suited for rapid, on-line, non-destructive measurements. Many successful techniques to measure properties *on-line*rely on the use of low-power electromagnetic radiation. Most notable among these is the use of near infrared (NIR) radiation. Although NIR has been employed successfully in many situations, it does suffer from some significant drawbacks. For instance, NIR spectra are often very complicated and require chemometric software to extract useful information. Additionally, NIR frequently suffers from poor sampling efficiency due to the shallow penetration depth of NIR waves and the small size of most fiber-optic probes.

Microwave dielectric sensing is an alternative to other, more common process monitoring techniques. Microwave sensors do not require chemometric software to decode their outputs and possess significantly improved sample penetration. However, as has been reported recently, microwave sensors, like NIR sensors, may be sensitive to particle size and other particle properties. This study was designed to examine these claims as well as investigate the effects of other particulate properties not already correlated to microwave sensor measurements. In total, over 15 particulate and bulk powder properties were investigated and their effects on microwave sensor measurements were elucidated.

(19d) Influence of Physical Factors and Powder Properties On the Accuracy and Precision of Multivariate Calibration Models Using NIR Spectroscopy

Leonel Quiñones, Chemical Engineering, University of Puerto Rico - Mayaguez Campus, Mayaguez, PR and Luis Obregon, Chemical Engineering, University of Puerto Rico Mayaguez, Mayaguez, PR

During the last decade the study and implementation of NIR spectroscopy has increased exponentially giving a valuable advance to the science especially in the study of powder behavior. The pharmaceutical industry is moving toward the use of continuous production enabling the use of continuous powder mixing. This last step involves the use of important sensors to measure homogeneity of mixing online which has been a great challenge. NIR spectrometer plays an important role in this matter and that is why a continuous deep study is necessary in order to be completely sure of the degree of prediction that can be achieved under different physical conditions. Considering this subject different calibration models were made in a complete range of concentration from 0 to 100% w/w for Naproxen as API. The excipient used was Lactose Monohydrate with Magnesium Stearate 1%. The mixtures were prepared with intervals of 1 % from 0 to 30% and with intervals of 5 % from 30% to 100%. These powders have different properties making the mixture a challenge for the NIR. The powder mixing was done initially in a V-blender and the mixing time for all the mixtures were the same. Then the powders were introduced in the continuous mixer rotated at different RPM and NIR spectra were taken with two different NIR spectrophotometers at the output stream and samples were taken to be analyzed and compared with UV spectrophotometer. Precision and exactitude results had variability that depends on the cohesion of the mixtures, average particle size and type of NIR.

(19e) The Modified Drucker-Prager/Cap Properties of Pharmaceutical Powder Blends: The Influence of MgSt and Acetaminophen Concentrations

Shrikant Swaminathan[1], **Ariel Muliadi**[2], Petrus Geldenhuis[2], Andrew S Yuk[2], Carl R. Wassgren[1] and James D. Litster[3], (1)Mechanical Engineering, Purdue University, West Lafayette, IN, (2)Chemical Engineering, Purdue University, West Lafayette, IN, (3)School of Chemical Engineering, Department of Industrial and Physical Pharmacy, Purdue University, West Lafayette, IN

In granulation processes involving applications of mechanical force, the mechanical properties of the powder being processed often dictate the key characteristics of the end-product. For that reason, the modified Drucker-Prager/Cap properties of common pharmaceutical blends were measured in this study. In particular, the influence of magnesium stearate (MgSt) and acetaminophen (APAP) concentrations on the DPC parameters was studied. It was found that adding up to 2% w/w MgSt to a pure excipient system (microcrystalline cellulose Avicel PH200) resulted in a considerable decrease in the powder's internal friction angle and cohesion, but had a corresponding increase in the hydrostatic yield stress and elastic parameters (Young's modulus and Poisson's ratio). The influence of MgSt appeared to asymptote, however, as adding the MgSt concentration past 2% w/w did little more to the above-mentioned parameters. Adding APAP to an Avicel system generally resulted in an increase in both cohesion and internal friction angle. The APAP's influence on the elastic parameters, however, was complex; increasing the APAP concentration up to 50% w/w resulted in a decrease in both the Young's modulus and Poisson's ratio, but increasing it past 50% w/w resulted in an increase in both parameters. Hypotheses for the observed trends will also be presented.

(19f) Quantitative Correlation of the Effect of Process Conditions On the Capping Tendencies of Tablet Formulations

Ilgaz Akseli[1], Adrian Stecula[2], Xiaorong He[1] and Nadia Ladyzhynsky[1], (1)R&D Pharmaceutical Development, Boehringer Ingelheim, Ridgefield, CT, (2)University of San Francisco, San Francisco, CA

Tablet compaction presses and compaction emulator were used to rank order capping tendencies of a diverse sample set which includes excipients, binary, tertiary blends and final compression blends of commercial products. The compression forces of 5 kN, 10 kN, 15 kN, 20 kN, and 25 kN were studied for oval and round tooling shapes, while 4 kN, 6 kN, 8 kN, 10 kN, and 14 kN were chosen for the beveled edge due to the force restrictions for this tooling. Compaction speeds of 25 rpm (24.8 ms dwell time), 40 rpm (15.5 ms dwell time), and 80 rpm (7.9 ms dwell time) were chosen as representative ranges for bench-to-pilotscale processing. Five replicate tablets were compressed at each of the factor levels to ensure repeatability. Tablets were tested by nondestructive ultrasonic method to obtain effective elastic (E) and shear (G) moduli in the axial and radial directions. The anisotropic properties of tablet formulations are represented by a dimensionless ratio (EG). The ultrasonic measurements revealed that effective elastic and shear modulus values vary with different testing orientations which indicated elastic and shear moduli anisotropy in tablets. It was found that as the EG ratio decreases; the compressed powder exhibited an increasing level of anisotropic nature, which led to a higher tendency for capping. Acceptance criteria of the EG ratio have been proposed based on manufacturing performance. It was shown that altering process conditions such as tooling shape, compression force and press speed significantly affect the capping tendency of tablet formulations. A systematic approach has been applied to develop predictive tools to assess capping tendencies of formulations. These tools have been shown to correlate well with actual capping performance of a diverse sample set. The developed tools are fast, material sparing and have potential to flag the risk of capping issues and provide insight into the performance of common materials during formulation and process development. Application of these tools may help reduce development time, API usage and facilitate development of a robust product.

(19g) Online Particle Size Monitoring During Roller Compaction

Rhye Hamey, PD, Bristol-Myers Squibb, new brunswick, NJ

Online particle size monitoring during roller compaction

Rhye Hamey, David Good, Dongsheng Bu, Ziyang Su, Chandra Vema-Varapu

Roller compaction can be a challenging process to monitor and control, often relying on offline measurements of ribbon thickness, and ribbon density to determine if the process is running properly. It is well known that granule particle size can significantly influence powder flow (in turn effects tablet weight), compactability, and dissolution. Yet there has been limited work demonstrating the use of online equipment to monitor or control particle size during granulation.

An online particle size measurement via Mettler Toledo focus beam reflectance measurement (FBRM) was incorporated into the process ruggedness design of experiment (DOE) in order to collect data necessary to establish a control strategy based on real-time particle size information. To demonstrate the applicability of online particle size measurements for monitoring and controlling the granulation process, correlations between particle size, ribbon density, and ribbon thickness were evaluated. The sensitivity on the granule particle size and size distribution as a function of varying roller pressures and gap settings was evaluated. The study used offline sieve analysis data of samples taken during the process to validate the online particle size measurements. A correlation of online particle size measurements to product attributes was achieved by relating particle size data to powder flow, tablet compactability, and dissolution data.

This study suggests the use of online particle size monitoring as a process control in place of or in conjunction with ribbon density and ribbon thickness. The added product information may reduce the amount of time to transfer a product to commercial manufacturing by reducing the number of batches required to demonstrate process control. Particle size data would allow mapping of a design space in real-time by observing the effect of changes in roller compaction process parameters on the particle size distribution. Ultimately, a feedback control utilizing particle size data may reduce product variability due to manufacturing.

(26a) Enantioselective Separation On Chiral Au Nanoparticles

Nisha Shukla, Institute for Complex Engineered Systems, Carnegie Mellon Univeristy, Pittsburgh, PA and Andrew J. Gellman, Chemical Engineering, Carnegie Mellon University, Pittsburgh, PA

Adsorption of chiral compounds on chiral surfaces is the initial step in enantioselective processes such as separations and catalysis. There has been a significant effort over the past decade aimed at the preparation of chiral nanoparticles based on metallic cores modified by chiral ligands. In principle, these can serve as the basis for enantioselective chemical processing. In this work we demonstrate a simple measurement of enantioselective adsorption on chiral metal nanoparticles using a method that can yield quantitative measures of the enantiospecific adsorption equilibrium constants [1].

The surfaces of chemically synthesized Au nanoparticles have been modified with D- or L-cysteine to render them chiral and enantioselective for adsorption of chiral molecules. Their enantioselective interaction with chiral compounds has been probed by optical rotation measurements when exposed to racemic propylene oxide. The ability of optical rotation to detect enantiospecific adsorption arises from the fact that the specific rotation of polarized light by R- and S-propylene oxide is enhanced by interaction Au nanoparticles. This effect is related to previous observations of enhanced circular dichroism by Au nanoparticles modified by chiral adsorbates. More importantly, chiral Au nanoparticles modified with either D- or L-cysteine selectively adsorb one enantiomer of propylene oxide from a solution of racemic propylene oxide, thus leaving an enantiomeric excess in the solution phase. Au nanoparticles modified with L-cysteine (D-cysteine) selectively adsorb the R-propylene oxide (S-propylene oxide). A simple model has been developed that allows extraction of the enantiospecific equilibrium constants for R- and S-PO adsorption on the chiral Au nanoparticles.

[1] N. Shukla, M.A. Bartel, A.J. Gellman "Enantioselective separation on chiral Au nanoparticles" *Journal of the American Chemical Society*, *132*(25), (**2010**), 8575–8580

(26b) Specific Targeting Stealth Nanoparticles Coated with Ultra-Low Fouling Peptide

Ann K. Nowinski, Lei Zhang, Wei Yang and Shaoyi Jiang, Chemical Engineering, University of Washington, Seattle, WA

Nanoparticle-based biotechnology is relevant to many biomedical applications including drug delivery, diagnosis, and biosensing. A large obstacle to these applications is nonspecific protein adsorption, which can result in cellular uptake, nanoparticle aggregation, or an immune system response. Here we present an ultra-low fouling all -natural peptide coating based on the design principle of alternating positively and negatively charged residues. These mixed charge peptides exhibit high resistance to nonspecific protein adsorption. Furthermore, they can be extended with a functional peptide sequence to achieve specific targeting. Integrating the ultra-low fouling peptide sequence with a functional peptide sequence allows replacement of nonfouling synthetic materials and avoids the complex chemistries used in bioconjugation. These peptide-coated gold nanoparticles display stability in salt solution, protein solutions, and even undiluted human blood serum. Particles lacking specific targeting moieties possess stealth properties, whereas particles containing specific targeting moieties (RGD) are specifically uptaken by cells. This talk will discuss the assembly conditions, characterization, stability, and versatility of the peptide-based gold nanoparticle system.

(26c) Development of Magnetic Field-Responsive Surfactants

Sun Hae Ra Shin[1], Rajiv Misra[2], Peter Schiffer[2] and Kyle J. M. Bishop[3], (1)Chemical Engineering, The Pennsylvania State University, University Park, PA, (2)Physics, The Pennsylvania State University, University Park, PA, (3)Chemical Engineering, Penn State University, University Park, PA

Development of magnetic field-responsive surfactants

Sun Hae Ra Shin[1], Rajiv Misra[2], Peter Schiffer[2], Kyle Bishop[1]*

[1]Department of Chemical Engineering, The Pennsylvania State University, University Park, PA 16802

[2] Department of Physics, The Pennsylvania State University, University Park, PA 16802

*Address correspondence to kjmbishop@engr.psu.edu

This talk presents the development of magnetic field-responsive surfactants that can be used for dynamic and continuous tuning of interfacial properties. The magnetic surfactants consist of inorganic nanoparticles with amphiphilic surfaces. Particles of magnetic materials such as cobalt ferrite exhibit large magnetic moments and can be oriented and positioned by externally applied fields. The preparation of amphiphilic surface is readily achieved by spontaneous self-organization of a mixed monolayer of hydrophobic and hydrophilic ligands on the particle surface. Strong coupling between the particles' internal magnetic moments and their asymmetric surface chemistry allows external magnetic fields to directly influence the properties of the interface through the competition of magnetic and surface forces. The field-induced response of liquid interfaces containing these responsive surfactants is investigated with pendant drop tensiometry.

(26d) Deposition of Coated Titanium Dioxide Nanoparticles with Controllable Wetting Properties

Mikko Aromaa[1], Janne Haapanen[1], Hannu Teisala[2], Mikko Tuominen[2], Jurkka Kuusipalo[2], Milena Stepien[3], Jarkko J. Saarinen[3], Martti Toivakka[3] and Jyrki M. Mäkelä[1], (1)Aerosol Physics Laboratory, Tampere University of Technology, Tampere, Finland, (2)Paper Converting and Packaging Technology, Tampere University of Technology, Tampere, Finland, (3)Laboratory of Paper Coating and Converting, Abo Akademi University, Turku, Finland

Liquid Flame Spray (LFS) method can be used in synthesis and deposition of nanoparticles. The deposited nanoparticle coating forms a functional surface on top of the substrate. We have previously demonstrated that the coating can be applied on paperboard in a roll-to-roll process in a pilot scale with the substrate speed up to 150 m/min [1]. Titanium dioxide nanoparticles are hydrophilic by nature but with a careful selection of the coating parameters the substrate, especially paper board, can participate in the process. Even without affecting the visual or mechanical properties of the substrate some combustion fumes might be released from the substrate. The combustion fumes condense on top of the LFS deposited titania nanoparticles forming a coating on top of the particles [2]. The coating layer is only a few nanometres thick and thus transparent. The photoactive properties of titanium dioxide are preserved as the coating on top is transparent and therefore UV-light can be used in switching the wetting properties of the surface from superhydrophobic to superhydrophilic [3].

[1] Mäkelä, J.M., Aromaa, M., Teisala, H., Tuominen, M., Stepien, M., Saarinen, J.J., Toivakka M. and Kuusipalo, J., Nanoparticle deposition from liquid flame spray onto moving roll-to-roll paperboard material, Aerosol Science and Technology, 45, 827-837, 2011.

[2] Aromaa, M., Arffman, A., Suhonen, H., Haapanen, J., Keskinen, J., Honkanen, M., Nikkanen, J.-P., Levänen, E., Messing, M.E., Deppert, K., Teisala, H., Tuominen, M., Kuusipalo, J., Stepien, M., Saarinen, J.J., Toivakka, M. and Mäkelä, J.M., Atmospheric Synthesis of Superhydrophobic TiO_2 Nanoparticle Deposits in a Single Step using Liquid Flame Spray, Journal of Aerosol Science, 17 pages, accepted 2012.

[3] Stepien, M., Saarinen, J.J., Teisala, H., Tuominen, M., Aromaa, M., Kuusipalo, J., Mäkelä, J.M. and Toivakka, M., Surface chemical analysis of photocatalytic wettability conversion of TiO_2 nanoparticle coating, Surface Coatings Technology, 17 pages, submitted 2011.

(26e) Nanoparticles Surface Design Process for Aggregation and Dispersion Behavior Control in Liquid to Apply Functional Material

Hidehiro Kamiya and Motoyuki Iijima, Institute of Engineering, Tokyo University of Agriculture and Technology, Koganei, Japan

Nanoparticle and nanomaterials dispersion and aggregation behavior control is one of the most important fundamental processes to apply the functional properties development of energy related devices, such as polymer composite and electrode of Li battery. Surface modification and treatment by using various surfactants and coupling agents with different molecular structure on prepared nanoparticle surfaces in organic solvents were applied to control surface interaction between particles. After surface treatment on nanoparticles in suspension, if surface structure on nanoparticles was almost completely covered and modified by surfactant or coupling agent, hard aggregate or agglomerate of nanoparticles were not formed after evaporating and drying process. Such completely surface modified nanoparticles were able to be re-dispersed into different organic solvent. Uniform nanoparticles dispersed polymer composite or electrode for battery were prepared from the above prepared nanoparticle's suspension without aggregation. Furthermore, molecular and nanometer scaled surface structure and interaction after various modification and treatment was characterized by using a colloid probe atomic force microscope, AFM, method. Based on the measurement results, the action mechanism of surface modification was discussed.

Reference

Iijima, M., Kobayakawa, M., Yamazaki, M., Ohta, Y., Kamiya, H., (2009). *Anionic Surfactant with Hydrophobic and Hydrophilic Chains for Nanoparticle Dispersion and Shape Memory Polymer Nanocomposites,* J. Am. Chem. Soc. 131(45) 16342-16343.

Iijima, M., Kamiya, H., (2010). *Layer-by-Layer Surface Modification of Functional Nanoparticles for Dispersion in Organic Solvents*, Langmuir, 26, 17943–48.

(45a) Reduction of a Mechanistic Drying Model of Pharmaceutical Granules for Inclusion in a Population Balance Model with Continuous Growth Term

Séverine T.F.C. Mortier[1,2], Ingmar Nopens[1], Thomas De Beer[2] and Krist V. Gernaey[3], (1)Department of Mathematical Modelling, Statistics and Bioinformatics, Ghent University, Ghent, Belgium, (2)Department of Pharmaceutical Analysis, Ghent University, Ghent, Belgium, (3)Department of Chemical and Biochemical Engineering, Technical University of Denmark (DTU), Kgs. Lyngby, Denmark

Nowadays, there is a trend to move from batch towards continuous production processes in the pharmaceutical industry (Leuenberger, 2001). To accomplish this transition properly, all consecutive unit operations executed during the continuous manufacturing should be understood in detail to design adequate control strategies for guaranteeing product quality at all time. Adequate control mechanisms should rely on in-process measurements of critical process and product parameters and real-time adjustment of input variables which are able to direct the process in the right way.

Pharmaceutical processes often consist of multi-phase systems with high complexity. Developing mechanistic models of these multi-phase systems is a means of mapping the detailed knowledge available for such a process (Mortier et al., 2011). Such a modelling approach has only recently started to become more popular in the pharmaceutical area. It has been recognized that the use of mechanistic models and system-based approaches are important for Process Analytical Technology (PAT) and a Quality By Design (QbD) (Sin et al., 2008; Gernaey et al., 2012).

The work presented here resides in the modeling of a continuous fluidized bed drying process, which is used to dry the wet granules originating from the continuous granulator. The ConsiGma™ from Collette™(GEA Pharma Systems) is a full continuous from powder to tablet manufacturing line, including a six-segmented fluidized bed drying unit. The drying behaviour of single pharmaceutical granules was previously unraveled in a modeling study including calibration and validation (Mortier et al., 2012). The drying of the wet granules occurs in two subsequent phases. The first drying phase includes the water at the free surface of the droplet. As soon as the radius of the droplet equals the radius of the dry particle, the evaporation rate slows down and the second drying phase begins.

Population Balance Modelling (PBM) is a modeling framework allowing to describe the behaviour of a population of particles with a distributed property. In the case under study, the latter is the moisture content and the evolution of the moisture content distribution for a batch of wet granules is envisioned. The general Population Balance Equation (PBE) can here be simplified into a homogeneous equation solely with a negative continuous growth term expressing the drying. A good candidate for this drying process model is that of a single granule. However, this model is too complicated to integrate it straight away into the PBM. Hence, a model reduction step is required. Different available methods described in the literature were evaluated, e.g. heuristic methods (Van Nes et al., 2005), mathematical based methods (projection based (Bernhardt, 2008), Krylov-based approximation (Antoulas, 2005), sensitivity based methods (Lawrie et al., 2007)). These methods are very case specific and could not be used here.

Therefore, a new strategy was developed for this particular case. It consists of a Global Sensitivity Analysis (GSA) followed by a model reduction step. The GSA is performed to determine the most important parameters and model inputs that need to be retained in the reduced model. Based on the GSA the gas temperature was found to be the most sensitive parameter, and was retained in the model reduction. Details of the GSA are the subject of a separate abstract.

The reduced model needs to be able to describe the decrease of the moisture content for a drying particle. Analysis revealed that the growth term of both drying phases displayed a different behaviour, and that the growth term is strongly dependent on the wet radius, which is a representation of the moisture content. Finally, a size dependent growth term for both drying phases was developed separately. The model reduction starts with collecting data by simulating the full model for a vast range of the selected degree of freedom, i.e. the gas temperature. Simulation at an intermediate value was chosen to reveal the model structure of the reduced empirical model. Subsequently, this structure is verified for all other simulations in the range to ensure the structure is able to describe the behaviour for all values of the degree of freedom in the range. The selected model structure contains a number of parameters, which are calibrated for each simulation by minimizing the Root Mean Square Error (RMSE). In a next step, dependencies (submodel structures) between these parameters and the selected degree of freedom (gas temperature) are determined. The latter was done in a step-wise approach, and upon introducing each new submodel the remaining coefficients are recalibrated. This eventually results in an empirical model solely function of gas temperature, and which can be implemented easily in a PBM. The reduced model consists of two sets of algebraic equations, one for each drying phase, and can be used to solve the Population Balance Equation (PBE) for different gas temperatures, different sizes of the granule and different initial conditions. For the first drying period a mean weighted relative error of 0.53% was obtained compared to the detailed mechanistic drying model, whereas for the second drying period 1.97% was obtained. The latter is an acceptable error given the reduction in complexity.

A.C. Antoulas, An overview of approximation methods for large-scale dynamical systems, Annual Reviews in Control 29 (2005) 181-190.

K. Bernhardt, Finding alternatives and reduced formulations for process-based models Evolutionary computation 16 (2008) 63-88.

K.V. Gernaey, A.E. Cervera-Padrell, and J.M. Woodley, A perspective on Process Systems Engineering in pharmaceutical process development and innovation. Comput. Chem. Eng. (2012) (in press).

J. Lawrie, J. Hearne, Reducing model complexity via output sensitivity Ecological Modelling, 207 (2007) 137-144.

H. Leuenberger, New trends in the production of pharmaceutical granules: batch versus continuous processing, Eur. J. Pharm. Biopharm. 52 (3) (2001) 289-296.

M. D. McKay, R. J. Beckman, and W.J. Conover, A comparison of three methods for selecting values of input variables in the analysis of output from a computer code. Technometrics, 21 (2) (1979) 239-245.

S.T.F.C. Mortier, T. De Beer, K.V. Gernaey, J. Vercruysse, M. Fonteyne, J. Remon, C. Vervaet, I. Nopens, Mechanistic modelling of the drying behaviour of single pharmaceutical granules, Eur. J. Pharm. Biopharm. 80 (3) (2012) 682-689.

S.T.F.C. Mortier, T. De Beer, K.V. Gernaey, J. Remon, C. Vervaet, I. Nopens, Mechanistic modelling of fluidized bed drying processes of wet porous granules: a review, Eur. J. Pharm. Biopharm. 79 (2011) 205-225.

G. Sin, P. Odman, N. Petersen, A. Lantz, K. Gernaey, Matrix notation for efficient development of first-principles models within PAT applications: integrated modeling of antibiotic production with streptomyces coelicolor, Biotechnol. Bioeng. 101 (1) (2008) 153-171.

E.H. Van Nes, M. Scheffer, A strategy to improve the contribution of complex simulation models to ecological theory, Ecological Modelling 185 (2005) 153-164.

(45b) A Systematic Approach for Quantifying the Multi-Particle Interactions in Dry Milling: Rational Function Type Effectiveness Factor

Maxx Capece[1], Rajesh Dave[2] and Ecevit Bilgili[1], (1)Chemical, Biological and Pharmaceutical Engineering, New Jersey Institute of Technology, Newark, NJ, (2)Department of Chemical, Biological and Pharmaceutical Engineering, New Jersey Center for Engineered Particulates, Newark, NJ

Non-linear population balance models (PBMs) have recently been shown to be superior to their linear counterparts for quantifying the breakage kinetics in dense-phase dry milling processes, where mechanical multi-particle interactions are significant. At the particle ensemble scale, particles of different sizes interact causing a population dependence of the specific breakage rate function. Non-linear PBMs explicitly account for these multi-particle interactions through a population-dependent functional called the effectiveness factor. Due to the inherent difficulty and complexity of proposing and selecting an appropriate form of the effectiveness factor, a general and systematic approach is desired to assess breakage kinetics in an unbiased manner. In this study, toward addressing the need for a general form of the effectiveness factor, we propose a rational function approximation. Computer generated data on the effectiveness factor and evolution of the particle size distribution during batch dry milling of quartz in a tumbling ball mill were fitted. Goodness-of-fit and statistical significance of the parameters estimated were evaluated to discriminate various forms of the effectiveness factor. It is concluded that a rational function approximation may replace specific forms of the effectiveness factor for a less restrictive analysis of the multi-particle interactions. Additionally, fitting a non-linear PBM using the rational function approximation to the effectiveness factor was able to accurately describe the dense-phase dry milling data. Accordingly, this study enables experimenters to quantify the impact of the multi-particle interactions in a robust, systematic, and unbiased manner via the rational function approximation to the effectiveness factor within the context of the non-linear PBM.

(45c) Theoretical Breakage Distribution Functions

Priscilla J. Hill, Dave C. Swalm School of Chemical Engineering, Mississippi State University, MS State, MS

A variety of different breakage distribution functions have been proposed for population balance models. Although particle breakage is dependent on both particle size and shape, initial functions only considered the particle size but not particle shape. Later approaches accounted for the volume of a crystal removed with each impact, assumed an initial particle shape (such as a cube) for all of the initial crystals, and/or assumed that all of the particles break in the same way. For example, the corners of cuboid particles were removed. The overall goal was to determine how the particle size and shape changed with repeated impacts.

Newly developed breakage distribution functions will be presented that remove the limitation that all particles break in the same way. The effect of the model parameters on the particle size and shape will be presented. In all cases the models conserve mass while accounting for the number of particles generated. Comparisons will be made with existing theoretical models. Although developed for modeling breakage of crystals during solution crystallization, these models are general and can be applied to other unit operations.

(45d) Optimization and Control of Growth Modifier Concentration Profile for Achieving Target Crystal Shape Distribution

Aniruddha Majumder and Zoltan K. Nagy, Department of Chemical Engineering, Loughborough University, Loughborough, United Kingdom

The crystal product quality, functionality and properties such as flow characteristics are greatly influenced by the crystal shape distribution. In order to obtain desired crystal shape, growth modifiers (GMs) are often used as additives [1]. The objective of this work is to determine the optimal concentration profile of the GMs that is required for the desired crystal shape distribution and subsequently use a simple control configuration for achieving the target crystal shape distribution. The optimal profile is achieved by formulating and solving an optimization problem. The formulation of the problem involves description of crystal shape distribution using multidimensional population balance model (PBM) in combination with Kubota-Mullin's model for the pinning mechanism of GM molecules on crystal face to take into account the possible hindering or promoting effect of GM on crystal growth [2,3]. The Kubota-Mullin's model uses surface coverage of the adsorbed species on the growing crystal and a proportionality constant denoting the effectiveness of the adsorbed species [2]. Finally, a simple feedback PI control configuration is implemented which manipulates the GM concentration profile in the crystallizer [4]. With this control implementation, it is demonstrated that the desired crystal shape distribution can be achieved.

References

[1] M. A. Lovette, A. R. Browning, D. W. Griffin, J. P. Sizemore, R. C. Snyder and M. F. Doherty. Crystal Shape Engineering. *Ind. Eng. Chem. Res.*, 47:9812-9833, 2008.

[2] N. Kubota. Effect of Impurities on the Growth Kinetics of Crystals. *Cryst. Res. Technol.*, 36:749-769, 2001.

[3] F. Fevotte and G. Fevotte. A method of characteristics for solving population balance equations (PBE) describing the adsorption of impurities during crystallization processes. *Chem. Eng. Sci.*, 65(10):3191-3198, 2010.

[4] D. B. Patience and J. B. Rawlings. Particle-Shape Monitoring and Control in Crystallization Processes. *AIChE J.*, 47(9):2125:2129, 200

Self-starting and Globally Convergent Parameters Estimation Method for Aggregation-Breakage Population Balance Model using Genetic Algorithm

Akinola Falola, Xue Wang, Antonia Borissova, Jingjing Liu
Institute of Particle Science and Engineering, School of Process, Environmental and Materials Engineering, University of Leeds, Leeds LS2 9JT, UK

ABSTRACT

Some of the main challenges to the estimation of parameters for the aggregation-breakage population equation from experimental data include the determination of a good initial point to ensure the convergence of the numerical procedure and also ensuring convergence to the global optimum not a local minimum. The initial point is usually determined by trial and error. This method is very time consuming and convergence to the global minimum is not guaranteed. In this write-up, a procedure is proposed for choosing good initial points for the optimisation using Genetic Algorithm and then employing a gradient searching optimisation method to speed-up convergence to the minimum point. The method is then validated with simulated data with and without artificial random errors.

1. INTRODUCTION

Population balance modelling (PBM) is the most useful tool for estimating the dynamic evolution of particle size distribution (PSD) of most particulate processes [1]. It has been widely applied in granulation, dry and wet milling, bubble breakage up, flocculation etc. The general population balance equation incorporates all the mechanisms that affect the properties and number of particles in a particulate processing. These mechanisms include aggregation, breakage, growth and nucleation. For the purpose of this work, only aggregation and breakage will be presented. Nevertheless, the method presented here is also applicable to the general PBM including growth and nucleation.

Given the parameters of the population balance equation (PBE), the solutions of the systems of integro-partial differential equation have been well studied. There are several methods for the solution of PBE including the Discretised Population Balance (DPB) methods, the method of moments and its variants (SMOM, QMOM, EMOM, etc.), Monte Carlo Methods, the finite difference method and finite element method [2, 3]. The determination of the population balance parameters given experimental size distributions is however not so trivial. Much work has been done in the estimation of breakage parameters from the time evolution of the particle size distribution but very little for the combined aggregation and breakage processes [4, 5]. Even for the breakage only population balance equation, the inverse determination of breakage parameters from the PSD is degenerate: there may be multiple solutions [5]. Bilgili and co-workers have made a lot of contribution to the determination of kinetics parameters from the size distribution of breakage only population balance equation using a Full Numerical Scheme [4, 6]. Two of the limitations of the Full Numerical Scheme are (1) It is not self-starting (i.e. it requires the specifications of suitable starting points for the optimisation) and (2) Convergence to the global minimum[1] is not guaranteed unless a good starting value is provided and this usually involved trial and error.

In this work, the Full Numerical Scheme is applied to the aggregation-breakage population balance equation. Since the optimisation problem is generally more complex and the number of parameters is higher than for a breakage only population balance equation, the trial and error method

[1] The procedure may not even converge at all in some cases.

for the determination of starting values for the optimisation procedure is less efficient and more time consuming. Therefore, it is important that the optimisation procedure is self-starting i.e. requires no starting values for the optimisation. One such procedure is presented in this work and validated with computer simulated data.

1. MODEL AND METHODS

2.1 Aggregation-Breakage PBE

The aggregation-breakage population balance equation for a batch system, assuming no variation in external coordinates and particle size as the only internal coordinate is [7]:

$$\frac{\partial n(v;t)}{\partial t} = \int_v^\infty S(\varepsilon)b(v\mid\varepsilon)n(\varepsilon;t)d\varepsilon - S(v)n(v;t) + \frac{1}{2}\int_0^v \beta(\varepsilon, v-\varepsilon)n(\varepsilon;t)n(v-\varepsilon;t)d\varepsilon \\ - n(v;t)\int_0^\infty \beta(v,\varepsilon)n(\varepsilon,t)d\varepsilon \qquad 1$$

where $n(v;t)$ is the particle number density, $S(\varepsilon)$ is the rate of breakage of particle of size ε, $b(v\mid\varepsilon)$ is the daughter size distribution function which is the number of particles of size v formed on breakage of particle of size ε, $\beta(\varepsilon, v-\varepsilon;)$ is the rate at which particles of sizes ε and $v-\varepsilon$ aggregate to form particle of size v. The first term on the RHS of Equation 1 is the birth due to particle breakage, the second term is the death due to breakage, the third term is the birth due to aggregation and the fourth term is the death due to aggregation.

Several expressions have been proposed for the for the breakage and aggregation kernels. For this work, the power-law breakage kernel according to Kapur [8] is used:

$$S(v) = \dot{S}v^{\lambda} \qquad 2$$

where $\dot{S}$ is the breakage rate coefficient and λ is the exponent of the power-law kernel. To allow for the simulation of several breakage mechanisms, the generalized breakage daughter size distribution according to Diemer and Olson [9] is employed:

$$b(v\mid\varepsilon) = p\frac{z^{s-1}(1-z)^{r-1}}{\varepsilon\cdot \mathrm{B}(s,r)} \qquad 3$$

$$p \geq 2,\ r > 0, s = r(p-1)$$

where p is the number of fragments for each breakage event, r determine the shape of the daughter size distribution function, z is the ratio of the daughter-to-parent size i.e. $z = v/\varepsilon$ and $\mathrm{B}(q,r)$ is the beta function [9]. In general, the theory of aggregation is more developed than that of breakage [10]. Therefore several expressions for different aggregation kernels are available in the literature. The turbulent shear kernel is used in this contribution and is written as [1, 10]:

$$\beta(v,\varepsilon) = \dot{\beta}(v^{1/3} + \varepsilon^{1/3})^3 \qquad 4$$

where $\dot{\beta}$ is the breakage kernel coefficient.

Analytical solution of Equation 1 are available only for special cases of aggregation and breakage kernels, therefore, numerical methods are mostly used. The Discrete Population Balance (DPB) methods, which are the most successful methods for the solution of aggregation-breakage population balance equation, will be employed for the solution of PBE in this work. The DPB methods were developed by Hounslow, Ryall [11], Litster, Smit [12] and Kumar and Ramkrishna [13].

In the DPB methods, the PBE is converted to a set of discrete equations in number concentration by defining the number concentration within each volume interval $[v_i, v_{i+1})$ as:

$$N_i(t) = \int_{v_i}^{v_{i+1}} n(v;t)dv \qquad 5$$

By integrating Equation 1 between each volume interval $[v_i, v_{i+1})$, the PBE can be written as:

$$\frac{\partial N_i}{\partial t} = \left.\frac{\partial N_i}{\partial t}\right|_{aggregation} + \left.\frac{\partial N_i}{\partial t}\right|_{breakage} \qquad 6$$

where the first and second terms on the RHS of Equation 6 is the change in number concentration due to aggregation and breakage respectively. In order to avoid high number of classes, non-uniform geometric grids are usually employed in the DPB methods which introduced internal inconsistency to the normal discretisation of the equation [14]. Several expressions have been proposed for the aggregation and breakage source terms in Equation 6 in order to resolve the internal inconsistencies. For the purpose of this work, the method by Litster, Smit [12] is used to resolve the aggregation term while the fixed pivot method of Kumar and Ramkrishna [13] is employed to resolve the breakage source term because both methods allow the use of variable geometric grids.

The discretization of the PBE in this work is of the form $v_{i+1} = 2^{1/q} v_i$ where q is a positive integer greater than 0. Using the approach of Litster, Smit [12], the aggregation source term in Equation 6 is

$$\begin{aligned}\left.\frac{\partial N_i}{\partial t}\right|_{aggregation} &= \sum_{k=1}^{i-Sum(q)-1} \frac{2^{(k-i+1)/q}}{2^{1/q}-1}\beta_{i-1,k}(t)N_{i-1}N_k + \sum_{k=2}^{q}\sum_{j=i-Sum(q-k+2)-k+1}^{i-Sum(q-k+1)-k} \frac{2^{(j-i+1)/q}-1+2^{-(k-1)/q}}{2^{1/q}-1}\beta_{i-k,j}(t)N_{i-k}N_j \\ &+ \frac{1}{2}\beta_{i-q,i-q}(t)N_{i-q}N_{i-q} + \sum_{k=2}^{q}\sum_{j=i-Sum(q-k+2)-k+2}^{i-Sum(q-k+1)-k+1} \frac{2^{(j-i)/q}}{2^{1/q}-1}\beta_{i-k+1,j}(t)N_{i-k+1}N_j \\ &- \sum_{k=1}^{i-Sum(q)} \frac{2^{(k-i)/q}}{2^{1/q}-1}\beta_{i,k}(t)N_iN_k - \sum_{k=i-Sum(q)+1}^{m}\beta_{i,k}(t)N_iN_k\end{aligned} \qquad 7$$

where $Sum(q) = \sum_{1}^{q} p$ and M is the number of classes or bins. Also, employing the approach by Kumar and Ramkrishna [13], the aggregation source term in Equation 6 is [3, 13]:

$$\begin{aligned}\left.\frac{\partial N_i}{\partial t}\right|_{aggregation} &= \sum_{k=1}^{M}\sum_{j=k}^{M}\left[\theta(v_{i-1} < v_j + v_k < v_i)\frac{(v_j+v_k)^a v_{i-1}^b - (v_j+v_k)^b v_i^a}{v_i^a v_{i-1}^b - v_i^b v_{i-1}^a}\right. \\ &\left. + \theta(v_i < v_j + v_k < v_{i+1})\frac{(v_j+v_k)^a v_{i+1}^b - (v_j+v_k)^b v_i^a}{v_i^a v_{i+1}^b - v_i^b v_{i+1}^a}\right] \\ &\times\left(1-\frac{\delta_{j,k}}{2}\right)\beta_{j,k}N_jN_k - N_i\sum_{k=i}^{M}\beta_{i,k}N_k\end{aligned} \qquad 8$$

Equation 4 can then be solved using suitable ODE solvers. For this work, the ode13s of MATLAB version R2012a is used to solve the ODEs with absolute and relative tolerances of $1\text{x}10^{-6}$ and $1\text{x}10^{-9}$ respectively.

2. PARAMETERS ESTIMATION METHOD

The breakage and aggregation model parameters can be obtained by fitting the size distribution obtained from the PBM above to experimental data. This yields the breakage and aggregation kernels parameters. The minimization problem is posed as:

$$\min_{p\geq 2,\ r>0,\dot{S}\geq 0,\beta\geq 0,} \left[F_{obj} = \frac{\sum_{i=1}^{N}\sum_{j=1}^{M}\left(m_{i,j}^{\exp} - m_{i,j}^{siim}\right)}{N-1} \right] \qquad 9$$

where $m_{i,j}^{\exp}$ and $m_{i,j}^{sim}$ are respectively the experimental and simulated cumulative volume fraction. The subscripts i and j refers to the discrete experimental sampling time and size class respectively and N is the total number of sampling times. The cumulative volume fractions are used in the objective function instead of the number concentrations because the number concentrations tend to give greater weight to the smaller particles and will therefore require a weighting factor in Equation 9. The use of cumulative volume fraction therefore eliminates the need for a weighting factor for each term in Equation 9 thereby simplifying it.

In this work, the constrained non-linear minimisation is carried out using the *fmincon* built-in optimiser in MATLAB 2012a. MATLAB implementation of *fmincon* is used to find the constrained minima (or maxima by using the negative of the objective function) of a function of several variables and it implements four different algorithms for finding the minima of a function namely: Sequential Quadratic Programming (SQP), Interior Point, Active Set and Trust Region Reflective algorithms [15]. In this work, the SQP algorithm is used for the optimisation. The termination tolerances for both the function and parameters are both set to $1\text{x}10^{-9}$. *fmincon* also requires the specification of an initial parameter set.

As pointed out by Capece, Bilgili [6], the convergence of the non-linear optimisation may be slow and is not always guaranteed especially if poor initial parameters are specified. Also, the optimisation problem can give a local minimum instead of the global minimum. The determination of the initial values is usually done by try and error and this can be very difficult especially for cases with large number of parameters. Even for the moderate number of parameters (3 parameters) considered in this work, determination of a good initial parameter set by trial and error can be very difficult and time consuming. Therefore, a method is required than can reliably estimate a good starting value for the optimisation procedure.

In this work, the MATLAB 2012a implementation of genetic algorithm (*ga*) is used to provide the initial parameter set for the optimisation. The genetic algorithm mimics the process of natural selection to find the optimal solutions for constrained and constrained optimisation [15]. At each step, the *ga* algorithm selects individuals (parameter sets) from the current population to be parents and generates the children from them for the next generation [15]. Over successive generations, the population evolves towards the optimal solution. Since the initial population is selected randomly, *ga* does not require the specification of an initial parameter set. Because the population at each step is generated randomly, the ga algorithm does not get trapped at a local minimum like other classical derivative based methods. The *ga* is also suitable for problem where the objective function is non-differentiable, stochastic or highly non-linear and for mixed-integer optimisation [15].

The *ga* method however is usually much slower than the SQP algorithm due to the higher number of function evaluations. Therefore, in this work, a combination of *ga* and SQP is used to determine the best set of parameters. The optimisation starts with the *ga* and the value of the parameters obtained from the *ga* is used as the starting parameters for the SQP algorithm in order to speed up the convergence to the best set parameters. The initial parameter passed to the SQP is good because it is obtained using the *ga* method; therefore the convergence of the method is guaranteed. For the genetic algorithm step, the termination tolerances for the function and parameters were both set to $1x10^{-3}$ while the maximum number of generations and function evaluations were 20 and 15000 respectively. The tolerances of the genetic algorithm step of the optimization should not be too large to allow the algorithm to get a good starting parameters for the optimization or too small in order to reduce the time taken for the optimization.

With enough computation resources, the genetic algorithm can be employed on its own to find the parameters directly especially for cases where the derivative based methods are un-suitable e.g. finding the number of daughter cells for each breakage event (mixed-integer optimisation).

3. METHOD VALIDATION

The accuracy and robustness of parameter estimation algorithm can be assessed using simulated data with or without artificial errors [6]. The artificial errors are introduced because experimental data are never perfectly correct and the robustness of the parameter estimation method is determined by how well the method is able to predict the right parameters given data with some level of errors.

The system used to validate this method was the shear induced flocculation and breakup of spherical Polystyrene studied experimentally by Oles [16] and numerically by Spicer and Pratsinis [17] and Sommer, Stenger [1]. The initial solution was mono-disperse with diameter of 2.17μm (volume=$5.35x10^{-12}cm^3$) and volume fraction of $5x10^{-5}$. This corresponds to a number concentration of $9.34x10^9\#.cm^{-3}$. Sommer, Stenger [1] fitted[2] the following sets of parameters to the experimental data of Oles [16]for shear rate of $50s^{-1}$: $\dot{S} = 2.429cm^{-3\lambda}s^{-1}, \dot{\beta} = 18.565s^{-1}, \lambda = 1/3$. Sommer, Stenger [1] assumed binary equisized daughter size distribution (*p=2, r=31*) to fit the data. To benchmark the DPB solution method programmed in MATLAB, the PBE with parameters above was solved.

The parameters were used to generate the simulated data using the DPB methods presented in section 2. The discretisation used for the solution is of the form $v_{i+1} = 2^{1/3}v_i$ with primary particle of size (v_1) of $3.34x10^{-13}cm^3$ and 132 classes/sections. Therefore, the system spans a range of particles of one to $2^{\wedge 43}$ primary particles i.e. the largest particle has a volume of $4.398x106cm^3$. Figure 1 shows the good agreements between the temporal variations of the predicted and experimental dimensionless volume weighted mean length. Randomly normalised errors of 5% and 10% were also introduced into the simulated data using the method highlighted by Capece, Bilgili [6].. After generating the number density, the algorithm of section 3 was then used to generate the parameters. Table 1 gives the parameters estimated from the simulated data without error and those with 5% and 10% random errors.

Table 1, Table 2 and Table 3 show the parameters estimated using the method for the case of no error, 5% and 10% errors respectively. For the case of no error, the parameters estimated have very little deviation from the actual parameter value. Even for the case of 5% and 10% errors, the parameters obtained were still within reasonable range of the true value i.e. the maximum percentage error of 3.31% was obtained in the aggregation kernel coefficient for the data with 10% error.

2 The fitting is only to the mean particle length rather than to the particle size distribution

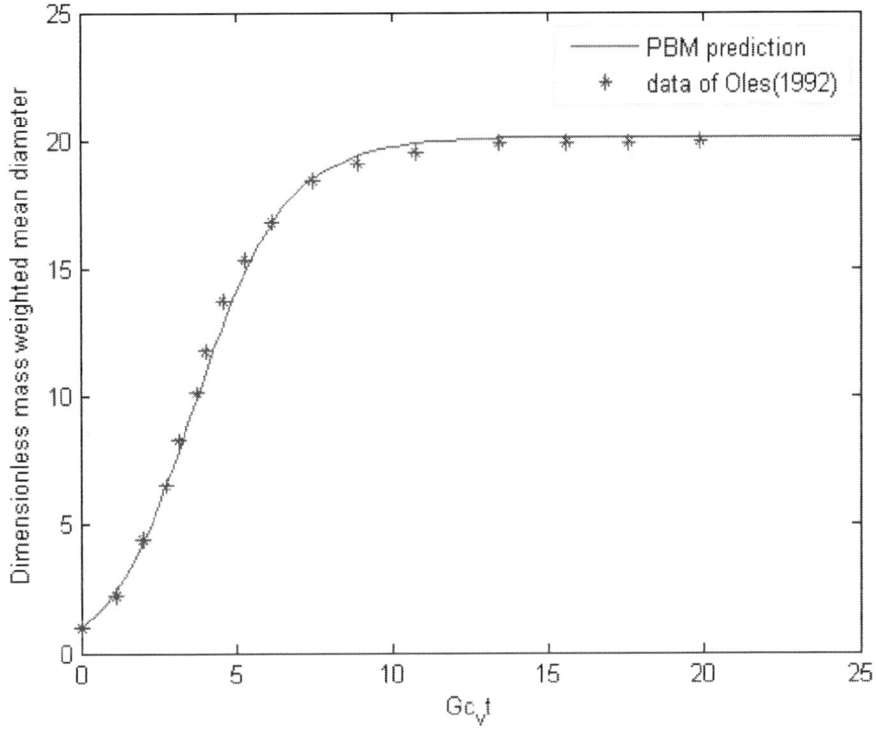

Figure 1 Time variation of dimensionless mass weighted mean diameter for PBM generated data and experimental data of Oles [16]

The tables also show the values of the parameters obtained at the genetic algorithm stage of the optimisation process which are then subsequently used as the initial values for the SQP stage. On the average, the genetic algorithm step of the optimisation took 9347 s with 420 function evaluations while the SQP step of the optimisation took 6955 s. The objective functions for the final values of the parameters in the table are $3.908\text{x}10^{-6}$, $3.240\text{x}10^{-4}$ and $6.227\text{x}10^{-4}$ respectively for the simulated data without error, for data with 5% error and for data with 10% error.

Table 1 Estimated parameter from simulated data without error

Parameter	Parameter Value	GA Predicted Parameters	Estimated Parameter	% |Error|
$\dot{S}$	2.429	0.846	2.423	0.247
$\dot{\beta}$	18.565	5.991	18.568	0.016
λ	1/3	0.447	0.333	0.04

Table 2 Estimated parameter from simulated data with 5% random error

Parameter	Parameter Value	GA Predicted Parameters	Estimated Parameter	% |Error|
$\dot{S}$	2.429	4.417	2.482	1.717
$\dot{\beta}$	18.565	6.800	18.579	1.607
λ	1/3	0.488	0.3346	0.38

Table 3 Estimated parameter from simulated data with 10% error

Parameter	Parameter Value	GA Predicted Parameters	Estimated Parameter	% |Error|
$\dot{S}$	2.429	1.497	2.400	1.194
$\dot{\beta}$	18.565	4.790	17.950	3.313
λ	1/3	0.688	0.3326	0.22

4. CONCLUSIONS

A method for the estimation of aggregation-breakage population balance is proposed and validated in this work. The method combines the genetic algorithm and classical derivative based method and does not require a starting parameter. Also, because of the automatic determination of the starting value for the optimisation, the method can avoid being trapped at a local minimum.

The method was validated with simulated data with and without artificial errors and was able to estimate the correct parameters to reasonable accuracy.

REFERENCES

1. Sommer, M, et al., *Agglomeration and breakage of nanoparticles in stirred media mills - a comparison of different methods and models.* Chemical Engineering Science, 2006. **61**(1): p. 135-148.
2. Marchisio, DL, RD Vigil, and RO Fox, *Quadrature method of moments for aggregation-breakage processes.* Journal of Colloid and Interface Science, 2003. **258**: p. 322 - 334.
3. Vanni, M, *Approximate Population Balance Equations for Aggregation-Breakage Processes.* Journal of Colloid and Interface Science, 2000. **221**: p. 143 - 160.
4. Bilgili, E, R Hamey, and B Scarlett, *Nano-milling of pigment agglomerates using a wet stirred media mill: Elucidation of the kinetics and breakage mechanisms.* Chemical Engineering Science, 2006. **61**(1): p. 149-157.
5. Williams, MC, TP Meloy, and M Tarshan, *Assessment of Numerical-Solution Approaches to the Inverse Problem for Grinding Systems - Dynamic Population Balance Model Problems.* Powder Technology, 1994. **78**(3): p. 257-261.
6. Capece, M, E Bilgili, and R Dave, *Identification of the breakage rate and distribution parameters in a non-linear population balance model for batch milling.* Powder Technology, 2011. **208**(1): p. 195-204.
7. Randolph, AD and MA Larson, *Theory of particulate processes; analysis and techniques of continuous crystallization*1971, New York,: Academic Press. xiii, 251 p.
8. Kapur, PC, *Self-Preserving Size Spectra of Comminuted Particles.* Chemical Engineering Science, 1972. **27**(2): p. 425-&.
9. Diemer, RB and JH Olson, *A moment methodology for coagulation and breakage problems: Part 3 - generalized daughter distribution functions.* Chemical Engineering Science, 2002. **57**(19): p. 4187-4198.
10. Marchisio, D, *Quadrature method of moments for aggregation–breakage processes.* Journal of Colloid and Interface Science, 2003. **258**(2): p. 322-334.
11. Hounslow, MJ, RL Ryall, and VR Marshall, *A Discretized Population Balance for Nucleation, Growth, and Aggregation.* Aiche Journal, 1988. **34**(11): p. 1821-1832.
12. Litster, JD, DJ Smit, and MJ Hounslow, *Adjustable discretized population balance for growth and aggregation.* Aiche Journal, 1995. **41**(3): p. 591 - 603.
13. Kumar, S and D Ramkrishna, *On the solution of population balance equations by discretization .1. A fixed pivot technique.* Chemical Engineering Science, 1996. **51**(8): p. 1311-1332.

14. Ramkrishna, D, *Population balances : theory and applications to particulate systems in engineering*2000, San Diego, CA: Academic Press. xvi, 355 p.
15. The Mathworks Inc *Global Optimization Toolbox User's Guide*. 2012.
16. Oles, V, *Shear-Induced Aggregation and Breakup of Polystyrene Latex-Particles.* Journal of Colloid and Interface Science, 1992. **154**(2): p. 351-358.
17. Spicer, PT and SE Pratsinis, *Coagulation and Fragmentation: Universal Steady-State Particle-Size Distribution.* Aiche Journal, 1996. **42**(6): p. 1612 - 1620.

(45f) Fast Monte Carlo Algorithm for Heterogeneous Population Balance Models: Effect of Reactor Dissipation Energy Distribution and Blending Time On Metal Nanoparticle Formation

Roberto Irizarry, Electronic Technologies, DuPont, Raleigh, NC

To study the scale-up of particulate processes from laboratory scale to industrial scale, the effects of heterogeneities on the final particle properties needs to be understood. The ability of Monte Carlo, MC, to track individual events/particles and to model complex multidimensional particulate processes makes it a very important tool for this task. Most of the existing MC strategies are developed for the homogeneous case and very little attention has been given to the heterogeneous case. In this work, I develop efficient time-driven and event-driven Monte Carlo methodologies for the simulation of population balance models under heterogeneous conditions. A new stochastic procedure called particle bundle flow (PBF) is utilized to model the transfer of particles [1]. To simulate the population dynamics inside each cell or compartment, the PEMC and t-PEMC algorithms are utilized [2, 3]. The resulting algorithm is very fast and robust.The algorithm is applied to study the formation of silver nanoparticles [4] in a large reactor utilizing compartment models. The parameters of the compartment models are extracted CFD simulations. The MC ability to track single events was utilized to study the impact of turbulence and stability factor on the generation of large particles. Controlling the production of oversized particles is very important in many areas like printable electronics. This information can be utilized to optimize the industrial reactors. [1] Irizarry R. (2012) submitted to Ind. Eng. Chem. Res..[2] Irizarry R. (2007) Chem. Eng. Sci. 63, 7665-7675.[3] Irizarry R. (2011) Chem. Eng. Sci. 66, 4059–4069.[4] Irizarry R. (2010) Ind. Eng. Chem. Res., 49, 5588–5602.

(54a) Computation of Minimum Bubbling Velocity for Large Particles

Dimitri Gidaspow and Fang Li, Chemical and Biological Engineering, Illinois Institute of Technology, Chicago, IL

A theory for the origin of bubbles in fluidized beds is described in chapter 6 in Gidaspow's book, Multiphase Flow and Fluidization, Academic Press, 1994. The theory shows that without the effects of solids pressure, the Ergun drag law leads to bubble formation for small particle Reynolds numbers and to dispersion for large Reynolds numbers. We have now discovered by numerical simulation using the codes in Gidaspow and Jiradilok 2009 book,Computational Techniques, Nova Science that for large particles there exists a region above minimum fluidization without bubble formation, similar to that for Geldart group A particles.

We have visually confirmed the existence of a bubble free region for fluidization of a mixture of 5 mm glass beads and 3.5 mm red glass beads in a 15 cm wide two dimensional glass fluidized bed. Similar to the computation the smaller red beads are near the top of the bed. When the bed width is decreased to about 5 cm, we see distinct slugging. Such behavior is well known and was computed . To obtain non-slugging behavior, the bed width needs to be sufficiently large.

The computations show that in the region between minimum fluidization and minimum bubbling the granular temperatures of the particles are sufficiently large for the dispersion coefficients to be high enough near minimum bubbling velocity. The computed wall to bed heat transfer coefficients are also sufficiently high to allow heating of the bed through the wall. Hence this new flow regime is practical and allows high production rates, due to the high gas velocities that can be achieved with the large particles.

(54b) Constitutive Model for the Fluid-Particle Drag Coefficient in Filtered Two-Fluid Models for Gas-Particle Flows

Chris Milioli[1], Fernando Milioli[1], William Holloway[2] and **Sankaran Sundaresan**[2], (1)Mechanical Engineering, University of Sao Paulo, São Carlos-SP, Brazil, (2)Chemical Engineering, Princeton University, Princeton, NJ

AIChE 2012 Annual Meeting

Special Session to Celebrate Tom O'brien's Career Long accomplishments

Constitutive model for the fluid-particle drag coefficient in filtered two-fluid models for gas-particle flows

Chris Milioli*, Fernando Milioli*, William Holloway, Sankaran Sundaresan

Department of Chemical and Biological Engineering, Princeton University, Princeton, NJ

Abstract

It is well known that gas-particle flows manifest fluctuations over a wide range of length and time scales. Filtered two-fluid model formulations, where the coarse flow structures are resolved and the effects of fine structures are modeled, are beginning to appear in the literature [1-5]. These filtered models require constitutive relations to capture the effect of sub-filter scale inhomogeneities on the effective drag force and the particle and fluid phase stresses. Such closures have been derived from the results obtained from highly resolved simulations [1-5]. Both Parmentier et al. [5] and Igci and Sundaresan [3,4] model the correction to the fluid-particle drag coefficient as a function of the filtered particle volume fraction and filter size. Such models average over all possible sub-filter scale meso-scale structures.

It is reasonable to expect that there should be additional variables characterizing the sub-filter scale meso-scale structure, which are not recognized in these models. In the present study we ask if one can formulate a more elaborate model for the filtered drag force containing additional variables characterizing the sub-filter scale meso-scale structures, thereby allowing us to differentiate between the filtered drag coefficients corresponding to appreciably different sub-filter scale structures. To this end, we have examined the drag coefficient values obtained by filtering the results of highly resolved kinetic theory based simulations. We have classified the filtered drag coefficient as a function of solid fraction and various additional filtered quantities. We found no systematic dependence of the filtered drag coefficient on the variances of the solid fraction, gas velocity, or the solid velocity. However, it is quite sensitive to changes in the filtered slip velocity; as the filtered slip velocity increases, the filtered drag coefficient decreases and approaches an asymptotic dependence on particle volume fraction at high slip velocities. This trend is opposite of what one would obtain when inertial effects become important in flow past individual particles. We rationalize it as a consequence of the formation of larger meso-scale structures at larger slip velocities. We also propose a model for the filtered fluid-particle drag coefficient in terms of the filtered solid fraction, filter size and filtered slip velocity.

References

1. K. Agrawal, P. N. Loezos, M. Syamlal, and S. Sundaresan, *J. Fluid Mech.*, **445**, 151 (2001).

2. D. Z. Zhang and W. B. VanderHeyden, *Int. J. Multiphase Flow*, **28**, 805 (2002).

3. Y. Igci and S. Sundaresan, *Ind. Eng. Chem. Res.*, **50**, 13190–13201 (2011).

4. Y. Igci and S. Sundaresan, *AIChE J.*, **57**, 2691-2707 (2011).

5. J.F. Parmentier, O. Simonin, and O. Delsart, *AIChE J.*, 2011, DOI: 10.1002/aic.12647.

Department of Mechanical Engineering, University of S‹o Paulo, S‹o Carlos-SP, Brazil

(54c) Predicting Gas-Solid Bubbling Bed Flows Using Continuum Modeling

Jennifer Sinclair Curtis and Deepak Rangarajan, Chemical Engineering, University of Florida, Gainesville, FL

A high speed gas jet injected into a bed of particles causes bubble formation that typically enhances mixing between the gas medium and particles. This favorable type of flow is employed in many industrial processes such as coal and biomass gasification. The aim of the current project is to develop an accurate and feasible method to model such flows. The Eulerian/Continuum framework, which treats particles as a continuous fluid-like phase is employed for versatility and computational feasibility. The major challenge in continuum modeling is in appropriate descriptions of terms requiring closure, which are particle stress due to translation and collision, gas-particle drag force, particle stress due to friction and gas-particle turbulence interaction. While the description of the former two terms are now more or less well-established, there is no consensus on the description of the latter two terms, which play a very important role in such dense turbulent flows. This talk will focus on evaluating currently available friction and turbulence interaction models by validating them against our own bubbling fluidized bed experiments with the aim of finding an appropriate description for the particle frictional stress and turbulence interaction.

(54d) Simulation of Mono- and Bi-Disperse Gas-Particle Riser Flow with Quadrature-Based Moment Methods

Alberto Passalacqua, Department of Chemical and Biological Engineering, Iowa State University, Ames, IA and **Rodney O. Fox**, Chemical & Biological Engineering, Iowa State University, Ames, IA

Gas-particle flows can be described by a kinetic equation for the particle phase coupled with the Navier-Stokes equations for the fluid phase through a momentum exchange term. The direct solution of the kinetic equation is prohibitive for most applications due to the high dimensionality of the space of independent variables. A viable alternative is represented by moment methods, where moments of the velocity distribution function are transported in space and time. A fully coupled third-order, quadrature-based moment method (Fox, 2008; Passalacqua et al., 2010) was applied in this work to the simulation of mono- (He et al, 2009a) and bi-disperse (Mathiesen et al., 2000) gas-particle flows in the riser of a circulating fluidized bed. Closures for the moment spatial fluxes and the source terms in the moment transport equations are provided by a quadrature-based representation of the particle velocity distribution function. A Bhatnagar-Gross-Krook collision model is used in the mono-disperse case, while the full Boltzmann integral is adopted in the bi-disperse case. The predicted values of mean local phase velocities, RMS velocities, and particle volume fractions are compared with the predictions of Euler-Euler multi-fluid models, with kinetic theory closures for the particle phase, Euler-Lagrange simulations (He at al, 2009a; 2009b), and experimental data from the literature (He at al, 2009a; Mathiesen et al, 2000). The grid-independence of the solution provided by Euler-Euler hydrodynamic models and quadrature-based moment methods is examined to illustrate that QMOM achieves grid-independent solutions, due to its ability of predicting particle-trajectory crossing without incurring in the formation of delta-shocks, and allowing converged statistics to be extracted from the simulation.

References

Fox, R. O. A quadrature-based third-order moment method for dilute gas-particle flows. J. Comput. Phys. 2008, 227, 6313–6350.

He, Y.; Deen, N.; Van Sint Annaland, M.; Kuipers, J. A. M. Gas-solid turbulent flow in acirculating fluidized bed riser: Experimental and numerical study of monodisperse particle systems. Ind. Eng. Chem. Res. 2009a, 8091 – 8097.

He, Y.; Deen, N.; Van Sint Annaland, M.; Kuipers, J. A. M. Gas-solid turbulent flow in a circulating fluidized bed riser: Numerical study of binary particle systems. Ind. Eng. Chem. Res. 2009b, 8098 – 8108.

Passalacqua, A.; Fox, R. O.; Garg, R.; Subramaniam, S. A fully coupled quadrature-based moment method for dilute to moderately dilute fluid-particle flows. Chem. Engng. Sci. 2010, 65, 2267 – 2283.

Mathiesen, V.; Solberg, T.; Hjertager, B. H. An experimental and computational study of multiphase flow behavior in a circulating fluidized bed. Int. J. Multiphase Flow 2000, 387 – 419.

(54e) A Constitutive Model for Intermediate, Quasi-Static and Inertial Regimes of Dense Granular Flow

Shankar Subramaniam and V. Vidyapati, Mechanical Engineering, Iowa State University, Ames, IA

Experiments [Powder Technol. (2003), vol. 131, pp. 23-39; Phys. Rev. E 78 (041306)] and discrete element simulations (DEM) [J. Fluid Mech. (2002), vol. 465, pp. 261-291; Powder Technol. (2006), vol. 169, pp. 10-21] have established that dense granular flow exhibit three different kinds of rheological behavior. In the quasi-static regime the granular shear stress is independent of strain rate, in the intermediate regime the shear stress has a power-law dependence on the strain rate with exponent between 0 and 2, and in the inertial regime the shear stress has a quadratic dependence on the strain rate. A recent assessment of several constitutive models with DEM data for homogeneous sheared dense granular flow [Chem. Eng. Sci. (2012), vol. 72, pp. 20-34] reveals that none of the models is able to capture the granular rheology in the intermediate regime. It has also been recently demonstrated through DEM calculations of silo discharge that the error in granular stress predicted by current constitutive models is directly implicated in accurate prediction of the discharge rate, and the spatial region of this erroneous stress prediction coincides exactly with the spatial extent of the intermediate regime. Analysis of the DEM data in the intermediate regime [Chem. Eng. Sci. (2012), vol. 72, pp. 20-34] also reveals that the dominant contribution to granular stress is from the contact (virial) contribution, while the streaming contribution is negligible. Based on these findings a constitutive model is derived by averaging the DEM expression for the contact stress. This contact stress model (CSM) naturally gives rise to the dependence of average contact stress on mesoscale flow descriptors such as the coordination number and the fabric tensor, as well as the average normal contact force. Appropriate closures for the coordination number and the fabric tensor are provided by solving their modeled evolution equation proposed by Sun and Sundaresan [J. Fluid Mech. (2011), vol. 682, pp. 590-616]. The predictive capability of the proposed contact stress model (CSM) is tested in homogeneous shear flow using DEM data corresponding to the intermediate, quasi-static and inertial regimes. A simple extension of CSM that is obtained by combining it with the refined OP (ROP) concept results in the ROP-CSM model that satisfactorily predict dense granular flow in all regimes: quasi-static, intermediate, and inertial.

(54f) Minimum Fluidization and Bubbling Properties - the Effects of Fines Content and Pressure

Allan S. Issangya, Roy C. Hays, S. B. Reddy Karri, Ray A. Cocco and Ted M. Knowlton, Particulate Solid Research, Inc., Chicago, IL

Gas-solids fluidized beds have wide industrial applications as they offer rapid solids mixing, relatively uniform temperature, good solids-gas contacting and heat and mass transfer. In the fluidization processes, generally once the velocity of the gas exceeds the minimum velocity necessary to suspend the solids, Umf, any excess flow passes through the bed in the form of gas bubbles or voids. Geldart's Group A powders (Geldart, 1986) are an exception in that they show an appreciable bubble-free fluidization range at gas velocities between the minimum fluidization velocity and the velocity at which bubbles start to appear, the minimum bubbling velocity. There is interest in knowing the bed densities, rhomf and rhomb, corresponding to the two fluidization points. Wells(2001) gas bypassing model suggested that the emulsion phase density in deep beds isn't constant but rather it increases from a value equal to rhomb at the bed surface to a value approaching rhomf near the bed bottom and that this increase in the emulsion phase density eventually leads to gas bypassing. Our earlier gas bypassing studies in deep beds of FCC catalyst particles showed that increasing the system pressure and/or addition of fines eliminated gas bypassing or reduced it significantly. were conducted in a 15-cm-diameter column to determine how the two factors affect the the minimum fluidization and bubbling densities of FCC catalyst particles. System pressures of up to 200 kPag and fines contents of up to 12% <44 microns were used. There was a significant decrease of the minimum bubbling density with fines content. For the pressure range tested there was only a slight decrease in the minimum bubbling density. There was no significant influence of pressure or fines content on the minimum fluidization density. The results are used to explain the effect of pressure and fines on gas bypassing in deep beds of FCC catalyt particles.

(76a) The Primary Particle Diameter of Aerosol Agglomerates & Aggregates From Mass-Mobility Characterization

Max Eggersdorfer[1], Arto J. Gröhn[1], C. M. Sorensen[2], Peter H. McMurry[3] and Sotiris E. Pratsinis[1], (1)Particle Technology Laboratory, Swiss Federal Institute of Technology (ETH Zurich), Zurich, Switzerland, (2)Physics, Kansas State Univ., Manhattan, KS, (3)Mechanical Engineering, University of Minnesota, Minneapolis, MN

The Primary Particle Diameter of Aerosol Agglomerates & Aggregates from Mass-Mobility Characterization

Gas-borne nanoparticles generated at high temperatures undergo coagulation forming agglomerates (physically-bound particles) and aggregates (chemically- or sinter-bound particles). The structure of such particles influences their transport, light scattering, effective surface area and density. Significant advances have been made in characterization of agglomerates (physically –bonded particles) by employing fractal theory and relating agglomerate structure to its generation pattern through the mass fractal dimension, D_f. The D_f values have been developed for agglomerates of *monodisperse* primary particles. For coagulating aerosols, however, this needs to be carefully examined as Brownian coagulation leads to polydisperse particles [1]. Furthermore, once coalescence or sintering starts between these primary particles, sinter necks are formed between them converting the agglomerates to aggregates.

Real-time characterization of nanoparticles is necessary for continuous monitoring of aerosol manufacturing and airborne pollutant particle concentrations, but is still challenging [2]. Mostly ex-situ methods have been used to characterize such structures in terms of agglomerate mass, mobility and radius of gyration, D_f, primary particle diameter and number and specific surface area (SSA). Figure 1 shows a sphere, an aggregate [3] and agglomerate [1] having the same mobility diameter, d_m, but different mass and surface area. So measuring only one particle property is not sufficient to characterize those structures.

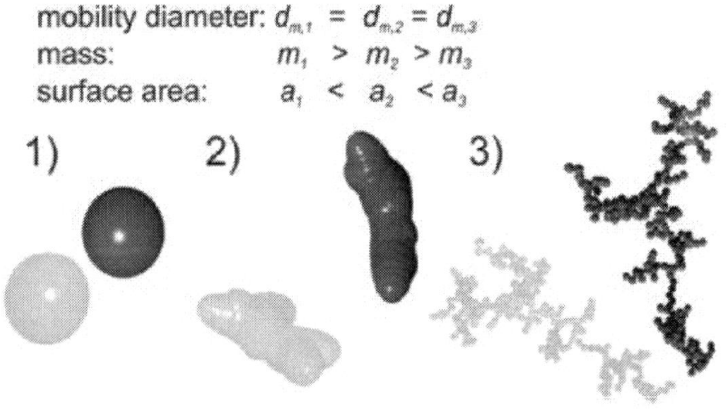

Figure 1. Snapshots of simulated nanoparticle structures: 1) sphere, 2) aggregate & 3) agglomerate.

Here, zirconia nanoparticles are generated by a scalable flame spray process and are characterized in almost real-time with their mass and mobility diameter. The mobility diameter is measured by a differential mobility analyser (DMA) and the mass by an aerosol particle mass (APM) analyser to determine the mass-mobility exponent (D_{fm}). Additionally, a new relation [4] between surface area mean primary particle diameter, aggregate/agglomerate mass and mobility diameter is used to extract the surface area mean primary particle diameter or SSA from these data. The effect of oxygen flow (Fig. 2) and precursor feed rate as well as precursor concentration on agglomerate/aggregate structure and primary particle diameter are investigated. Good agreement between ex-situ nitrogen adsorption (BET), transmission electron microscopy (TEM) and on-line DMA-APM is found for all investigated process conditions.

1. Eggersdorfer, M.L. and Pratsinis, S.E. (2012) *Aerosol Sci. Technol.* **46**, 347-353.

2. Scheckman, J.H., McMurry, P.H. and Pratsinis, S.E. (2009) *Langmuir* **25,** 8248-8254.

3. Eggersdorfer, M.L., Kadau, D., Herrmann, H.J. and Pratsinis, S.E. (2011) *Langmuir* **27**, 6358-6367.

4. Eggersdorfer, M.L., Kadau, D., Herrmann, H.J. and Pratsinis, S.E. (2012) *J. Aerosol Sci.* **46**, 7-19.

(76b) Combined On-Line Differential Mobility and Particle Mass Analysis for Determination of Size Resolved Particle Density and Microstructure Evolution

Qin Liu, **Xiaofei Ma** and Michael Zachariah, Department of Mechanical Engineering and Chemistry, University of Maryland, College Park, MD

A new application of on-line tandem differential mobility-particle mass analysis (DMA-APM) is used to obtain the density distribution, detect the reaction evolution, and investigate the mechanism of porous particle formation. The expected uncertainty of the density measurement is within 5%. The method has several advantages over absorption methods such as BET in which pore model is required. The methodology also enables a size resolved measurement to understand how particle porosity varies over a given particle population. The DMA-APM method is materials independent, and can measure the density even in a particle with inaccessible pores. In this paper we demonstrate the method in the synthesis of mesoporous iron oxide and copper oxide. For example, in the formation of iron oxide we found that small particles (~50 nm) in the size distribution had a higher density (~2.6 g/cm^3) relative to larger 130 nm particles with a measured density of 2.2 g/cm^3. Syntheses at higher temperatures lead to high density particles which were also less size sensitive.

(76c) Roughness Effects On Particle Adhesion

Nyah V. Zarate[1], James D. Litster[2] and Stephen P. Beaudoin[1], (1)Chemical Engineering, Purdue University, West Lafayette, IN, (2)School of Chemical Engineering, Department of Industrial and Physical Pharmacy, Purdue University, West Lafayette, IN

Surface roughness is influential in many solid processing variables, such as choice of excipients and flowing parameters (compression, pneumatic conveying, etc.) that affect the quality of the final product. Surface roughness is one of the most important parameters in surface/particle engineering.

In this work, we have utilized our experimentally-validated simulation scheme as a tool to estimate the adhesion forces between nanosized silicon nitride and ultra-fine particles (d<10 μm); silica, alumina and alumina coated with nano-silica particles, against a silica surface.

This study focuses on the importance of different physical parameters such as the material, size and surface roughness of the interacting bodies on the force of adhesion between them, with particular attention to particle roughness as the particle size and roughness is varied from the micro- to the nanoscale. The effective Hamaker constant (a force constant for the van der Waals force) for nanoscale particle-substrate systems was derived from experimentally obtained force measurements between a model particles and the substrate of interest, silica surface (polished and unpolished).

By using surface features (geometry and roughness) examined by AFM and SEM imaging of particles, it was determined that surface roughness generally reduces adhesion of both micro- and nanoscale particulates. However, the effect of roughness on the adhesion force is greatly dependent on the particle probe asperities, with asperity diameters below 50 nm, observed in silco. Also, through simulation the adhesion force distribution was related to the distribution of asperities where normal distribution asperities (randomly generated surface) resulted in lognormal adhesion force distributions for all sizes (4 μm, 8 μm, 10 μm, 50 μm). The particle size and geometry can directly influence the adhesion force by altering the area of interaction between the particle and substrate. As a result, it remains an important area of study for particulate systems.

(76d) Using Inverse Gas Chromatography to Assess Passivation of High Surface Energy Sites of Milled Pharmaceutical Crystals Via Dry Coating of Nano-Silica

Xi Han, Otto H. York Department of Chemical, Biological and Pharmaceutical Engineering, New Jersey Institute of Technology, Newark, NJ, Laila Jai Jallo, New Jersey Center for Engineered Particulates, New Jersey Institute of Technology, Newark, NJ, Daniel To, Chemical Engineering, New Jersey Institute of Technology, Montclair, NJ, Chinmay Ghoroi, Department of Chemical, Biological and Pharmaceutical Engineering, New Jersey Center for Engineered Particulates, Newark, NJ and **Rajesh N. Dave**, Chemical, Biological and Pharmaceutical Engineering, New Jersey Institute of Technology, Newark, NJ

In this work, we examine the influence of simultaneous micronization and surface modification via dry coating of nano-silica particles on the surface energy of ibuprofen crystals milled down to different sizes to be able to explain higher surface energy of milled ibuprofen and mitigation thereof via dry coating. Fluid energy mill was used to micronize ibuprofen particles to sizes from 5 to 30 µm with or without simultaneous dry coating with nano silica. Bulk powder properties such as bulk density, compressibility, unconfined yield strength were characterized using Freeman FT4 rheometer, and powder dispersibility was measured using Rodos/Helps system by Sympatec. Surface properties were characterized using the Surface Energy Analyzer, via Inverse Gas Chromatography (IGC) to be able to explain the bulk flow improvement due to dry coating.

As received, uncoated, micronized ibuprofen samples show an increase of dispersive surface energy (from 40.7 to 54.4 mJ/m^2) as the particle size decreases with increasing milling intensity. On the other hand, dry coated milled ibuprofen powders show an overall decrease in the dispersive surface energy. The detailed surface energy maps reveal that pure micronized ibuprofen surface is highly heterogeneous, while dry coating reduces the heterogeneity. All of the measured bulk powder properties showed an improvement due to dry coating.

Overall, it was found that improved powder flow and packing characteristics may be obtained when milling includes simultaneous surface modification via dry coating of nano-silica, thus eliminating disadvantages of micronization. These improvements are attributed to decreased cohesion indicated by reduced dispersive surface energy. Increased surface energy, hence cohesion for ordinary milled ibuprofen may be due to creation of high energy sites during milling, where as the detailed surface energy measurements indicate that dry coating with nano-particles quench the high energy sites and making ibuprofen surface less heterogeneous and having reduced surface energy comparable to that of the nano-particles used for dry coating.

(76e) Electrically and Thermally Conductive Oxide Particles Functionalized by Metallic W Atomic Layer Deposition

Staci A. Van Norman[1], Joseph W. Tringe[2], John Falconer[1] and Alan W. Weimer[1], (1)Chemical Engineering, University of Colorado at Boulder, Boulder, CO, (2)Lawrence Livermore National Lab, Livermore, CA

Functionalization of insulating particles by Atomic Layer Deposition (ALD) could provide a unique range of electrical conductivity and outperform metal doped polymers for applications as replacements for metal electrodes in devices. The characterization and method of applying conductive metal ALD coatings on insulating materials to selectively produce precisely controlled properties is discussed.

In this study, functionalization of four types of oxide and polymer particles, 18 μm diameter or larger, was achieved by ALD of conformal metallic W thin films. Hollow and solid oxide particles as well as high and low surface area porous polymers were examined for conductivity range with increasing W film thickness. Particles were characterized for electrical and thermal conductivity using a four point probe planar van der Pauw test geometry. Measurable decreases in resistance over bare oxide support particles have been seen for film thicknesses of 25 nm or greater, additional decreases observed by increasing W thickness. Combining resistance characterization with ALD of conductive films on non-conductive particles allows for the development of a novel technology to selectively control electrical and thermal properties for defined applications.

(76f) Phase Separation and Reaction Mechanism During Agcu Particle Formation by Cosolvent Assisted Pyrolysis Technology

Kai Zhong[1], George Peabody[1], Elizabeth Blankenhorn[2], Howard Glicksman[3] and **Sheryl H. Ehrman**[1], (1)Department of Chemical and Biomolecular Engineering, University of Maryland, College Park, MD, (2)Chemical and Biomolecular Engineering, University of Maryland, College Park, College Park, MD, (3)DuPont Electronic Technologies, Research Triangle Park, NC

The spray pyrolysis process was used for the generation of micron sized metal particles because of its advantages of making adjustable and narrowly distributed particle sizes. In our process, a copper nitrate and silver nitrate mixture solution was used as the precursor for the generation of AgCu particles, of interest as ingredients in metal pastes and in isotropic conductive adhesives (ICA). Either 40 vol% ethanol or 40 vol% ethylene glycol was used as the cosolvent, to produce a reducing environment inside the reactor, resulting in metallic powder production without additional hydrogen added as a process gas. XRD and SEM were used for the characterization of AgCu particles. Phase separation was observed during the generation of AgCu particles, and particles mainly consisted of a Ag-rich solid solution phase and a Cu-rich solution phase. Short residence time experiments were used to investigate the particle formation process. With a cosolvent of ET, Ag precipitated during the evaporation of the solvent, and agglomerated on the surface of the particles. $Cu(NO_3)_2$ decomposed to $Cu_2(OH)_3NO_3$ first, then to CuO and Cu_2O. CuO and Cu_2O were reduced to Cu in the reducing atmosphere resulting from the use of the ET cosolvent. When EG was used, nano-sized Ag particles were generated during the evaporation process. Layered structures of $Cu_2(OH)_3NO_3$ and Cu_2O with embedded Ag particles were observed during the formation of the particles. Understanding the mechanism of bicomponent particle formation, including role of co-solvent, is important for the optimization of the process and to obtain the desired morphology.

(88a) Synthesis of Visible Light-Sensitive TiO2 Coated Silica Beads by Cu(II) or Fe(III) Grafting

Junichi Ida[1], Kazumasa Tokumoto[2], Sho Suzuki[2], Soshi Watanabe[1], Tatsushi Matsuyama[1] and Hideo Yamamoto[1], (1)Department of Environmental Engineering for Symbiosis, Soka University, Hachioji, Tokyo, Japan, (2)Department of Environmental Engineering for Symbiosis, Soka University, Hachioji, Japan

Visible light sensitive TiO_2 coated silica beads were prepared by grafting Cu(II) or Fe(III) ion onto TiO_2. In the experiment, the surface of silica beads (D=0.5μm) was coated by TiO_2 by applying direct hydrolysis reaction of Ti-tetraisopropoxide(TTIP) in ethanol solution. Scanning electron microscopy observation confirmed that TiO_2 was coated on the silica beads in the form of amorphous nano particles uniformly by this method. When the amount of supplied TTIP solution increased during coating process, the amount of coated TiO_2increased. After calcined at high temperature the amorphous TiO_2 nano particles changed into crystalline $TiO_2$and their diameter also increased to few ten~few hundred nm. When the coated amount increased, the ratio of rutile/anatase also varied even calcined at the same temperature.

To synthesize visible light-sensitive TiO_2 by Cu(II) or Fe(III) grafting, it is known that rutile phase TiO_2 shows better photocatalytic activity than in the case of using anatase phase TiO_2. Therefore, in this experiment, the TiO_2 coated beads (TCB) with higher rutile/anatase ratio was used for grafting. When Cu(II) or Fe(III) ion was grafted by the impregnation method to TCB, it showed visible light absorption (>400nm). The photocatalytic activity of the Cu(II) of Fe(III) grafted TCB were evaluated by the photocatalytic decomposition of acetic acid as a model substance under the visible light (410-530nm) irradiation using Xe lamp. The result showed that Cu(II) or Fe(III) grafted TCB demonstrated higher activity compared to the non grafted TCB. The result indicated that visible light sensitive TiO_2 coated silica beads were successfully prepared by this method.

(88b) Experimental Studies in the Characterization Occurring At the Catalytic Surfaces of Silver On Titania Adsorbents

Zenda D. Davis and Bruce J. Tatarchuk, Chemical Engineering, Auburn University, Auburn, AL

Solid catalysts have been used in the transformation of hydrocarbons to useful products for many years. The majority of adsorbents/catalysts used consists of a non metallic support promoted by transition metallic oxides. It has been well established that particle size and dispersion is directly related to the efficacy of adsorbents in heterogeneous catalysis. Thus, information on these two properties in particular is crucial to the development of adsorbents/catalyst with greater efficiency. Gaseous adsorption, XRD, and electron microscopy have been routinely applied in heterogeneous catalysis research for particle size determination. This work will detail the use of XPS analysis in the investigation of Ag nanometer sized particles dispersed on TiO_2 support. XPS will be shown to be a more reliable technique for the small particle size determination than the traditional particle size estimation techniques. The limitations of the three techniques will be discussed in their application to particle size determination of Ag/TiO_2 adsorbents. A correction factor will be introduced to chemisorption based on XPS analysis that will allow more accurate particle size determination.

(88c) Preparation and Assembly of Anisotropically Polymer-Grafted Silica Nanospheres Via Adsorption On Charged Surface

Dieu Huong Pham, Ayae Sugawara-Narutaki, Atsushi Shimojima and Tatsuya Okubo, Department of Chemical System Engineering, The University of Tokyo, Tokyo, Japan

Anisotropic particles that possess non-uniformity in surface structure or function have attracted great attention. Owing to the anisotropy in their inter-particle interaction, these particles are expected to become an important element of functional nanomaterials if they can be assembled into useful structures. However, up until now, the preparation of anisotropic particles less than 100 nm in diameter has been a difficult challenge. In this study, a method to synthesize anisotropically polymer-grafted silica nanospheres was developed by utilizing their adsorption to a charged surface. First, Carboxyl-terminated polystyrene (Mw = 37800) was synthesized via atom transfer radical polymerization (ATRP). Next, spherical silica nanospheres (SNSs) 50 nm in diameter were mixed together with hydrotalcite (HT), a plate-like layered crystals, in water at pH 11.5. Under this pH condition, SNSs are adsorbed on the surface of HT by electrostatic interaction since the surface of SNSs is negatively-charged while that of HT is positively-charged. Next, SNSs-adsorbing HT were treated with 3-aminopropyltriethoxysilane to introduce amino groups on the exposed surface of the SNSs. Coupling reaction between amino group on SNSs surface and carboxyl-terminated polystyrene was carried out at room temperature using coupling agents (DCC and DMAP). HT substrates were removed by being dissolved in acidic water to give anisotropically polymer-grafted SNSs. Mixture of these particles, water and chloroform was emulsified to formed water-in-oil microemulsion where polymer-grafted SNSs are located at the water-oil interface. Carefully removing the liquid phase produced hollow particles of several hundred nm in size with their wall constructed from polymer-grafted SNSs.

(88d) Metal:Metal Oxide Complexes Via Organic Coating Techniques: Understanding Electro-Optical Properties

Diane M. Steeves[1], Jason W. Soares[1], Jisun Im[2], Jagdeep Singh[2], Yiwen Wang[2] and James E. Whitten[2], (1)U.S. Army Natick Soldier Research Development and Engineering Center, Natick, MA, (2)Department of Chemistry and Centers for Advanced Materials and High-Rate Nanomanufacturing, University of Massachusetts Lowell, Lowell, MA

Metal oxide nanoparticles are finding use in various applications ranging from catalysis to optoelectronics. In particular, nano-crystalline zinc oxide (nano-ZnO) is an important electro-optical material due to its wide bandgap and bimodal photoluminescence (PL) spectrum. Nano-ZnO represents a promising semiconductor for future hybrid devices. To facilitate the creation of ZnO hybrid complexes, a simple thiol modification approach has been developed, wherein organothiol surface modifiers are covalently attached to the surface of ZnO nanoparticles. Reaction under ambient conditions creates a thiol monolayer, while elevated temperatures result in complete encapsulation of the ZnO nanoparticles. While previous studies have demonstrated that alkanethiols adsorb on ZnO, the present work is the first to demonstrate encapsulation of ZnO nanoparticles. The thiol modification procedure has been extended to other metal oxides, in particular CuO and Cu_2O nanoparticles. CuO nanoparticles demonstrated a similar thiol uptake to that seen with ZnO; however, Cu_2O thiol uptake is significantly reduced. In addition to the influence of the organothiol on the inherent optical properties of the nanoparticles, single crystal studies to elucidate the interfacial interactions responsible for the differential thiol uptake between CuO and Cu_2O will be discussed. The thiol modification process has been used as a basis to form hybrid metal:metal oxide complexes. More specifically, implementation of dithiol surface modifiers has facilitated the creation of hybrid systems with gold nanoparticles (AuNPs) tethered to the functionalized metal oxides nanoparticles. The creation of AuNP-ZnO and AuNP-CuO hybrid complexes and the influence on the electro-optical properties of the metal oxides were examined. The information derived enables the formation of future metal oxide:AuNP complexes with increasing control of their electro-optical properties, thereby generating a series of hybrid materials with specific, tailored optical responses. The creation of hybrid metal:metal oxide complexes through the described thiol modification approach could have significant implications on future chemical sensors, as well as photovoltaic devices. Additionally, because a large selection of functionalized thiols is available, it is possible to surround metal oxide nanoparticles with a variety of chemical functional groups for subsequent interactions, or to facilitate dispersion within polymer matrices.

(88e) Surface Modification of Carbon Hard Sphere Via Surface Initiated Atom Transfer Radical Polymerization

Lakhinder Kamboj, Chemical and Biomolecular Engineering, Tulane University, metairie, LA

Lubricants play an integral role in the operation of several technologies and in biology also, ranging from moving parts in machinery to the biolubrication of artificial joints. We have found that a colloidal dispersion consisting of easily synthesized highly spherical and uniform carbon particles coupled with surface grafting of polymers that result in a stable dispersion and provides low friction coefficients and excellent surface wear protection. The covalent functionalization with the polymer graft is achieved through the surface initiated atom transfer radical polymerization (SI-ATRP). We hypothesize that these particles would decrease the overall friction coefficient compared to ungrafted particles. These particles use a rolling mechanism similar to nano –or microscale ball bearing under confinement. The effect of several parameters of lubrication including the thickness of the grafted polymer layer, type of polymer grafted, along with the size of the carbon particles will be presented. These parameters listed would be optimized to generate the lowest frictional coefficient. Broader applications beyond just lubrication include adsorbents, and reinforcement in composite materials

(88f) Adsorption of Acrylic/Sulfonic Acid Copolymer-Coated Iron Oxide Nanoclusters On Silica Microspheres At High Divalent Salinity

Hitesh G. Bagaria[1], Zheng Xue[1], Andrew Worthen[1], KiYoul Yoon[1], Bethany Neilson[2], Victoria Cheng[1], JaeHo Lee[1], Christopher W. Bielawski[2] and Keith P. Johnston[1], (1)Chemical Engineering, The University of Texas, Austin, TX, (2)Chemistry, The University of Texas, Austin, TX

Minimizing the adsorption of engineered nanoparticles (NPs) on rock surface is paramount for feasibility of sub-surface applications and environmental fate of NPs. Natural reservoir brines often have high salinity including the presence of significant divalent cations including calcium that lead to strong NP adsorption. We recently identified a class of acrylic /sulfonic acid copolymer-coated iron oxide (IO) nanoclusters that display enhanced colloidal stability in the presence of divalent cations. Here, we report a combinatorial approach to study batch adsorption of IO coated with different polymers on silica microsphere models at various salinity and pH. The isotherms provide the adsorption capacity and equilibrium adsorption constants for various conditions. We find that a low adsorption capacity from these batch experiments is a necessary condition for successful transport in flow experiments. Thus, in addition to providing the equilibrium adsorption capacity, the batch isotherms also save time for guiding the more laborious/expensive flow experiments.

(114a) Role of Aeration with Standpipe Flow

Ray Cocco, Ted Knowlton and S. B. Reddy Karri, Particulate Solid Research, Inc. (PSRI), Chicago, IL

Standpipes provide the pressure balance in a circulating fluidized bed and thereby dictate the solid flux into the riser. A higher pressure build in a standpipe provides the higher pressure needed to move more solids in the riser. Thus, in today's FCC units with Geldart Group A catalyst powder, the standpipe is most likely the bottleneck in increasing plant capacity. Aeration can help with this pressure build, but how it helps is somewhat contentious. For small-scale standpipes, aeration appears to provide fluidization of the standpipe core, especially at high aeration rates [1]. Pressure drop is translated from shear and wall stresses to increasing the pressure build in the standpipe. However, this may not be the mechanism for large-scale standpipes. It seems unlikely that aeration jets can penetration deep into a one-meter diameter standpipe, a size typically used in industry.

Several CFD models were developed using Barracuda® to understand the role of aeration in large-scale standpipes. Modeling results suggest that jet penetration from the aeration ports does not provide sufficient aeration of the standpipe core. Aeration appeared to be providing a "lubricating" effect for wall shear. Hence, having more aeration around the perimeter of the standpipe may be equally important as the axial aeration position for these commercial-scale standpipes.

[1] A. Srivastava, S. Sundaresan, K. Agrawal, S. R. Karri, and T. Knowlton, "Dynamics of gas-particle flow in circulating fluidized beds," Powder Technology, vol. 100, no. 2, pp. 173–182, 1998.

(114b) Agglomeration and De-Agglomeration of Wetted Solids: A Summary of Recent Findings

Christine M. Hrenya[1], Carly M. Donahue[2] and Robert H. Davis[2], (1)Chemical and Biological Engineering, University of Colorado at Boulder, Boulder, CO, (2)Chemical and Biological Engineering, University of Colorado, Boulder, CO

Liquid-coated, solid particles occur in a wide variety of applications – filtration, agitated drying, and granulation, to name a few. Much of the past work on these systems has focused on the agglomeration behavior of two-particle collisions that occur in the normal (head-on) direction. Over the past several years, we have extended this work by considering two cases in details: (i) the normal collision of 3 wetted particles, and (ii) the oblique collision of 2 wetted particles. The experiments are rich in surprising results, and have shed light on important physical mechanisms previously believed to be negligible. Corresponding advances in the theoretical description have ensued, including the important role of glass transition, added resistance upon rebound from existing liquid bridges, capillary forces even for high Capillary-number flows, and a new de-agglomeration mechanism characterized by a dimensionless number relating centrifugal forces to capillary forces. An overview of these findings will be given.

(114c) A Novel Mppic Methodology for Numerical Simulations of Gas-Solids Flows

Rahul Garg[1,2], Jean-Francois Dietiker[2,3] and Mehrdad Shahnam[2], (1)URS Corp, Morgantown, WV, (2)National Energy Technology Laboratory, Morgantown, WV, (3)West Virginia University Research Corporation, Morgantown, WV

CFD simulations of gas-solids flows are done either using continuum representation for both gas and solid phases (Two-Fluid Method) or using continuum representation for gas phase and a discrete representation for solid phase (Continuum-Discrete Method). The CFD simulations of dense multiphase flow applications suffer from very high computational turnaround times. This is either due to severe numerical instabilities in the frictional stress regime observed in two fluid method or due to very small time steps in fully resolved collision operators (such as soft spring model) in continuum discrete method using real particles. In the MPPIC (**M**ulti **P**hase – **P**article **In** **C**ell) methodology, continuum representation is used for the gas-phase. The solids phase is represented by discrete entities that map the solids loading. These discrete entities are generally referred to as computational particles or parcels or notional particles. In this approach, multiple particles are clubbed together to form one parcel. The parcels trajectories are tracked in a Lagrangian frame and a frictional stress term is added in order to prevent solids loading over the theoretically allowed limits. So far, available MPPIC models implemented into commercial CFD solvers provide limited description of the exact form of frictional stress models used. In this study we present an MPPIC method with a novel frictional stress term that has been implemented in the open source reactive multiphase flow CFD solver MFIX (Multiphase Flow with Interface eXchange). The newly developed frictional stress model is based on a combination of a coloring function and four simple post-collision rules. The results of MPPIC simulations obtained from MFIX for chosen fluidized bed applications are compared with more accurate continuum discrete method and also the two-fluid method in order to assess the relative accuracy and computational efficiency of different numerical simulation methodologies.

(114d) Implementation and Verification of Population Balance Model Using FCMOM in CFD Codes

Emadoddin Abbasi, Wanger Institute for Sustainable Energy Research (WISER),Chemical and Biological Engineering Department, Illinois Institute of Technology, Chicago, IL and Hamid Arastoopour, Wanger Institute for Sustainable Energy Research (WISER) Chemical and Biological Engineering Department, Illinois Institute of Technology, Chicago, IL

Successful computational fluid dynamics (CFD) simulations of poly-disperse flows must include modeling of particulate processes, i.e. nucleation, growth, agglomeration and breakage. In addition, in poly-disperse reactive flows heterogeneous chemical reactions would affect the particulate phase property distribution and also hydrodynamics of the system (1). A computationally attractive way of computing the temporal and spatial changes in the particulate phase property distribution function is utilizing the population balance equation (PBE). However, Integro-differential nature of the PBE demands efficient numerical techniques for the solution of PBE and its implementation in CFD codes (2).

Finite size domain complete set of trial functions method of moments (FCMOM) is a novel solution method in the "method of moment" family for population balance equation developed by Strumendo and Arastoopour (3, 4). The method poses unique properties including fast convergence to the exact solution and providing the solution of PBE in terms of the moments of the distribution and the reconstructed distribution function itself, which makes it distinct from other available approaches (5, 6). In addition, the method has been validated against cases with available analytical solution and the comparisons were excellent (2, 3, 4). However, the applicability of the method in simulations of a complex system using multiphase CFD approach has not been studied. This work is an effort to provide a coupled CFD-PBE model based on FCMOM and probe its applicability for CFD simulations of poly-disperse flows.

The method was employed in CFD simulations of an oil-water emulsion flow in a backward facing step geometry including different particulate processes as proposed by Silva et al. (6). The results were numerically verified against the simulation results obtained by QMOM model.

References:

1- Abbasi, E., Arastoopour, H., "CFD Simulation of CO_2 Sorption in a Circulating Fluidized Bed Using Deactivation Kinetic Model", Proceeding of the Tenth International Conference on Circulating Fluidized Beds and Fluidization Technology, CFB-10, edited by T. M. Knowlton, ECI, New York (2011): 736-743

2- Abbasi, E., Arastoopour, H., "Numerical Simulation of CO_2 Removal Process Using Solid Sorbent In a Fluidized Bed: A CFD-PBE Model", *2011 AIChE Annual Meeting*, Minneapolis, MN, USA (2011).

3- Strumendo, M., Arastoopour, H., "Solution of PBE by MOM in Finite Size Domains", *Chemical Engineering Science*, 63.10(2008): 2624-2640.

4- Strumendo M., Arastoopour H., "Solution of population balance equations by the FCMOM for in-homogeneous systems", *Industrial and Engineering Chemistry Research*, 49.11 (2010): 5222–5230.

5- Marchisio, D. L.; Pikturna, J. T.; Fox, R. O.; Vigil, R. D., "Quadrature Method of Moments for Population-Balance Equations", *AIChE Journal.*, 49 (2003a): 1266.

6- Silva, L.F.L.R. Damian, R.B., Lage, P.L.C., "Implementation and analysis of numerical solution of the population balance equation in CFD packages", *Computers & Chemical Engineering*, 32- 12 (2008): 2933–2945

(114e) DNS of Particle-Laden Gas Flow with Four-Way Coupling

John B. McLaughlin, Chemical and Biomolecular Engineering, Clarkson University, Potsdam, NY, Goodarz Ahmadi, Department of Mechanical and Aeronautical Engineering, Clarkson University, Potsdam, NY and Hojjat Nasr, MAE, Clarkson University, Potsdam, NY

This study is concerned with the effects of particle–particle collisions and the two-way coupling on the dispersed and carrier phase turbulence fluctuations in a channel flow. The time history of the instantaneous turbulent velocity vector was generated by the two-way coupled direct numerical simulation of the Navier–Stokes equations via a pseudo-spectral method. The particle equation of motion included the wallcorrected nonlinear drag force and the wall-induced and shear-induced lift force. The effect of particles on the flow was included in the analysis via a feedback force that acted on the computational grid points. Several simulations for different particle relaxation times and particle mass loadings were performed, and the effects of particle–particle collisions, particle feedback force and inter-particle interactions on the particle deposition velocity, fluid and particle fluctuating velocities, and particle concentration profiles were determined. The effect of particle aerodynamic interactions was also examined for certain cases. The simulation results indicated that when particle–particle collisions were included in the computation but two-way coupling effects were ignored, the particle normal fluctuating velocity increased in the wall region causing an increase in the particle deposition velocity. When the particle collisions were neglected but the particle– fluid two-way coupling effects were accounted for, the two-way coupling and the particle normal fluctuating velocity decreased near the wall causing a decrease in the particle deposition velocity. In the case of the four-way coupling in which both inter-particle collisions and two-way coupling effects were present, it was found that the particle deposition velocity increased compared with the one-way coupling case. When the particle aerodynamic interactions were added to the four-way coupled case (termed six-way coupled case), no significant changes in the mean fluid and particle velocities and the fluid and particle fluctuating velocities were obtained. The results for the particle concentration profile indicated that the inclusion of two-way coupling or inter-particle collisions into the computation reduced the accumulation of particles near the wall. It was also observed that particle–particle collisions and two-way coupling weakened the preferential distribution of particles. -

(114f) ECVT Imaging of 3-D Flow Structures of Irregular Geomteric Multi-Phase Flow Systems

Aining Wang, Qussai Marashdeh and L. - S. Fan, William G. Lowrie Department of Chemical and Biomolecular Engineering, The Ohio State University, Columbus, OH

ECVT IMAGING OF 3-D FLOW STRUCTURES OF IRREGULAR GEOMTERIC MULTI-PHASE FLOW SYSTEMS

Aining Wang, Qussai Marashdeh and Liang-Shih Fan*

William G. Lowrie Department of Chemical and Biomolecular Engineering

The Ohio State University

140 West 19th Avenue, Columbus, Ohio 43210, USA

*To whom correspondence should be addressed (fan@chbmeng.ohio-state.edu)

Measuring the dynamic flow structure in multi-phase flow systems requires a 3-D technique that is capable of sensing the flow field in real-time. Electrical Capacitance Volume Tomography (ECVT) is a newly developed technique that can provide such measurement. In ECVT, a set of non-invasive capacitance sensors are placed around a measurement section of the flow field. Sensors interact with each other collecting real-time measurements that can be related back to phase concentrations in the section. The technique is based on reconstructed volume images from capacitance signals acquired from sensors with inherent 3D features. The attractiveness of the technique is in its low profile sensors, fast imaging speed and scalability to different section sizes, low operating cost, and safety. Moreover, the flexibility of ECVT sensors enables them to be designed around virtually any geometry of the columns. The feasibility and accuracy of the ECVT technique for volume imaging have been demonstrated for various multiphase flow systems including those of a large circular column of 60 inches, with internals, at high temperatures, and with varied shapes.

In this work, an ECVT sensing system is utilized to examine gas-solid flows in irregular shaped conduits of a circulating fluidized bed. Specifically, volume images will be acquired and analyzed of elbow, cylindrical, and converged sections of the flow system. The results will be analyzed quantitatively. Results of volume solid holdup distributions will be reported also. The multi-phase flow system under interrogation will be examined at various flow conditions.

(124a) Multi-Scale Modeling and Validation of Twin Screw Granulation Processes

Dana Barrasso, Chemical Engineering , Rutgers University , Piscataway , NJ and Rohit Ramachandran, Department of Chemical and Biochemical Engineering, Rutgers University, Piscataway, NJ

Process robustness and controllability is of particular importance in the highly-regulated pharmaceutical industry. While batch processing is typically used in pharmaceutical manufacturing operations, continuous processing has numerous potential advantages in cost, efficiency, scalability, and controllability. In wet granulation, a liquid binder is sprayed onto a powder mixture of active pharmaceutical ingredient (API) and excipient particles, which aggregate to form larger granules. Granulation processes are often inefficient because granule size and composition can be widely distributed and difficult to control. Continuous granulators that have been developed or investigated for use in pharmaceutical processing include fluidized bed granulators, twin screw granulators, and high shear granulators. In this study, a multi-scale model for twin screw granulation was developed and validated with experimental data.

Wet granulation is governed by three rate processes: wetting and nucleation, consolidation and aggregation, and attrition and breakage. Population balance models are used track the number of particles with a given set of characteristics, such as size and composition, taking these rate processes into account. A three-dimensional population balance model was developed to simulate concurrent distributions in size, composition, and liquid binder content. It has been demonstrated that these properties strongly influence the aggregation rate. In order to represent the continuous granulator, spatial coordinates were added to the population balance model, along with an inflow and outflow of particles. This model was coupled with discrete element modeling (DEM). Axial and radial velocities within the twin screw granulator were determined from DEM simulations, and these particle fluxes were input into the population balance model. Experimental data was used to select values for adjustable parameters in the aggregation and breakage rate kernels. Using these rate constants, simulations were performed under varying process parameters, such as liquid binder spray rate. These results were compared to experimental data, demonstrating the predictive capability of the model.

Application of Euler-Lagrange Approach to Predict the Multiphase Flow with Simultaneous Heat and Momentum Transfer, and Solidification Predicted by a Shrinkage Unsolidified Core Model

Guilherme A. N. Ricardo[*1], *Dirceu Noriler*[1], *Waldir P. Martignoni*[2], *Henry F. Meier*[1]
[1]*Regional University of Blumenau - Blumenau, Santa Catarina, Brazil*
[2]*PETROBRAS/AB-RE/TR/OT - Rio de Janeiro, Rio de Janeiro, Brazil*
[*]*e-mail: guilherme_novelletto@yahoo.com.br*

Abstract

This work aims to the mathematical formulation and numerical simulation of a gas-liquid flow with simultaneous heat and momentum transfer, and solidification of droplets. The Euler-Lagrange approach was used in the mathematical formulation. The finite volume method present in a commercial Computational Fluid Dynamics (CFD) code was used as the discretization method. The specific heat of the discrete phase as a function of temperature was estimated through models based on contributions groups. The solidification and melting phenomena are modeled and evaluated by a shrinkage unsolidified core model. The model calculates the solid thickness formed within the liquid droplets during solidification or the disappearance of the solid thickness during the melting. The model that describes these phenomena was programmed in C language and linked in a commercial CFD code through subroutines. A numerical case was performed to analyze the physical consistency of the results. The analysis of the results showed that the models have physical consistency, and now can be applied to an industrial case. Thus, it is possible to analyze some important aspects of the phenomena involved in the granulation process, and especially the influence of the solidification in the heat transfer between the phases, and their impact on the quality of the product.

Introduction

In granulation process, the melt is sprayed in the top of a cylindrical tower where droplets are formed and they fall downward in counter-current with ambient air, which cools and solidifies the droplets, and the solid particles are collected in the bottom of the tower (Perry, 2008). In practice, the main problem is the deficiency in the solidification period, which can cause operational problems. When incomplete solidification occurs, the formed grains can break hitting the walls of the tower, or by collision between each other. This creates a kind of "cake" on the walls of the tower. When the solidification phenomenon starts, a solid thickness is formed inside the liquid droplets, so an additional conductivity resistance appears and decreases the heat transfer rate between the gas and the droplets. The most important aspect in solidification is the heat transfer, because it is necessary remove the heat for it to occur. The same occurs with the melting, but heat must be added. This way, seeks to a better understanding of these phenomena so that these problems can be solved.

Mathematical Modeling and Numerical Solution

The mathematical modeling uses an Euler-Lagrange approach and the theoretical basis about these approaches can be found in authors like Balachandar (2010), Cross *et al.* (2006), Hoef *et al.* (2008), Toschi (2009), Yeung (2002) and Chrigui (2010). The model takes into account some hypotheses: (1) the continuous phase is modeled on an Euler framework; (2) the discrete phase (droplets) is modeled on a Lagrange framework; (3) the turbulence is treated by the Reynolds Averaged Navier-Stokes (RANS) equations; (4) there are no chemical reactions; (5) there are only exchange of heat and momentum between the phases; (6) a two-way coupling is used to represent the interactions between the phases; (7) the solidification phenomenon occurs in constant temperature; (8) the droplets are considered spherical; (9) there are only heat transfer by convection and conduction (during solidification) mechanism; (10) a shrinkage unsolidified core model (Yuan, 2007) is used to represent the solidification phenomenon, where the solidification starts from the outside surface until the center of the droplets and the melting starts from the center of the droplets until the outside surface (Fig. 1); (11) there is no temperature profile inside the liquid droplet, so the Biot number $<< 0.1$.

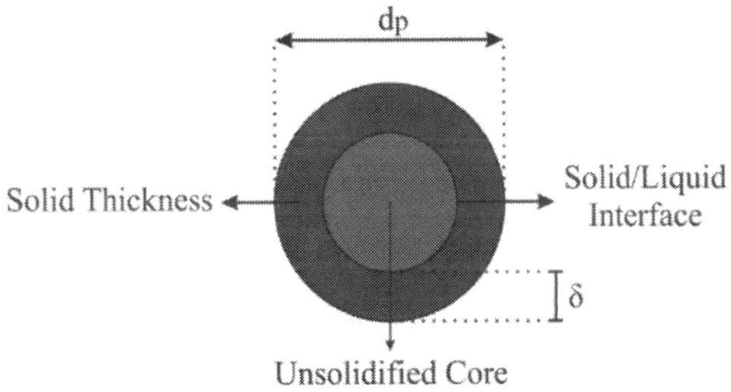

Figure 1. Shrinkage unsolidified core model (Adapted from Yuan, 2007).

The thermal behavior of droplets passes through three periods: (i) cooling of the droplets until the solidification temperature is reached; (ii) solidification of the droplets, where the resistance to the heat transfer due to the formation of the solid thickness inside the droplets cannot be neglected; (iii) cooling of the solids particles.

Additional equations are necessary to represent the interphase heat and momentum transfer, and the turbulence. For the interphase momentum transfer it is considered only the drag force which represents the two-way coupling. The thermal exchange between the phases takes into account the convective heat transfer from the droplets surface to the gas, and the conductive heat transfer due to the formation of a solid thickness inside the droplets during solidification, and these terms represents the two-way coupling. An average procedure is applied to transform the conservative equations, which are instantaneous in time and space, in average equations, resulting in the Reynolds Averaged Navier-Stokes (RANS) equations (Wilcox, 1994). The turbulence is modeled with standard k-ε model. The discretization method used is the Finite Volume Method (FVM) present in the software FLUENT 13.0 from ANSYS®. From Tab. 1 to Tab. 3, it is presented the synthesis of the model, which includes the Euler-Lagrange conservative equations, the turbulent equations and the auxiliary equations.

Table 1. Conservative equations of the discrete phase.

Discrete Phase	
Momentum Conservation	$\frac{d\mathbf{v_p}}{dt} = \frac{\mathbf{g}(\rho_g - \rho_p)}{\rho_p} + \frac{\beta(\mathbf{v_g} - \mathbf{v_p})}{\rho_p}$
Energy Conservation – 1st period	$\frac{dT_{p_l}}{dt} = -\frac{6h\left(T_{p_l} - T_g\right)}{\rho_l C_{p_l} d_p}$
Energy Conservation – 2nd period – Solid Thickness	$\frac{d\delta}{dt} = \frac{U d_p^2 (T_{pf} - T_g)}{\Delta H_f \rho_{p_s} (d_p - 2\delta)^2}$
Energy Conservation – 3rd period	$\frac{dT_{p_s}}{dt} = -\frac{6U\left(T_{p_s} - T_g\right)}{\rho_{p_s} C_{p_s} d_p}$

Table 2. Conservative and turbulent equations of the continuous phase.

Continuous Phase	
Continuity Equation	$\frac{\partial}{\partial t}(\rho_g) + \nabla \cdot (\rho_g \mathbf{v}_g) = 0$
Momentum Equation	$\frac{\partial}{\partial t}(\rho_g \mathbf{v}_g) + \nabla \cdot (\rho_g \mathbf{v}_g \mathbf{v}_g) = -\nabla p_g - \nabla \cdot (\mathbf{T}_{eff}) + \rho_g \mathbf{g} + \sum_{i=1}^{n_p} \beta(\mathbf{v}_g - \mathbf{v}_p)$
Energy Equation	$\frac{\partial}{\partial t}\left(\rho_g C_{p_g} T_g\right) + \nabla \cdot \left(\rho_g \mathbf{v}_g C_{p_g} T_g\right) = \nabla \cdot (\lambda_{eff} \nabla T_g) + \sum_{i=1}^{n_p} U a_p (T_{pf} - T_g)$
Turbulent Kinetic Energy Equation	$\frac{\partial}{\partial t}(\rho_g k_g) + \nabla \cdot \left[\rho_g \mathbf{v}_g k_g - \left(\mu_g + \frac{\mu_g^t}{\sigma_k}\right) \nabla k_g\right] = P_g - \rho_g \varepsilon_g$
Dissipation of Kinetic Energy	$\frac{\partial}{\partial t}(\rho_g \varepsilon_g) + \nabla \cdot \left[\rho_g \mathbf{v}_g \varepsilon_g - \left(\mu_g + \frac{\mu_g^t}{\sigma_\varepsilon}\right) \nabla \varepsilon_g\right] = \frac{\varepsilon_g}{k_g}(C_1 P_g - C_2 \rho_g \varepsilon_g)$

Table 3. Auxiliary equations for the model.

Auxiliary Equations	
Effective Stress Tensor	$\mathbf{T}_{eff} = \mu_{eff}\left[\nabla \mathbf{v}_g + (\nabla \mathbf{v}_g)^T - \frac{2}{3}(\nabla \cdot \mathbf{v}_g)\boldsymbol{\delta}\right]$
Effective Viscosity	$\mu_{eff} = \mu_g + \mu_g^t$

Table 3. Auxiliary equations for the model (concluded).

Turbulent Viscosity	$\mu_g^t = \rho_g C_g \frac{k_g^2}{\varepsilon_g}$
Effective Thermal Conductivity	$\lambda_{eff} = \lambda_g + \lambda_g^t$
Turbulent Thermal Conductivity	$\lambda_g^t = \frac{\mu_g^t}{Pr_g^t}$
Nusselt Number	$Nu = \frac{hd_p}{k_g} = 2{,}0 + 0{,}6Re^{1/2}Pr^{1/3}$
Reynolds Number	$Re = \frac{\rho_g d_p \lvert v_g - v_p \rvert}{\mu_g}$
Prandlt Number	$Pr = \frac{\mu_g C_{p_g}}{k_g}$
Overall Heat Transfer Coefficient – During Solidification/Fusion	$U = \frac{1}{1/h + \delta/k_s}$
Overall Heat Transfer Coefficient – After Complete Solidification	$U = \frac{1}{1/h + d_p/2k_s}$
Interface Coefficient for Drag Force	$\beta = \frac{18\mu_g C_D Re}{\rho_p d_p^2 24}$

Results and Discussions

To analyze if the mathematical model and the subroutines capture the solidification/melting phenomena and its physical consistency, numerical cases were performed. A 3-dimensional main case was performed to validate the physical consistency of the models. A numerical grid is made for the case (Fig. 2) that has two inlets for cold and hot air and two outlets. The geometry has 60 meters in height and 2 by 2 meters in base. The droplets (1.2 millimeters in diameter) were injected on top of the tower. The case is configured so that the droplets are cooled to solidify, and then heated to melt. Cold air enters in the upper inlet of Fig. 2 and is directed to the top of the tower, and hot air enters the lower inlet which is directed to the bottom of the tower.

In Fig. 3 (a) it can be noted that, initially, the droplet analyzed is only cooled. Then, the solidification starts, and the solid thickness starts to grow until a maximum value and then, due to the heating of the droplet, the melting starts, and solid thickness starts to decrease until zero, so the droplet it is fully liquid again at the end of the tower. The Fig. 3 (b) shows the liquid diameter inside the droplet, which starts constant during the cooling, decreases during solidification, increases during the melting, and ends constant again.

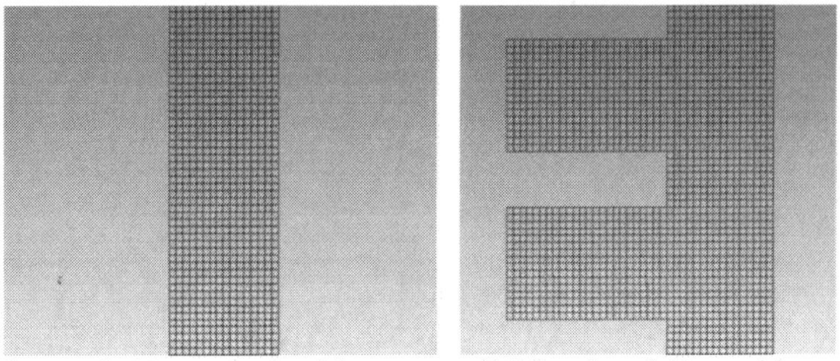

Figure 2. Numerical grid to test solidification followed by melting.

In Fig. 4 (a), the coefficient initially decreases due to the velocity reduction when the droplet is injected in the tower (during cooling period). In solidification period, it decreases due to the formation of the solid thickness, which produces a conductivity resistance to the heat transfer and in melting period, it stars to increase due to disappearance of the solid thickness. Finally it becomes constant, because in this period there is only convective resistance, which became constant because the particle has reached its terminal velocity. Figure 4 (b) shows the temperature behavior of the droplet/particle.

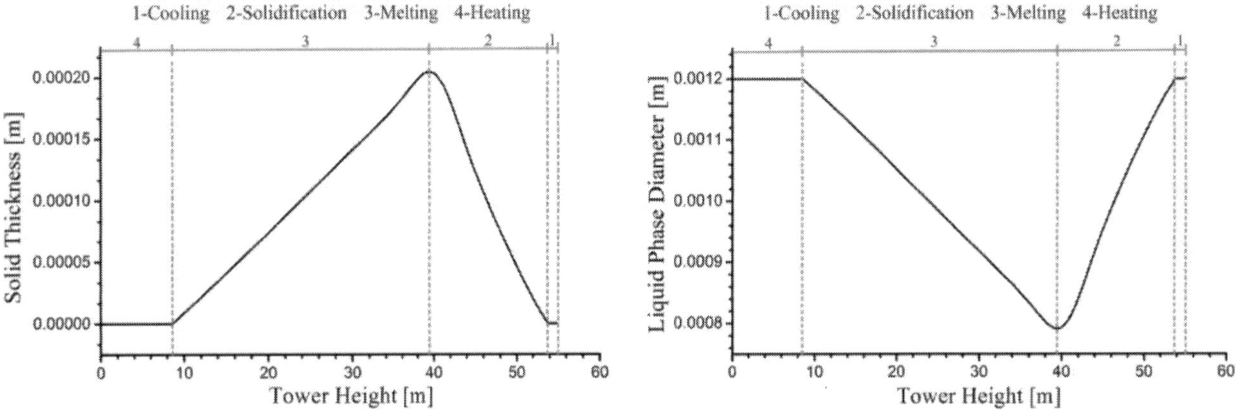

(a) (b)

Figure 3. (a) Solid thickness behavior of the droplet along the tower height; (b) liquid phase diameter of the droplet along the tower height.

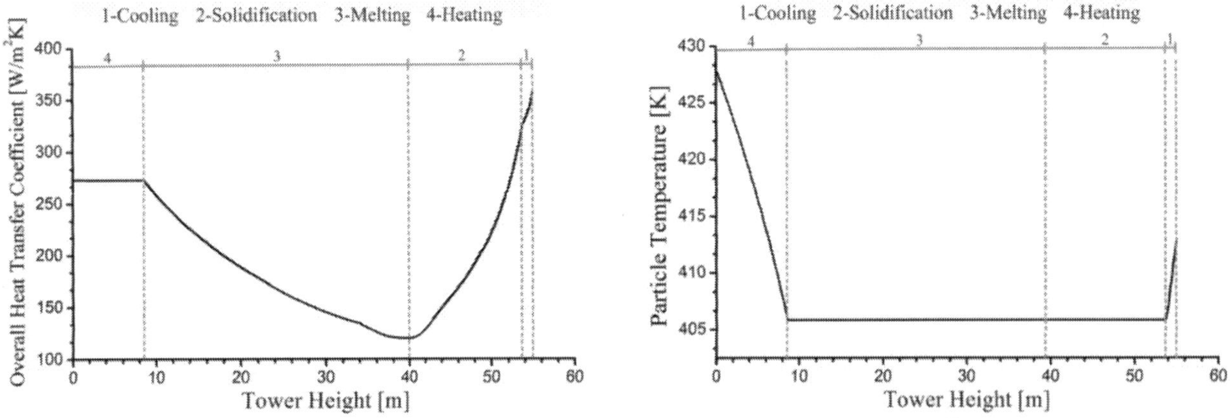

(a) (b)

Figure 4. (a) Behavior of the overall heat transfer coefficient along the tower height; (b) droplet/particle temperature along the tower height.

Conclusions

A 3-dimensional simulation was carried out to prove physical consistency of the model. The results showed that the model is able to predict the solidification and melting phenomena and has physical consistency. The temperature behavior of the droplets is qualitatively consistent with the numerical data presented by Yuan (2007). Now the model can be applied to an industrial case.

References

1. Balachandar, S., Eaton, J.K.. Turbulent Dispersed Multiphase Flow. **Annual Review Of Fluid Mechanics,** [s. L.], v. 42, p. 111-133. Jan. 2010.
2. Chrigui, M., Sadiki, A., Ngoma, G.D. **Unsteady turbulent two-phase flow using Euler/Lagrange approach devoted to two-way coupling conditions.** International Conference on Multiphase Flow, 7th, 2010, Tampa. International Conference On Multiphase Flow (icmf), 2010. 6 p. Avaliable in: <http://ufdc.ufl.edu/UF00102023/00463/1j>. Access: 30 Mar. 2012
3. Cross, M. *et al.* Computational modelling of bubbles, droplets and particles in metals reduction and refining. **Applied Mathematical Modelling,** [s. L.], v. 30, p. 1445-1458. Apr. 2006.
4. Hoef, M.A. Van Der *et al.* Numerical Simulation of Dense Gas-Solid Fluidized Beds: A Multiscale Modeling Strategy. **Annual Review Of Fluid Mechanics,** [s. L.], v. 40, p. 47-70. Jan. 2008.
5. Perry, R.H, Green, D.W. Perry's chemical engineers' handbook.8th ed. New York : McGraw-Hill, c2008. 1v. il.
6. Toschi, F., Bodenschatz, E. Lagrangian Properties of Particles in Turbulence. **Annual Review Of Fluid Mechanics,** [s. L.], v. 41, p. 375-404. Jan. 2009.
7. Wilcox, D.C. Turbulence Modeling for CFD. 2. ed. La Cañada: DCW Industries, 1994.
8. Yeung, P.K.. Lagrangian investigation of turbulence. **Annual Review Of Fluid Mechanics,** [s.L.], v. 34, p. 115-142. Jan. 2002.
9. YUAN, Wu; CHUANPING, Bao; YUXIN, Zhou. An innovated tower-fluidized bed prilling process. **Chinese Journal Of Chemical Engineering,** Wuhan, p. 424-428. jun. 2007.

Nomenclature

Latin Letters

a	Surface area per volume unit $[m^2/m^3]$
C_1	Constant of the $k-\varepsilon$ model
C_2	Constant of the $k-\varepsilon$ model
C_D	Drag force constant
C_p	Specific heat $[W/kg \cdot K]$
d	Diameter $[m]$
ΔH_f	Melting heat $[J/kg]$
$\mathbf{g}$	Gravitational acceleration $[m/s^2]$

h	Convective heat transfer coefficient $[\mathrm{W/m^2 \cdot K}]$
k	Thermal conductivity $[\mathrm{W/m \cdot K}]$ or Turbulent kinetic Energy $[\mathrm{m^2/s^2}]$
Nu	Nusselt number
p	Pressure [Pa]
P	Production of turbulence due to viscous and gravitational forces
Pr	Prandtl number
Re	Reynolds number
t	Time [s]
T	Temperature [K]
U	Overall heat transfer coefficient $[\mathrm{W/m^2 \cdot K}]$
v	Velocity vector $[\mathrm{m/s}]$

Greek Letters

ρ	Density $[\mathrm{kg/m^3}]$
μ	Viscosity $[\mathrm{kg/m \cdot s}]$
T	Stress tensor $[\mathrm{N/m^2}]$
δ	Unitary tensor
δ	Solid thickness [m]
β	Momentum transfer coefficient due to the drag force $[\mathrm{kg/s \cdot m^3}]$
λ	Thermal conductivity of continuous phase $[\mathrm{W/m \cdot K}]$
ϵ	Dissipation of kinetic energy $[\mathrm{m^2/s^3}]$
σ_ε	Constant of the $\mathrm{k} - \varepsilon$ model
σ_k	Constant of the $\mathrm{k} - \varepsilon$ model

Subscripts

eff	Effective
f	Melting or solidification
g	Continuous phase
l	Liquid phase
p	Discrete phase
s	Solid phase

Superscripts

t	Refers to turbulence

(124c) Prediction of Single Drop Granule Formation Mechanisms for Regime Separated Granulation

Heather N. Emady, School of Chemical Engineering, Purdue University, West Lafayette, IN, Defne Kayrak-Talay, Dow Chemical, William C. Schwerin, Honeywell, Des Plaines, IL and James D. Litster, School of Chemical Engineering, Department of Industrial and Physical Pharmacy, Purdue University, West Lafayette, IN

Wet granulation is the process of adding a liquid binder to a fine powder in order to get larger granules for improved particle properties. This process is used in a variety of industries, including pharmaceuticals, food, agricultural chemicals, and detergents. In typical granulation equipment, many of the granulation rate processes (wetting and nucleation, consolidation and growth, and breakage and attrition) occur simultaneously, making it notoriously difficult to control and predict product properties, such as size and shape. Recently, a new granulation approach, regime separated granulation, has been proposed as a way to physically separate the different rate processes to get dramatically better control of the product granule attributes.

The first and most important stage in any regime separated granulation process is nucleation, where new granules are formed by the addition of liquid to the powder bed. Drop controlled nucleation, where one drop forms one granule, is the most desirable operating regime. In order to apply regime separated granulation in practice, more knowledge is necessary regarding the mechanisms by which granules are formed from drop impact and penetration into a static powder bed.

Single drop granule experiments with a syringe and a dish filled with powder are used to simulate drop controlled nucleation in static beds. From high speed camera videos of drop impact and penetration, three different granule formation mechanisms were identified: *Tunneling*, *Spreading*, and *Crater Formation*. Each mechanism produced distinctly different granule shapes. To quantify the conditions under which each mechanism will occur, dimensional analysis was performed and a new regime map was created that plots the powder bed porosity (ε) against the modified granular Bond number ($Bo_g{}^*$), which is a ratio of the capillary force acting on a particle to the gravitational force acting on a particle. Subsequently, a mechanistic model based on a force balance was derived for the *Tunneling* mechanism and compared to the regime map.

The results of this study have important implications for controlling the granule formation mechanism via formulation and process properties to produce the desired granule shape.

(124d) Optimization and Scale up of High Shear Wet Granulation Process

Saurabh Sarkar, Apurva More and Bodhisattwa Chaudhuri, Pharmaceutics, University of Connecticut, Storrs, CT

High shear wet granulation is traditionally defined as a unit operation which produces bigger granules from fine powder under the influence of a binder. The bigger granules have better flow and compressibility ensuring better mixing and content uniformity while retaining the same effective surface area for dissolution and minimizing solvent resistance and inhalational hazards. Despite widespread use in pharmaceutical, chemical, agriculture, food, detergent and mineral industries; this process is often characterized by a lack of well-defined end point due to limited process understanding. Different parameters that affect how well the product is mixed include: liquid addition rate, impeller speed, chopper speed, fill volume, binder volume, viscosity and surface tension of the liquid binder. Current study focusses on optimization and scale up of this process in a standard granulator (KG5, Key International Inc.) in lab scale. The performance of granulation is quantified by recording the granule size distribution and dynamic strength of granules as a function of time. Particle size distributions were measured through Mastersizer 2000E. Parametric studies on 1L and 5L vessels were conducted with a 3 level factorial design using spray dried lactose (d_{50} = 107 μm) and microcrystalline cellulose (MCC, d_{50}= 54 μm) particles with water, hydroxypropyl cellulose (HPC) and sodium lauryl sulfate (SLS) as binders. Fixed liquid volume percentage and maximum power consumption were used to determine end point. Scale up studies on vessels of defined sizes were done using Buckingham pi theorem.

Torque curves gathered during high shear wet granulation experiments are found to allow reasonable end point control of the process. Statistical analysis showing the significance of individual parameters and interaction between two parameters were done using MINITAB 16.0 using Principal Component analysis (PCA) Methodology. The influence of chopper speed had relatively little influence on both dynamic power and granule size especially in the smaller vessel for both lactose and MCC systems. Impeller speed was found to have a relatively greater impact for distribution of viscous binders. However, increasing impeller speed shortens granulation time but increases adhesion producing large lumps. Adhesion to the walls was much greater for lactose compared to MCC. In general, order of wall adhesion was water> HPC>SLS. This was also correlated with the maximum impeller power consumption being less reproducible in smaller vessel with material sticking to the walls. Increased content of liquid binder corresponded to increased size and strength but also increased wall adhesion. Decreased surfaces tension led to more granule collisions leading to quicker granulations but relatively weak granules were formed. A 3D Discrete Element Method (DEM) based model comprising microscopic particle interactions in agglomeration behavior within a shearing granular bed is validated in smaller scale.

(124e) The Effect of Intra-Granular Binder Distribution On Single Granule Strength

Steven Dale, Department of Industrial and Physical Pharmacy, Purdue University, West Lafayette, IN, James D. Litster, School of Chemical Engineering, Department of Industrial and Physical Pharmacy, Purdue University, West Lafayette, IN and Carl Wassgren, Mechanical Engineering, Purdue University, West Lafayette, IN

Granule strength is important for further processing of granules in the pharmaceutical industry as it influences granule attrition rates and the strength of tablets that can be made from granules (Horisawa, Komura, Danjo, & Otsuka, 1995). The current study evaluates the effect that the spatial distribution of binder solid bridges has on granule strength. Granules are formed through single drop tests on a static powder bed. These granule are excavated and dried at an elevated temperature. Granule structure is measured through the use of x-ray microtomography for model systems in which x-ray attenuation coefficients can be used to distinguish primary particles from dried binder bridges. This allows segmentation of the x-ray images and quantification of the spatial distribution of binder. Once the binder distribution is quantified, granules are crushed to measure the breaking strength of the granules. Implications on desired granule structure and process design are discussed.

Horisawa, E., Komura, A., Danjo, K., & Otsuka, A. (1995). Effect of granule strength on compressed tablet strength. *Chemical & Pharmaceutical Bulletin*, *43*(12), 2261-2263.

(124f) Evaluation of the Effect of Blade Angle On a High Shear Wet Granulation

David A. Mota-Aguilar, Chemical Engineering Department, University of Puerto Rico - Mayaguez Campus, Mayaguez, PR and Carlos Velázquez, Department of Chemical Engineering, University of Puerto Rico at Mayaguez, Mayaguez, PR

Wet granulation is a widely used intermediate process in solid dosage manufacturing. It improves flowability of in-process materials, decreases segregation, reduces dust and modifies the appearance and properties of final product.

As well as other granulation methods, high shear wet granulation is very sensitive to operational and design parameters that have not been fully studied nor understood. The effectiveness of this process, particle growth behavior and dynamics are the result of different mechanisms that may take place simultaneously inside a granulator (e.g. wetting, coalescence, attrition), and these mechanisms are strongly affected by the energy provided to the system. In a high shear granulation the impeller plays a fundamental role since it determines the way in which the energy will be transmitted to the material. Therefore, the aim of this work is to study the effect of impeller design in the growth kinetics of a high shear granulation.

Experiments were performed in a laboratory-scale high shear granulator using a lactose-based formulation as model system and a diluted Povidone solution as binder. A factorial design was implemented using three-bladed impellers at different blade angles (42°, 90° and 127°) and agitation velocities (450 rpm, 620 rpm and 807 rpm). Early-stage growth kinetics was determined performing granulations at 6 different massing times from 15 s to 150 s after wetting stage. The study was supported using a Laser Diffraction Method (Insitec T, Malvern Inc.) and conventional sieving for particle size analysis and Powder Rheometry (FT4, Freeman Technology) to characterize final flowability of the material and other properties such as: compressibility and stability.

Experimental results show that as blade angle increase there is a tendency to decrease particle growth kinetics. Thus, a model was proposed to describe and quantify the influence of the blade angle and its interaction with the impeller velocity.

(143a) Boron Composite Nanoparticles for Enhancement of Biofuel Combustion

Neng Wang, Chemical engineering, Louisiana State University, Baton Rouge, LA, Jacob Hanberry, Audubon Engineering Operations, LLC , Metairie, LA, Srinibas Karmakar, Mechanical Engineering, Louisiana State University, Baton Rouge, LA, Sumanta Acharya, Department of Mechanical Engineering, Louisiana State University, Baton Rouge, LA and Kerry M. Dooley, Chemical Engineering, Louisiana State University, Baton Rouge, LA

Boron (in small quantities) is a candidate fuel additive for biofuels of low energy density. Its own energy density is very high on both a volumetric and gravimetric basis. However, burning boron can be difficult, because the B_2O_3 oxide shell melts at 722 K, far lower than the boron core (2350 K), and it often adheres to the particles. This liquid B_2O_3 shell can significantly delay boron combustion. The addition of either a catalytic (e.g., CeO_2) or easily ignitable metal (e.g., Fe) as a shell to boron nanoparticles is hypothesized to solve this problem.

In this study, we have synthesized such core (boron)-shell nanboparticles. A simple mechanical ball milling method was sufficient to coat CeO_2 or rare earth oxide (REO) mixtures to the boron core. We examined three synthesis methods for coating with Fe: (a) cold milling in dry ice; (b) borohydride reduction of Fe salts; (c) H_2 reduction of precipitated $Fe(OH)_2$. Simple mechanical ball milling of Fe and B nanoparticles failed because it resulted in some intermetallic $(Fe_9B)_{0.2}$, which is relatively inert to combustion. Using emission measurements at various wavelengths, it was shown that the addition of these rare earth oxides and Fe to boron did enhance its burning behavior, reducing the ignition delay and providing greater overall heat release. The boron was completely oxidized in the ethanol-fueled, swirl-type combustor.

(143b) Morphology and Molecular Weight Control of Core-Shell Polymer Nanoparticles

Ibrahim A. El-Hedok, Chemical & Biological Engineering, Iowa State University, Ames, IA and Jennifer M. O'Donnell, Chemical and Biological Engineering, Iowa State University, Ames, IA

Colloidally stable core-shell polymer nanoparticles can be simply synthesized using microemulsion polymerization through the semi-continuous addition of a second monomer to form the shell around seed polymer particles. However, there are two main limitations of traditional semi-continuous microemulsion polymerization. The first disadvantage is the lack of control over the molecular weight and polydispersity. While the second limitation is the challenge in maintaining the location of the shell forming polymer, since the morphology of the core-shell polymer is determined by equilibrium thermodynamics that is dictated by the polymer chemistry and composition. For instance, if the added polymer thermodynamically favors to be in the core more than the first seed polymer, phase inversion occurs between the core and shell polymers. In this work, we aim to produce core-shell polymer nanoparticles by employing a controlled polymerization technique, i.e. reversible addition-fragmentation chain transfer (RAFT), in semi-continuous microemulsion polymerization which enables us to obtain predefined molecular weight polymers and architectures that are inaccessible by current synthetic techniques.

RAFT is a successfully implemented method for controlling homogenous polymerization for producing polymers with low polydispersity and predefined molecular weight. The RAFT technique relies on a chain transfer agent (CTA) which controls the propagation of the polymer chains. Performing RAFT in heterogeneous microemulsion polymerization has been proven to facilitate the synthesis of latex nanoparticles with predetermined molecular weight and low polydispersity. In recent work, we have investigated enhancing the polymerization control in RAFT microemulsion polymerization by fixing the location of the CTA at the surface of the polymer particle through attaching a surface-active moiety to the CTA structure. Here, we extend our work to investigate the use of surface-active CTA for the synthesis of core-shell polymer nanoparticles. There are several advantages for using surface-active CTA in RAFT semi-continuous microemulsion polymerization:

(1) **Block copolymer shell formation:**The polymer propagation is constrained to the corona of the particle, which promotes the formation of a block co-polymer shell with further second monomer addition.

(2) **Prevention of phase inversion:**The covalent attachment of the surface-active CTA to the end of the polymer introduces an energetic penalty for transferring the shell polymer to the core of the particle, and thus diminishes the driving force for phase inversion with the right concentration of surface-active CTA per particle.

(3) **Surface functionality:**The surface-active CTA permits further surface functionalization by simple chemical modification of the CTA.

In this work we investigate the core-shell synthesis of n- butyl acrylate (BA) and styrene (St) using RAFT microemulsion with surface-active chain transfer agent, *2-((11-(((benzylthio)carbonothioyl)thio)undecanoyl)oxy)-N,N,N-trimethylammonium iodide* (B11T). In the literature, it has been shown experimentally and theoretically that at 50:50 St to BA ratio, St favors to form the shell while BA favors to be in the core, irrespective of monomer addition sequence. We hypothesize that, with enough B11T chain transfer agents, phase inversion can be prevented during St/BA semi-continuous microemulsion polymerization.

The number average molecular weight, polydispersity, and rate of polymerization are analyzed during the semi-continuous microemulsion polymerization to assess polymerization control and kinetics, while the final morphology of the polymer particles will be directly visualized by transmission electron microscopy. The results will be compared against core-shell synthesis using traditional semi-continuous microemulsion polymerization, and shall provide an understanding of the requirement to prevent phase inversion.

Thermogravimetric Analysis of the Kinetics of the Reaction of Alkoxysilane with Silica

Yuk Yu Law[1], Donald L. Feke[1] and Ica Manas-Zloczower[2]
(1) Department of Chemical Engineering, (2) Department of Macromolecular Science and Engineering, Case Western Reserve University, Cleveland, OH

The surface functionalization of silica by alkoxysilane can improve its dispersion into polymers by beneficially modifying interparticle and particle-polymer interactions. Thermogravimetric analysis (TGA) has been used to study the kinetics of binding a bi-functional alkoxysilane onto silica over the temperature range of 50 to 110°C. Derivative TGA curves of the functionalized silica show distinct peaks which correspond to the alkoxysilane bound to one or two surface silanol sites. Binding at two silanol sites is postulated to occur in a two-step series reaction model. An Arrhenius analysis for the alkoxysilane binding reactions shows that the activation energies of the two reaction steps are similar, which is expected since the alkoxysilane contains two identical binding groups. This work demonstrates the suitability of the TGA technique to investigate the reaction kinetics for modifying the surface of silica.

(143d) Atom Transfer Radical Polymerization Synthesis to Yield Copper-Free Block Copolymers

Justin R. Adams, Chemical & Biological Engineering, Iowa State University, Ames, IA and Surya K. Mallapragada, Chemical and Biological Engineering, Iowa State University, Ames, IA

We report a method of atom transfer radical polymerization (ATRP) for the synthesis of copper-free poly(2-diethylaminoethyl methacrylate (PDEAEM) pentablock copolymers. ATRP is a controlled radical polymerization technique for the development of functional polymers with complex and diverse architectures. Biocompatibility is an increasing concern with growing applications for these polymers synthesized by ATRP in the fields of biology and medicine because of the residual copper catalysts. The reported ATRP synthesis method utilizes novel copper(I) oxide nanoparticles with a polyvinylpyrrolidone coating as catalysts, that can be easily removed after polymerization. We demonstrate this through synthesis of PDEAEM-based pentablock copolymers with pH- and temperature sensitive gelation behavior using a pluronic macrointiator. The copper(I) oxide nanoparticles are easily removed using a column of basic alumina with x-ray spectroscopy (XPS) showing no residual copper in the final product. This ATRP method allows for the synthesis of polymers with low toxicity for in vivo applications such as drug and gene delivery.

(143e) Engineered Quantum Dots for Light Selective Polymer Films with High Photoluminescence and Photostability

Abdul Mumin[1], Golam Moula[1], Jenna Allan[1] and Paul A. Charpentier[2], (1)Chemical and Biochemical Engineering, University of Western Ontario, London, ON, Canada, (2)Chemical and Biochemical Engineering, Western University, London, ON, Canada

This work focused on optimizing the optical properties of polymer films using quantum dot (QD) nanoparticles to manipulate and control light transmission for applications in light selective nanofilms. A new approach was undertaken to synthesize Quantum Dots (QDs) with highly luminescent properties and photostability for solar energy applications. Highly luminescent Cd- based and non Cd-based QDs: CdS, ZnS, ZnSeS, and CdS/ZnS core/shell QDs with 5 nm size were synthesized using a facile approach based on pyrolysis of the single molecule precursors. After capping the CdS QDs with a thin layer of ZnS to reduce toxicity, the photoluminescence and photostability of the core-shell QDs was significantly enhanced. The ZnS shell was found to passivate the surface defects of the CdS dots while confining the electron-hole pair to the CdS core region.

To make both the bare and core/shell structure QDs more resistant against photochemical reactions, a mesoporous silica layer was grown on the QDs through a reverse microemulsion technique based on hydrophobic interaction and surface silanization using 3-mercaptopropyl trimethoxysilane as the coupling agent. The average size of the mesoporous silica nanoparticles (MSNs) was 60 nm with tuneable QDs core from 10-35nm depending on the concentration of QDs loaded. This encapsulation improved the quantum yield 40% and photostability 30% compared to the bare QDs by providing much stronger resistance to oxidation and Oswald ripening of QDs. Various silane coupling agents are being examined to chemically attach these silica covered QDs to poly(ethylene vinyl acetate) polymer that provides simultaneous UV protection, infrared retention, and light selectivity for controlling plant growth, ultimately working to improve biomass production and reduce energy consumption in commercial greenhouses. A new type of environmental friendly quantum dot-sensitized photovoltaic device was also developed by decorating QDs using supercritical carbon dioxide ($scCO_2$). With a liquid electrolyte as the hole transport medium, quantum-dot-sensitized nanowire photovoltaic cells exhibited short-circuit currents ranging from 1 to 2 mA/cm^2 and open-circuit voltages of 0.5–0.6 V, when illuminated with 100 mW/cm^2 simulated AM1.5 spectrum. Internal quantum efficiencies as high as 30–50% were also obtained.

(143f) Self-Cleaning and Anti-Microbial Polyurethane Coatings

Adria F. Lotus, Yixing Tang, Koosha Azhie and Paul A. Charpentier, Chemical and Biochemical Engineering, Western University, London, ON, Canada

Self-Cleaning and Anti-Microbial Polyurethane Coatings

A. F. Lotus, Y. Tang, K. Azhie, and P. A. Charpentier*

Department of Chemical and Biochemical Engineering, Faculty of Engineering,

Western University, London, Ontario Canada, N6A5B9

Abstract

Tremendous interest exists for self-cleaning and anti-microbial coatings due to their wide range of potential applications, spanning from surface coatings to textiles. This work has devoloped a new synthesis method for preparing self-cleaning hydrophilic coatings using nano-titanium dioxide (TiO_2) with step-growth polyurethane polymers. The photocatalytic activity of the coatings is attained from a transparent layer of TiO_2 which is made using a 'grafting from' polymerization technique with various bi-functional molecules that are coordinated to nano-TiO_2 while providing functionalities for subsequent "grafting from" polymerization. 2,2-bis (hydroxymethyl) propionic acid (DMPA) was chosen for the functionalization with $nTiO_2$, which is a bifunctional agent with one carboxyl group that can coordinate to $nTiO_2$, and two hydroxyl groups that can react with diisocyanate terminated pre-polyurethane through step-growth polymerization to obtain functionalized $nTiO_2$/polyurethane composite coatings. Different physico-chemical characterization techniques were used to verify the coordination reaction between DMPA and $nTiO_2$ including SEM, TGA, ATR-FTIR, and EDX. The $nTiO_2$/PU coatings can chemically break down organic dirt and microbes by photocatalysis when exposed to solar irradiation. The antimicrobial behavior was investigated using both gram negative (*E. coli*) and gram positive (*M. luteus*) bacteria at different exposure time using a UV lamp and Solar Simulator. The self cleaning property was evaluated by water contact angle measurements on the coating surfaces. The self-cleaning property was also examined using stearic acid and excess DMPA as model compounds for organic dirts, and monitoring the degradation of "dirt" on the surfaces using FTIR spectroscopy analysis. The functionalized $nTiO_2$-polyurethane composite coatings showed promising self-cleaning property and antibacterial activity against both gram-positive and gram-negative bacteria under UV and solar irradiation.

**To whom it all correspondence should be addressed: Email: pcharpentier@eng.uwo.ca;*

Tel: (519) 661-3466 ; Fax: (519) 661-3498

(143g) Magnetoresistive Conductive Polyaniline - Barium Titanate Nanocomposites with Negative Permittivity

Xi Zhang, Integrated Composites Laboratory (ICL), Dan F Smith Department of Chemical Engineering, Lamar University, Beaumont, TX, Suying Wei, Chemistry and Biochemistry, Lamar University, Beaumont, TX and Zhanhu Guo, Dan F. Smith Department of Chemical Engineering, Lamar University, Beaumont, TX

Conductive polyaniline (PANI)-based polymer nanocomposites (PNCs) with barium titanate ($BaTiO_3$) were synthesized by a surface initiated polymerization (SIP) method. Two different particle sizes (100 and 500 nm) were studied. By varying the loading level, size of $BaTiO_3$ nanoparticles (NPs) and stirring method, a series of PNCs were obtained and the effects of these parameters on the crystalline structure, thermal stability, morphology, electrical conductivity and dielectric permittivity were systematically studied. FT-IR analysis indicated a strong interaction between the formed PANI and the $BaTiO_3$ NPs, and TEM observations showed that the $BaTiO_3$ NPs are well coated with a this PANI layer, however, the thickness of the PANI layer decreased with increasing the $BaTiO_3$ particles loading. The XRD reflection patterns indicated that the crystallinity of PANI in the PNCs depends on the nanoparticle loading. However, the resistivity does not increase with increasing the crystallinity, and the temperature dependent resistivity result reveals a 3-d variable range hopping (VRH) electron transport mechanism. The resistivity variation with $BaTiO_3$ loading is mainly attributed to the ferroelectric nature of $BaTiO_3$ and the space charge on the interface between $BaTiO_3$ and PANI. Compared with the positive real permittivity for the PNCs prepared from physical mixing, all the chemical synthesized PNC samples show negative dielectric permittivity and the permittivity change is related to the instinct metallic state in PANI. In addition, positive magnetoresistance (MR) is observed in all kinds of PNCs and analyzed theoretically from the wave-function shrinkage model.

(172b) COAL Direct Chemical Looping Process: Metallurigical Coke and PRB COAL Conversions

L. - S. Fan, William G. Lowrie Department of Chemical and Biomolecular Engineering, The Ohio State University, Columbus, OH

As a novel oxy-combustion approach for CO_2 capture from coal power plants, the chemical looping processes have been extensively studied. The chemical looping process can easily extract highly concentrated CO_2 and efficiently recover heat, which has exhibited the promising economic and technical feasibility to capture CO_2 from the coal conversion processes.

At the Ohio State University, Coal-Direct Chemical Looping (CDCL) process using iron-based oxygen carriers has been developed for the application to the power generation with in-situ CO_2 capture. The iron-based oxygen carrier particles can undergo extensive characteristic and performance tests for the application in the chemical looping processes. These oxygen carrier particles are consisted of reactive metal oxides, Fe_2O_3 and inert supports. The study of particular oxygen carrier particle shows that the particle can be regenerable under reduction-oxidation conditions for more than 100 cycles. Also, these particles are capable to convert various fuels such as methane, syngas and solid fuels to generate hydrogen and/or heat. In this study, a 2.5 kW_{th} bench unit and a 25 kW_{th} sub-pilot demonstration unit for the CDCL process are presented. Design, construction and operation of CDCL sub-pilot unit are discussed. The sub-pilot unit was successfully operated in the integrated and continuous manner with different types of solid fuels such as metallurgical coke and PRB coal.

(172c) Important Length Scales in Dense Gas-Particle Flows

Stefan Radl[1], Chris Milioli[2], Fernando Milioli[2] and **Sankaran Sundaresan**[3], (1)Institute for Process and Particle Engineering, Graz University of Technology, Graz, Austria, (2)Mechanical Engineering, University of Sao Paulo, São Carlos-SP, Brazil, (3)Department of Chemical and Biological Engineering, Princeton University, Princeton, NJ

There is a long history of studies on application of dimensional analysis to develop scaling relations that can be employed for scale-up and scale-down of gas-particle fluidized beds. It is now well established that two-fluid models (TFM), with constitutive models for the stresses deduced from the kinetic theory of granular materials, are able to capture the formation inhomogeneous structures over a wide range of time and length scales. As resolving all these structures in simulations of industrial scale devices is impractical, filtered models that average over small-scale structures are currently being developed. This has brought to the surface the issue of relevant length scales in gas-particle flows.

In this talk, we will review the various length scales that naturally enter in gas-particle fluidized beds and discuss how an understanding of these scales is being used to construct simplified filtered models. A dimensional analysis of these filtered models then suggests a simple set of dimensionless groups that are important to capture the dynamics of such devices when simulated on coarse computational grids. Finally, we present simulation results, using both a TFM and an Euler-Langrange model, which support our choice of dimensionless parameters.

(172d) CFD Simulation of Pharmaceutical Particle Drying in Bubbling Fluidized Bed Dryers At Different Scales

Hamid Arastoopour, IIT, Chicago, IL and **Jungkee Jang**, Chemical and Biological Engineering, Illinois Institute of Technology, Chicago, IL

CFD Simulation of Pharmaceutical Particle Drying in Bubbling Fluidized Bed Dryers at Different Scales

Jungkee Jang and Hamid Arastoopour

Department of Chemical and Biological Engineering, Illinois Institute of Technology, USA

Drying is one of the major unit operations in the manufacturing of solid pharmaceuticals. The optimum design of this process will significantly enhance the rate and reliability of the production and, in turn, decrease the cost of the pharmaceutical products. In this study, the gas-solid mixing and drying processes of pharmaceutical particles in bubbling fluidized bed dryers at different scales were simulated using a two-fluid granular model. Our mathematical model is composed of the continuity, momentum, energy, and species transfer equations for water vapor to simulate the flow pattern and heat and mass transfer for the pharmaceutical bubbling fluidized bed dryer process. User defined scalar (UDS) transport equations were added to extend the FLUENT code capability to account for moisture transport and to simulate the drying rate and extent of the solid particles.

The simulations were performed at different particle diameters, bed heights, inlet gas velocities, and inlet gas temperature distributions, respectively. Our simulation results were validated and refined by comparing with Duquesne University laboratory-scale experimental data on solid particles moisture content and outlet gas temperature. Then, our CFD model was used to predict larger scale (kilo- and 10-kilo scale units) Abbott Laboratory data for the drying process of the same material. Our simulation results compared well with the experimental data for the drying process for all three different scales.

(172e) Dynamics of Particle Cratering Using High Speed Gas Jets

Casey LaMarche[1], **Jennifer Sinclair Curtis**[2], Shibo Kuang[3] and Aibing Yu[4], (1)University of Florida, Gainesville, FL, (2)Chemical Engineering, University of Florida, Gainesville, FL, (3)University of New South Wales, Sydney, Australia, (4)The University of New South Wales, Sydney, Australia

Future space exploration necessitates improved gas-solid and solid-solid interaction models. Impinging rocket plumes from spacecraft interact with the lunar or Martian surfaces and release a high velocity particle spray that is hazardous to surrounding surface structures and affects the stability of spacecraft landings. Understanding and predicting the dynamics of this particle cratering and erosion can aid in mitigating these hazards. Ground-based experiments performed at the University of Florida provide detailed data on the particle cratering behavior as a function of time. Data were obtained for the cratering of 2mm glass beads using a turbulent subsonic jet at various velocities. It is shown that the particle cratering behavior can be successfully described using a combined CFD-DEM model developed at UNSW. The model incorporates gas turbulence, gas-particle drag, and particle-particle contact forces. In addition, the model enables the determination of the primary mechanism for particle cratering under the experimental conditions studied. Particle erosion is confined to a thin shearing region along the crater boundary; hence, viscous erosion is the primary cratering mechanism.

Particle Scale Modelling and Analysis of Fluid Flow and Heat Transfer in Fluidised Beds

Zongyan Zhou, Qinfu Hou and Aibing Yu

Laboratory for Simulation and Modelling of Particulate Systems
School of Materials Science and Engineering,
The University of New South Wales
Sydney, NSW 2052, Australia

Extended Abstract

Fluid bed reactors have been extensively used in chemical processes due to their high heat transfer efficiency. Typical examples are ironmaking blast furnace which involves complicated multiphase flow, heat transfer and chemical reactions in a packed bed, and fluidized bed combustors whose performance heavily depends on the hydrodynamics and thermal-chemical behavior of particles in interaction with gas. To achieve optimal design and control of such a fluid bed reactor, it is important to understand the flow and heat transfer characteristics. Particle scale studies, achieved by the use of approach of coupled discrete element method (DEM) and computational fluid dynamics (CFD) [1, 2], are useful to achieve this goal. In this approach, the motion of discrete particles is obtained by solving Newton's second law of motion, and the flow of continuum fluid by solving the Navier-Stokes equations based on the concept of local average as used in CFD, with the coupling of CFD and DEM through particle-fluid interaction forces. This approach has recently been extended to study heat transfer in gas fluidization [3]. It considers different heat transfer mechanisms in detail, including, for example, particle-fluid convection, particle-particle conduction, and particle radiation. The proposed model offers a useful numerical technique to elucidate the fundamentals governing the heat transfer in packed/fluidized beds at a particle scale. Here our major application of this approach is summarized below.

- *Heat transfer in gas fluidization with non-cohesive particles [3]*

In physical experiments, the temperature of hot spheres is measured using thermocouples connected to the spheres [4]. The cooling process of such hot spheres can be easily traced and recorded in the CFD-DEM simulations. It is shown that there is a general tendency for the HTC (heat transfer coeffcient) of hot sphere increasing first with gas superficial velocity U in the fixed bed ($U<U_{mf}$), and then remaining constant, independent of the gas superficial velocities for fluidized beds ($U>U_{mf}$). This is consistent with physical experiments. The HTC-U relationship is affected significantly by the thermal conductivity of bed particles. The higher the k_p, the higher the HTC of hot spheres. For example, when k_p=30 W/(m·K), the predicted HTC in the fixed bed is so high that the trend of HTC-U relationship is totally changed. The HTC decreases with U in the fixed bed, then may reach a constant HTC in the fluidized bed. But when the thermal conductivity of particles is low, the HTC always increases with U, independent of bed state. Generally, the convective HTC increases with U; but conductive HTC decreases with U. For a proper particle thermal conductivity, the two contributions (convective HTC and conductive HTC) could compensate each other, then the total HTC is nearly constant after the bed is fluidized. So HTC independence of U is valid under this condition.

The model can then be used to analyse the sub-heat transfer mechanisms. The relative contributions by different heat transfer paths can be quantified. For example, when k_p=0.08 W/(m·K), particle-fluid-particle conduction always contributes more than particle-particle contact, but both vary with gas superficial velocity. For particle-fluid-particle conduction, particle-fluid-particle heat transfer with two contacting particles is far more important than that with two non-contacting particles in the fixed bed. This can be further explained by the local structures, e.g. the hot sphere contacts about 6 particles when $U<U_{mf}$. But such a feature changes in the fluidized bed ($U>U_{mf}$), where particle-fluid-particle conduction between non-contacting particles is relatively more important. This is because most of particle-particle contacts with an overlap are gradually destroyed with increasing gas superficial velocity, which significantly reduces the contribution by particle-fluid-particle between two contacting particles. However, particle-particle conduction through the contacting area becomes more important with an increase of particle thermal conductivity.

Further, the HTCs of all the bed particles can be examined, and it was found that the features are similar to those observed for hot spheres. The similarity illustrates that the hot sphere approach can, at least partially, represent the general features of particle thermal behaviour in a particle-fluid bed. Overall, the particles in a uniformly fluidized bed behave similarly. But a particle may behave differently from another at a given time.

- *Effective thermal conductivity in a packed bed [5]*

Effective thermal conductivity (ETC) is an important parameter describing the thermal behaviour of packed beds with a stagnant or dynamic fluid. Various mathematical models have been proposed to help solve this problem, but they are often limited by the homogeneity assumption in a continuum model. The CFD-DEM model has shown a promising advantage in predicting the ETC under the different conditions. Young's modulus is an important parameter affecting the particle-particle overlap, hence the particle-particle heat transfer. However, in the DEM, it is computationally very demanding to carry out the simulation using a real Young's modulus (often at an order of 10^3~10^5 MPa), particularly when involving a large number of particles. It is therefore very helpful to have a method that can produce accurate results but does not have a high computational cost.

The calculation of heat fluxes for conduction heat transfer mechanisms is related to an important parameter: particle-particle contact radius r_c. The overestimation of r_c in DEM based on the soft-sphere approach significantly affects the calculation of conductive heat fluxes. To reduce such an over-prediction, a correction coefficient c is introduced, and then the particle-particle contact radius used to calculate the heat flux between particles through the contact area is written as $r_c' = c \cdot r_c$ where $c = r_{c,0}/r_c = (E_{ij}/E_{ij,0})^{1/5}$, where E_{ij} is the value of Young's modulus used in DEM and $E_{ij,0}$ is the real value of Young's modulus of the materials considered. The otained correction coefficeints have shown applicable by some cases. Further, effects of some parameters on ETC are examined, and revealed tha ETC is not sensitive to particle-particle sliding friction coefficient. ETC increases with the increase of bed average temperature. The predicted ETC at 1475°C can be about 5 times larger than that at 75°C. The effect of particle size on ETC is shown to be more complicated. At low thermal conductivity, the ETC varies little with particle size from 250 μm to 10 mm. But for high thermal conductivity ratios, ETC increases with particle size. The main reason could be that the particle-particle contact area is relatively large for large particles, and consequently, the increase of thermal conductivity enhances the conductive heat transfer between particles.

The modified approach has also been applied to gas fluidization to test its applicability. The results are very comparable and consistent. The significance of proposed modified model is to save computational cost. For the case shown, the use of a low Young's modulus significantly reduces the computational time, i.e. 4~5 times faster with 16,000 particles. Such a reduction becomes more significant for a larger system involving a large number of particles.

- *Heat transfer between a fluidized bed and an insert tube [6]*

Fluidization is widely used in industries as a major flow mode in chemical processes. Temperature control is one of the important concerns in these applications. Immersed tubes are usually adopted to control the flow and heat transfer. The CFD-DEM approach is used here to study the heat transfer in a fluidized bed with a horizontal tube. The tube exchanges heat with its surrounding particles and fluid. The distribution and magnitude of HTC are two factors commonly used to describe the heat transfer in such a system. Gnerally speaking, the local HTC is higher at sides of the tube around 90° and 270° while it is lower at the upstream and downstream of the tube around 0° and 180°. The local HTC increases firstly with the increases of gas velocity and then decreases. The local HTC increases with the increase of tube position within the bed static height.

The heat transfer between an immersed tube and a fluidized bed depends on many factors, such as the contacts of particles with the tube, porosity and gas flow around the tube. These factors are affected by many variables related to operational conditions, and some of them are examined as described below. (i) *Gas velocity*: the heat flux by convection is dominant and increases with the increase of gas velocity while the heat flux by conduction decreases; (ii) *Particle thermal conductivity:* local HTC increases with the increase of k_p from 1.10 to 100 W/(m·K). However, there is no significant increase of HTC from 100 to 300 W/(m·K). This variation is expected because heat transfer between the fluidized bed and the tube is mainly through convection while the k_p only affects the conductive heat transfer; (iii) *Young's modulus E:* E affects the deformation of a particle and the tube when they are colliding with each other, but has insignificant effect on local HTC. The effect of E on this kind of heat transfer is limited and the percentages of heat fluxes by different modes change insignificantly. However, it does affect the conductive heat transfer. The total conductive heat flux decreases a little with the increase of E because of the decrease of particle-tube contact time and area; (iv) *Tube temperature*: the heat transfer by radiation is important in the considered system because the increase of environmental temperature of the tube, and its significance has already been pointed in the literature. The increase of gas thermal conductivity with temperature is one of the main reasons for the increase of HTC. The radiative heat flux increases more quickly than the convective heat flux according to the fourth power law of the temperature difference. The radiative heat flux becomes larger than that of conductive heat flux around T_s = 300°C and then, larger than that of the convective heat flux around T_s = 1200°C. These show that the radiation is an important heat transfer mode with high tube temperatures.

- *Effects of material properties on heat transfer in gas fluidization [7]*

Heat transfer in a fluidized bed is closely related to the gas-solid flow characteristics which depend on material properties. In general, powders can be classified into groups A, B, C and D according to the gas and solid properties of a fluidized system. Compared to other groups, group A powders are characterized by an expanded bed flow regime that occurs at gas velocity u_f between the minimum fluidization velocity u_{mf} and the minimum bubbling velocity u_{mb}. Thus, three flow regimes exist for group A powders, including fixed, expanded and fluidized beds, as reproduced in the CFD-DEM simulations [8]. Under given conditions, different powders may give different flow behaviors and hence different heat transfer behaviors. The heat transfer characteristics of group A powders different from those of group B powders are depicted. Convective heat transfer is dominant for group A powders, similar to that of group B powders. However, conductive heat transfer can contribute over 20% of the overall heat transfer for typical operations. This mechanism should not be ignored for group A powders. Radiation can be safely ignored at low temperatures but it should be considered if the bed temperature is high enough (e.g. >500°C).

The effects of many pertinent variables on flow and heat transfer can be investigated by the model through controlled numerical experiments. Two have been studied in details. (i) *Hamaker constant H_a*: the interparticle cohesive force plays a critical role in the transition of different flow regimes and the classification of powders. This force is largely determined by H_a. The heat transfer varies complicatedly and significantly with the increase of the Hamaker constant. With the increase of the Hamaker constant at a given gas velocity ratio, convective heat flux increases to a maximum at the point of minimum bubbling velocity, then decreases to a minimum at the point of minimum fluidization velocity, finally increases slightly. Conductive heat flux increase to a maximum at the point of minimum bubbling velocity, then decreases a bit with small variation; (ii) *Particle size d_p*: d_p is another important material property affecting the flow and hence heat transfer characteristics of fluidized beds. A fixed bed with larger particles needs a significant longer time to reach the temperature of the inlet air while the time periods are similar for fluidized beds. Given a velocity ratio u_f/u_{mf}, the contribution of conductive heat transfer decreases and that of convective heat transfer increases with the increase of particle size. A longer time is often needed for larger particles to reach the temperature of the inlet gas under given fluidization conditions.

The macroscopic heat transfer process can be quantified by $ln(1 - Ts) = -Ct$ where $T_s=(T_{solid}-T_0)/(T_{in}-T_0)$ and T_{solid} is the bed average temperature. $C=\tilde{h}A/m\tilde{c}_p$ is a lumped parameter. It indicates the heating rate of a bed, which is mainly related to material properties and gas-solid flow structures such as local gas flow and gas-solid contact area. The equation suggests that there is a linear relationship between $ln(1-T_s)$ and t, and C can be determined from this relationship. Parameter C, mainly determined by the convective heat flux, varies with the increase of the Hamaker constant complicatedly, and decreases exponentially with the increase of particle size. The effect of gas velocity ratio on C is linear in the fixed and fluidized bed flow regimes for the studied cases, and a transitional stage is observed in the expanded bed flow regime.

The examples presented have demonstrated that the CFD-DEM approach is very promising in quantifying the role of various heat transfer mechanisms in packed/fluidized beds, which is useful to the optimal design and control of fluid bed ractors.

References

1. H.P. Zhu, Z.Y. Zhou, R.Y. Yang, and A.B. Yu. Discrete Particle Simulation of Particulate Systems: A Review of Major Applications and Findings. Chem. Eng. Sci. 63 (2008) 5728-5770.
2. H.P. Zhu, Z.Y. Zhou, R.Y. Yang, and A.B. Yu. Discrete Particle Simulation of Particulate Systems: Theoretical Developments. Chem. Eng. Sci. 62 (2007) 3378-3396.
3. Z.Y. Zhou, A.B. Yu, and P. Zulli. Particle Scale Study of Heat Transfer in Packed and Bubbling Fluidized Beds. AIChE J. 55 (2009) 868-884.
4. A.P. Collier, A.N. Hayhurst, J. L. Richardson, and S.A. Scott. The Heat Transfer Coefficient between a Particle and a Bed (Packed or Fluidised) of Much Larger Particles. Chem. Eng. Sci. 59 (2004) 4613-20.
5. Z.Y. Zhou, A.B. Yu, and P. Zulli. A New Computational Method for Studying Heat Transfer in Fluid Bed Reactors. Powder Technol. 197 (2010) 102-10.
6. Q.F. Hou, Z.Y. Zhou, and A.B. Yu. Computational study of the heat transfer in bubbling fluidized beds with a horizontal tube. AIChE J. 58 (2012) 1422-1434.
7. Q.F. Hou, Z.Y. Zhou, and A.B. Yu. Computational Study of the Effects of Material Properties on Heat Transfer in Gas Fluidization. Ind. Eng. Chem. Res. 51 (2012) 11572-11586.
8. Q.F. Hou, Z.Y. Zhou, and A.B. Yu. Micromechanical modeling and analysis of different flow regimes in gas fluidization. Chem. Eng. Sci. (2012) DOI: 10.1016/j.ces.2012.08.051.

(172g) Measurement of Density Fluctuations in a Vertically Oscillated Granular Bed At the Onset of Vibrofluidization

Hugo Caram, Chemical Engineering, Lehigh University, Bethlehem, PA, James F. Gilchrist, Department of Chemical Engineering, Lehigh University, Bethlehem, PA and Kenneth J. Ford, Merck, Inc., Bethlehem, PA

Through measurements of vane power and density, three granular bed states, the static granular state, the pseudo-plastic state and vibro-fluidized state, along with the transitions between the states are identified. A granular bed state phase diagram is constructed, mapping the effects of vibration frequency and intensity input. The pseudo-plastic state contains two sub-classifications. For vibration force less than the force of gravity at low frequency and all vibration force levels at high frequency and moderate acceleration, vibration has no measurable effect on the granular bed. For the pseudo-plastic state subjected to low frequency vibration force greater than the force of gravity, the vane power, extrapolated zero-shear rate torque, and density change penetration into the bed are a strong function of the vibration force and frequency. Measurements demonstrate hysteresis with changes in vibration acceleration and frequency. The transition to a vibro-fluidized state at a critical dimensionless vibration acceleration is characterized by dramatic decreases in vane power, dynamic zero-shear rate torque and measurable fluctuations in density throughout the bed. In the vibrofluidized state, the vane power and extrapolated zero-shear rate torque are weakly influenced by the vibration force and frequency in the vibro-fluidized state. Within the range tested, particle diameter had little affect on the measurements of noncohesive granular media.

(172h) Quantitative Predictions of Instabilities in Solid-Particle Flows

Peter P. Mitrano and **Christine M. Hrenya**, Chemical and Biological Engineering, University of Colorado at Boulder, Boulder, CO

The presence of clustering instabilities in gas-solid fluidized beds has been well documented experimentally and known to be a critical aspect of unit operation. Predictions of such clusters via kinetic-theory models was first reported about two decades ago, with a plethora of studies since then demonstrating the qualitative ability of such models. The quantitative ability of such models, however, has received relatively little attention. In this work, we describe a series of studies designed to critically probe the accuracy of kinetic-theory models to predict instabilities in solid-particle flows. The full, nonlinear solution of the transient form of the governing equations is solved. A linear stability analysis of the same starting equations is also considered. Predictions of the critical length scales obtained from both of these methods are compared with molecular dynamics simulations. The results reveal the impact of friction, nonlinearities, and Knudsen effects on the formation and evolution of instabilities. Overall, the quantitative performance of kinetic-theory models at predicting instabilities is quite encouraging.

(187a) Layer Melt Crystallization of Long-Chain Aldehydes From Hydroformylation Processes

Thorsten Beierling, Department of Biochemical and Chemical Engineering, Laboratory of Thermodynamics, Technische Universität Dortmund, Dortmund, Germany and Gabriele Sadowski, Department of Biochemical and Chemical Engineering, Laboratory of Thermodynamics, Technische Universität Dortmund, Dortmund, Germany

The hydroformylation of long-chain olefins, an industrially highly relevant reaction, produces a mixture of isomeric long-chain aldehydes. The purified substances are used as intermediates for further bulk and fine chemicals, such as diluent or for the flavour and fragrance industry. Conventional methods like stand-alone distillation tend to be costly or even infeasible, due to a decreasing difference in boiling points with increasing chain lengths. Layer melt crystallization in contrast is a technically proven and selective method for separating isomeric compounds and has generally the advantages of an easy scale up, facile solid-liquid separation and process handling. Despite these advantages, the promising method of separating isomeric long chain aldehydes with melt crystallization has not been investigated yet.

In this work the applicability of layer melt crystallization for various isomeric long chain aldehyde systems with different chain lengths was evaluated. Necessary parameters like related phase-equilibria were measured and modeled. Layer melt crystallization experiments were conducted, in which achievable growth rates and the corresponding purities were studied. The growth rates were modeled using the physical properties of the aldehyde systems as a function of process conditions like concentration, temperature and cooling rate. Additionally, the efficiency of sweating as a post-purification step was investigated. The sweating behavior can mathematically be described in high accordance to the experimental values as a function of temperature, crystal purity and layer thickness.

Summing up we show that melt crystallization should, due to its high efficiency, be considered as the method of choice for the separation of isomeric long-chain aldehydes. Furthermore our results reveal useful methods for the extrapolation of melt crystallization including the sweating step to systems with different chain-lengths, notably without conduction of further experiments.

(187b) Hybrid Filtration-Drying-Dissolution Device for Compact Pharmaceutical Manufacturing System

Shin Yee Wong, Jie Chen and Allan S. Myerson, Department of Chemical Engineering, Massachusetts Institute of Technology, Cambridge, MA

Pharmaceutical manufacturing is typically done in batch vessels. The recent interest in continuous pharmaceutical manufacturing had triggered a lot of interest in both academic and industrial communities, driven by the many benefits of the transformation. To bring the idea of continuous manufacturing further, we are developing a bench-top sized Pharmacy-on-Demand (PoD) system, capable of producing formulated drugs, corresponding to a daily production rate of few kilograms of Active Pharmaceutical Ingredients (API). The realization of the compact PoD system would require innovation in all unit operations (chemical synthesis, crystallization, and downstream processes), and high level of controls and automations. In this study, a hybrid device that integrated all post-processing operations after the crystallization process was developed.

This device consisted of five major components: 1) top cover consisted of inlets for slurry (from crystallizer) and liquid excipients; 2) steel mesh (0.5μm) holder; 3) lower barb end connected to vacuum; 4) wall heating component; 5) side drain valve. The filtration, drying and dissolution operations were conducted in semi-batch manner. Once the crystallization process was completed, the slurry was pumped into the hybrid device. Then, the refined API was separated from the mother liquor through the steel mesh under vacuum condition. After the filtration, the API was dried under vacuum condition at the temperature controlled by power supplied to the wall heating element. Once dried, the API was re-dispersed in the liquid excipients and drained from the side valve. Initial testing had shown high separation and drying efficiency. More than 98% of the crystallized API was successfully recovered after re-dissolution with the liquid excipients. The hybrid device had successfully transformed the slurry obtained from the crystallization process into formulated drug products in a single integrated device.

In short, the invention of this hybrid device had essentially eliminated the need for multiple stand-alone units for filtration, drying, blending and dissolution. On top of its integrated function, the compact (max dimension = 90mm) device had also significantly simplified the implementation of the downstream purification process of the PoD system.

Design and Characterization of a Continuous Stirred Tank Crystallizer

G. Morris, G. Hou, M. Barrett, B. Glennon

Solid State Pharmaceutical Cluster, The School of Chemical and Bioprocess Engineering, University College Dublin, Belfield, Dublin 4, Ireland

Keywords: Mixed Suspension Mixed Product Removal Crystallizer, Benzoic Acid, Cooling Crystallization, Process Analytical Technology

Paper ID#: 270744

Introduction

This study presents the development of a continuous crystallization platform consisting of a single stirred tank Mixed Suspension Mixed Product Removal (MSMPR) crystallizer. A photo and schematic drawing of the MSMPR setup appear in Figures 1 and 2 respectively. A recirculation loop is incorporated such that the withdrawn product is recycled to a dissolution vessel which serves as the feed tank to the crystallizer. Two OptiMax™ workstations (Mettler-Toledo) are used as both the feed/dissolution tank and the MSMPR crystallizer. Difficulties can often arise in the continuous removal of representative, non-classified product from lab-scale MSMPRs as a result of the relatively low volumes involved and consequently slow flowrates and removal velocities observed **[1]**. Intermittent withdrawal, whereby a certain percentage volume of the MSMPR is periodically removed at a high velocity so as to aid the potential for iso-kinetic removal and mitigate classification has been applied previously and is adopted in this work **[2, 3]**. This quasi-continuous transfer of the MSMPR product to the feed/dissolution vessel is achieved using vacuum with 7.1% of the suspension volume removed per transfer.

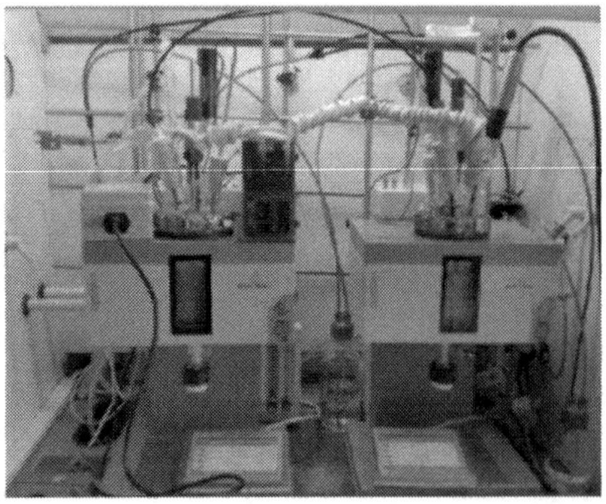

Figure 1: Image of the MSMPR setup showing feed/dissolution vessel on the left and MSMPR on the right with FBRM in-situ

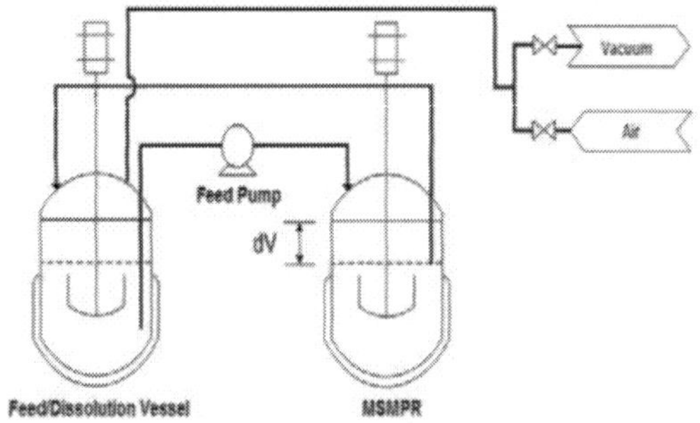

Figure 2: Schematic drawing of the MSMPR setup showing application of vacuum in recycling material from MSMPR to feed/dissolution vessel

In-situ process analytical technologies (FBRM, PVM), together with periodic sampling of the withdrawn product at steady-state and during the dynamic period of the process, have been applied in order to characterize the continuous cooling crystallization of benzoic acid from water and ethanol. Comparison is drawn between *high magma density* crystallizations (feed saturated at 40°C with the crystallizer at 18°C) and *low magma density* crystallizations (feed saturated at 15°C with the crystallizer at 0°C). For the low magma density runs the influence of residence time on affecting the steady-state particle size was investigated. For each of these studies, continuous operation was started with a suspension in the MSMPR at time zero by first batch cooling the saturated feed to the desired MSMPR temperature. In addition, a further set of experiments was carried out to investigate whether the mode of start-up bears any influence on the final steady-state product obtained and the route to steady-state.

Characterization with FBRM and Periodic Sampling

The crystallization is monitored in-situ using FBRM and PVM. This allows real time tracking of the changes in the particle size distribution and attainment of steady-state. Figure 3 shows the dynamic response in the FBRM total counts and mean chord length trends for a high-magma density crystallization proceeding to steady-state.

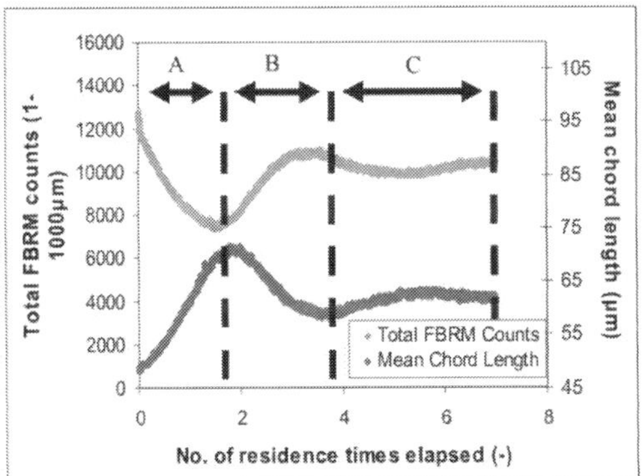

Figure 3: Tracking MSMPR to steady-state with FBRM for a high magma density run of 60 minute residence time. Showing regions: (A) wash-out period, (B) nucleation response and (C) attainment of steady-state

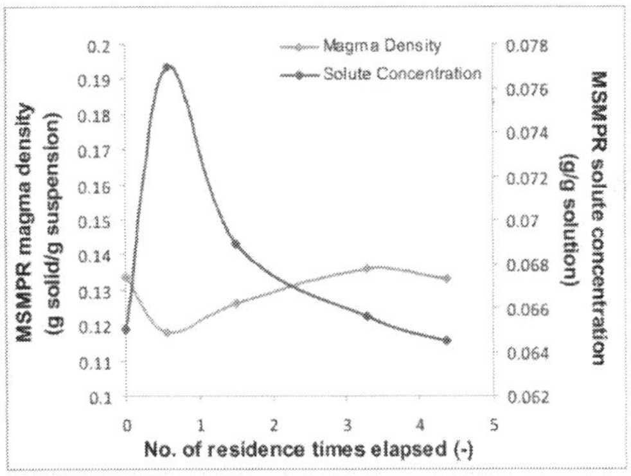

Figure 4: MSMPR solute concentration and magma density trends from start-up to steady-state for a 15 minute residence time high magma density run

As the crystallization proceeds initially there is a wash-out period (A) where particles are withdrawn in the absence of nucleation as the supersaturation builds. Over this period crystal growth is observed which is confirmed by an increase in the FBRM size trends and in-situ PVM images (Figure 5). Supersaturation increases to a maximum just before region (B), where the kinetics of the system force a period of constant nucleation which steadily consumes some of the supersaturation and slightly shifts the CSD back towards the fine end. Eventually the rate at which new crystals are formed by nucleation balances the rate at which they are withdrawn from the MSMPR and the system tends towards a steady-state (C).

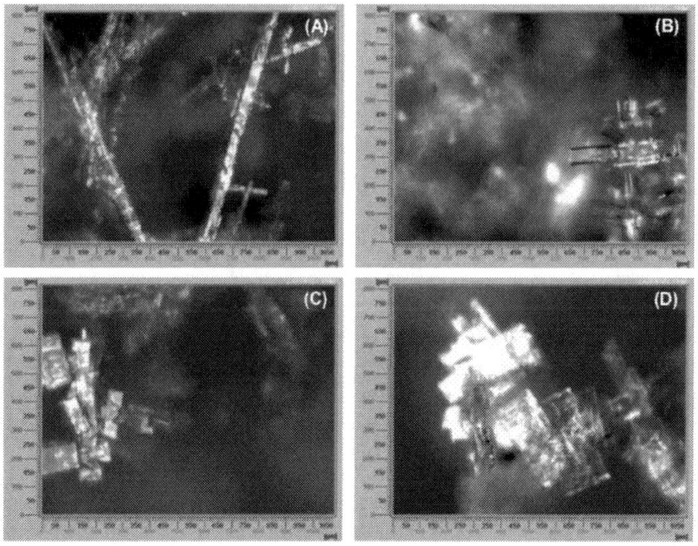

Figure 5: In-situ PVM images from a low magma density run of 15 minute residence time showing the change in particle size during MSMPR start-up. Taken at: (A) time zero after batch cooling at 1.5°C/min, (B) at 1 residence time, (C) at 2 residence times and (D) at 3 residence times

High & Low Magma Density Crystallizations

Comparing high and low magma density crystallizations it was observed that the level of the nucleation response in the dynamic route to steady-state is reduced when the slurry density in the crystallizer is relatively low (Figure 6), demonstrating the dependency of the nucleation rate on magma density. Consequently for the high magma density process the steady-state chord length distribution is shifted more towards the fine end (Figure 7).

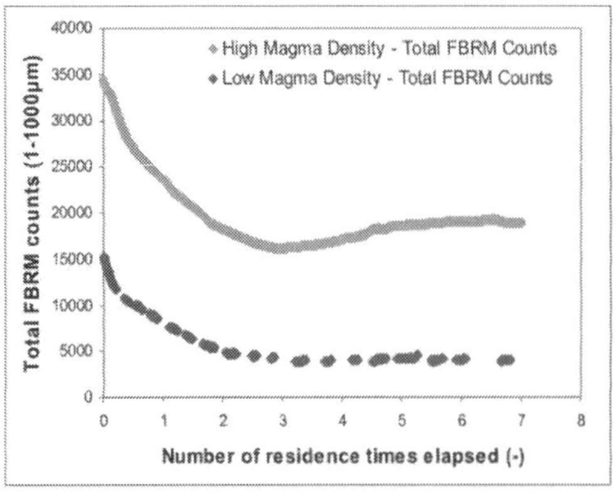

Figure 6: FBRM total counts trend for high and low magma density MSMPR runs of 15 minute residence time. For the low magma density run FBRM fouling was an issue so periodic cleaning of the probe was adopted

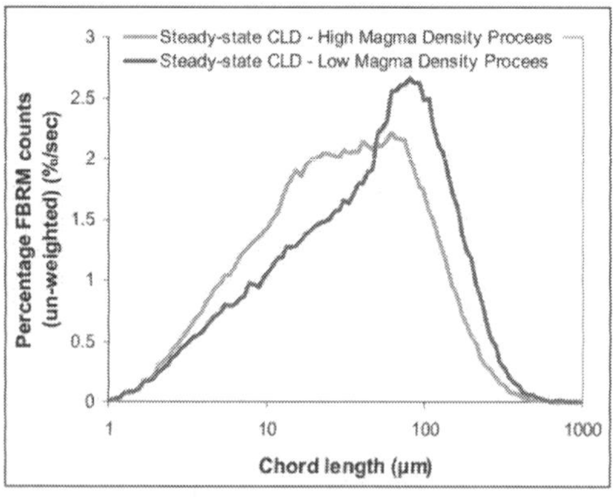

Figure 7: Final steady-state (un-weighted) chord length distributions (CLDs) for both high and low magma density processes

Evolution of Particle Size with Increasing Residence Time

In an MSMPR, for a given temperature drop between feed and crystallizer, it is possible to access different positions on the phase diagram by varying the residence time and as a result manipulate the steady-state supersaturation. The speed of the crystallization kinetics and the meta-stable zone width will ultimately dictate the supersaturation range accessible, and whether changing residence time will be favourable in terms of manipulating particle size advantageously. The benzoic acid system used in this study has a small meta-stable zone width and is more growth dominated. Consequently, the accessible range of supersaturation is limited but particle size has been shown to grow with residence time (Figure 8). Figure 9 demonstrates that steady-state particle size increases whilst MSMPR solute concentration and hence supersaturation deceases with increasing residence time.

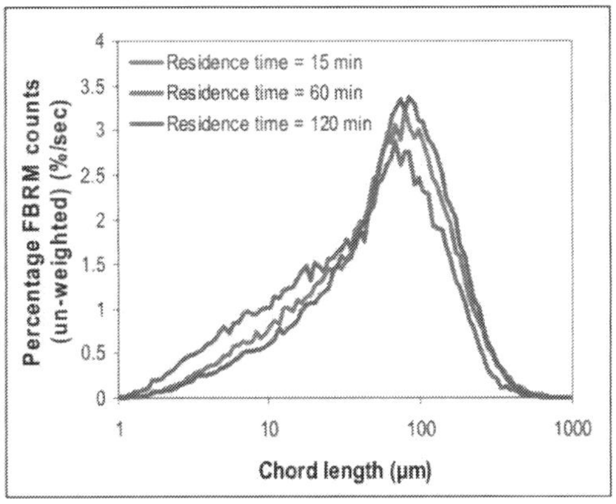

Figure 8: Steady-state CLDs (un-weighted) for low magma density runs of varying residence time

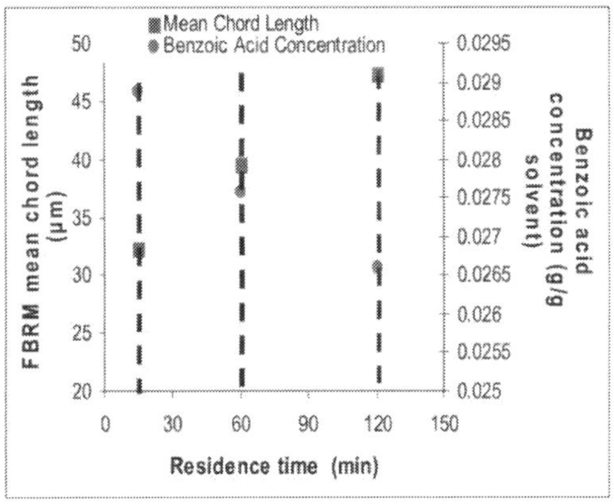

Figure 9: Demonstrating increase in particle size with residence time whilst solute concentration decreases

Influence of Start-Up Mode on Attainment of Steady-State

Several start-up regimes for running an MSMPR could potentially be employed. To examine whether particular choices in this area can have a significant bearing on the steady-state particle size obtained by the system, three methods of starting the MSMPR were investigated whilst keeping all other conditions the same:

(i) *Batch cooling the saturated feed to the desired MSMPR operating temperature, therefore allowing continuous operation to be started with an equilibrium suspension in the MSMPR at time zero*

(ii) *Starting with a clear solution which is saturated at the MSMPR operating temperature*

(iii) *Seeding a solution that is saturated at the MSMPR operating temperature with isolated steady-state product from a previous run, therefore creating a suspension of the final product at time zero*

All start-up methodologies were assessed through high magma density benzoic acid crystallizations, where the feed to the MSMPR is saturated at 40°C and the crystallizer is operated at 18°C. The resulting steady-state chord length distributions from each start-up regime are seen to be similar in all cases (Figure 10). This suggests that the final steady-state condition attained by the system is largely unaffected by the start-up methodology adopted.

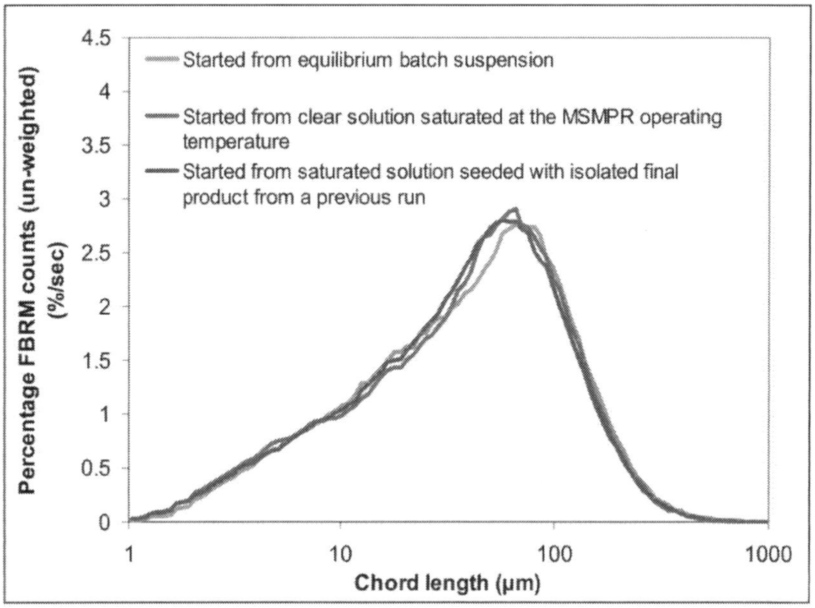

Figure 10: Steady-state CLDs (un-weighted) for high magma density crystallizations with different start-up regimes

Comparison of the FBRM total counts trends during the dynamic route to steady-state suggests that starting with saturated solution that is seeded with final product material may offer the quickest route to steady-state (Figure 11). Unlike the equilibrium batch start-up, in this case the initial particle size distribution is the same as the final and as a result the system does not require periods of significant oscillation in the size distribution to develop the steady-state CSD for the continuous process. This therefore may represent an optimum start-up strategy for MSMPRs, as the time required to reach steady-state is reduced and hence wastage of material whilst the final product is potentially out of spec is minimised.

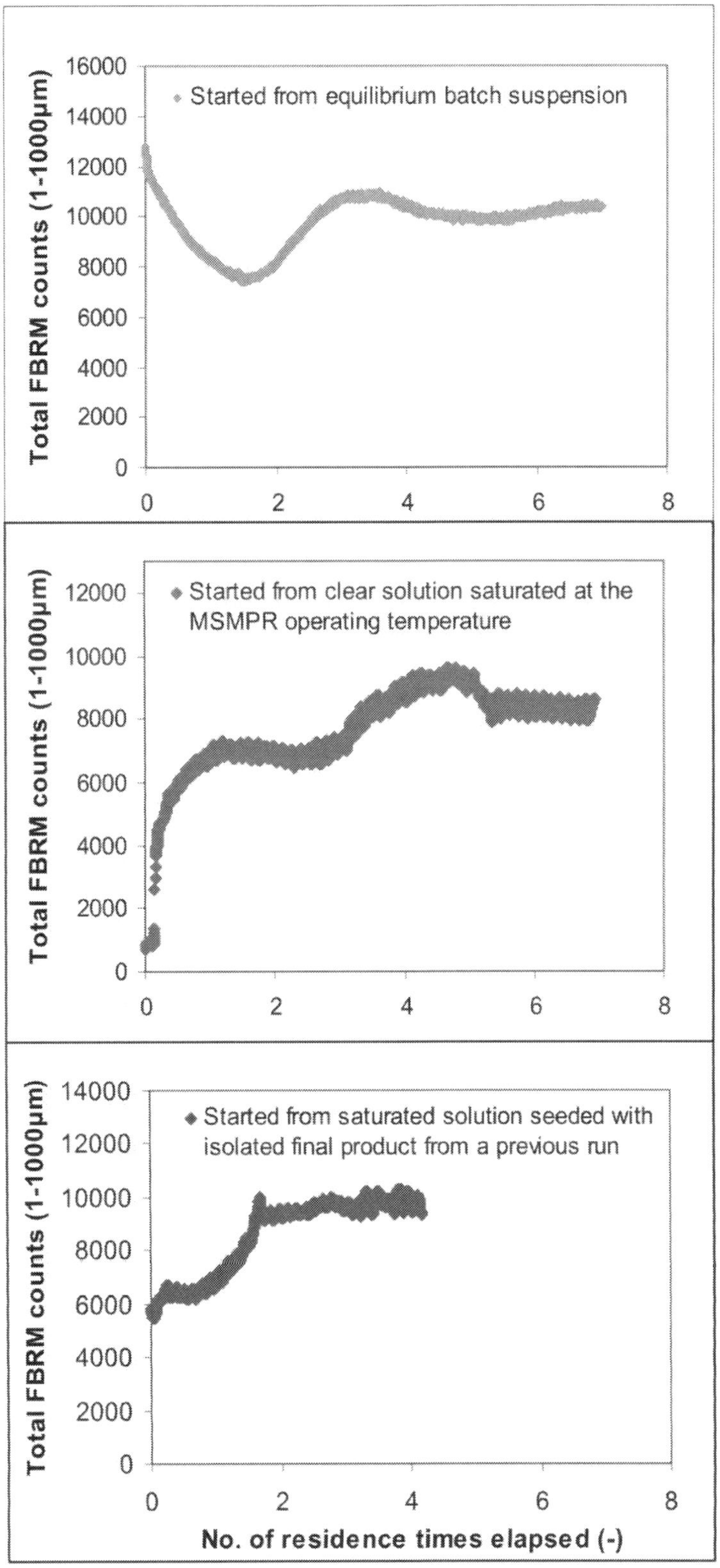

Figure 11: Comparison of the dynamic response in the FBRM total counts trends as the system proceeds to steady-state for all three start-up methodologies investigated

References

[1] A.D. Randolph, M.A. Larson, Theory of Particulate Processes, 2nd edition, Academic Press, New York, 1988.
[2] V.A. Chan, H.M. Ang, Journal of Crystal Growth 166 (1996) 1009-1014
[3] E. Kougoulos, A.G. Jones, K.H. Jennings, M.W. Wood-Kaczmar, Journal of Crystal Growth 273 (2005) 529–534

(187d) X-Ray Compatible Microfluidic Platforms for On-Chip Structure Determination of Membrane Proteins

Sudipto Guha, Ashtamurthy Pawate, Sarah L. Perry and Paul J.A. Kenis, Department of Chemical & Biomolecular Engineering, University of Illinois at Urbana-Champaign, Urbana, IL

Microfluidics has found extensive use in fields as diverse as energy conversion, medical and clinical diagnostics, point-of-care medical devices, proteomics and protein crystallization. Microfluidics allows for ease of sample preparation, small sample sizes, ease of scale-up, as well as integration of various sample analysis techniques on a single platform[1,2].

Here, we focus on the application of microfluidic platforms for membrane protein crystallization. Membrane proteins (MPs) play a crucial role in many important biological processes including energy and material transduction across cellular membranes, molecular recognition and immune response. They are amphiphilic, structurally diverse, and extremely responsive to the surrounding environment. Efforts to understand the structure-function aspects of these proteins through X-ray crystallography have been hampered by difficulty in obtaining high quality crystals[3]. Identifying the optimal crystallization condition(s) requires systematic screening of a huge chemical space (salts, buffers and precipitants), which requires a lot of protein sample that is generally not available. Additionally, current methods of structure determination are not optimized for handling small and fragile crystals that MPs tend to form.

We present an array-based, X-ray compatible microfluidic platform comprised of cyclic olefin copolymer (COC) and a thin polydimethylsiloxane (PDMS) membrane needed for valve actuation. The array-based design allows for screening for upto 96 conditions using $< 6\mu L$ of protein solution. After screening, a similar chip is used for structure determination of the protein by merging data from multiple crystals grown on-chip. The use of X-ray compatible materials allows *on-chip* analysis, thus preventing damage to fragile protein crystals that might occur during manual harvesting and mounting of crystals. As proof of concept, we have screened crystallization conditions for a soluble protein, phosphonoacetate hydrolase (PhnA). After obtaining the optimal condition and using anomalous data collected solely on-chip, we determined its structure to a resolution of 1.99 Å.

We are now in the process of validating our platform by screening and solving the structure of two MPs - photosynthetic reaction center (RC) and cytochrome ba_3-type oxidase. We have collected data on RC to 3.0 Å and are currently optimizing the crystallization conditions and chip design to get higher quality data.

(1) Hansen, C.; Quake, S. R. *Current Opinion in Structural Biology* **2003**, *13*, 538-544.

(2) Li, L.; Ismagilov, R. F. *Annual Review of Biophysics* **2010**, *39*, 139-158.

(3) Stevens, R. C. *Current Opinion in Structural Biology* **2000**, *10*, 558-563.

(187e) Investigation of a Counter-Current Crystallization Process for Separation of Systems Forming Solid Solutions

Erik Temmel[1], Sebastian Wloch[1], Uwe Müller[2], Detlef Grawe[3], Robert Eilers[4], Heike Lorenz[5] and Andreas Seidel-Morgenstern[6], (1)Max Planck Institute for Dynamics of Complex Technical Systems, Magdeburg, Germany, (2)Hapila Gera GmbH, Gera, Germany, (3)Jesalis Pharma GmbH, Jena, Germany, (4)Hapila Gera GmbH, Magdeburg, Germany, (5)Physical and Chemical Foundations of Process Engineering, Max Planck Institute for Dynamics of Complex Technical Systems, Magdeburg, Germany, (6) Physical and Chemical Foundations of Process Engineering, Max-Planck Institute for Dynamics of Complex Technical System, Magdeburg, Germany

Investigation of a counter-current crystallization process for separation of systems forming solid solutions

Erik Temmel[1], Sebastian Wloch[1], Uwe Müller[2], Detlef Grawe[3], Robert Eilers[2], Heike Lorenz[1], Andreas Seidel-Morgenstern[1,4]

[1]*Max Planck Institute for Dynamics of Complex Technical Systems, Magdeburg, D-39108, Germany*

[2]*HAPILA GmbH, Gera, D-07552, Germany*

[3]*Jesalis Pharma GmbH, Jena, D- 07745, Germany*

[4]*Otto von Guericke University Magdeburg, Chair of Chemical Process Engineering, Magdeburg, D-39108, Germany*

The occurrence of complete or partial solid solutions in solution crystallization is still the focus of several recent investigations. One reason is that the underlying mechanisms of the formation of mixed crystals are not clarified in every detail. On the other hand this special solid liquid equilibrium (SLE) limits the applicability of commonly used crystallization processes and complicates separation tasks. Especially for pharmaceuticals this can lead to a huge economical drawback.

Process design for crystallization from solution is mainly based on the solid-liquid equilibria (SLE) of the substances of interest [1]. For the, in case of ternary mixtures, commonly appearing simple eutectic and intermediate compound-forming systems several opportunities for crystallization based separation from a solvent are known and are well understood, e.g. processes based on applying preferential crystallization in case of two dissolved enantiomers. Besides, the tie lines of systems involving solid solutions do not lead to a corner of the pure substance in the ternary phase diagram (figure 1). Therefore it is not possible to separate these mixtures within one crystallization step and there is no option to apply the commonly used crystallization processes. For many examples such a miscibility behavior is published including inorganic systems like salts containing one common ion [1, 2] or simple organic substances like amino acids [3]. For more complex chiral molecules the occurrence of complete mixed crystal formation seems to be rare [4]. However, there are more cases reported where partial solid solutions are present at the pure enantiomer side of conglomerate and racemic compound-forming chiral systems.

Some theoretical studies suggest a separation opportunity based on fractional crystallization [5-7]. By means of repeated dissolution and crystallization it is possible to separate systems forming solid solutions. This principle can be realized by exploiting a cascade of batch crystallizers. Furthermore, it was shown that a counter-current between the solid and liquid phase in such a process would lead to optimal process performance in terms of yield and productivity. Nevertheless, realization and automation of a counter-current crystallization cascade is quite complicated. Between every crystallization stage a solid/liquid separation is required. Additionally, the automatic transport of the crystalline solid phase between the stages is difficult and would lead to huge instrumental effort.

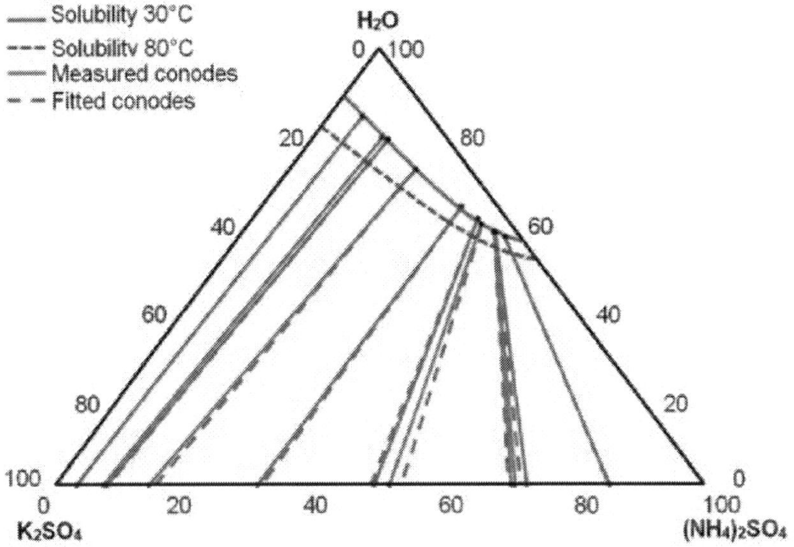

Figure 1: Ternary phase diagram of the system K_2SO_4 / $(NH_4)_2SO_4$ / H_2O for 30°C and 80°C with conodes indicated.

Recently an automatic counter-current crystallization cascade was developed by the HAPILA Gera GmbH [8]. In this plant exclusively liquid phases are transported. The solid phase is dissolved after the crystallization to avoid the expensive transport of the crystalline phases in the solid state. Hence, it is possible to proof this up to now only theoretically considered separation opportunity for systems involving solid solutions.

In the contribution the SLE of two different systems forming complete solid solutions will be shown. Furthermore, the mathematical description of the process of interest will be explained and verified by a manually operated counter-current crystallization in a rotary evaporator. The validated model was then used to run a semi-automatic pilot plant followed by a comparison of theoretical and experimental results. Finally a partial solid-solution forming system was theoretically investigated in a simulation study.

References

[1] Mullin, J. W., *Crystallization*, Boston, Butterworth-Heinemann (1997)

[2] Dejewska, D.; Szymański, T., *Cryst. Res. Technol.*, 1998, 33, 757-765

[3] Kurosawa, I. W.; Teja, A. S.; Rousseau, R. W., *Fluid Phase Equilib.*, 2004, 224, 245-249.

[4] Jaques, J.; Collet, André; Wilen, *Enantiomers, racemates, and resolutions*, S. H., Florida, Krieger (1994)

[5] Lin, S. W; Ng, K. M.; Wibowo, C., *Comput. & Chem. Eng.*, 2008, 32, 956–970

[6] Balawejder, M.; Galan, K. ; Elsner, M. P.; Seidel-Morgenstern, A.; Piatkowski, W.; Antos, D., *Chem. Eng. Sci.*, 2011, 66, 5638–5647

[7] Matz, G., *Chem. Ing. Tech.*, 1980, 52, 562-570

[8] HAPILA GmbH, Offenlegungsschrift DE 10 2008 023 833 A1, 2009

(187f) Designing a Fit-for-Purpose Agglomeration Method for Pharmaceutical Crystals

Chenchi Wang, Daniel Hsieh, Mario Hubert and Chiajen J. Lai, Research and Development, Bristol-Myers Squibb Company, New Brunswick, NJ

Designing a Fit-for-Purpose Agglomeration Method for Pharmaceutical Crystals

Chenchi Wang, Daniel Hsieh, Mario Hubert, Chiajen Lai.

Research and Development, Bristol-Myers Squibb Company, New Brunswick, New Jersey

In pharmaceutical development various particle engineering techniques are commonly employed to alter crystal morphology, improve crystal size, or increase purity of the active pharmaceutical ingredient (API). Crystal agglomeration is an alternative method to improve slurry and powder properties, especially for materials with a crystal habit or size that is difficult to change by classical crystallization techniques. In a crystal agglomeration process, the primary crystals can be directed to form agglomerates in a multi-solvent system. The formation of agglomerates can significantly improve slurry filterability or bulk powder properties compared to the primary crystals. A "fit-for-purpose" method will be presented that can be used to identify suitable conditions to encourage crystal agglomeration. Multiple case studies will be presented in detail to demonstrate how specific slurry or powder properties can be attained through the crystal agglomeration approach.

(187g) Utilization of Population Balance Models to Develop a Continuous Crystallization Process; Determination of Crystallization Kinetics

Christopher L. Burcham[1], Michael A. Lovette[2], Christopher S. Polster[2], Sean K. Bermingham[3] and Hassan Mumtaz[3], (1)Chemical Product Research and Development; Particle Design Laboratory, Eli Lilly and Company, Indianapolis, IN, (2)Chemical Product Research and Development, Eli Lilly and Company, Indianapolis, IN, (3)Process Systems Enterprise, London, United Kingdom

Due to material and time constraints, experimentation alone often cannot provide a means for design optimization of continuous crystallization processes within the pharmaceuticals industry. While several conventional process design software packages offer the ability to handle process optimization, such software packages often lack a robust means for predicting the effect of changing process variables on the corresponding particle size distributions (PSD). In the crystallization of active pharmaceutical ingredients, specific PSD requirements are often a crucial to the manufacturability and bioavailability of the final drug (solid dosage form). We use a combined population balance and process design software (e.g., gCRYSTAL, Process Systems Enterprises) as a means for process optimization for continuous crystallization. In performing this process optimization, we use rate constants for nucleation and growth that were obtained from regressed experimental data using the same software package. Furthermore, we are able to perform this process optimization with the assertion of constraints on the PSD for the optimal solution.

In this talk, experiments are conducted to provide the necessary data needed to regress kinetic parameters associated with crystallization; nucleation, growth and agglomeration. This talk will discuss both the experimental and computation aspects of the work and subsequent refinements as continuous crystallization results are collected later in development. This is the first of two discussions at this meeting, illustrating our proposed "universal" work flow for the design of continuous crystallizations.

(187h) Utilization of Population Balance Models to Develop a Continuous Crystallization Process; Process Design Using Optimization

Christopher L. Burcham[1], Michael A. Lovette[2], Christopher S. Polster[2], Sean K. Bermingham[3] and Hassan Mumtaz[3], (1)Chemical Product Research and Development; Particle Design Laboratory, Eli Lilly and Company, Indianapolis, IN, (2)Chemical Product Research and Development, Eli Lilly and Company, Indianapolis, IN, (3)Process Systems Enterprise, London, United Kingdom

Due to material and time constraints, experimentation alone often cannot provide a means for design optimization of continuous crystallization processes within the pharmaceuticals industry. While several conventional process design software packages offer the ability to handle process optimization, such software packages often lack a robust means for predicting the effect of changing process variables on the corresponding particle size distributions (PSD). In the crystallization of active pharmaceutical ingredients, specific PSD requirements are often a crucial to the manufacturability and bioavailability of the final drug (solid dosage form). In this talk, we use a combined population balance and process design software (e.g., gCRYSTAL, Process Systems Enterprises) as a means for process optimization for continuous crystallization. In performing this process optimization, we use rate constants for nucleation and growth that were obtained from regressed experimental data using the same software package (the details of this regression are contained in the talk "Utilization of Population Balance Models to Develop a Continuous Crystallization Process; Determination of Crystallization Kinetics," which is also being given at this conference). Furthermore, we are able to perform this process optimization with the assertion of constraints on the PSD for the optimal solution. This talk concludes with our insights regarding the implementation of process optimization as a part of a proposed "universal" work flow for the design of continuous crystallizations.

(208a) Improved Mixing in Dry Catalyst Impregnation Using a Double Cone Blender: An Experimental and Computational Approach

Yangyang Shen, Rutgers University, Piscataway, NJ and M. Silvina Tomassone, Chemical and Biochemical Engineering, Rutgers University, Piscataway, NJ

In the manufacture of heterogeneous catalysts, the impregnation of active metals onto a porous catalyst support is a crucial preparation step that may significantly affect the activity and selectivity of the resulting catalysts. In a typical dry impregnation (pore filling or incipient wetness impregnation) process, metal solutions are sprayed over a powder bed in a mixing vessel until the pore volume is reached. Traditionally, these systems are studied experimentally, but such experiments can be difficult to examine quantitatively, and the analysis is often inaccurate due to frequent disturbance to the powder bed. In this work, Discrete Element Method (DEM) simulations coupled with a novel algorithm allowing the transfer of fluid to and between particles were used in combination with geometrically equivalent experiments to model dry catalyst impregnation with the goal of improving the mixing and content uniformity of the resulting catalyst powder. While the developed model and corresponding algorithm have previously shown excellent agreement with matching experiments, the double cone blender itself does not establish an optimal content uniformity alone. As a result various improvements were simulated and subsequently analyzed within the system including baffle and nozzle configurations. In this study, it is shown that the use of both two and four baffles significantly reduced the mixing time for a double cone blender, thus increasing overall content uniformity while maintaining other important process parameters. No noticeable difference was evident between two and four baffles. Additionally, matching one-to-one experiments with baffles were also conducted at three distinct fill levels to validate that the addition of as little as two baffles can significantly improve mixing and content uniformity, as discovered in the DEM simulations. In addition, particle tracking techniques unique to the simulations have better quantified the poor axial mixing in the double cone, and further established the improved mixing generated by adding baffles. The culmination of this work has been the development of simulations capable of analyzing dry catalyst impregnation in limitless parameters, geometries, and conditions for improving overall mixing, content uniformity, and product quality.

(208b) DEM Validation of Tablet Motion in a Vibratory Coating Pan

Rahul Kumar and Carl Wassgren, Mechanical Engineering, Purdue University, West Lafayette, IN

Discrete Element Method (DEM) modeling holds great promise for simulating tablet coating processes. This study presents a DEM model for simulating the movement of true biconvex tablet shapes in a vibrating coater pan, along with the experimental methods used to validate the model. High speed video imaging is used to measure the circulation speed of tablets in the experiments and comparisons with DEM simulations are made for a range of vibration amplitudes and frequencies. Inputs to the DEM model, like tablet-tablet and tablet-wall friction coefficients are measured directly using a variation on the pin-on-disk apparatus and the coefficient of restitution values were obtained from literature. In addition, the sensitivity of the model to tablet-level parameters, including friction coefficients, coefficients of restitution, and contact stiffnesses, are presented. The tablet-wall friction coefficient was found to have a very strong influence on the circulation speed with the circulation speed increasing as the friction coefficient is increased. The coefficient of restitution was also found to have a significant effect on the circulation speed.

(208c) Axial Dispersion of Non-Spherical Particles in Horizontal, Rotating Cylinders

Guang Lu, James Third and **Christoph R. Müller**, Department of Mechanical and Process Engineering, ETH Zurich, Zurich, Switzerland

Rotating cylinders or kilns are widely employed in industry, ranging from the pharmaceutical, food and agricultural to the chemical industry. Therefore, a fundamental understanding of the underlying physics of such systems is not only of academic interest, but also relevant for industrial applications. One specific area that is currently only poorly understood, but at the same time particularly relevant for industrial applications, is the axial dispersion of non-spherical particles in horizontal, rotating cylinders. So far, most studies have only concentrated on spherical particles in such systems (Zhu et al., 2008). Granular materials are typically opaque, making the acquisition of experimental measurements in such systems, e.g. using conventional high speed optical imaging techniques, very challenging (Müller et al., 2010; Müller et al., 2011). As an alternative, computational techniques, such as the discrete element method (DEM), can be applied as a tool to investigate the underlying physics of particle interactions and dynamics within a bed of grains (Third et al., 2010b; Third et al., 2011). The DEM can provide both macroscopic and microscopic 'measurements' in granular system and, in addition, model non-spherical particles of various shapes (Lu et al., 2012). First, as a reference case we study the axial dispersion of monodisperse spheres within a horizontal, rotating cylinder. Experiments using glass spheres have shown that axial dispersion in rotating cylinders obeys Fick's second law, i.e. the mean square deviation of the particle positions is linearly proportional to time (Hogg et al., 1966; Parker et al., 1997). These observations are in agreement with previous DEM simulations (Third et al., 2010a; Taberlet and Richard, 2006) and the results obtained here. However, the corresponding information from non-spherical particle systems, e.g. beds composed of cubes, is currently not available. In this work, non-spherical particles were modeled using the super-quadric equation, viz. $(x/a)^m + (y/b)^n + (z/c)^p = 1$ (a, b and c are the lengths of the principle axes and m , n and p are indices controlling the sharpness of the edges). In the simulations performed here 11220 particles were placed in a horizontal, rotating cylinder. The aspect ratio of the non-spherical particles was varied, but the volume of the non-spherical particles was kept constant, and was equal to the volume of a sphere of diameter 3 mm, leading to an approximately constant fill level of 26% in the DEM simulations. It was found that the particle shape had a strong influence on the rate of axial dispersion. For example, cubic particles of aspect ratio 1 were found to disperse much faster than spherical particles of equal volume. Furthermore, the rate of axial dispersion increased with increasing blockiness of the particles.

References Hogg, R., Cahn, D.S., Healy, T.W., and Fuerstenau, D.W., 1966. Diffusional mixing in an ideal system. Chemical Engineering Science, 21, 1025-1038. Lu, G., Third, J.R., and Müller, C.R., 2012. Critical assessment of two approaches for evaluating contacts between super-quadric shaped particles in DEM simulations. submitted to Chemical Engineering Science. Müller, C.R., Holland, D.J., Sederman, A.J., Dennis, J.S., and Gladden, L.F., 2010. Magnetic resonance measurements of high-velocity particle motion in a three-dimensional gas-solid spouted bed. Physical Review E, 82, 050302. Müller, C.R., Holland, D.J., Third, J.R., Sederman, A.J., Dennis, J.S., and Gladden, L.F., 2011. Multi-scale magnetic resonance measurements and validation of discrete element model simulations. Particuology, 9, 330-341. Parker, D.J., Dijkstra, A.E., Martin, T.W., and Seville, J.P.K., 1997. Positron emission particle tracking studies of spherical particle motion in rotating drums. Chemical Engineering Science, 52, 2011-2022. Taberlet, N. and Richard, P., 2006. Diffusion of a granular pulse in a rotating drum. Physical Review E, 73, 041301. Third, J.R., Scott, D.M., and Müller, C.R., 2011. Axial transport within bi-disperse granular media in horizontal rotating cylinders. Physical Review E, 84, 041301. Third, J.R., Scott, D.M., and Scott, S.A., 2010a. Axial dispersion of granular material in horizontal rotating cylinders. Powder Technology, 203, 510-517. Third, J.R., Scott, D.M., Scott, S.A., and Müller, C.R., 2010b. Tangential velocity profiles of granular material within horizontal rotating cylinders modeled using the DEM. Granular Matter, 12, 587-595. Zhu, H.P., Zhou, Z.Y., Yang, R.Y., and Yu, A.B., 2008. Discrete particle simulation of particulate systems: A review of major applications and findings. Chemical Engineering Science, 63, 5728-5770.

(208d) Application of Enhanced DEM Simulations for the Improvement of a Given Tablet Coating Process

Gregor Toschkoff[1], Sarah Just[2], Georg Scharrer[1], Klaus Knop[2], Peter Kleinebudde[2], Dejan Djuric[3], Andreas Altmeyer[3], Adrian Funke[4] and Johannes G. Khinast[5], (1)Research Center Pharmaceutical Engineering, Graz, Austria, (2)Institute of Pharmaceutics and Biopharmaceutics, Heinrich Heine University, Düsseldorf, Germany, (3)L.B. BOHLE Maschinen + Verfahren GmbH, Ennigerloh, Germany, (4)Bayer HealthCare Pharmaceuticals, Berlin, Germany, (5)Institute for Process and Particle Engineering, Graz University of Technology, Graz, Austria

An important unit operation in the pharmaceutical industry is the coating of tablets. A common technique is drum coating, where the tablets are placed in a rotating drum and a coating liquid is sprayed onto the moving tablets. For most applications, the uniformity of coating is of great importance. This includes the even distribution of coating on single tablets, and the uniformity of coating mass between the different tablets of a batch . The demand for process understanding shows in the increasing number of investigations in recent years, both experimentally[1] and using the Discrete Element Method (DEM)[2-4].

The aim of this work is to improve the inter-tablet uniformity of a given coating process, and to predict the scale-up behaviour. To this end, exhaustive investigations based on DEM simulations are performed. Methods established in literature were applied, and new models developed. For the simulations, the real geometry of the apparatus was used. The material properties came from measurements, the operational parameters were taken from a real coating process. To capture the spray itself, two methods for the modelling of a spray in DEM simulations were implemented (Fig. 1): : a user module for the DEM software that models the spray as droplets (Fig. 1), and a ray-traced algorithm that operates on the output data of a prior DEM simulation.

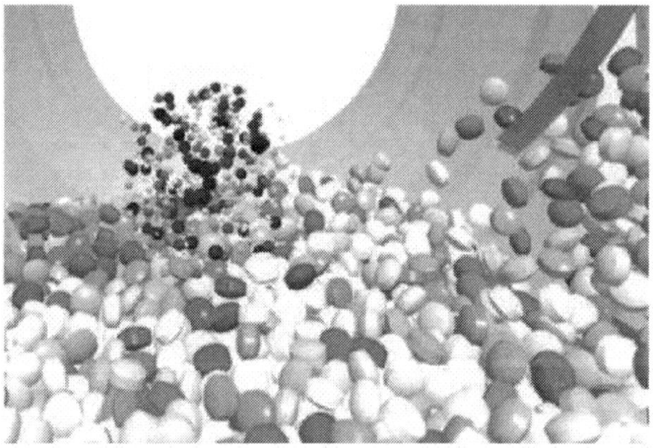

Figure 1: DEM simulation of a tablet coating process, including the coating spray. The spray droplets are colored in blue. The tablets are colored according to coating mass, from light yellow (no coating mass) to red (high coating mass).

The basis for the investigation is a full factorial 2^3 Design of Experiments for the lab-scale coater (BFC5, L.B. Bohle, Ennigerloh, Germany). The two spray models were applied to extract detailed information on the spray process, such as coating mass distribution or Single Visit Residence Time. First, a statistical analysis of how the DoE parameters affect the product quality was done. Second, the influence of the spray setup was studied in detail. This includes the geometry of the spray zone, the number of nozzles, and the placement of them. Finally, similar investigations were done for the pilot-scale to determine the scale-up performance.

The results of the DEM simulation give a deep understanding of a realistic coating process from different angles. They point out the main influence factors, and show feasible ways to improve the coating uniformity.

[1] R. Mueller and P. Kleinebudde, "Prediction of tablet velocity in pan coaters for scale-up," *Powder Technology,* vol. 173, pp. 51-58, 2007.

[2] M. Kodam, J. Curtis, B. Hancock, and C. Wassgren, "Discrete element method modeling of bi-convex pharmaceutical tablets: Contact detection algorithms and validation," *Chemical Engineering Science,* vol. 69, pp. 587-601, Feb 2012.

[3] A. Dubey, R. Hsia, K. Saranteas, D. Brone, T. Misra, and F. J. Muzzio, "Effect of speed, loading and spray pattern on coating variability in a pan coater," *Chemical Engineering Science,* vol. 66, pp. 5107 - 5115, 2011.

[4] B. Freireich, W. R. Ketterhagen, and C. Wassgren, "Intra-tablet coating variability for several pharmaceutical tablet shapes," *Chemical Engineering Science,* vol. 66, pp. 2535-2544, 2011.

(208e) A Renewal Theory Approach to Understanding Inter-Particle Coating Variability

Ben Freireich, Core R&D - Solids Processing, The Dow Chemical Company, Midland, MI and Jianfeng Li, School of Chemical Engineering, Purdue University, West Lafayette, IN

Many models, both analytical and numerical, have been derived for predicting coating mass variability for particle coating operations (e.g. tablet pan coating, Wurster coating). These existing models require lengthy simulation times or artificial discrete time scales, diminishing their utility. In this work, a simple alternative is proposed. Here an alternating dependent renewal theory model is presented to predict the inter-particle coating mass distribution for spray coating processes. By accounting for the variability of coating mass per visit to the spray and cycle time in the mixing vessel, the new model gives an explicit equation for the coating mass distribution. Comparisons are made against other modeling approaches including Bernoulli trial models, the Monte-Carlo models, and population balance models. It is shown that all existing modeling methods can be more easily and compactly expressed under the modeling approach developed in this work. Furthermore, using this approach, the same results may be obtained without introducing artificial discrete time scales or running lengthy simulations.

(208f) Sphero-Cylindrical Particle Stresses in a Vertical Axis Mixer

Xia Hua, School of Mechanical Engineering, Purdue University, West Lafayette, IN, Jennifer Sinclair Curtis, University of Florida, Gainesville, FL, Bruno C. Hancock, Pfizer Inc., Groton, CT, William R. Ketterhagen, Pfizer Global Research and Development, Pfizer Inc., Groton, CT and Carl R. Wassgren, Mechanical Engineering, Purdue University, West Lafayette, IN

A discrete element method model is used to examine the stresses in sphero-cylindrical particles agitated in a vertical axis mixer. The model is validated against experimental measurements of the rotating shaft torque. Using this model, the effects of particle aspect ratio from one to approximately nine, on blade torque, flow patterns, and average particle internal stress are investigated. Increasing particle aspect ratio generally decreases the flow rate of particles over the blade and increases the height of the heap in front of the blade. The mean particle hydrostatic and Mises stresses are found to increase with increasing bed depths ranging from one to two times the blade height and impeller speeds from 0 to 90 rpm. The stress distributions also appear to follow two different exponential distributions, depending on the stress magnitude. Increasing the particle aspect ratio increases the mean Mises stress due to an increase in the bed height in front of the mixer blade, as well as due to the interference of particles as they flow past one another. Increasing either the inter-particle friction coefficient from 0.11 up to 0.66 or the particle elastic modulus from 1 MPa up to 100 MPa results in larger internal stresses. However, increasing the particle-drum friction coefficient from 0.17 up to 0.68 has a much weaker influence.

(208g) Experimental Validation of a 3-D Finite Element Method Roll-Compaction Model

Ariel R. Muliadi, Chemical Engineering, Purdue University, West Lafayette, IN, Carl R. Wassgren, Mechanical Engineering, Purdue University, West Lafayette, IN and James D. Litster, School of Chemical Engineering, Department of Industrial and Physical Pharmacy, Purdue University, West Lafayette, IN

This study compares the FEM-predicted density distribution of roll-compacted ribbons to those measured experimentally. The following steps were undertaken to ensure one-to-one comparisons of model and measured results: (i) an air powered piston feeder configuration that applied a known, uniform stress on the powder was manufactured to replace the roll compactor feed auger; (i) all model boundary conditions, such as the powder/roll and powder/cheekplates friction angles, were based on experimentally-measured values; and (iii) density-dependent stress-strain constitutive parameters were used in the simulations to describe the powder mechanical behavior. Results show that model results agree well with the experimentally measured counterparts. Discrepancies between the two quantities were the result of cantilevered roll shaft supports that caused the experiments to be performed with non-parallel rolls. This effect made the experimentally-measured ribbon densities slightly asymmetrical against the roll half-width. Such trends were not captured by the simulations since the model assumed perfectly parallel rolls. Model shortcomings include its inability to resolve yield paths when the roll gap is too small (< 1.00 mm) and the ribbon's springback response downstream of the roll minimum gap width.

(212a) A Favre-Averaged Turbulence Model for Gas-Particle Flows

Thomas J. O'Brien, Retired, Boulder, CO

Computational fluid dynamics (CFD) simulations are routinely used to analyze a wide range of industrial and natural processes, mostly for single phase flows. In spite of their many well known shortcomings, RANS equations are used for the majority of these single phase CFD calculations. These are usually based on a k-ε turbulence model, although many other types of turbulence models have been developed. However, multiphase CFD calculations are very costly since they are invariably transient; no adequate turbulence model has been developed for the gas-particle two-fluid model which would allow a direct calculation of the time-steady behavior of the flow. A long transient calculation must be performed from which the time-steady behavior is extracted by averaging.

The two-fluid equations are themselves derived by an averaging procedure which introduces several constitutive relations, the most important being the phase-interaction (drag) term. This presents an additional complication for a multiphase RANS theory since many different forms of these relations have been presented in the literature. For the purposes of this initial development of a RANS model for gas-solid flow, the full equation set of MFIX (Syamlal, *et al.*, 1993) will be truncated to describe isothermal, non-reactive flow. Further, the granular pseudo-temperature equation will not be included; the granular stress will be described by a fixed viscosity as suggested by Miller and Gidaspow (1992), appropriate for dense, vertical gas-solid flow in a pipe.

For the case of multiphase flow, a RANS theory has not been extensively developed. Because of the many non-linear terms in the two-fluid theory equation set, a formal RANS theory presented in an early paper of Elghobashi and Abou-Arab (1983) contained 38 "closure" terms and the equations for multiphase turbulence kinetic energy (k equations) contained 64 such terms. In order to reduce the required number of closure relationships, Besnard and Harlow (1988) used an extension of Favre averaging which was introduced to address this same issue for compressible single phase flows (Favre, 1965; 1992). This approach has been followed for the description of bubbly flows, mainly motivated by nuclear reactor safety issues (Lahey, 2005; Krepper *et al.*, 2005, 2009). However, an equivalent development has not occurred to describe gas-solid flows. In this paper, these ideas are developed for the two-fluid equation set of gas solid flow, as used in the open-source software MFIX, developed at the National Energy Technology Laboratory (https://mfix.netl.doe.gov). The Favre-like average of the phasic velocities are introduced and useful identities are derived from these definitions. The relationship between the Favre-averaged velocity and the Reynolds-average is discussed. The averaged continuity and momentum equations are then derived for each phase. The continuity equations have the identical form as their transient counterparts. However, the momentum equations contain many (but much less than 38) terms that require closure, including a specific turbulent Reynolds stress for each phase. Specific turbulent kinetic energies can be introduced as the contraction of the Reynolds stresses. As in single phase turbulence theory, transport equations have been derived for these turbulent stresses and turbulent kinetic energies, which include dissipative terms. It is shown that the energy of the flow can be partitioned into two quadratic terms, one involving the square of the Favre-averaged velocity and the other being the turbulent kinetic energy. Thus the energy of the flow can be partitioned between a term involving the mean flow and the fluctuations associated with volume fractions and the turbulent kinetic energy which includes the energy of the remaining fluctuations. It is clear from these equations that the turbulent energy cascade is much more complicated than single phase flow since fluctuation energy can be exchanged between phases. Most likely, energy introduced in the fluid (carrier) phase is transferred to the granular phase and dissipated there. (A very important dissipative channel, inelastic particle collisions, is not open in this initial theory since the granular temperature equation is not included.) Some suggestions are introduced for calculating the closure terms, although these are very preliminary.

References:

Besnard, D.C., and F .H. Harlow, "Turbulence in multiphase flow," *Int. J. Multiphase Flow* **14**, 679-699, 1988.

Elghobashi, S.E., and T.W. Abou-Arab, "A two equation turbulence model for two-phase flows," *Phys. Fluids* **26**, 931-938, 1983.

Favre, A., "Équations des gaz turbulents compressibles I.- Formes générales," Journal de Mécanique **4**, 361-390, 1965.

Favre, A., "Formulation of the statistical equations of turbulent flows with variable density," in *Studies in Turbulence*, T.B. Gatski, S. Sarkar and C.O. Speziale, eds., Springer-Verlag, 324-341, 1992.

Krepper, E., D. Lucas and H.-M. Prasser, "On the modelling of bubbly flow in vertical pipes," *Nucl. Eng. Des.* **235**, 597–611, 2005.

Krepper, E., M. Beyer, T. Frank, D. Lucas and H.-M. Prasser, "CFD modelling of polydispersed bubbly two-phase flow around an obstacle," *Nucl. Eng. Des.* **239**, 2372–2381, 2009.

Lahey Jr., R.T., "The simulation of multidimensional multiphase flows," *Nucl. Eng. Des.* **235**, 1043-1060, 2005

Miller, A., and D. Gidaspow, "Dense, vertical gas-solid flow in a pipe," *AIChE J.* **38**, 1801-1815, 1992.

Syamlal M., W. Rogers and T.J O'Brien, "MFIX documentation: theory guide," Technical Note, DOE/METC-94/1004, NTIS/DE94000087, Springfield, VA: National Technical Information Service, 1993.

(212b) L-Valve Operation Under High Temperature Conditions

Dawei Wang, Hyung Kim, Samuel Bayham and L.-S. Fan, William G. Lowrie Department of Chemical and Biomolecular Engineering, The Ohio State University, Columbus, OH

The L-Valve as a solids flow control device has been used extensively in fluidized beds and circulating fluidized beds for solids recycling. Despite extensive research on the experimental characterization and correlative equations conducted under cold flow conditions, reports for the operation of the L-Valve under high temperature condition which are the normal practice in the industry are very rare. This paper presents an experimental study on the characterization of the L-Valve operation under high temperature condition. The tests were performed in a circulating fluidized bed system with various solids flow rates and temperatures. The relationships among the solids flow rate, aeration rate and pressure drop were investigated and compared with those under cold condition. It is found that a linear relationship between aeration gas flow rate and solids flow rate still holds with different coefficients than those under cold flow model. Some potential factors, such as agglomeration and sintering, which may cause the malfunction of the L-Valve under high temperature conditions, are also discussed.

STUDY OF DISTRIBUTION AND STABILITY OF SOLID PARTICLES IN A SPOUTED BED

Ghanim M. Alwan[1], Muthanna Al-Dahhan[2], Stoyan Nedeltchev[2] and Shreekanta Aradhya[3]
(1) Chemical Engineering Department, University of Technology, Baghdad, Iraq.
(2) Department of Chemical and Biological Engineering, Missouri University of Science and Technology, Rolla, MO.
(3) Chemical and Nuclear Engineering Department, Missouri University of Science and Technology, Rolla, MO.

Abstract

The present work is a part of scale-up methodology. The selected key parameter of the scale-up process is the solid particles' concentration distribution in the spouted bed. Steady-state measurements were carried out to study the solid particle's behaviors in the 60° conical shape spout-air bed with the diameter of 3 inches and height of 36 inches. Radial concentration of particles (glass and steel beads) at various elevations of the bed under different flow patterns were measured using sophisticated optical probes. The superficial air velocity and particles' densities were selected as the effective variables on the particles' distribution and stability into the spouted bed. The hydrodynamic parameters of the spouted process are difficulty to be measured at the unstable conditions. A stable bed is obtained when the beads can fluidize homogenously. The stable spouting is distinguished by the formation of the stable fountain. Unstable spouting is characterized by swirling and pulsation of the spout. The homogeneity and stability of solid particles appeared at the annular region within the position of 5.5 inches above the conical base. The results showed that decreasing in superficial air velocity could promote the solids' concentration, uniformity and stability in the annular region of the bed. In addition, it has been found that high beads' concentration and stable flow can be obtained with high-density particles.

Keywords: Spouted bed, Solid particles' distribution, Stability.

Nomenclature

C	Relative concentration of solid particles
dp	Diameter of solid particle, [mm]
P	Pressure into spouted bed, [Kpa]
V_i	Voltage for particular run, [volt]
V_{min}	Min voltage for particular run, [volt]
V_{max}	Max voltage for particular run, [volt]
UI	Uniformity index of solid particles
Vg	Superficial velocity of gas, m/s
Greek Symbols	
ρs	Density of solid particles, Kg/m^3
€	Porosity of bed, [-]
Ø	Sphersity of the solid particle

1. Introduction

Gas-solid spouted beds are either cylindrical bed with cone base or the whole bed is in a cone shape where the gas enters as a jet. The gas forms a spout region that carries the solids upward in a diluted phase which forms a fountain at the top of the bed where the solids fall down and move downward in the annular region. These beds have found wide and efficient applications as reactors, dryers, coaters, etc.

A spouted bed is a special case of fluidization. It is an effective means of contacting gas with coarse solid particles. There is increasing application of spouted as reactors such as; coating, desulfurization, CO capture, combustion and gasification of coal and biomass [1].The

*Corresponding author. Tel.: +964 7714437368
E-mail: ghnm_mag@yahoo.com

spouted bed is a kind of high performance reactor for fluid-solid particles reaction, also it is a hybrid fluid-solid contacting system [2].

It is better to develop the design of the spouted bed to overcome the large pressure drop and instability of the operation, and improves the uniformity of the products resulting from the chemical or physical treatment due to the elimination of the back mixing [3].

2. Content

2.1. SCOPE OF THE WORK

The objective of the present work is to study the behavior of the spouted bed under steady state conditions using different flow patterns. The selected process variables are superficial gas velocity and solid particle's density. The distribution of solid particles concentration is good indication for the uniformity of the solid particles across the bed .The uniformity of solid particles enhances the gas flow and the porosity distributions that improve the mass and heat transfer through the bed. The stability conditions can be limited depends on the concentration profile curves of solids across the bed. All the hydrodynamic parameters of the process are a difficulty to be measured at unstable conditions.

2.2. Experimental set-up

The present work is a part of scale-up methodology which is investigate and develop into the Multiphase and Multi-scale processes Laboratory of the (MMPL) of Chemical and Biological Engineering Department, Missouri University of Science and Technology, MO, USA. The experimental set-up was designed and constructed in the best way to collect the data as explained in Figure (1).The cylindrical spouted bed made of Plexiglas was employed in this study as shown in Figure (2) which was the major part of the experimental set-up. The bed is (3 inches in diameter and 36 inches in height), on which 20 holes (0.5 inch in diameter) are perforated horizontally .At the bottom of the bed, there is a 60° cone-shaped base (3 inches in height).The spouting orifice was made of Pyrex with 0.25 inch in diameter locates in the center of the conical base.

The spouted gas is air using the air compressor where the pressure was controlled by the manual regulator. The flow of air was measured by the rotameter (Figure 1). The solid particles used were steel and glass beads with different average diameters and properties (Table 1).

Holes are drilled at vertical intervals of (1.86 inch) along the column wall in which the optical probe was placed at different radial positions: 1.5, 1.25, 1.0, 0.75, 0.5,and 0.25 inch also at axis positions:7.5, 5.5 and 3.5 inches above the conical base

The newly optical probe (manufactured by the Institute of Chemical Metallurgy, Chinese, and Academy of Sciences) can be used to measure both solids concentration and solids velocity and their fluctuation.
The velocity of the solids into the bed cannot be zero as they will be in constant circular motion. The new optical probes are of the sizes: 1.93mm, 1.97 mm, 2.12mm and 4.15mm to cover a wide range of solid particles dimensions and properties.

The Particle Velocity analyzer (PV6) which manufactured by the 'Institute of Chemical Metallurgy, Chinese, Academy of Science' consists of; photoelectric converter and amplifying circuits, signal pre-processing circuits, high-speed A/D interface card and its software PV6, is adapted to the optical probe .

2.3. Measuring of solid particles concentration

The PV6 particle velocity analyzer is used for measuring the relative (local dimensionless) concentration of solid materials. Regarding the technical sheet of the optical probe, the concentration is evaluated by using Equation (1).

The optical probes are calibrated for measurement of the local particles concentration. For empty-bed state, the material concentration is equal to zero correspond to zero voltage of probe signal. Under the bulk concentration state (material concentration equal to 1.0), the output signal from the probe is equal to 4.5 voltages. Two or three bundles of optical fibers with diameter of 0.2-0.3 mm are arranged at certain interval according to different sizes of particles to be measured by the optical probe. To choose the appropriate probe, the average particle diameter should not exceed the diameter of the optical fiber in the probe.

The pressure drop across the bed is measured by the pressure transceducer (Type: PX309-002G5V) with the range of (0-2.0 psig) manufactured by omega company. The pressure into the spouted bed adjusting within the desired values by using the inverted circular stabilizer, 60 mm in diameter was installed at the top of the bed column. This was preventing the spout fountain from swaying. The selected process variables affecting the solid concentration and stability into the bed are; gas velocity and particle density. The operating conditions of the process are explained in Table (2).

2.4. Result and Discussion

Figure (2-a) illustrates the solid concentration distribution of steel beads for the fountain region. The system behaves unstable within the fountain region (7.5 inches above the conical base) for different superficial velocities of air (0.74 to 1.0 m/s).These are due to the vortex and revolution of the particle's bulk [4 & 5].The concentration profile curve has the triangular pulse shape and not uniform. The maximum peak of pulse is appeared for all air velocities at 0.75 inches of radial position that represents the middle location between the center and the wall of the bed. It has high vortex motion and particle's interaction compared to others' regions. This behavior indicates to instability conditions. Unstable spouting is characterized by swirling and pulsation of the spout [6]. The concentration of solid particles is higher at the wall (0.25 inch in radial position) then decreased forward the center of the bed (1.5 in radial position). The solids concentration profile at low velocity (0.74 m/s) is higher when compared with high velocity (0.95 and 1.0 m/s) as shown in Figure (2-a).

Figure (3-a) illustrates the solids concentration profiles of the steel beads at the cylindrical region (5.5 inches above conical base) for different air velocities. The solid concentration profile curve has the uniform exponential shape. The stability conditions appeared within the behavior of the system since the particles in the bed fluidized homogenously, and they are not discharged from the apparatus [6]. This position located into the lower gas motion region of the spouted bed (cylindrical part).Furthermore, the profile of Vg of 0.74 m/s has high value when compared with the others' velocities of (0.95 & 1.0 m/s).The packed bed and stable spouting are distinguished by the formation of the stable spout. Figure (3) explains that the solid concentration is much less in the lean bed region in the center than in the dense bed region near the wall [5].

For glass beads system, the same behavior obtained as in the steel beads system which are explained in Figures (2-b and 3-b).The fountain region is unstable while the stability appeared at cylindrical region for different flow regimes .The solid concentration profile curve with the air velocity of 0.74 m/s has higher value when compared to that of the air velocities of 0.95&1.0 m/s. Figures(2 and 3) explain that the glass particles concentration profiles system are low when compared to steel particles system for the same flow regimes..This behavior is due to the high uniformity of solid particles across the spouted bed can be obtained with the high density system. At the fountain region (7.5 inches above the conical base), the system is changed from packed bed to spouted bed, then unstable spouting.

The gas velocity is more effective variable on the solid concentration distributions and the stability of the spouted bed. Stable condition at the bed is obtained when the particles are fluidized homogenously. The process variables are a difficulty to measure also, the performance of the spouted bed is dropped at the unstable conditions.

2.5. Equations, Formulas, Symbols, Units

$$C_i=(V_i-V_{min})/(V_{max}-V_{min}) \quad (1)$$

2.6. Illustrations

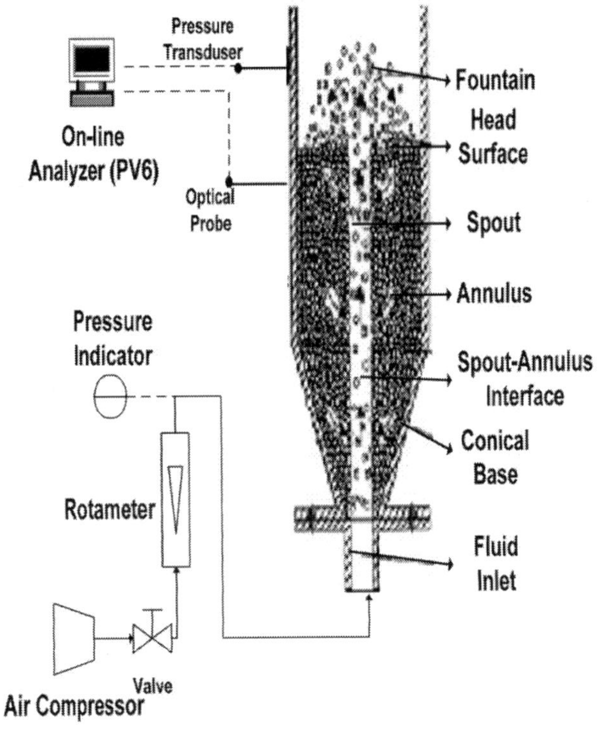

Fig.1.EXPERIMENTALSET-UP.

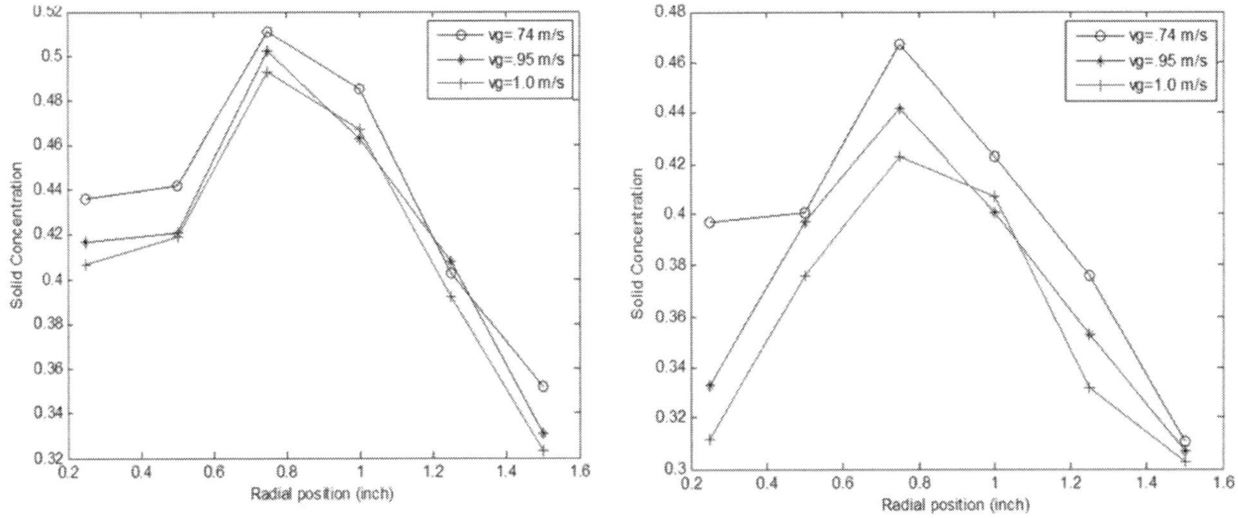

(a) (b)

Fig.2. Distribution of solid concentration into fountain region for ;(a) steel beads, (b) glass bead.

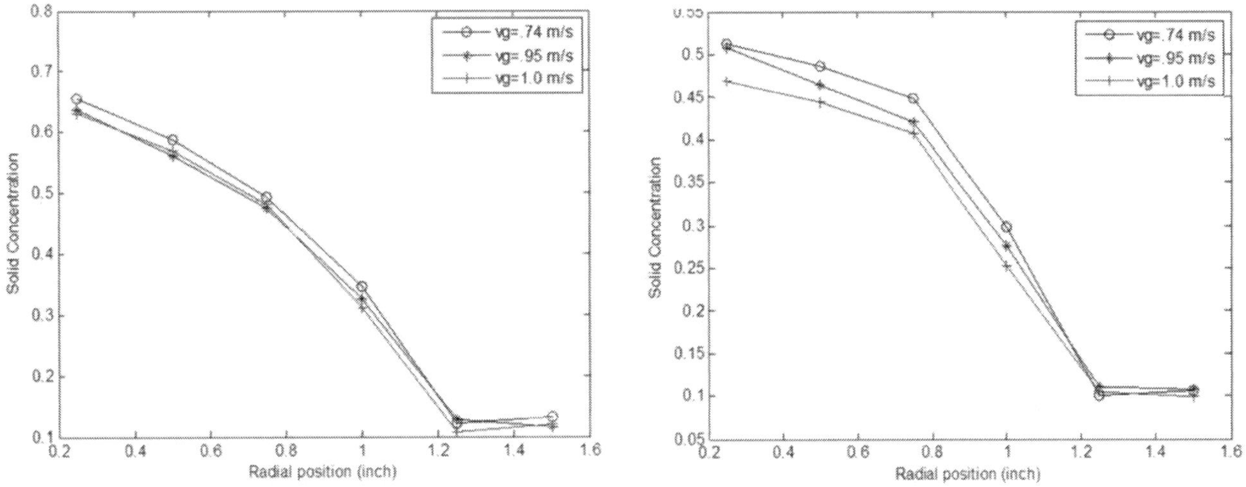

(a) (b)

Figure .3. Distribution of solid concentration into annulus region for; (a) steel beads, (b) glass beads.

2.7. Tables

Table 1. Properties of the particulate materials.

Material	Dp (mm)	ρs (Kg/m^3)	€	Ø
Steel beads	1.09	7400.0	0.42	1.0
Glass beads	1.09	2450.0	0.42	1.0
Glass beads	2.18	2400.0	0.41	1.0

Table 2. Operating conditions of the system.

Air velocity (m/s)	Bed pressure (Kpa)	Temperature (K)
0.74	101	298
0.95	101	298
1.0	312	298

3. Conclusion

1. Distributions of solid particle's concentration have found the suitable portrait of the particles' behavior into the spouted bed.

2. Solids' concentration is much less in the lean bed region in the center than in the dense bed region near the wall for different flow regimes.
3. The stable region into spouted bed is observed when the particles are fluidizing homogenously while unstable spouting is characterized by pulsation of the spout.
4. Stable spouted region is obtained at position within 5.5 inches above the conical base of the bed.
5. Superficial gas velocity has found the more effective variable on the stability of spouted bed.
6. High solid concentration and uniformity can attained with low gas velocity and high particle's density.

Acknowledgments

We thank all the participants to the Chemical and Biological Department-Missouri University of S & T, Rolla, Missouri (USA).

References

[1] Limtrakul, S., Boonsrirat, A.and Vatanathm, T.'DEM Modeling and Simulation of a Catalytic Gas-Solid Fluidized Bed Reactor: a Spouted Bed as aCase Study', Dept.of Chem.Engng.kasetsar University, Bangkok, Thailand, Elsevier Ltd, (2004).

[2] Wang, Z., Chen, P., Li, H., Wu, C.andChen, Y.'Study on the Hydrodynamics of a Spouting-Moving Bed', Ind.Engng.Chem.Res, 40, pp.4983-4989, (2001).

[3] Prachayawrakorn, S., Reuengnarong, S.and Soponronnarit, S.'Characteristics of Heat Transfer in Two-Dimensional Spouted Bed', School of Energy and Materials, University of Technology, Bangkok, Thailand,(2005).

[4] Zhong, W., Zhang, M., Jin, B.and Xiao, R.,'Experimental and Model Investigations on Gas Mixing Behaviors of Spout-Fluid Beds,' International Journal of Chemical Reactor Engineering, vol.5, Article A32, (2007).Zhang, Y.,

[5] Zhang, W. and Jin, B.,'New Method for the Investigation of Particle Mixing Dynamic in a Spout-Fluid Bed', Powder Technology, 24, January, (2011).

[6] Xu,J.,Bao,X.Wei,W.,Bi,H.T.,Grace,J.R.,and Lim,C.J.'Chaotic Characteristics of Pressure Fluctuation in a Gas Spouted Bed', the Canadian Journal of Chemical Engineering, vol.87, April, (2009).

(212d) A Novel Bubble-Emulsion Model for Modeling Gas-Solid Bubbling Fluidized Beds

Xi Gao, Li-jun Wang, Cheng Wu, You-wei Cheng and Xi Li, Department of Chemical and Biological Engineering, Zhejiang University, Hangzhou, China

The enlargement and enhancement of fluidized bed reactor represents the tendency of current technology development. Fluidization hydrodynamics is the basis of the reactor design and scale up. However the present gas-solid two fluid model (TFM) based on kinetic theory of granular flow seems incapable of scale-up simulation for bubbling fluidized bed. This project proposed an analogous research program, according to the analogies in flow regime and mechanism between fluidized bed and bubble column, that aims to develop a novel and unified two phase hydrodynamic model applicable both to fluidized bed and bubble column for mathematical scale up of the reactors. The method will benefit the present fluidized bed model in getting rid of the inadequacies of TFM by implanting some novel and effective modeling tools and results that recently developed for bubble column into fluidized bed model. The research content includes a complete description of the constitutive relationship for inter-phase forces around the bubbles, a comprehensive investigation of correlations for average voidage, bubble and emulsion phase properties in fluidized bed, simulation of large scale industrial fluidized bed reactor, and the related experimental and computational works. Both the methodology and research results are important in application and scientific areas.

Particle Segregation in a Liquid Fluidized Bed Incorporating Inclined Channels Subjected to Centrifugal Forces

K.P. Galvin and J. Dickinson

Centre for Advanced Particle Processing and Transport, Newcastle Institute for Energy and Resources, University of Newcastle, Callaghan, NSW 2308, Australia

ABSTRACT

A liquid fluidized bed was used to elutriate and hence separate particles on the basis of size and also density, with the ratio of the superficial fluidization velocity, U, to particle terminal velocity, u_t, (terminal velocity at g=9.8 ms^{-2}) exceeding 1000 fold. This ratio is extraordinary because the centrifugal acceleration involved was much lower at 73g. The fluidized bed system incorporated a system of parallel channels inclined at an angle of 20° to the direction of the centrifugal force. Inclined channels have been deployed previously in a centrifugal field. However, this is the first study to also introduce fluidization, allowing efficient separation to be achieved. Moreover, it was shown that the combination of the inclined channels and the centrifugal force helped to almost fully suppress the effects of particle size over a wide size range.

The processing of ultrafine particles less than 0.1 mm is often limited by their exceedingly low terminal settling velocity. This is especially true when the particles need to remain dispersed in order to facilitate their classification, either by size or by density. Centrifugal forces, with an acceleration G times the normal gravitational value, offer one approach for increasing the particle settling velocity by a significant factor, up to a maximum of G. The particle Reynolds number then increases significantly, ultimately leading to a change in the settling regime from the Stokes' to Intermediate, and hence dependence of the terminal velocity on the particle diameter, from d^2 to d^1. This reduction in the dependence on the particle size also leads to improved separation on the basis of the particle density, hence centrifugal separations in mineral processing are referred to by the term, *Enhanced Gravity*.

Inclined channels have also been used to achieve a significant increase in the segregation rate of particles, exploiting the so-called "Boycott Effect" (Boycott, 1920) by presenting an increased settling area to the suspension of particles. Galvin et al (2009) and later Galvin and Liu (2011) investigated the segregation achieved using closely spaced inclined channels designed to promote a laminar flow profile and a high shear rate at the planar surfaces. They applied an elutriation condition that equated the local fluid velocity acting on the particle at the wall with the terminal velocity of the particle. In classifying particles with a terminal velocity dependent on d^n, the superficial velocity was found to scale with d^{n-1}, hence there was a full decade reduction in the dependence on the particle diameter. Further, the ratio of the superficial fluidization velocity, U\`, through the channel relative to the terminal velocity, $u_{tT}$, in the tangent direction of the channel, was found to be approximately $U`/u_{tT}=z/(3d)$.

This study was concerned with the combined effect of applying a centrifugal force to particles fluidized and then conveyed into planar channels inclined to the centrifugal force. The key question was whether the benefits of the inclined channels and the centrifugal force would multiply, delivering a significant throughput advantage. The Coriolis Force is known to impact negatively on the benefits of the inclined channels rotating in a centrifuge (Schaflinger, 1990), however, by utilizing a confined fluidization housing, it was possible to overcome this problem. A further objective of the study was to examine, for the first time, whether the benefits of the inclined channels and the centrifugal force could combine to eliminate the effect of particle size on the separations, potentially from d^2 to almost d^0, thereby fully suppressing the effect of particle size, and hence amplifying the significance of the particle density.

Indeed, both of these effects were observed, with $U'/u_{tT}=gz/(3d)$. The superficial fluidization velocity within the inclined channels was as high as 0.34 m/s in some experiments involving separations of silica in water as fine as 7.5 microns. Here the value of U'/u_{tT} reached 3300 at $g=73$. Moreover, relatively low density particles were observed to elutriate, with the dependence of the superficial fluid velocity on the particle diameter scaling with $d^{0.3}$. A powerful mechanism for separating particles on the basis of density was established, utilising the combined effects of inclined channels and a centrifugal force.

References

Boycott, A.E., 1920, Sedimentation of Blood Corpuscles. Nature, 104, 532

Galvin, K.P., Walton, K., and Zhou, J., 2009, How to Elutriate Particles According to their Density, Chemical Engineering Science, 64, 2003-2010

Galvin, K.P., Walton, K., Zhou, J., 2010, Application of Closely Spaced Inclined Channels in Gravity Separation of Fine Particles, Minerals Engineering, 23 326-338

Galvin, K.P., and Liu, H., 2011, Role of Inertial Lift in Elutriating Particles According to their Density, Chemical Engineering Science, 66 3686-3691

Galvin, K.P., and Dickinson, J., 2012, Particle Transport and Separation in Inclined Channels Subject to G Forces, in preparation

King, M.R., and Leighton, D.T., 1997, Measurement of the inertial lift on a moving sphere in contact with a plane wall in shear flow, Physics of Fluids 9(5), 1248-1255

Nakamura H. and Kuroda K., 1937. La Cause de l'acceleration de la Vitesse de Sedimentation des Suspensions dans les Recipients Inclines. Keijo Journal of Medicine, 8, 256– 296

Ponder P., 1925. On sedimentation and Rouleaux formation. Quarterly Journal of Experimental Physiology. 15, 235–252

Schaflinger, Uwe, 1990, Review Article. Centrifugal Separation of a Mixture, Fluid Dynamics Research, 6, 213-249.

(212f) An Analysis Method for Fluidized Bed Flow Behavior Based On Equivalent Circuit Model

Masanori Ishizuka[1], Yuji Yoshie[2], Yasuki Kansha[1] and Atsushi Tsutsumi[1], (1)Institute of Industrial Science, The University of Tokyo, Tokyo, Japan, (2)School of engineering, The University of Tokyo, Tokyo, Japan

In this research, we proposed a novel method for analyzing the fluidized bed systems. We applied the analogy between flow behavior of fluidized bed and electric analog circuit with correlations of pressure to voltage, solid mass flux to electric current and hydraulic resistance to electric resistance. Equivalent circuit model were generally used in electronic equipment designs, especially integrated circuit (IC) and print circuit board (PCB). Usually, equivalent circuit model is translated to "SPICE" model for simulating its electric circuit behavior. Circuit simulation program SPICE is widely used, in this program SPICE model is solved to design analog circuits and digital circuits. SPICE program can calculate large scale modeling problem such as VLSI and complex analog circuit such as PCB. Then by using SPICE, we will be able to solve the complex and large size problem of flow behavior of a fluidized bed. By applying the method of equivalent circuit model and SPICE simulation, it will become possible to gain new methodology about the scale up problem, startup and shutdown procedure, stability of operation condition, etc.

Our final goal is to develop the novel analysis and design a model for the fluidized bed reactor. We will report the first result where the equivalent circuit method is applied to the fluidized bed flow behavior.

(212g) A Numerical Model for the Flow Including Dense Solids with Large Size Difference

Takuya Tsuji, Kyohei Higashida, Yoshitomo Okuyama and Toshitsugu Tanaka, Department of Mechanical Engineering, Osaka University, Suita, Osaka, Japan

In a large number of applications of fluidized bed such as gasification, combustion, separation and granulation, the size of solids is not uniform and large objects are coexisting with small emulsion solids. The size-difference induces characteristic motion of solids and results in segregation and mixing phenomena. It is difficult to treat these systems by existing numerical models and, in the present study, a numerical model which expresses the motion of large objects in gas-fluidized beds is proposed. The present model is based on DEM-CFD mesoscopic model and inspired by volume penalization method (Kolomenskiy & Schneider, J. Comput. Phys., 2009), one of the methods for the direct simulation of solid-flow systems. The interaction force between fluid and large object is estimated by assuming that large object is consisted of dense small particles, fictitiously. In this study, the motion of a large sphere in a bubbling fluidized bed is demonstrated and results are compared with experiments.

(231a) Atomic Layer Deposition As a Catalyst Synthesis Technique for Nickel Nanoparticles

Troy D. Gould, Alia M. Lubers, John L. Falconer, J.Will Medlin and Alan W. Weimer, Chemical and Biological Engineering, University of Colorado at Boulder, Boulder, CO

Efforts to design new catalysts for renewable energy applications have revolved around increasing selectivity & activity by exploring advantages of catalysis at the nanoscale, mainly due to geometric & electronic effects created by nanostructures or bimetallic structures. This work investigates the synthesis of Ni nanoparticle catalysts through the scalable gas phase synthesis route of Atomic Layer Deposition (ALD). Typically ALD deposits Ni as a NiO film when using oxygen as the secondary precursor. However; by using hydrogen as the secondary precursor, metallic Ni nanoparticles were synthesized with particle diameters of 2 nm with 1 ALD cycle. The size of the nanoparticles can be increased by performing more ALD cycles to give the ALD synthesis an element of particle size control. The catalysts were characterized via chemisorption, temperature programmed desorption (TPD), transmission electron microscopy (TEM), inductively coupled plasmon mass spectroscopy (ICP-MS), energy dispersive x-ray spectroscopy (EDS) & compared to traditional incipient wetness Ni catalysts. The incipient wetness Ni catalysts were also used as a base comparison of catalytic activity for alkene hydrogenolysis reactions using propylene as a probe reactant molecule. Hydrogenation turnover frequencies were at least one order of magnitude higher for the nanoscale ALD catalysts than the incipient wetness catalysts. In tuning the size of the nanoparticles, certain sized nanoparticles were found to have hydrogenolysis activity in addition to the typical hydrogenation. To further explore the benefits of Ni ALD, bimetallic ALD catalyst nanoparticles were investigated for a NiPt bimetallic system. Pt ALD has been an established technique for creating Pt nanparticles. However, by combining the novel Ni ALD with Pt ALD, nanoscale bimetallic catalysts were created. To address issues of specific Pt deposition on Ni particles, temperature programmed reduction (TPR) and chemisorptions were used to characterize the Ni & Pt depositions separately and together to determine the extent of bimetallic interaction.

(231b) Microfluidic Synthesis of Electro-Catalytically Active Nanoparticles

Anand Kumar and Klavs F. Jensen, Chemical Engineering, Massachusetts Institute of Technology, Cambridge, MA

Metallic nanoparticles, owing to their exceptional properties have gained a lot of research interest. Potential applications of these nanoparticles are well known in catalysis, pigments, electronic and magnetic materials, and drug delivery. Recent advances in colloidal synthesis methods[1] have enabled us to synthesize nanoparticles of various shapes and sizes. However the conventional batch processes give a broad size distribution because of poor mixing and temperature fluctuations. Consequently, they require further size classification step to obtain uniform particles. In our laboratory these materials are being synthesized by using microfluidic reactors which offer rapid mixing and heat transfer effects resulting in precise size control. Development of high pressure and high temperature micro-reactor techniques[2] has provided us with unique capability of synthesizing metallic/bimetallic nanoparticles in a fast, continuous fashion with enhanced homogeneity and monodispersity along with enabling special synthesis conditions such as segmented gas liquid flow[3].

In this work, we demonstrate the use of microfluidics for the continuous synthesis of Pt nanodendrites (Pt-ND) and the effect of synthesis parameters on the electrocatalytic activity of the Pt-ND. Our results indicate that the synthesized Pt-ND are much more active as compared to Pt-black obtained from Sigma Aldrich, as indicated in by the cyclic voltammetry curve (Fig 1 (a), potential wrt standard Ag/AgCl electrode) even though the amount of Pt-ND loading used is approximately one tenth of Pt-black. We attribute the effect to dendritic structure (Fig 1 (b)) giving high surface area of the active sites for the reaction. We discuss the detailed synthesis method, electrocatalytic and TEM characterizations in order to understand the effect of synthesis parameters on the electrocatalytic activity of Pt-ND.

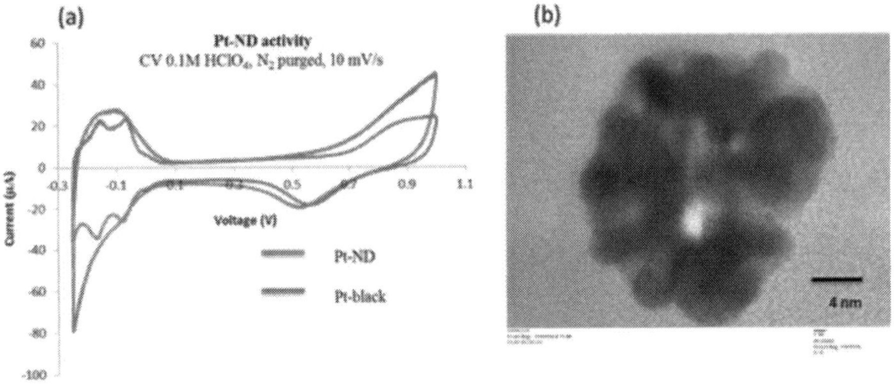

Figure 1: (a) Electrocatalytic activity of synthesized Pt-ND wrt to Pt-black, (b) HRTEM of Pt-ND

[1] Y. Xia, Y. Xiong, B. Lim and S. E. Skrabalak, *Angew. Chem. Int. Ed.* **2009**, 48, pp. 60-103

[2] S. Marre, A. Adamo, S. Basak, C. Aymonier, K. F. Jensen, *Ind. Eng. Chem. Res.* **2010**, 49 (22), pp 11310-11320

[3] V. Sebastian, S. Kuhn, A. A. Kulkarni, K. F. Jensen, *Langmuir* **2011**, 27, pp. 6519-6527

(231c) High-Precision Manufacturing of Core-Shell Nanoparticles for Catalysis

J. Ruud van Ommen[1], Barbara M. Kampa[1], David Valdesueiro[1], Gabrie Meesters[1,2], Freek Kapteijn[1] and Michiel T. Kreutzer[1], (1)Chemical Engineering, Delft University of Technology, Delft, Netherlands, (2)DSM, Delft, Netherlands

Many catalysts consist of nanometer-sized particles dispersed on a high-surface area support. In several cases, these nanoparticles are not just made of a single material, but have a more complex morphology such as a core-shell nature. A high degree of control over the synthesis of such particles could make catalysts substantially more efficient. In the current paper, we will discuss reaction engineering approaches for rational manufacturing of such catalysts. We propose two different ways to make such particles: one in the liquid phase and one in the gas phase using.

Our liquid-phase approach is based on multiphase-flow microfluidic devices. The synthesis of nanoparticles takes place inside aqueous droplets dispersed in oil. Each droplet acts as a batch reactor with a high level of control over reagent flow rates, residence time, mixing and temperature. Moreover, we are able to study kinetics of the reaction in a real time by applying optical fiber ports at different positions of the microchannel.

We use the gas-phase method Atomic Layer Deposition to deposit ultrathin coatings on nanoparticles through a layer-by-layer growth mechanism. The precursors, which are fed into the reactor in subsequent steps, react on the surface of the nanoparticles to deposit a monolayer of the coating. The thickness of the coating is controlled by repeating this cycle a desired number of times. This process is carried out in a fluidized bed reactor, which offers an excellent contact between the solids and reactants and a good scale-up potential.

(231d) Effects of Synthetic Variations On the Dispersion of Pt Catalysts Supported On Aluminum-Modified TiO2

Rebecca E. Olsen, Department of Chemistry and Biochemistry, Brigham Young University, Provo, UT, Calvin H. Bartholomew, Chemical Engineering, Brigham Young University, Provo, UT and Brian F. Woodfield, Chemistry and Biochemistry, Brigham Young University, Provo, UT

TiO_2 is a popular catalyst support because it is stable, versatile, non-toxic, and relatively inexpensive. We have focused on producing high surface area Al-stabilized anatase TiO_2 supports of controlled pore diameters with surface areas up to 490 m^2/g. While TiO_2 is commonly used to support Pt, it is not well understood how synthetic variations affect the final structure and dispersion of the Pt. Therefore, previous studies have used laborious and often ineffective 'trial and error' methods to develop stable support materials and catalysts. Development during the past 3-4 decades of sophisticated surface and high energy spectroscopy techniques has enabled effective approaches based on scientific understanding of the structure of doped nanomaterials, which is the key to efficient design of stable high surface area materials. Through statistically designed experiments and EXAFS/XANES analysis, we have examined the effects of varying synthetic parameters on the bonding of Pt metal clusters dispersed on Al-TiO_2. Here we present insight into why some methods of preparation lead to better dispersion of the Pt, with dispersions ranging from 12%-53%. Understanding how sensitive the metal cluster structure and metal cluster-support interactions are to synthetic variations will help us predict the efficiency of dispersing Pt in TiO_2 supports and enable us to rationally design better TiO_2-supported Pt catalysts.

Polymer and Surfactant Templated Synthesis of ZnS (II-VI) Semiconductor Nanocrystals in a Spray Pyrolysis Reactor

Munish K. Sharma[1], Parham Rohani[1], and Mark T. Swihart[1]*

[1] Department of Chemical and Biological Engineering, The University at Buffalo (SUNY), Buffalo, NY, 14260

ZnS is a direct (II-VI) wideband gap semiconductor with 3.68 eV for cubic phase and 3.80 eV for hexagonal phase. It is used in many applications such as phosphor in luminescent devices due to strong emission peak at 410 nm, photo catalyst for water splitting, solar cells for energy conversion and in anti-reflection coatings. We present here single step poly (ethylene glycol) methyl ether (PEG) mediated approach to synthesize ZnS nanocrystals using spray pyrolysis reactor at 500°C and 700°C. The atomizer produces a fine mist of aerosol droplets within size range of 25-300 nm which enters in a 4s residence time rector maintained at a constant temperature. Inside the reactor, solvent evaporates and solute comprising of polymer and metal precursor diffuses at the core of droplet. ZnS starts precipitating outside the core of polymer as polymer tries to minimize surface free energy by occupying core. PEG has a melting point at 365°C and hence it decomposes into gaseous species which escapes leaving behind a hollow core and ZnS shell. ZnS nanospheres were characterized via X-ray diffraction, scanning electron microscopy, transmission electron microscopy with single area electron diffraction, and energy dispersive spectroscopy. Thermal gravimetric analysis of ZnS nanospheres were carried out for different synthesis conditions under inert atmosphere. It was found that higher temperatures and lower polymer concentration are ideal conditions for highly crystalline ZnS nanospheres. Temperature has strong effect on the formation of ZnS nanospheres. These hollow nanospheres can be of potential use in increasing the photocatalytic efficiency due to their large specific surface areas confirmed by Brunauer Emmett Teller (BET). Also they can be of potential use as contrast agents for bioimaging for drug delivery applications.

Use of a cationic surfactant such as Cetyl trimethylammonium bromide (C-TAB) mediated synthesis of ZnS nanocrystals is underway. Surfactant templated assembly proceeds via micelle formation with polar head groups pointing outwards in aqueous region. These micelles decomposes under high temperature inside reactor and paves the way of producing nanocrystals with uniform pore size distribution unlike PEG which has a non-uniform pore size distribution confirmed by scanning electron microscopy. In this presentation, we will discuss spray pyrolysis reactor and reaction scheme in brief followed by polymer and surfactant templated synthesis of ZnS nanocrystals.

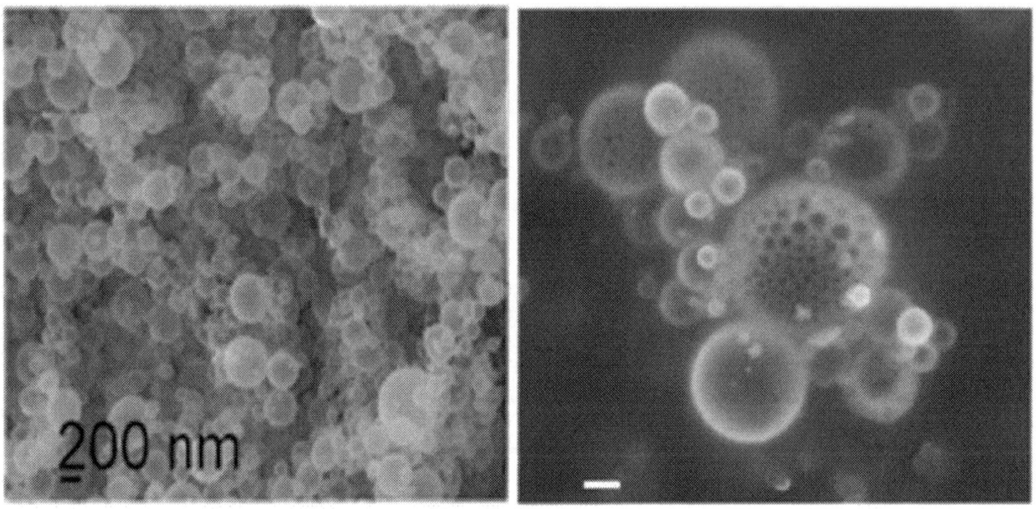

Figure 1. SEM images of ZnS micro and nano- spheres synthesized with the addition of PEG into ZnS precursors: (a) 20KX magnification SEM image of particles synthesized at 500°C with 0.75 g PEG added; (b) 60 KX magnification SEM image of particles synthesized at 500°C showing the surface pores.

(231f) Synthesis and Phase Transfer of Oxidation Resistant Copper Nanoparticles to Various Organic Solvents

Manish Niranjan[1], Jayanta Chakraborty[2] and Aabid Shaikh[1], (1)Indian Institute of Technology Kharagpur, Kharagpur, India, (2)Chemical Engineering, Indian Institute of Technology, Kharagpur, Kharagpur, India

Metallic copper nanoparticles are difficult to synthesize because of its low resistance to oxidation in compared to other noble metals. In this work we show a new and facile way to synthesize metallic and oxidation resistant copper nanoparticles in aqueous and organic phase. The aqueous phase oxidation resistant copper nanoparticle has been synthesized by using borohydride reduction of copper salt and it has been shown that excess borohydride maintains a reducing atmosphere in which copper particles remains metallic without any significant oxidation for at least two hours. Even blowing air could not destabilize/oxidize those particles.

Such particles could readily be transferred into a variety of organic solvents including hexane, toluene and chloroform. The efficiency of transfer depends on various properties of the solvent. The phase transfer process has been optimized for various organic solvents and the effeciency has been quantified. Interesting particle pattern forms on the glass substrate during phase transfer which also been investigated.

(270a) Towards a Refined Theory for Liquid Transfer in Dense and Dilute Particle Beds

Stefan Radl[1], Matthew Girardi[2], Charles Radeke[3], Johannes G. Khinast[1] and Sankaran Sundaresan[2], (1)Institute for Process and Particle Engineering, Graz University of Technology, Graz, Austria, (2)Department of Chemical and Biological Engineering, Princeton University, Princeton, NJ, (3)Research Center Pharmaceutical Engineering (RCPE), Graz, Austria

Existing research has focused on forces connected to liquid bridges, yet there is much less theory concerned with the process of liquid transfer upon particle-particle and particle-wall collisions. Previous work on liquid transfer [1–3] used either a solution of the Navier-Stokes equation in a simplified (i.e., axisymmetric) setup, or simpler models that often neglect fluid inertia. These previous studies focused exclusively on the rupture process. One of the missing links for a more general model for liquid transfer seems to be a dynamical description of the process of bridge formation.

Starting with the algorithm documented by Shi and McCarthy [3], we calculate the driving pressure difference that causes a drainage of free liquid from the wetted surfaces into a liquid bridge. This process is coupled with the evolution of the film thickness on wetted surfaces via a simple mass and force balance. This allows us to calculate the instantaneous liquid bridge volume as a function of the amount of liquid present on the surfaces in contact. A fit of the solution of Shi and McCarthy model [3] for the liquid transfer ratio upon rupture completes our liquid transfer model.

Finally, we study the effect of various liquid transfer models on the liquid distribution for various test cases using our in-house Discrete Element Method-based code. Interestingly, already one of our simplest models predicts an imbibition front that progresses with $l^2 \sim t$, i.e., yields Washburn's law.

References

[1] P. Darabi, T. Li, K. Pougatch, M. Salcudean, and D. Grecov, Modeling the evolution and rupture of stretching pendular liquid bridges. Chem. Eng. Sci. 65 (2010) 4472-4483.

[2] S. Dodds, M. Carvalho, and S. Kumar, Stretching liquid bridges with moving contact lines: The role of inertia. Phys. Fluids 23 (2011) 092101.

[3] D. Shi and J.J. McCarthy, Numerical simulation of liquid transfer between particles. Powder Technol. 184 (2008) 64-75.

(270b) Responsive Discrete Element Model for Preparation of Advanced Materials by Mechanical Milling Using an Attritor Mill

Priya R. Santhanam and Edward L. Dreizin, Otto H. York Department of Chemical, Biological, and Pharmaceutical Engineering, New Jersey Institute of Technology, Newark, NJ

Various research groups [1, 2, 3] have identified mechanical milling as a suitable technique to prepare advanced reactive materials. With growing demand for such mechanically alloyed and nanocomposite materials, there is need for predictive models enabling one to transfer synthesis from the laboratory to larger scale (and, possibly, operation principle) milling devices. Due to lack of mechanistic models in literature, a trial and error approach for scale-up is the current methodology, which is both imprecise and inefficient. In order to overcome this inadequacy, a concept of milling dose was suggested to describe milling progress in different mills [4,5]. The milling dose D_m is defined as the rate of energy transfer from the milling tools to the powder, E_d, multiplied by the milling time, t, and divided by mass of the milled powder, m_p: $D_m=(E_d.t)/m_p$

In our previous work, the concept of milling dose was considered for three milling devices: shaker, planetary and attritor mill. The shaker and planetary mills are used in laboratory scale preparations and attritor is more suited for commercial scale manufacture. In experiments, the same starting powders were used in each mill. The milling time, t, for each mill was measured while the final product with the desired yield strength (serving as an indicator of material refinement) was prepared. Theoretical models of the three mills were set up using discrete element method (DEM) [6] and the energy dissipation rate, E_d, for the milling dose was computed. The milling dose values for the shaker and planetary mills were close to each other, as expected. However, the computed energy dissipation rate for the attritor was too high, resulting in a higher than expected milling dose.

Detailed analysis of the milling tool interactions in the attritor mill showed that from time to time, balls are predicted to jam, resulting in very strong forces exerted onto the milling media from the impeller moving at a constant, pre-set speed. In experiments, similar jamming events can also occur; however, they cause small changes in the impeller speed. Thus, the variations in the force are attenuated. Initially, to account for the computationally predicted events associated with unrealistically high forces, a screening scheme was proposed, in which any events involving such high forces were discarded [6]. The resulting, corrected milling dose for the attritor was much closer to that predicted for other milling devices.

The objective of this effort is to modify the DEM model for the attritor mill to imitate the response of impeller to the jamming events, i.e., to allow the instantaneous changes in its speed, depending on the packing of the milling balls. To develop a modified model, the impeller is described in the code as a complicated particle, or an element similar to milling balls. It is rotated using several driving cylinders placed at its top, around its circumference. The rotation rate is thus fixed for the driving cylinders, but not for the impeller. When a jamming event occurs in the mill, impeller can slip relative to its driving cylinders, so that its rotation speed is reduced momentarily. Once the ball packing is adjusted inside the mill, the impeller speed returns to its pre-set value.

In order to compare predicted and actual variations in the impeller rotation speed, the measurements are made using a controller equipped with data acquisition available for the attritor mill. Details of model development and its experimental validation will be presented. The implications for the preparation of advanced reactive materials using mechanical milling will also be discussed.

REFERENCES:

1. Shteinberg A.S., Lin Y.S., Son S.F., Mukasyan, A.S., "Kinetics of high temperature reaction in NI-Al system:

Influence of mechanical activation", Journal of Physical Chemistry A, Volume 114, 2010, pp 6111-6116.

2. Bazyn T., Lynch P, Krier H, Glumac N., "Combustion of fuel-rich aluminum and molybdenum oxide nano-composite mixtures", Propellants, Explosives, Pyrotechnics, Volume 35, 2010, pp 93-99.
3. Umbrajkar S.M., Seshadri, S., Schoenits, M., Dreizin E.L., "Aluminum-rich Al-MoO3 nanocomposite powders prepared by arrested reactive milling", Journal of Propulsion and Power, Volume 24, 2008, pp 192-198.
4. Cocco G., Delogu F., Schiffini L., "Toward a quantitative understanding of the mechanical alloying process", Journal of Materials Synthesis and Processing, Vlume 8, 2000, pp 167-180.
5. Ward T.S., Chen W., Schoenitz M., Dave R.N., Dreizin E.L., "A study of mechanical alloying processes using reactive milling and discrete element modeling", Acta Materialia, Volume 53, 2005, pp 2909-2918.
6. Santhanam P., Dreizin E.L., "Predicting conditions for scaled-up manufacturing of materials prepared by ball milling", Powder Technology, Volume 221, 2012, pp 403-411.

(270c) Assessment of Collision-Less Vs. Collisional Simulation of Lunar Regolith Ejection During Spacecraft Landing

Kyle Berger, Chemical and Biological Engineering, University of Colorado Boulder, Boulder, CO, Philip Metzger, Granular Materials and Regolith Operations Laboratory, Kennedy Space Center, FL and Christine M. Hrenya, Chemical and Biological Engineering, University of Colorado at Boulder, Boulder, CO

When a spacecraft lands on the Moon (or other extraterrestrial body; i.e. Mars, asteroids, etc.), the rocket exhaust causes regolith (soil-like material) to be ejected. Due to the lack of atmosphere and corresponding drag on regolith motion, the ejection can be extremely hazardous to objects both close and far from the landing point. For example, the Apollo 12 landed ~160 meters from the deactivated Surveyor III lander. An examination of the Surveyor revealed that it had been sandblasted by the ejection of fine regolith (order of microns at speeds up to 2 km/s) during the Apollo landing. Current models used to describe such ejection involve single-particle trajectories and thus do not consider particle collisions. Here, we seek to critically assess the impact of a collision-less vs. collisional description on regolith erosion. To accomplish this goal, the discrete element method (DEM), which incorporates collisions directly, is used to simulate the many-particle system. The system examined here is located 6m from the impingement point. Lift and drag forces from the plume exhaust are included in the simulation, as well as the Moon's gravity. A one-way coupling is used such that the particles are affected by the plume but the plume is unhindered by particles. Two versions of the DEM are utilized, namely one which resolves collisions using a soft-sphere model and the other which ignores collisions and allows for the (unrealistic) crossing of particle trajectories. Moreover, two cases of the collisional description are also considered – one with non-dissipative collisions and the other with dissipative (inelastic and frictional) collisions. Somewhat surprisingly, the erosion rate of the non-collisional case lies between that of the dissipative and non-dissipative collisional cases. Specifically, the non-dissipative collisional case exhibits the highest erosion rate since it allows for highly-effective collisional momentum transfer from the interior of the regolith layer to the regolith surface. In addition, a sensitivity analysis to collisional input parameters is also performed. Finally, a polydisperse system is examined and compared to the results of the monodisperse system.

(270d) Development of an Accurate Granular Normal Contact Force Model Based On a Non-Newtonian Liquid Filled Dashpot

Ebrahim Alizadeh, Francois Bertrand and Jamal Chaouki, Department of Chemical Engineering, Ecole Polytechnique de Montreal, Montreal, QC, Canada

It is well known that normal contact force models from the literature may fail to predict adequately particle collisions. For instance, they regularly predict inaccurately the energy restitution behavior when the normal impact velocity is increased. For example, most non-linear models yield a non-zero attractive force between two impacting particles just before they break apart, which is unrealistic. Such limitations have motivated the development of a novel normal contact force model that better predicts the unfolding of a collision between two particles. This model brings into play a Hertz elastic force and a dissipative force the value of which depends on the motion of a non-Newtonian liquid in a dashpot. The model parameters are set using experimental restitution data for particle/particle and particle/wall contacts. In this work, the measurement of energy restitution for particle/wall collision was done using several materials for a wide range of impact velocities, whereas particle/particle collision data were taken from the literature. Model predictions for microscopic (e.g. particle velocity) and macroscopic (e.g. collision time) quantities will be presented and compared to experimental data and results obtained from other non-linear models. It will be shown that the model adequately predicts the coefficient of restitution and that it yields a much smaller attractive force at the end of a particle collision.

(270e) Population Density Dependence of Mechanical Multi-Particle Interactions

Maxx Capece, Chemical, Biological and Pharmaceutical Engineering, New Jersey Institute of Technology, Newark, NJ and Rajesh Davé, Chemical Engineering, New Jersey Institute of Technology, Newark, NJ

Mechanical interactions between particles are important in many processes involving granular materials. Dense phase dry milling processes in particular are sensitive to mechanical interactions at the particle ensemble scale. During milling, particles transmit forces amongst themselves and cause breakage if the resultant local stresses are high enough. However, breakage is not only dependent on intrinsic particle properties, but also on the population density due to mechanical multi -particle interactions. To elucidate the effects of mechanical multi -particle interactions, numerical experiments were conducted by means of discrete element modeling (DEM). Mono- , binary- , and poly - dispersed mixtures of particles were subjected to vibration in order to determine their influence on the distribution of impact energies and contact forces. The nature of the multi- particle interactions can be used to explain some interesting breakage kinetics. Results may also have a broader relevance to many other processes including that of particle dry-coating.

(270f) Multi-Grid Monte Carlo Algorithm for the Simulation of Multi-Component Population Balance Models

Roberto Irizarry, Electronic Technologies, DuPont, Raleigh, NC

In this work a new multi-grid strategy is utilized to accelerate the computational speed of the PEMC [1, 2] method for the case of multidimensional population balance models (PBM). The PEMC is a fast MC method to solve PBMs, increasing the simulation speed by order of magnitude without affecting accuracy (for 1D case). This method is based on discretizing the internal variables to create a dual structure consisting of pseudospecies and subensembles. The advantages of this method are: (1) A reduced search space that allows very fast sampling for next event (thus fast simulation speed). (2) The "chemical structure" allows applying simulations techniques available in chemical kinetics (for example tau leap methods and multiscale methods [2]). (3) The ability to track individual particle/events history like the standard MC method. (4) The kernel structure is preserved and complex kernels can be compressed into a small matrix.

For multidimensional PBM, the discretization can generate a very large number of events reducing the computational speed of PEMC. This problem is solved introducing a new multigrid strategy for the "next event" search. The multigrid strategy makes the computational speed of multidimensional PBM almost the same as a 1D case (orders of magnitude improvement in computational speed). This is because in this "divide and conquer" type strategy the search scales linearly with dimensions instead of exponential. This new strategy makes the PEMC a very fast method for any type of PBM (single dimension or multidimensional).

The new method is discussed in details and tested with kernels of industrial relevance including wet granulation used in pharmaceutical industry. In this unit operation, the primary solid particles are mixed with a liquid binder. The physics of the process is governed by the aggregation kernel that is a function of the size and binder composition of the two colliding particles. A multidimensional population balance model is utilized to tracks the evolution of total granule size and binder size within granules.

[1] Irizarry R. (2007) *Chem. Eng. Sci.* **63**, 7649-7664.

[2] Irizarry R. (2011) *Chem. Eng. Sci.* **66**, 4059–4069.

(273a) Pressure Signal Analysis and Bubble Behavior in a 3D Fluidized Bed. Comparison of Experimental Data and Computational Results

Alireza Abbasi[1], Mohammad Ashraful Islam[2], Paul E. Ege[1] and Hugo H. De Lasa[2], (1)Reactech Process Development Inc., Markham, ON, Canada, (2)Chemical Reactor Engineering Centre, University of Western Ontario, London, ON, Canada

The present study compares computational fluid dynamic (CFD) simulation of a cylindrical fluidized bed with experimental pressure and optical probe measurements in a bed of close dimensions and operating under the same conditions. The applied CFD model is a hybrid of the continuum and discrete approaches. It allows to describe using particle size distributions, the dense solid phase, the bubble behavior and the bed dynamics. The main objective of this study is to establish a methodology for validating CFD simulation with experimental data, The fluidized bed is 0.1016m internal diameter and 1m height. The bed is filled with Geldart B silicon particles to a packed bed height of 0.28 m. The particle size distribution is between 25 to 500micron. With proper tuning of the CFD model, both bed dynamics as measured by wall mounted pressure probes and bubble characteristics, retrieved from the experimental optical signals, can predict both particle fraction and pressure variations. Bubble characteristics compare fairly well at different radial and axial positions. This validation demonstrates that CFD modeling can be validated with measurements establishing confidence in the simulation for fluidized bed reactor units.

THE COMPARISON OF LARGE SQUARE CYCLONE AND CENTRIFUGAL CYCLONE BY NUMERICAL SIMULATION

Dr. Zhihong Liu, Dr. Toshiyuki Suda

IHI Corporation, Japan

Introduction

Cyclone has been widely used in separation of gas and particle. Square cyclones are applied to the Circulating fluidized bed (CFB), due to their low-cost and facilitate of compacting with square riser. The mechanism of separation and characteristic of square cyclone have not been well studied. In this paper, numerical study by a commercial code was carried out to study the difference of square cyclone and centrifugal cyclone.

First, the calculated results were compared with measurement data taken from the literature, to evaluate the code. The comparisons showed the numerical simulation was reasonable. Second, the large square and centrifugal cyclone, which were with the same hydraulic diameter, the same vortex finder, the same area of inlet, the same velocity of gas and the same velocity of particle, have been numerically studied. The comparisons were made on the separation mechanism between square and centrifugal cyclones. The comparisons were also made on the gross separation efficiency, grade separation efficiency and pressure drop between the square cyclone and centrifugal cyclone. It was found that there was large difference between square cyclone and centrifugal cyclone. The mechanism of particle separation for centrifugal cyclone was the eccentricity force, whereas the mechanism of particle separation for square cyclone was both the eccentricity force and collision with wall. Last, the effect of inlet gas velocities and effects of inlet particle load were numerically studied for square cyclone and centrifugal cyclone. It was found that the square cyclone usually has lower separation efficiency than the centrifugal cyclone, but the difference of separation efficiency would become smaller as the inlet gas velocity and/or particle load decreased.

Numerical Method and its Validation

Numerical Method in this Study

There are two kinds of approach to study hydrodynamics of cyclone by numerical calculation. One is Euler-Lagrange approach[1], in which fluid is described by Euler approach and particles are described by Lagrangian approach. Another is Euler-Euler approach[2], which takes particles as continuous phase and both of fluid and particles are described by the Euler approach. In this paper, we used BARRACUDA's Multiphase Particle in a Cell approach (Euler-Lagrange Hybrid)[3] to calculate the behavior of hydrodynamics of cyclone.

Validation of the Numerical Simulation

Firstly, in order to evaluate the code, the results calculated by BARRACUDA have been compared with measurement data taken from the literature[4]. The shape and size of the test cyclone was shown in Fig.1 and table 1. The diameter of cyclone was 0.2m, the particle density was 2750 kg/m^3, and the inlet particle load was $10g/Nm^3$. The particle size distribution was shown in Fig.2. The numerical calculations by BARRACUDA were carried out with the same conditions with experiments.

Table 1 The Dimension of Small Test Cyclone

a/D	b/D	De/D	h/D	H/D	B/D	S/D
0.46	0.203	0.307	1.310	3.795	0.399	0.9

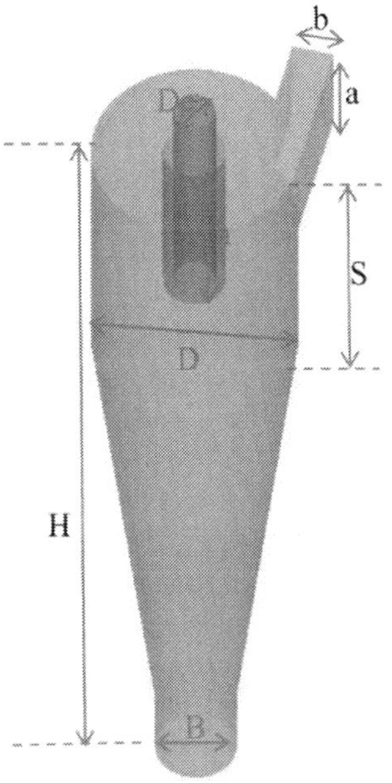

Fig.1 Geometry of small cycle

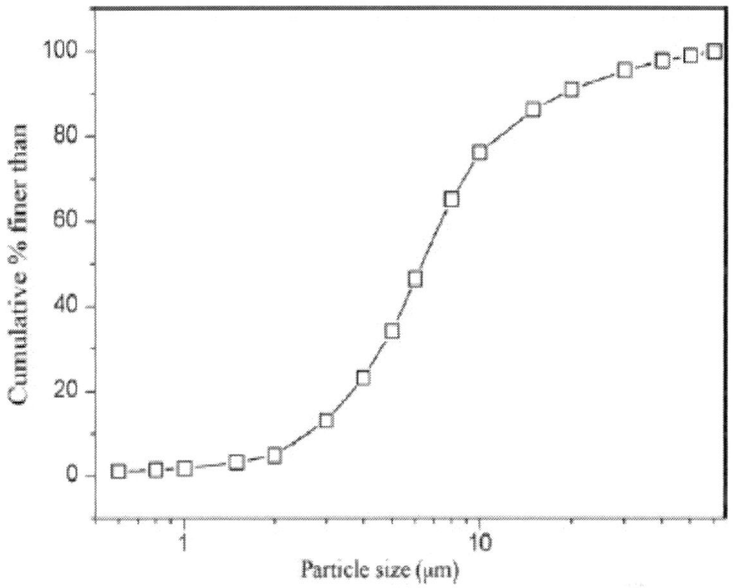

Fig.2 Particle size distribution of the test dust

The comparison of calculated results and data of experiments on the separation efficiency and pressure drop were shown in Fig.3 and Fig.4.It was shown that the results of BARRACUDA were reasonable.

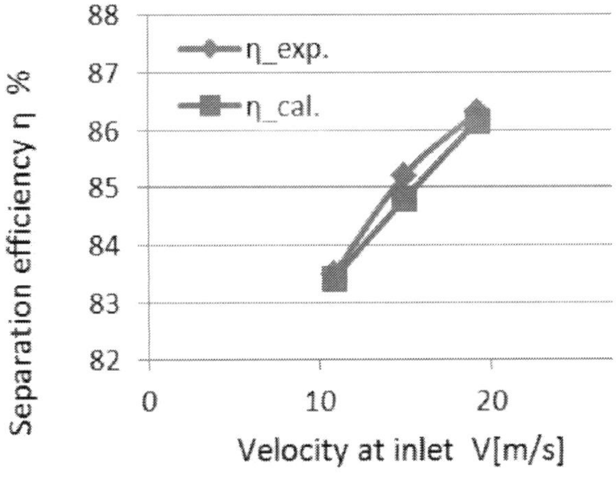

Fig.4 Comparison of sepration efficicy

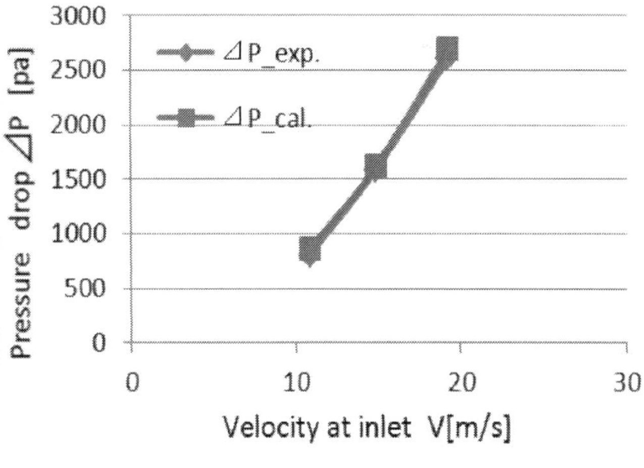

Fig.5 Comparison of pressure drop

The Conditions of Calculations

The large square and centrifugal cyclones have been numerically studied. The shape and size of the square and centrifugal cyclone are shown in Fig.6 and table 2. They had same hydrodynamic diameters. The areas at throat were the same. The other sizes were also the same.The size distribution of particle was shown in Fig.7. Density of particle was 2610kg/m^3. Inlet gas velocity at throat and particle load were the same respectively, shown in table 3. Temperature was 300K.

Table 2 The size of square and centrifugal cyclone (unit: m)

	D		De	b		a	H	h	S	B
Square Cyclone	Ds	hydrodynamic Diameter	1.8	bs	b	4	12.5	1.8	6.0	1.0
	4.431	5.0		2.083	0.8					
Centrifugal Cyclone	5.0		1.8	0.8		4	12.5	1.8	6.0	1.0

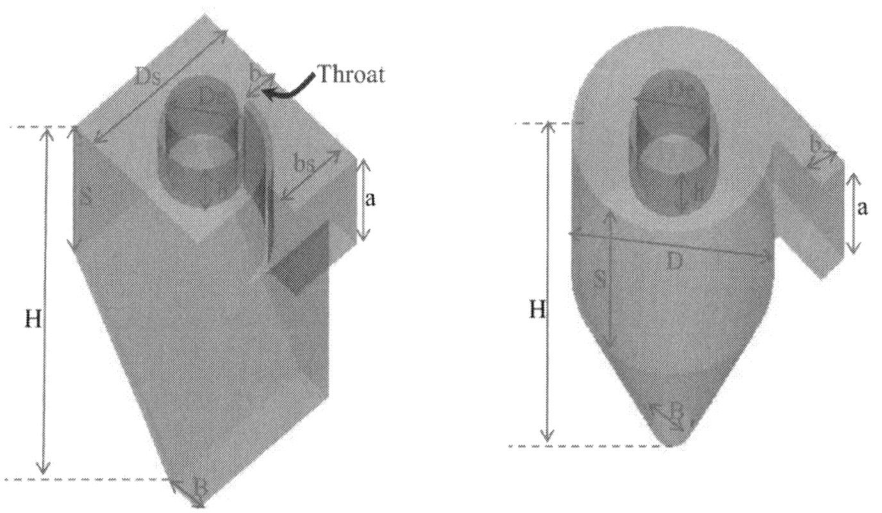

Fig.6 Geometry of Square and centrifugal cyclone

Table 3 Condition of inlet gas and particle

Case	Cyclone Type	Gas Velosity at Throat [m/s]	Particle Load [kg/s]
S26_100	Square	26	100
S26_50	Square	26	50
S13_100	Square	13	100
S13_50	Square	13	50
C26_100	Centrifugal	26	100
C26_50	Centrifugal	26	50
C13_100	Centrifugal	13	100
C13_50	Centrifugal	13	50

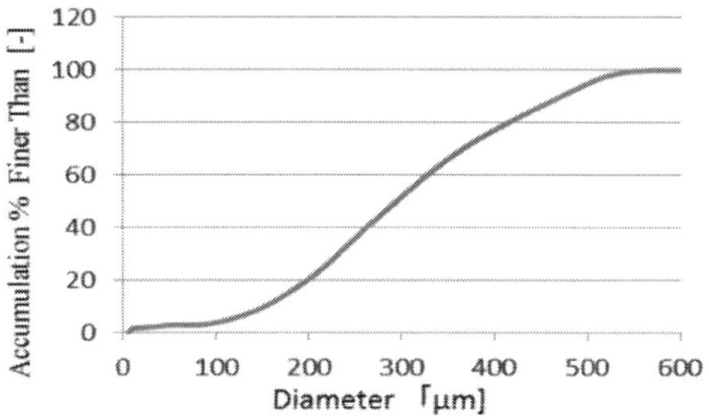

Fig.7 Size Distribution of Particle

Mechanism of Particle Separation of Square and Centrifugal Cyclone

The comparisons of square and centrifugal cyclone were based on numerical results. The numerical simulation was an unsteady calculation. After transient period, values were time-averaged.

Using the following time-averaged values, evaluations were made for square and centrifugal cyclone in this paper.

The following comparisons were under conditions of S26_100 and C26_100. The X-Y slice section was taken at the surface a little lower than the bottom of vortex finder. The X-Z slice section was taken at central of the cyclone.

Comparison in Terms of Particle Velocity

Firstly, the particle velocity vectors of square cyclone and centrifugal cyclone were compared. Fig.8 showed the particle velocity vectors of whole body, and the Fig.9 showed that in X-Y slice section.

From Fig.8 and Fig.9, it was observed that the particles in square cyclone conflicted with the wall and the speed of particle slowed down, therefore, some part of the particles were separated by collision with the wall, the rest particles were accelerated by the gas flow and then conflicted with the second wall, whereas the speed of particles in centrifugal cyclone decreased gradually due to the friction between the particle and wall. The tangent component of particle velocity of square cyclone was much lower than that of centrifugal cyclone. Based on the above observation, it could be said that the mechanism of particle separation of centrifugal cyclone was eccentricity force, whereas the mechanism of particle separation for square cyclone was both the eccentricity force and collision with the wall.

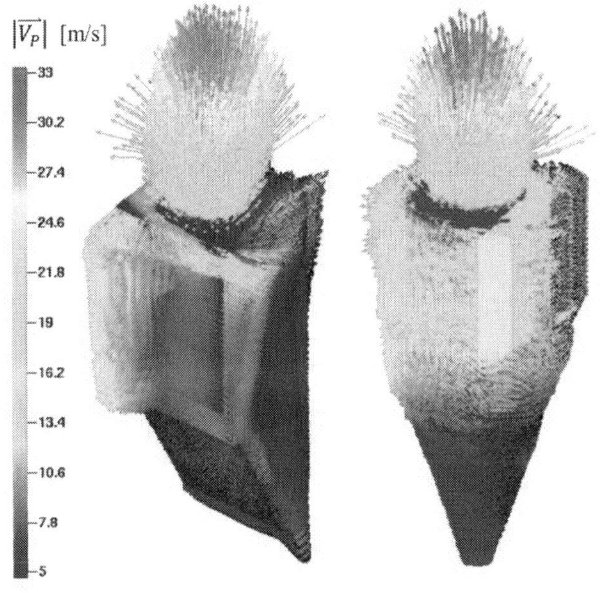

Fig.8 Particle Velocity Vector of Whole Body

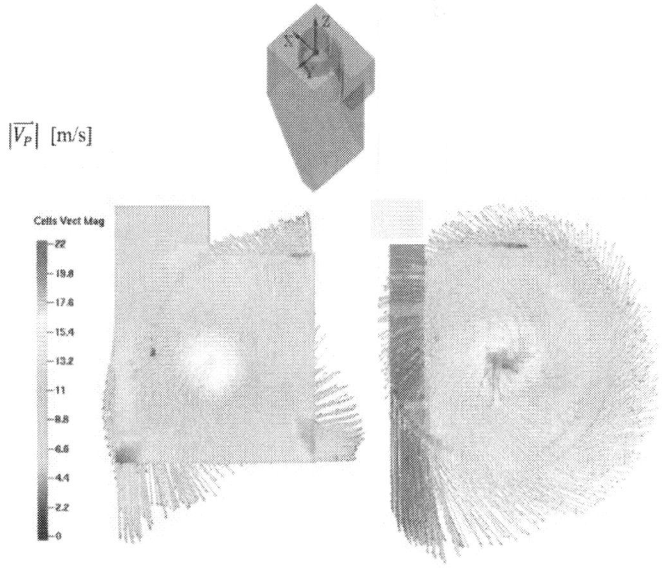

Fig.9 Particle Velocity Vector in X-Y Section

Comparison in Terms of Particle Mass Flux

Secondly, the particle mass flux vector of square cyclone and centrifugal cyclone were compared. Fig.10 showed the particle mass flux vector of whole body, the Fig.11 showed that in X-Y slice section, Fig.12 showed Z-direction component of Particle Mass Flux Vector in X-Y Section, and Fig.13 showed Y-direction Component of Particle Mass Flux Vector in X-Z Section.

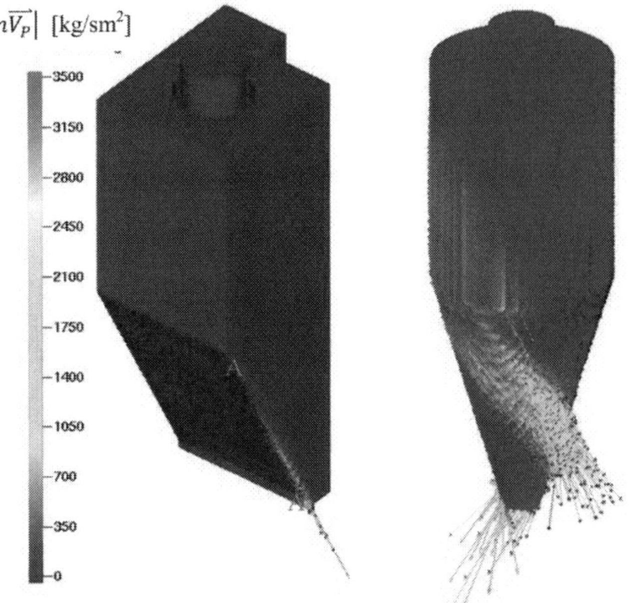

Fig.10 Particle Mass Flux Vector of Whole Body

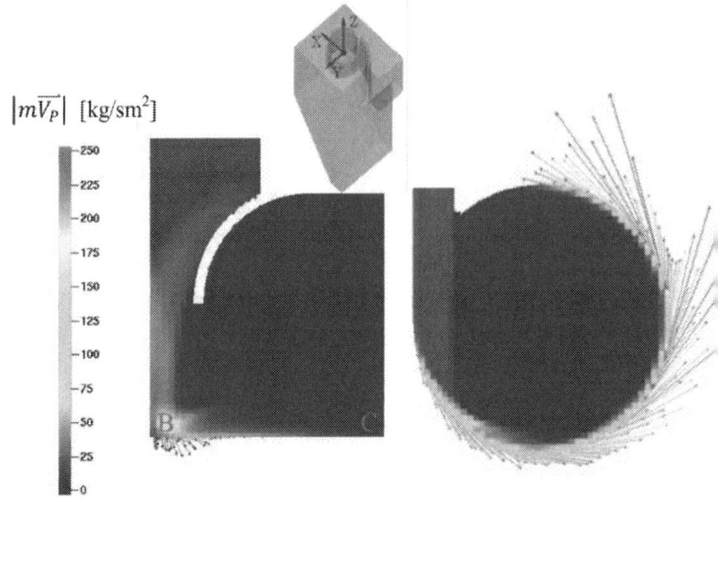

Fig.11 Particle Mass Flux Vector in X-Y Section

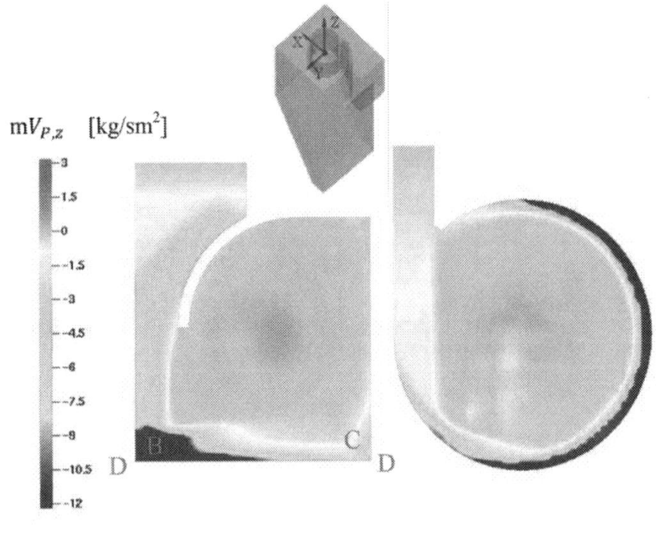

Fig.12 Z-direction component of Particle Mass Flux Vector in X-Y Section

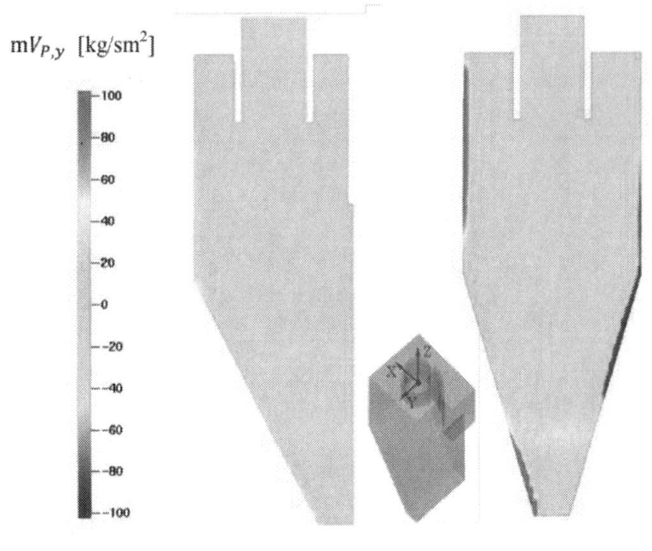

Fig.13 Y-direction Component of Particle Mass Flux Vector in X-Z Section

Observing the centrifugal cyclone in Fig.10, it was found that mass flux vector agreed with the envisagement of real cyclone. It was also found that the mass flux of square cyclone was very small except A-A edge, whereas the mass flux of centrifugal was large, and its direction was down-ward, it could be expected that the separation efficiency of centrifugal cyclone was higher than that of the square cyclone in this condition.

Observing Fig.11, the mass flux of square cyclone was small, only a little larger in the first corner B, whereas the mass flux of centrifugal cyclone was large along the whole wall. Observing Fig.12, for square cyclone, the down-ward mass flux was mainly in the in the first corner B, and then the second corner C and the wall D-D, while for centrifugal cyclone, the down-ward mass flux was on the whole wall. Observing Fig.13, absolute value of mass flux on the wall of square cyclone was near zero, while that of centrifugal cyclone was much larger, which means that the rotation of particle around the Z axis of square cyclone was much weaker than that of centrifugal cyclone. The eccentricity force acted on particles of square cyclone was much weaker than that of centrifugal cyclone.

Based on above observation, it was found again that the mechanism of particle separation of square cyclone was collision and eccentricity force, while that of centrifugal cyclone was eccentricity force. The particle separation mainly occurred near the wall D-D, especially near the corner B for square cyclone.

Comparison in Terms of Particle Volume Fraction

Finally, the particle volume fraction ε_P of square cyclone and that of centrifugal cyclone were compared. Fig.14 showed the particle volume fraction in X-Y slice section, and Fig.15 showed particle volume fraction in X-Z Section.

Observing the distribution of particle volume fraction ε_P in Fig.14, it could be found that the high particle concentration in the square cyclone was only on the wall D-D, especially on the corner B and C, while the high concentration in the centrifugal cyclone was on the whole wall. On the other hand, concentration in the inner space of the square cyclone was higher than that of centrifugal cyclone. Observing Fig.15, it could be seen that the high particle concentration zone in the square cyclone was closer to the bottom of vortex finder than that of centrifugal cyclone. It could be expected again that the separation efficiency of square cyclone was lower than centrifugal cyclone.

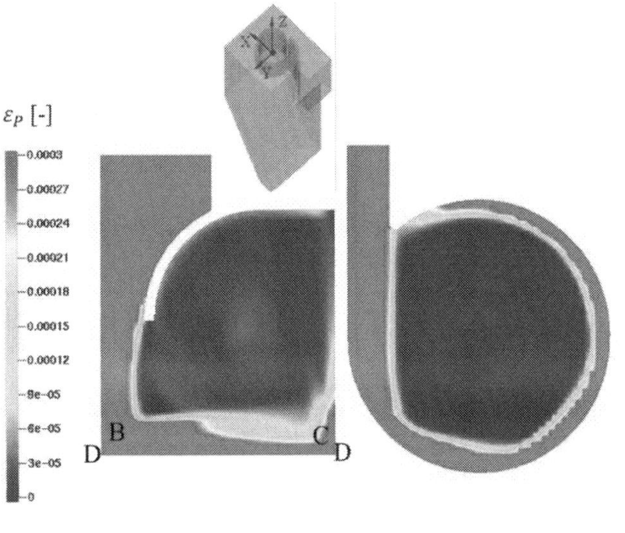

Fig.14 Volume Fraction of Particle in X-Y Section

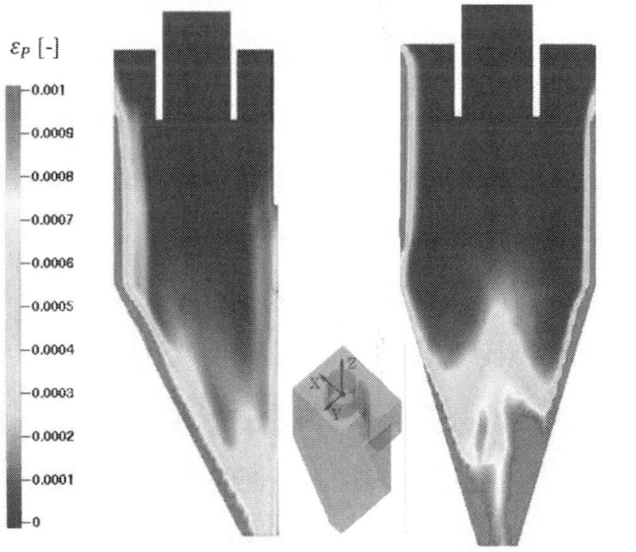

Fig.15 Volume Fraction of Particle in X-Z Section

Particle Separation Efficiency

Both of the effects of particle load and the effects of inlet gas velocity were compared between square cyclone and centrifugal cyclone. The computation conditions were listed in Table3.

The gross particle separation efficiency was shown in Fig. 16. The grade particle separation efficiency was shown in Fig. 17. It could be observed from Fig.16 that the particle separation efficiency of centrifugal cyclone was larger than that of square cyclone. It could be found that the particle separation efficiency increased when the particle load decreased, both for square and centrifugal cyclone, but the square cyclone would largely increase the separation efficiency than centrifugal cyclone. The effect of inlet gas velocity on the particle separation efficiency was different for square cyclone and centrifugal cyclone. When inlet gas velocity decreased, particle separation efficiency of centrifugal cyclone decreased, since the eccentricity force decreased. On the other hand, when inlet gas velocity decreased, particle separation efficiency of square cyclone increased. As the above observation, the mechanism of particle separation was collision with wall and eccentricity force. The collision with wall had a large contribution to the particle separation efficiency. In the calculated case, the decrease of inlet gas velocity improved the separation by the collision with wall and then improved the total separation effect, even though the separation by eccentricity force became weaker. It could be observed again from Fig.17 that the grade separation efficiency decreased when inlet gas velocity decreased for centrifugal cyclone, whereas the grade separation efficiency increased when inlet gas velocity decreased for square cyclone, even for small particles it was also true. It was believed that the ability of carrying particles by gas decreased as inlet gas velocity decreased, as the results, the separation efficiency increased for square cyclone.

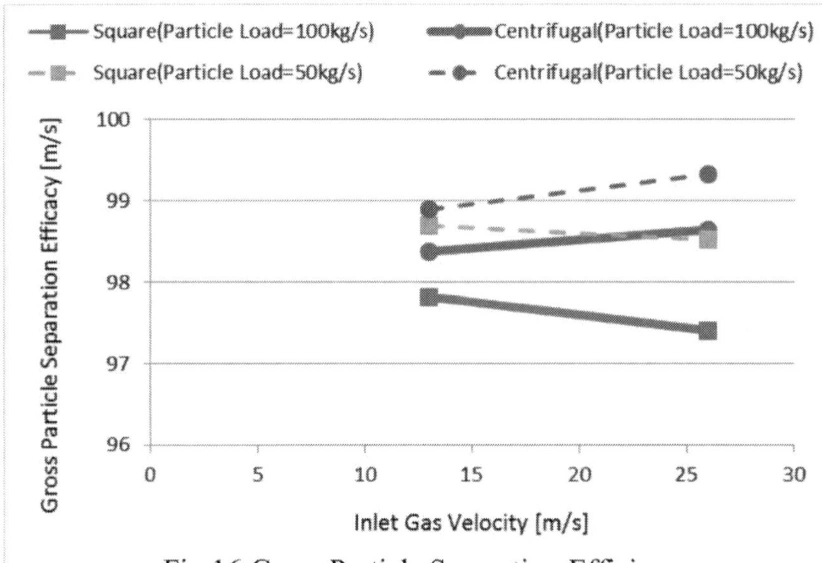

Fig.16 Gross Particle Separation Efficiency

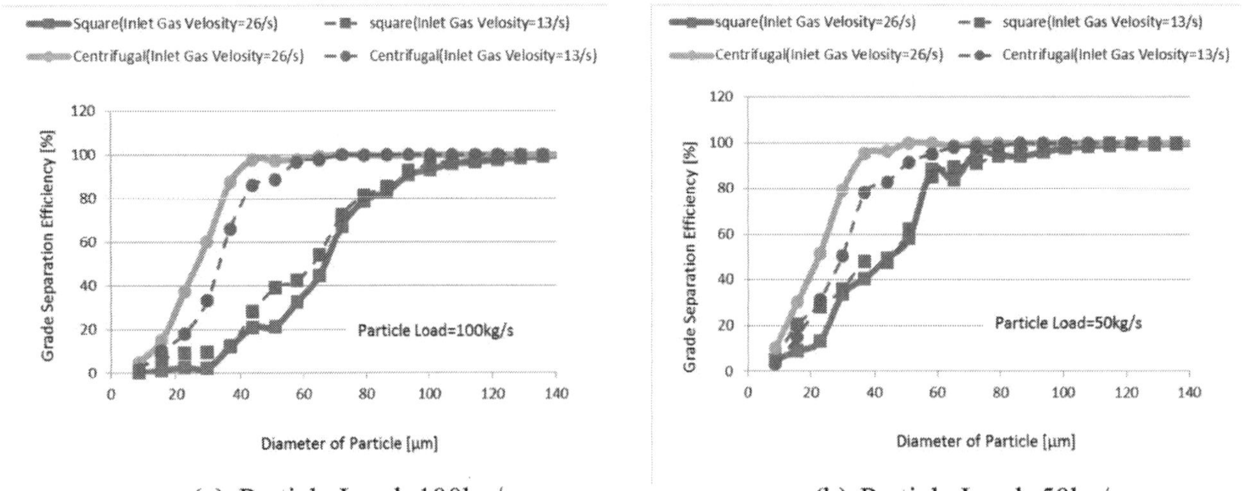

(a) Particle Load=100kg/s (b) Particle Load=50kg/s

Fig.17 Grade Particle Separation Efficiency

Pressure Drop

The pressure drop is important for the design of cyclone. The pressure drop was defined as the pressure difference between the outlet of cyclone and throat of inlet in this paper. The pressure drop was normalized by $\frac{1}{2}\rho V^2$, where V was the inlet gas velocity at throat. The pressure drop was shown in Fig.18. It would be found from Fig18 that the pressure drop of square cyclone was smaller than that of centrifugal cyclone, when they have the same hydraulic diameter and same inlet gas velocity. It could be found that the dimensionless pressure drop decreased when inlet gas velocity decreased. It was found that the pressure drop was decreased when the load of particle increased under this condition. This trend was also found in experiment [5] and numerical analysis [6]. It could be explained as reduction of the tangential gas velocity, caused by an increase in wall friction.

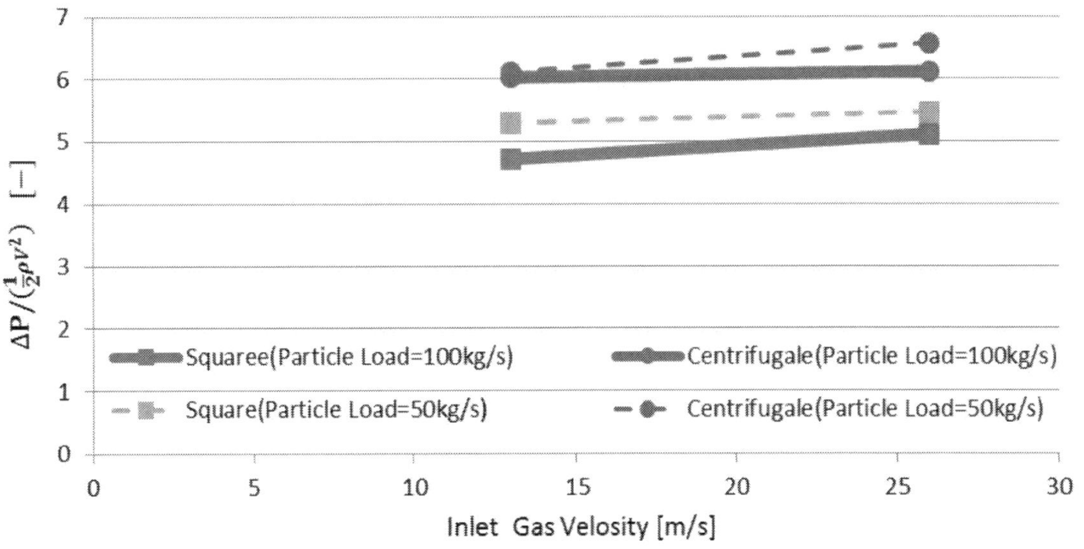

Fig.18 Pressure Drop

Conclusion Remarks

1. The comparisons were made between square cyclone and centrifugal cyclone by numerical calculation. By comparing the particle velocity vector, particle mass flux vector and particle volume fraction, it was found that the mechanism of separation was different. Centrifugal cyclone separated particle by eccentricity force, whereas square cyclone by both the eccentricity force and collision with wall.
2. Usually, the centrifugal cyclone had higher separation efficiency than square cyclone, but their difference on separation efficiency would become smaller when inlet gas velocity or/and particle load decreased. The square cyclone would be suitable for low inlet gas velocity and low particle load.
3. The pressure drop of square cyclone was lower than centrifugal cyclone under same conditions. The dimensionless pressure drop decreased when inlet gas velocity decreased. The pressure drop increase when particle load decrease under the conditions of this study.

References

1. K.W. Chu, B.Wanga, D.L.Xu, Y.X.Chen, A.B.Yu "CFD–DEM simulation of the gas–solid flow in a cyclone separator", Chemical Engineering Science 66834–847, (2011)
2. Y.Su, A. Zheng, B.Zhao, "Numerical simulation of effect of inlet configuration on square cyclone separator performance", Powder Technology 210, 293–303, (2011)
3. D.M.Sinder, P.J.O'Rourke, M.J.Andrews,"Sediment flow in inclined vessels calculated using multiphase particle-in cell model for dense particle flow", Int. J. Multiphase Flow, 24, pp.1359-1282, (1998)
4. Fuping Qian, Yanpeng Wu, "Effects of the inlet section angle on the separation performance of a cyclone" , chemical engineering research and design 8 7 1567–157 (2009)
5. Chen, J. Y., Luo, X. L., & Shi, M. X. "Pressure drop calculation on PV type cyclone under dust-bearing condition". Petro-Chemical Equipment Technology, 18, 1–3. (1997).
6. Fuping Qian, Zhijia Huang, Guang Chen, Mingyao Zhang, "Numerical study of the separation characteristics in a cyclone of different inlet particle concentrations" Computers and Chemical Engineering 31 pp. 1111–1122 (2007)

(273c) Filtered Drag Models for Euler-Lagrange Simulations of Gas-Solid Flow

Stefan Radl, Institute for Process and Particle Engineering, Graz University of Technology, Graz, Austria and Sankaran Sundaresan, Department of Chemical and Biological Engineering, Princeton University, Princeton, NJ

It is well known that fluidized gas-particle mixtures manifest meso-scale structures such as clusters and bubbles. Resolving small meso-scale structures in device-scale simulations is too expensive and this has led to the development of filtered Euler-Euler (EE) two-fluid models and related constitutive models [1,2]. In this contribution, we ask how the constitutive models for, say, fluid-particle interaction force, should be filtered in Euler-Lagrange (EL) simulations (using the averaged equations for the gas phase and Discrete Element Method for the particle phase – sometimes referred to as the CFD-DEM approach).

To address this question, we performed CFD-DEM simulations of $O(10^6)$ particles that freely sediment in a periodic box, and recorded statistics of the domain-averaged slip velocity, as well as of numerous spatially-averaged (i.e., filtered) quantities. Variations in the microscopic drag law, grid resolution, domain size, and mapping schemes show that our results are robust with respect to these settings.

We define two sets of filtered slip velocities and corresponding drag laws: First, we filter only the fluid velocity and define a filtered slip velocity as the difference between a filtered fluid velocity and the instantaneous particle velocity. The corresponding "fluid-only" filtered drag model is relevant for simulations where particles (or groups of particles) are tracked and coarse fluid grid cells have to be used (e.g., MP-PIC [3], or other parcel-based approaches [4]). Second, we define the filtered slip velocity based on the filtered fluid velocity and an average particle velocity in the filter region. The corresponding "particle-and-fluid" filtered drag model is relevant for coarse-grid EE simulations. Finally, we discuss the differences between the two classes of filtered drag laws, and provide closed expressions for their inclusion in parcel- and EE-based coarse-grid simulations. Variations in the particle size and density were used to extract the correct reference length scale that makes our filtered model applicable to gas-particle systems with sufficiently low particle Reynolds numbers.

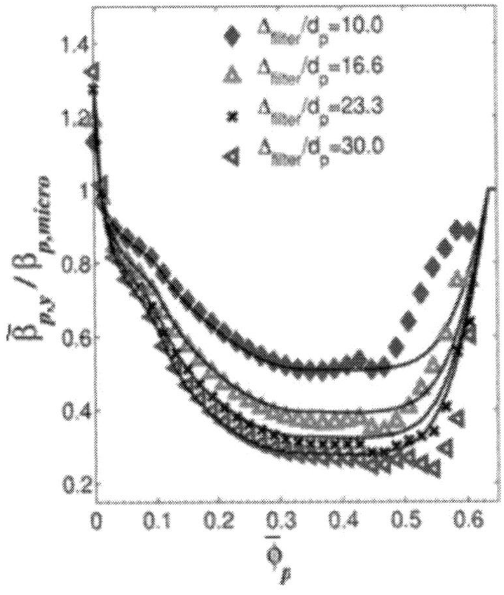

Dimensionless filtered drag coefficient for various filter sizes and volume fractions (symbols: data from EL simulations; black lines: model prediction)

References

[1] Y. Igci and S. Sundaresan, Constitutive Models for Filtered Two-Fluid Models of Fluidized Gas-Particle Flows. Ind. Eng. Chem. Res. 50 (2011) 13190-13201.

[2] J.-F. Parmentier, O. Simonin, and O. Delsart, A functional subgrid drift velocity model for filtered drag prediction in dense fluidized bed. AIChE J. 58 (2012) 1084-1098.

[3] D.M. Snider, An incompressible three-dimensional multiphase particle-in-cell model for dense particle flows. J. Comput. Phys. 170 (2001) 523-549.

[4] N.A. Patankar and D.D. Joseph, Modeling and numerical simulation of particulate flows by the Eulerian-Lagrangian approach. Int. J. Multiphase Flow 27 (2001) 1659-1684.

(273d) Effects of Pressure and Fines Content On Bubble Diameter in a Fluidized Bed Studied Using Fast X-Ray Tomography

J. Ruud van Ommen, Gerrit Brouwer, Evert Wagner and Robert F. Mudde, Chemical Engineering, Delft University of Technology, Delft, Netherlands

Using a fast X-ray tomography setup the bubble size and velocity has been measured in a 25 cm diameter fluidized bed of Geldart B powder and a 24 cm bed of Geldart A powder. The average bubble size has been determined for a measurement period of 60 seconds. The resolution of this setup is about 4.5 mm per pixel at a rate of 250 reconstructions per second. It is possible to detect bubbles as small as 2.2 cm. The Geldart B powder consisted of polystyrene particles with an average diameter of 607 mm and a bulk density of 625 kg/m^3. This bed was studied at pressures ranging from 1 to 5 bar. The superficial gas velocities varied from 12 to 32 cm/s for the atmospheric pressure measurements; for the highest pressure from 10 to 15 cm/s. The bubble size is significantly reduced at higher pressures for similar gas flows. The Geldart A powder consist of a base of aluminum oxide particles with an average diameter of 76 mm and a bulk density of 680 kg/ m^3. Varying amounts of fines were added to these base particles. The fines consist of aluminum oxide particles with an average diameter of 38 mm and a bulk density of 620 kg/ m^3. An increase in the fines content from 0 to 50% $_{\text{weight}}$ leads to a decrease in the average spherical equivalent bubble diameter by 20%.

(273e) Heat Transfer From Hot Particles to Cold Particles in Downward-Flow Binary Solids Systems

Yuji Yoshie[1], Masanori Ishizuka[2], Guoqing Guan[3], Chihiro Fushimi[4] and Atsushi Tsutsumi[2], (1)School of engineering, The University of Tokyo, Tokyo, Japan, (2)Institute of Industrial Science, The University of Tokyo, Tokyo, Japan, (3)North Japan Research Institute for Sustainable Energy (NJRISE), Hirosaki University, Hirosaki, Japan, (4)Department of Chemical Engineering, Tokyo University of Agriculture and Technology, Tokyo, Japan

Coal is the major source of energy for the generation of electricity throughout the world, however burning of coal in conventional coal-fired power plants is one of the major contributors to climate change and global warming. To cope with this issue, various Clean Coal Technologies (CCTs) such as integrated coal gasification combined cycle (IGCC) have been widely developed. IGCC can increase the efficiency of power generation from coal, and hence reduce the pollutants emissions.

For further improvement of the energy efficiency of IGCC, Advanced IGCC (A-IGCC) system, which makes use of the exergy recuperation concept; utilizing the exhausted heat from the gas turbine for steam gasification, has been proposed. For the A-IGCC system, a triple-bed combined circulating fluidized bed (TBCFB) gasifier, composed of a downer pyrolyzer of coal, a bubbling fluidized bed gasifier of char, and a riser combustor of unreacted char to produce heat, has been considered to replace the conventional high temperature gasifiers. The TBCFB gasifier is expected to enhance the efficiency of char gasification by separating the volatiles from coal soon after rapid pyrolysis process to avoid inhibition effects of volatiles to char gasification. In this system, it is essential to transfer the heat carried by inert particles from the riser combustor immediately and effectively to the coal particles injected into the downer pyrolyzer.

For the TBCFB gasifier, the design of the downer coal pyrolyzer with high performance is necessary. For this purpose, it is important to investigate the changes of heat transfer coefficients in binary solids systems (inert particle and coal particle) with different flow structures in the downer. In this work, an novel experimental technique was developed to measure the temperatures of coal particles in the binary solids systems via heat balance method. Simultaneously, local solids holdup and particle velocities were also measured. The relationship between heat transfer coefficient and flow structure was analyzed and discussed.

(273f) Experimental and Numerical Investigations On the Electrostatics in the Downer Reactor

Yongpan Cheng, Chemical and Biomolecular Engineering, National University of Singapore, Singapore, Singapore, Darren Yan Jun Lau, Department of Chemical and Biomolecualr Engineering, National Universtiy of Singapore, Singapore, Singapore, Guoqing Guan, North Japan Research Institute for Sustainable Energy (NJRISE), Hirosaki University, Hirosaki, Japan, Chihiro Fushimi, Department of Chemical Engineering, Tokyo University of Agriculture and Technology, Tokyo, Japan, Atsushi Tsutsumi, Institute of Industrial Science, The University of Tokyo, Tokyo, Japan and Chi-Hwa Wang, Department of Chemical and Biomolecular Engineering, National University of Singapore, Singapore, Singapore

In fluidization processes and granular flow system, tribo-electrification is an inevitable phenomenon due to the continuous and random collisions between particulates and walls with different materials. In this study the electrostatic performance in the downer reactor of a triple-bed combined circulating fluidized bed was investigated through both experiments and numerical simulations. In numerical simulation Discrete Element Method (DEM) was adopted to simulate the electrostatic charge generation and transfer occurring in the downer. Both experimental and numerical results showed that the averaged induced currents caused by electrostatics increased with increasing solids mass flux in the downer, and the electrostatics was strongest near the entrance of the downer because of the highest frequency of collisions between particles and wall caused by the largest solids holdup at that location. Numerical studies also revealed that the averaged induced currents increased with increasing initial particle velocity and particle size, and the averaged induced currents caused by the charge transfer was much larger than those by charge generation through tribo-charging. This is the first few studies in trying to reveal the tri-electrification in the fluidized bed, and may help us better understand the mechanism of electrostatic phenomena, and cope with challenges and problems caused by electrostatics.

(273g) Predicting Minimum Fluidization and Bubbling Velocities for Geldart Type A Particles

Janine Galvin, U.S. DOE National Energy Technology Laboratory, Albany, OR and Sofiane Benyahia, U.S. DOE National Energy Technology Laboratory, Morgantown, WV

The fluidization behavior of Geldart A particles is investigated using soft-sphere discrete particle simulations. The inclusion of cohesive forces permits ready distinction between the minimum fluidization velocity (U_{mf}) and the minimum bubbling velocity (U_{mb}) that are typically observed with these type of particles. The simulation predictions are consistent with experimental observations detailed in fluidization textbooks (e.g. Kunii and Levenspiel, Fluidization Engineering). Specifically, the predictions demonstrate a spike in pressure drop near U_{mf} as well as the formation of small voids that do not coalesce into fast bubbles and greatly increase the bed expansion until reaching U_{mb}. The results also clearly show the onset of bubbling coincides with the breaking of cohesive links between particles once the bed has expanded considerably. This occurs at a velocity U_{mb} that can be two to three times larger than U_{mf}. Quantitatively, the predicted values of these design-critical velocities depend on the size and geometry of the system as well as other considerations, such as polydispersity.

While cohesive van der Waals forces are easy to incorporate in a discrete particle model, the present simulations predict a very sticky bed which cannot be fluidized when realistic values of the Hamaker constant are employed. This issue was resolved by considering the effects of asperities on the particle's surface, which tend to diminish the strength of cohesive forces. Finally, an explanation is proposed to address the observed discrepancies between the results of this study and those of previous publications studying systems under similar conditions.

(275a) Hybrid Bilayer Hollow Submicron Particles with Carbon and Silica Nanoshells Fabricated Through a Template Free Process and Their Applications As Multifunctional Materials

Vijay T. John[1], Yingqing Wang[2], Jingjing Zhan[2] and Bhanukiran Sunkara[1], (1)Chemical and Biomolecular Engineering, Tulane University, New Orleans, LA, (2)Department of Chemical and Biomolecular Engineering, Tulane University, New Orleans, LA

We present a facile aerosol based method to create a hollow particulate system with a composite bilayer shell. The external layer is silica while the internal layer is carbon, thus creating an amphiphilic shell with an intrinsically hydrophilic external surface and a hydrophobic internal surface. The particles are obtained using a new concept of forming a silica shell extremely rapidly and locking in other precursors in the interior of an aerosol droplet. This is done by exploiting salt bridging concepts to confine a surfactant (CTAB) and carbon precursors together with iron oxides in the interior of a droplet while a silica shell is allowed to form on the droplet surface. Subsequent pyrolysis results in a buildup of internal pressure forcing carbon formation as a second layer adjoining the silica shell. The synthesis method is expected to be a general approach to fabricate a variety of hybrid double layer hollow microspheres with unique properties. In addition, the incorporation of magnetic iron oxide into the shells opens up opportunities in external stimuli responsive materials. Hollow nano and microparticles have a variety of applications in encapsulation, catalysis, separation and controlled drug release. The aerosol based process is intrinsically scalable and overcomes many of the drawbacks of traditional methods of making hollow structures through multiple steps of forming layers over sacrificial particle cores.

Creating Conductive Copper-Silver Bimetallic Nanostructured Coatings using a High Temperature Reducing Jet Aerosol Reactor

Munish K. Sharma[1], Raymond D. Buchner[1], William J. Scharmach[2], Vasilis Papavassiliou[2], and Mark T. Swihart[1]*

[1] Department of Chemical and Biological Engineering, The University at Buffalo (SUNY), Buffalo, NY, 14260
[2] Praxair, Inc., Research and Development, 175 E Park Dr. Tonawanda, NY 14150

We present here the high temperature flame-based synthesis of non-oxide nanoparticle hybrid films using a high temperature reducing jet (HTRJ) reactor. We synthesized bimetallic copper-silver nanostructured films by direct deposition of nanoparticles produced within the HTRJ reactor. The HTRJ process allows us to decouple the flame chemistry from the nanoparticle formation chemistry. The alloy and bi-component nanoparticles have potential applications in conductive ink formulations for printed electronics, in antibacterial coatings, and in heat-transfer fluids. Recent advances include synthesis of hybrid structures such as bimetallic nanoparticles and alloys. Interesting properties arise in bimetallic structures through synergistic effects of their components. Silver nanoparticles have been widely used in conductive nanoink formulations as well as in RFID technology. This is not only because silver has high room temperature electrical conductivity (~6×10^7 S/m), but also because its oxide has appreciable conductivity (of order 10^6 S/m). However, due to high cost of silver metal, researchers are looking for alternatives. Copper has electrical conductivity similar to that of silver (~6×10^7 S/m) and is much lower in cost than silver, but oxidizes rapidly in the presence of atmospheric oxygen.

In the present work, we report the single step direct synthesis of copper-silver nanostructured particles and films using the HTRJ flame-based process. The objective of this work was to synthesize a bimetallic coating using copper and silver with minimal use of silver to lower the cost and to meet the desired industrial window of electrical conductivity (> 100 S/m) for the aforementioned applications. The advantages of HTRJ technology stem from the use of low-cost, water soluble nitrate precursors, and a low-cost energy source in a continuous and environmentally benign process. The metal films were thermophoretically deposited on soda lime glass. In this presentation, we will describe the HTRJ process briefly, and then focus on the synthesis and characterization of bimetallic copper-silver films on glass substrates. The hybrid copper-silver coating yielded in very high bulk electrical conductivity of 10^4 S/m at 40 wt% Ag. The effect of silver concentration in the mixed copper-silver films on the electrical conductivity of these coatings was studied systematically.

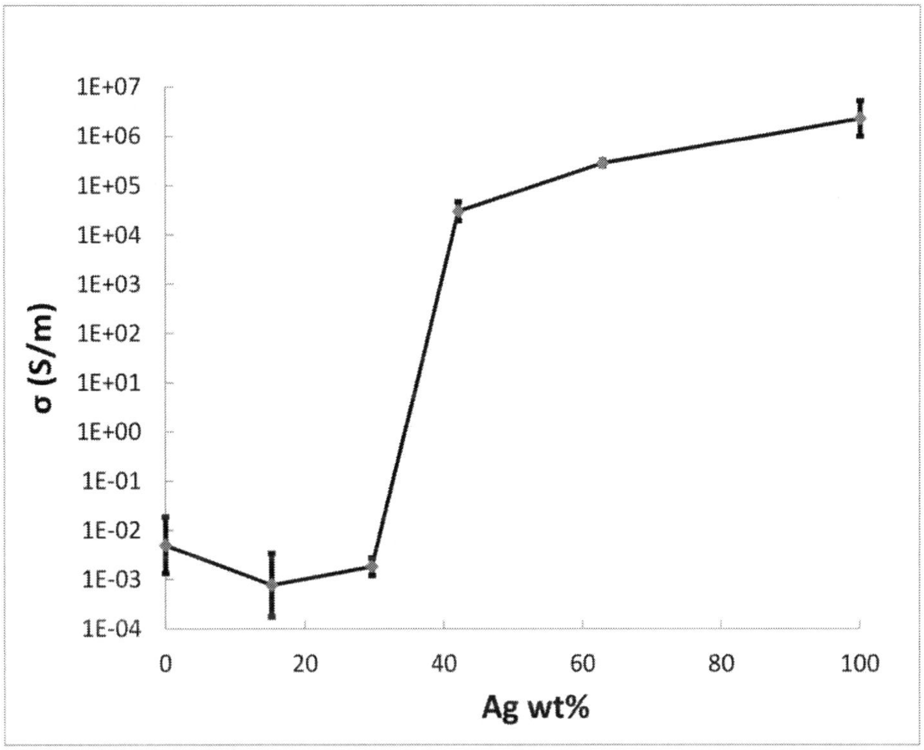

Figure 1. Electrical conductivity as a function of Ag wt%. Error bars are plus or minus one standard deviation of log(σ).

Combustion-Driven One Step Synthesis of Palladium Hybrid Films in a High Temperature Reducing Jet Reactor for Hydrogen storage and Purification

Munish K. Sharma[1], Di Qi[1], Raymond D. Buchner[1], and Mark T. Swihart[1]*

[1]Department of Chemical and Biological Engineering, The University at Buffalo (SUNY), Buffalo, NY, 14260

In recent years, hydrogen storage and purification became an important area of research for scientific community. There is an urgent need of a low cost, highly mechanical stable composite metallic membrane. Palladium and palladium based alloys have become a promising membrane material in this area due to high selectivity and mobility of hydrogen in Pd lattice. But pure palladium is expensive, low mechanical stability, and prone to hydrogen embrittlement below 573 K. All these issues can be handled by alloying palladium with comparatively less expensive metals such as copper and silver. Alloying palladium with copper and silver prevents hydrogen embrittlement and results in higher hydrogen permeation fluxes and enhances mechanical stability. Besides change in physical properties at nanoscale, these membranes are resistant to surface poisoning impurities such as CO, H_2S, and CH_4.

We present here the high temperature flame-based synthesis of non-oxide nanoparticle hybrid films utilizing a high temperature reducing jet (HTRJ) reactor developed by Scharmach et al (2010). We synthesized bimetallic palladium-copper, palladium-silver and ternary palladium-copper-silver hybrid films by alloying palladium with copper and silver nanoparticles in-situ using the hydrogen rich flame inside the HTRJ reactor. The HTRJ process allows us to decouple the flame chemistry from the nanoparticle formation chemistry. We synthesized these hybrid films in one step by thermal decomposition of low-cost, water soluble, and environment friendly route. The as-prepared films were sintered at 200°C inside the HTRJ reactor, eliminating the need of any post treatment methods to anneal and sinter the films. The metal nanoparticles in nm range were thermophoretically deposited on a 25 micron average pore size SS-316 discs leading to formation of these Pd based porous films. Hydrogen permeation experiments at different pressure and temperatures using a disc membrane module are in progress. The hybrid films were characterized by powder X-Ray diffraction, SEM and EDX for surface topology and elemental composition. In this presentation, we will describe the HTRJ process briefly, and then focus on the synthesis and characterization of bimetallic Pd-Cu, Pd-Ag, and ternary Pd-Cu-Ag films along with hydrogen permeation results using these low cost membranes.

Figure1: (a) SEM micrograph showing ternary Pd-Cu-Ag coating showing a porous network on porous sintered stainless steel disc; (b) SEM micrograph showing a cross-sectional view of as-prepared films on substrate

(275d) Pilot-Scale Flame Spray Aerosol Reactor for Manufacture of ZrO_2 Nanoparticles

Arto J. Gröhn, Department of Mechanical and Process Engineering, ETH Zurich, Zurich, Switzerland, Sotiris E. Pratsinis, Department of Mechanical and Process Engineering, ETH Zürich, Zurich, Switzerland and Karsten Wegner, Particle Technology Laboratory, ETH Zurich, Zurich, Switzerland

Pilot-scale Flame Spray Aerosol Reactor for Manufacture of ZrO_2 Nanoparticles

Arto J. Gröhn, Sotiris E. Pratsinis and Karsten Wegner

Particle Technology Laboratory, Swiss Federal Institute of Technology (ETH) Zurich, Sonneggstrasse 3, CH-8092 Zurich, Switzerland.

Flame Spray Pyrolysis (FSP) aerosol reactors have high potential to be employed for industrial nanoparticle manufacture for their versatility and relatively low production costs (Strobel et al., 2007; Wegner, 2011). To date, extensive research has been carried out on these reactors at the laboratory scale and a plethora of single and multicomponent nanoparticles has been developed (Teoh et al., 2010).

In this work FSP operation at the pilot-scale with up to 2 kg/h zirconia production rate is investigated. Therefore, the effects of reactor geometry and operating conditions on product particle properties are studied experimentally. Fourier-transform infrared spectroscopy, Phase-Doppler Anemometry and nanoparticle sampling in-situ the flame are employed for characterizing the temperature, velocity and species profiles of the flame reactor. The obtained results are used to determine optimized spray nozzle design. Finally, the potential of replacing oxygen dispersion gas by air and increasing the metal content of the starting raw materials for further reducing production costs in an industrial environment is explored.

References:

Strobel, R. and Pratsinis, S.E., Flame aerosol synthesis of smart nanostructured materials. *Journal of Materials Chemistry*, 2007. 17, 4743-4756.

Wegner, K. Schimmoeller, B. Thiebaut, B. Fernandez, C. and Rao, T.N., Pilot Plants for Industrial Nanoparticle Production by Flame Spray Pyrolysis. *KONA Powder and Particle*, 2011. 29, 251-265.

Teoh, W.Y. Amal, R. and Madler, L. Flame Spray Pyrolysis: An Enabling Technology for Nanoparticles Design and Fabrication. *Nanoscale*, 2010. 2, 1324-1347.

(275e) Spray Granulation of Nanopowders

Arto J. Gröhn and **Karsten Wegner**, Particle Technology Laboratory, Swiss Federal Institute of Technology (ETH Zurich), Zurich, Switzerland

Spray Granulation of Nanopowders

Arto Gröhn and Karsten Wegner

Dry synthesis of nanoparticles with flame reactors can yield large quantities of nanopowders at rather low cost, as has been shown by pilot-scale flame spray pyrolysis (FSP, Bickmore et al., 1996; Wegner et al., 2011). Products recovered from the process at the bottom of the bag-house filter have a very low packing density and are easily fluidized. This can impose problems on nanoparticle handling with standard equipment and techniques and may result in the release of nanoparticles into the workspace.

Here, spray granulation of flame-made nanoparticles is investigated at the pilot- scale by the examples of silica and zirconia. The effect of process parameters such as slurry flow rate, solids fraction, and temperature on the characteristics of the spray granulator as well as the product granule characteristics is studied. Properties of the granules as well as the bulk powders are described with one emphasis being placed on aerosolization and dust formation.

References:

Bickmore, C.R., Waldner, K.F., Treadwell, D.R. and Laine, R.M. (1996), *J. Am. Ceram. Soc.* **79**, 1419-1423.

Wegner, K., Schimmoeller, B., Thiebaut, B., Fernandez, C. and Rao, T.N. (2011), *KONA* **29**, 251-265.

Brownian dynamics simulation of recognition kinetics between lock and key colloid particles

Jun Liu[1], *Ronald G. Larson*[1]

Department of Chemical Engineering, University of Michigan, Ann Arbor, Michigan 48109, the United States

Abstract

We adopt Brownian dynamics (BD) simulation to study the recognition kinetics between lock-and-key colloidal particles. A general resistance matrix (12*12) to describe the translation, rotation and translation-rotation coupling for any orientation connecting two sphere centers is established. Meanwhile, the general form of hydrodynamic interactions (HIs) between two spheres, adopted from the literature (Jeffrey and Onishi, J. Fluid. Mech, 1984, 139, 261), has been verified to incorporate both the far and near-field HIs, and an error in this important literature has been corrected. Our results indicate a critical depletant concentration exists for stable non-specific binding. The simulated translational and rotational diffusion coefficient agree well with the theoretical predictions in the far-field case where HI plays a negligible role. Generally the simulation methodology developed here can be readily used in those situations where the effect of HIs on the translation and rotation needs to be accurately accounted for.

(337b) Measurement of Crystal Morphology: A New Method Based On Reflection of Light From Crystal Faces

Jayanta Chakraborty, Chemical Engineering, Indian Institute of Technology, Kharagpur, Kharagpur, India, Debasis Sarkar, Indian Institute of Technology, Kharagpur, Kharagpur, India and Abhishek Singh, Indian Institute of Technology, Kharagpur

Measurement of morphology of 3D faceted crystals: a new method on the basis of light reflection from faces.

Jayanta Chakraborty, Debasis Sarkar and Abhishek Singh

Department of Chemical Engineering, Indian Institute of Technology Kharagpur. Kharagpur, India.

Measurement of morphology of faceted crystals is a challenging task. Most of the examples of morphological measurements are applicable for flat crystals where image is taken from top and it is analyzed for edge detection. In this work we show how this simple strategy of image capture and analysis can be used to obtain the morphology of a 3D faceted crystal using our novel method of image capturing.

The main idea comes from the fact that the crystal faces being smooth, regular reflection of light dominates the irregular one. Hence a proper arrangement of optics will make one face *shining* where other faces will be dull. One such picture is shown in Fig. 1a for $CuSO_4$, $5H_2O$. This novel strategy makes image processing a trivial task and the face vertices can be detected readily. The vertex location, which is 2D information about a certain face can then be integrated with the angular orientation of faces (which is invariant for a given material and polymorph) to obtain the detail 3D morphology of the crystal. One example of such reconstruction is shown in Fig. 1b.

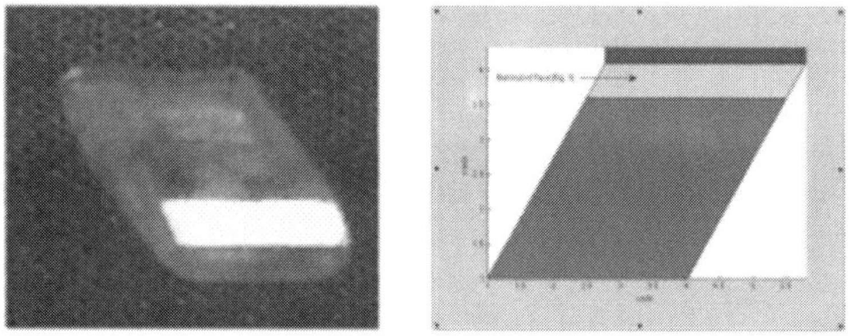

Figure 1: (a) The proposed illumination strategy has found a face. (b) The face features has been extracted using image analysis and the crystal is reconstructed using our software. The top view of the reconstructed crystal is shown.

(337c) Metal Nanoparticle Growth by Molecular Dynamics

Beat Buesser[1,2] and Sotiris E. Pratsinis[2], (1)Department of Chemical Engineering, Massachusetts Institute of Technology, Cambridge, MA, (2)Department of Mechanical and Process Engineering, ETH Zurich, Zurich, Switzerland

Metal Nanoparticle Growth by Molecular Dynamics

B. Buesser[1,2] and S.E. Pratsinis[1]

[1]Particle Technology Laboratory, Institute of Process Engineering, Department of Mechanical and Process Engineering, ETH Zürich, 8092 Zürich, Switzerland

[2]Department of Chemical Engineering, Massachusetts Institute of Technology, Cambridge, MA 02139, USA

Metal nanoparticles are attractive in catalysis, magnetic separations[1] or sensors[2], to name only a few applications. The performance of these particles, however, depends considerably on their primary particle size and structure. Gas-phase processes allow making economically such particles in large quantities with close control of their size and extent of aggregation[3]. In gas-phase reactors these two characteristics are determined mostly by particle sintering. The detailed understanding of sintering is crucial for the development and scale-up of such reactors to target product particle size and morphology at maximal yield especially when precious metals are involved.

Silver nanoparticles are one of the most studied noble nanomaterials. For example Shimada *et al.*[4] proposed a sintering rate for silver nanoparticles in gas-phase. They found their particle dynamics model in good agreement with the measured particle size evolution in their hot wall reactor, where primary particle sizes bigger than d_p = 8 nm were observed. The evolution of smaller primary particles (d_p < 8 nm) is increasingly difficult to determine experimentally although this would be the key size range where nanoparticle exhibit their extraordinary performance and exciting new properties.

Molecular dynamics (MD) simulations are reaching this range of particle sizes and sintering time scales with the proliferation of high-performance computer hardware. Sintering of metallic[5], metalloid[6] and ceramic[7] nanoparticles has been investigated, but often the surface area evolution, a key quantity in reactor design for particle synthesis, has been neglected.

Here, sintering of silver nanoparticles is investigated using MD simulations accelerated by graphical processing units (GPU) in the range of d_p = 2 – 5 nm. The sintering rate is determined by calculating the surface area evolution comparable to BET surface area measurements. First observations of the atom trajectories reveal that surface atoms exhibit a much higher mobility than bulk ones indicating that sintering by surface diffusion dominates[7] at these particle sizes and temperatures (Figure 1). The dependence of the sintering rate on particle morphology has been investigated during sintering of straight chains, triangles and stars of three and four particles.

An expression for the sintering rate as function of primary particle size and temperature has been extracted from MD, filling the gap of knowledge between clusters of a few atoms up to particles of several nanometers. This sintering rate will facilitate the design of large scale manufacture and processing of these small nanoparticles based on phenomenological models[8] or allow engineering estimations of the particle morphology by comparing it to the coagulation rate[3].

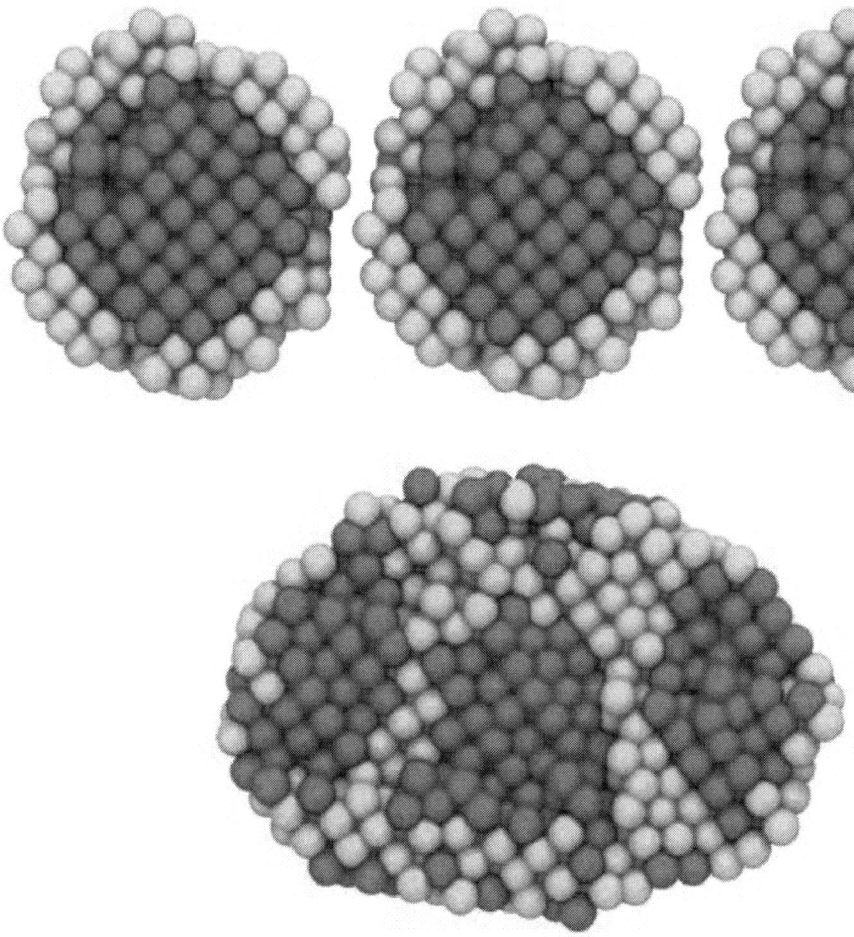

a)

b)

Figure 1 Snapshots of 3 silver nanoparticles with a diameter of d_p = 3 nm sintering at T = 800 K and time t = a) 0 ns and b) 100 ns. The atoms are colored green at the surface and red in the bulk at t = 0 ns. It is fascinating to see that these surface atoms largely move to the concave areas between these particles at t = 100 ns, revealing the dominance of surface diffusion during their sintering or coalescence.

Financial support from the European Research Council is gratefully acknowledged.

1. Grass RN, Athanassiou EK, Stark WJ. Covalently Functionalized Cobalt Nanoparticles as a Platform for Magnetic Separations in Organic Synthesis. *Angewandte Chemie International Edition.* 2007;46(26):4909-4912.

2. Sotiriou GA, Sannomiya T, Teleki A, Krumeich F, Vörös J, Pratsinis SE. Non-Toxic Dry-Coated Nanosilver for Plasmonic Biosensors. *Advanced Functional Materials.* 2010;20(24):4250-4257.

3. Buesser B, Pratsinis SE. Design of Nanomaterial Synthesis by Aerosol Processes. *Annual Review of Chemical and Biomolecular Engineering.* 2012;3:103-127.

4. Shimada M, Seto T, Okuyama K. Size change of very fine silver agglomerates by sintering in a heated flow. *Journal of Chemical Engineering of Japan.* 1994;27(6):795-802.

5. Arcidiacono S, Bieri NR, Poulikakos D, Grigoropoulos CP. On the coalescence of gold nanoparticles. *International Journal of Multiphase Flow.* 2004;30(7-8):979-994.

6. Zachariah MR, Carrier MJ. Molecular dynamics computation of gas-phase nanoparticle sintering: a comparison with phenomenological models. *Journal of Aerosol Science.* 1999;30(9):1139-1151.

7. Buesser B, Gröhn AJ, Pratsinis SE. Sintering Rate and Mechanism of TiO_2 Nanoparticles by Molecular Dynamics. *J. Phys. Chem. C.* 2011;115:11030-11035.

8. Koch W, Friedlander SK. The effect of particle coalescence on the surface area of a coagulating aerosol. *Journal of Colloid and Interface Science.* 1990;140(2):419-427.

(337d) Interfacial Tension in Multiphase, Highly Supersaturated Aqueous Solutions for CaCO3 Polymorphs

Ben Schroeder, Chemical Engineering, Institute for Clean and Secure Energy, University of Utah, Salt Lake City , UT, Sean T. Smith, Chemical Engineering, Institute for Clean and Secure Energy, University of Utah, Salt Lake City, UT and Derek D. Harris, Chemical Engineering, BYU, Provo, UT

A method of modelling the interfacial energy of CaCO3 particles in a supersaturated, aqueous solution will be presented. Four polymorphs of CaCO3 will be considered, each with its own distinct surface properties. The purpose of this model is for the application to nucleation, growth, and agglomeration kernels in a population balance in order to reproduce experimental data from the literature [Ogino et al., Geochimica et Cosmochimia Acta, 1987]. The system described by the literature varied in temperature and the supersaturation of the measured solutions changed over time.

All three of the physical processes effecting the evolution of the particle distribution are functions of and sensitive to the interfacial tension of the particle-solution interface. Interfacial tension itself is a function of both temperature and the ionic composition of the fluid surrounding the particle, thus making this a dynamic system. The interfacial tension between phases can generally be modelled using the van Oss & Good equation, which breaks up the surface tension of each phase into polar and non-polar components and then combines these phases components to find the interfacial tension. The effect of the temperature and varying amounts of multiple ionic compounds within the aqueous solution will be considered and incorporated in the components water's surface tension.

(337e) Fabrication of Highly Ordered and Close Packed Colloidal Crystals From Large Microparticles

Melissa H. Lash[1], Kaladhar Kamalssanan[1,2,3], Shu Li[1], Joseph McCarthy[1] and Steven R. Little[4], (1)Chemical and Petroleum Engineering, University of Pittsburgh, Pittsburgh, PA, (2)Immunology, University of Pittsburgh, Pittsburgh, PA, (3)The McGowan Institute for Regenerative Medicine, University of Pittsburgh, Pittsburgh, PA, (4)Department of Chemical and Petroleum Engineering, University of Pittsburgh, Pittsburgh, PA

Colloidal crystals have been explored in the literature for applications in molecular electronics, photonics, sensors, and drug delivery. However, much of the research on colloidal crystals has been focused on nano-sized particles with limited attention directed towards building blocks with dimensions ranging from tens to hundreds of microns. This can be attributed, in part, to the fact that large-scale particles are less prone to assemble in an organized fashion due to the relative absence of thermalizing forces. Nevertheless, ordered arrays of large particles are of interest both as a basis for tissue engineering scaffolds as well as a first step toward large-volume production of micron-scale "patchy particles" (anisotropically-labeled particles). In this work, ultrasonic agitation is being explored as a means of artificially "thermalizing" these particles in order to overcome kinetic barriers to packing in the creation of close packed, highly ordered, crystalline structures from large microparticles (18-100um). The film thickness is adjustable from a monolayler to a multilayer structure by changing the concentration of the solution and through a layer-by layer addition of particles. For this process, we are investigating the significance of substrate and solution properties like surface tension, viscosity, and particle concentration both experimentally and computationally. The interactions between the constituent particles within these colloidal formations can be estimated and analyzed using simulation-techniques. Simulations will not only inform as to what is occurring and ultimately achievable, but may also be used to direct the size, thickness, and quality of the overall crystal. This analysis and control over processing will ideally provide insight into, and control over the formation of a desired crystal structure.

(337f) Small and Wide-Angle X-Ray Scattering for in Situ Monitoring of Particle Size Evolution During Platinum Nanoparticle Synthesis

Samuel St. John, Naiping Hu, Dale W. Schaefer and **Anastasios Angelopoulos**, School of Energy, Environmental, Biological, and Medical Engineering, University of Cincinnati, Cincinnati, OH

The kinetics of metal synthesis at the onset of nanocrystal formation is of great interest due to reports of the unusual properties when matter is confined to the nanoscale. Particle structure in this critical size region impacts magnetics, electrocatalysis, optics, and heterogeneous synthesis. Despite significant progress regarding the characterization of size and shape of Pt nanocrystals, data concerning nucleation and growth has proven elusive. Current approaches relate equilibrated structures measured ex situ to initial synthesis parameters, missing the key nucleation and growth processes. Here we use real-time, in-situ small-angle and wide-angle x-ray scattering (SAXS and WAXS) and a Pt synthesis scheme with slower particle growth relative to the use of strong reducing agents to observe the formation of Pt critical nuclei on a length scale approaching 0.6 nm and with a dynamic resolution of 20 seconds. We correlate the autoreduction kinetics of precursor inorganic complexes with nanoparticle morphology, number density, polydispersity, and Pt volume fraction, thereby establishing the mechanism of particle size control at the onset of atomic coalescence. In addition, a hierarchical sub-structure is identified for the first time in the wide-angle scattering data related to the atomic structure of the growing Pt nanoparticles.

(338a) A Numerical Study of Granular Shear Flows of Rod-Like Particles Using the Discrete Element Method

Yu Guo, Chemical Engineering, University of Florida, GAINESVILLE, FL, Carl R. Wassgren, Mechanical Engineering, Purdue University, West Lafayette, IN, William R. Ketterhagen, Pfizer Global Research and Development, Pfizer Inc., Groton, CT, Bruno C. Hancock, Pharmaceutical Development, Pfizer Worldwide Research and Development, Groton, CT, Benjamin D. James, Chemical Engineering, University of Florida, Gainesville, FL and Jennifer Sinclair Curtis, University of Florida, Gainesville, FL

The effect of particle aspect ratio and surface geometry on granular flows is assessed by performing numerical simulations of rod-like particles in simple shear flows using the Discrete Element Method (DEM). The effect of particle surface geometry is explored by adopting two types of particles: glued-spheres particles and true cylindrical particles. Compared to frictionless spherical particles, smaller stresses are obtained for the glued-spheres and cylindrical particle systems in dilute and moderately dense flows due to the loss of translational energy, which is partially converted to rotational energy, for the non-spherical particles. For dilute granular flows of non-spherical particles, stresses are primarily affected by the particle aspect ratio rather than the surface geometry. As the particle aspect ratio increases, the effective particle projected area in the plane perpendicular to the flow direction increases, so that the probability of the occurrence of the particle collisions increases, leading to a reduction in particle velocity fluctuation and therefore a decrease in the stresses. Hence, a simple modification is made to the kinetic theory for granular flows to describe the stress tensors for dilute flows of non-spherical particles by incorporating a normalized effective particle projected area to account for the effect of particle collision probability. For dense granular flows, the stresses depend on both the particle aspect ratio and the surface geometry. Sharp stress increases at high solid volume fractions are observed for the glued-spheres particles with large aspect ratios due to the bumpy surfaces, which impede the flow. However, smaller stresses are obtained for the true cylindrical particles with large aspect ratios at high solid volume fractions. This trend is attributed to the combined effects of the smooth particle surfaces and the particle alignments such that the major/long axes of particles are aligned in the flow direction. In addition, the apparent friction coefficient, defined as the ratio of shear to normal stresses, is found to decrease as the particle aspect ratio increases and/or the particle surface becomes smoother at high solid volume fractions.

(338b) Experimental Validation of a Fast, 3D Physics-Based Cellular Automata Model of Granular Shear Flows

Martin C. Marinack Jr. and C. Fred Higgs III, Mechanical Engineering, Carnegie Mellon University, Pittsburgh, PA

Granular flows continue to be a complex problem in nature and industrial sectors where solid particles exhibit solid, liquid, and gaseous behavior in a manner which is often difficult to predict both locally and globally. In solids processing applications such as pharmaceutical production, food processing, and lawn fertilizer production, the ability to accurately predict particle flows can improve process efficiency and product effectiveness. The "gold standard" for the modeling and prediction of granular flows is the discrete element method (DEM), which provides a rigorous physical treatment of particle interactions in many granular systems. One possible supplement to DEM-based design or engineering analysis is lattice-based cellular automata (CA). The CA modeling framework employed by the authors provides a platform for obtaining fast predictions by employing both rule-based mathematics and high-fidelity physics to rapidly model physical processes, such as granular flows. Several prior physics-based CA models have been developed by the authors with each model's predictive capabilities being assessed by comparisons to experiments. The present work introduces a three-dimensional (3D) physics-based CA framework which aims to model all three granular flow regimes (kinetic, transitional, and frictional) effectively, include treatments for force chains, and increase the modeling accuracy of CA for granular flows. The predictive capabilities and quantitative accuracy of this CA model have been tested and validated against experimental shear flow results. Results from this work suggest that the model accurately predicts local flow properties. Parametric tests varying the particle-particle coefficient of restitution, macroscopic surface roughness, and global solid fraction have also been performed.

(338c) Slurry Flow Modeling Using Eulerian/Lagrangian CFD

Kuochen Tsai, Innovation R&D, Shell International E&P, Houston, TX

An attempt has been made to develop a general model for slurry flow transport in pipes using a Eulerian/Lagrangian CFD code. The main characeteristics to be modeled are the solid concentration profile and pressure drop. Similar efforts can be found with Eulerian/Eulerian approaches. However, several drawbacks hinders the extensive use of the Eulerian/Eulerian approach, namely, the difficulty in describing the particle size distribution (PSD) and the solution sensitivity to mesh resolution and numerical schemes. With the Eulerian/Lagrangian approach, the PSD can be naturally included and the solution is in general much less sensitive to numerical scheme and grid resolution. CPFD's Barracuda software is adopted for the modeling work in this study. Although Barracuda has been extensively validated for gas-solid flows, its applications to slurry flows have been limited. In this work, we attempt to devise a general model for solid-solid interactions that can be used to correctly account for solid concentration profiles and pressure drops under various pipe diameters, flow velocities, solid loadings and PSDs observed in many experiments. The developed model can be used for applications in many areas, including oil sand processing and hydraulic fracturing for oil extraction, and slurry catalyst cracking for heavy oil in the refineries.

(338d) Density and Velocity Distribution in Pneumatic Conveying System

Ankita Deshmukh and Manaswita Bose, Energy Science and Engineering, Indian Institute of Technology Bombay, Mumbai, India

Pneumatic conveying is one of the most preferred methods of solid transportation in industries handling large amount of particulate matter. Despite several advantages, pneumatic conveying system suffers from a few disadvantages including high specific power requirements, which is directly related to the inhomogeneity of the particulate flow in the system. As a general practice, pneumatic transport systems operate at a gas velocity much higher compared to the optimal value to ensure the homogeneity of the solid flow and causes high power consumption.

The superficial gas velocity at which the particles start settling at the bottom of the pipe, is defined as the "saltation velocity" [1]. It has been experimentally shown that for coarse particles the pressure gradient is lowest when the system operates at the saltation condition, however, it is also observed that the saltation occurs at a higher velocity in case of fine particles. It is, therefore, of interest to accurately predict the saltation condition. The onset of the inhomogeneity in pneumatic conveying systems, in other words the saltation condition, has been extensively studied in literature and there are several correlations that relate the saltation velocity to the solid conveying rate, properties of the particles such as size and density and those of the carrier gas such as specific gravity and viscosity. However, there has always been a need to develop fundamental understanding of the dynamics in pneumatically conveyed particles, especially during saltation.

In a recent series of articles published by the group of Prof. Kalman [2 and references therein], the influence of various fluid solid interaction on the saltation condition has been analysed. They have considered the fluid interaction on a single particle which is a good approximation for flows in very dilute regime. Kuan et al have reported the velocity and density profile of particles in a 90° bend obtained from their numerical simulations on dilute conveying system [3]. In both the works, interactions between particles have been neglected. However, the inter-particle interactions play crucial role in the pneumatic conveying system as in a typical operation, the Stokes number defined as the ratio of the particle relaxation time to the collision time, is much higher than unity.

In the present work, we have numerically investigated the velocity distribution of the particles pneumatically conveyed in a two dimensional channel where the inter-particle as well as wall-particles interactions are accounted for. The fluid phase is described by the continuum description of mass and momentum conservation, whereas the dynamics of solid phase are determined by solving the Newton's second law of motion for individual particles. Inter-particle as well as wall-particle interactions are modelled using the well known "spring dashpot" description, where the spring constant represents the stiffness of the material and is expressed in terms of modulus of elasticity. The viscous coefficient accounts for the dissipation of energy during collisions. The model equations are solved using the algorithms of computational fluid dynamics coupled with the discrete element method. Open source software MFIX [4] is used for the simulation.

Kinetic energy of the particle phase is monitored during the simulation. Data is collected and processed once the kinetic energy of the system has reached the equilibrium value. The density distribution of the particle phase is obtained by counting the number of the particles per unit volume of the collection bin at different channel heights and taking the average of the volume of the solids weighted based on the position of the particle in the bin. The influence of the ratio of the particle to fluid densities is then, investigated. The density of the particles, is observed to be higher close to the bottom wall which decreases monotonically along the height of the channel for the cases with moderate average number density of particles which are heavier than the carrier fluid. Also, string like clusters of particles are observed to form at the bottom of the channel. The density variation in the cross stream direction reduces as the system becomes more and more dilute. Simulations are extended for the neutrally buoyant particles as well. The modified Froude number (Fr), defined as $(V/\sqrt{(g\, d_p)}\sqrt{(\rho_p - \rho_f)}/\rho_f$, where V is the mean velocity of the fluid, D is the particle diameter, ρf is the fluid and ρP is the particle density respectively, is selected as the parameter to study the influence of gravity on the saltation condition, in addition to the density number defined as the ratio of the particle to the fluid density (ρ_P / ρ_f) . In the limit, where the Fr = 0, i.e., for the neutrally buoyant particles, the density profile has a maxima at the centre of the channel even when the system is not very dilute ruling out the possibility of saltation. However, in case of the neutrally buoyant particles, small clusters of particles are observed in the channel as opposed to the string like structures at the bottom wall, mentioned earlier.

The mean velocity of the solid phase is obtained by dividing the channel into several segments in the cross-stream direction and averaging the velocities of the individual particles. The velocity of the solid phase is observed to be nearly symmetric in case of neutrally buoyant particles. However, for particles heavier than the fluid, the mean velocity of the particle phase is higher near the top wall than the bottom of the channel.

We have next, obtained the stream-wise and the cross stream velocity distribution of the particles in the channel. The distribution functions are non-Gaussian in nature. Also, anisotropy in the "granular temperature" defined as the mean fluctuating kinetic energy of the particles, is observed. Non-Gaussian velocity distribution is obtained for neutrally buoyant but inelastic particles, as expected.

We have also determined the solid shear stress and the strain rate from the simulation. Influence of the particle properties such as the diameter and density of the particle on the transport coefficients, namely viscosity of the solid phase will be discussed during the meeting. The influence of the particle size on the onset of the clustering instability will also be presented at the conference.

Reference
[1] Clayton Crowe, ed. 2006 Multiphase flow handbook. CRC Taylor-Francis

[2] E Rabinovich , & H Kalman,2010 Phenomenological study of saltating motion of individual particles in horizontal particle-gas systems. Chem. Engng Sci. 65, 739–752
[3] K UAN , B, YANG , W & S CHWARZ , M P 2007 Dilute gas-solid two-phase flows in a curved 90◦ duct bend: CFD simulation with experimental validation. Chem. Engng Sci. 62, 2068–2088
[4] M Syamlal, W Rogers and T J O'Brien, 1993, MFIX Documentation Theory Guide, U.S. Department of Energy Office of Fossil Energy, Morgantown Energy Technology Center, Morgantown, West Virginia

(338e) Electrostatic Charging in Granular Flow

Daniel J. Lacks, Chemical Engineering, Case Western Reserve Unversity, Cleveland, OH and R. Mohan Sankaran, Chemical Engineering, Case Western Reserve University, Cleveland, OH

When granular materials flow, the particles rub against one another and become electrostatically charged. The electrostatic charge on the particles can greatly alter how the material flows; this can have important consequences, such as disrupting industrial operations. We use a combined experimental-theoretical approach to address the electrostatic charging in granular materials with a polydispersity in particle sizes. We have developed a methodology to investigate the electrostatic charging due strictly to particle-particle interaction, such that the particles being examined have no contact with any surface besides those of other particles. Our experiments show that under these conditions smaller particles tend to charge negatively and larger particles tend to charge positively. Results on a range of granular materials give the same result regarding the particle size dependence of electrostatic charging. We believe that this size dependence is universal (i.e., independent of material properties) in single component granular systems, and we describe a simple theory based on statistical mechanics to explain why this universality occurs. The implications of this bipolar charging on the flow of the material are discussed.

(338f) Quantitative Validation for Frictional Granular Flows

Adel F. Alenzi[1], Jiaming Cheng[1], Martin C. Marinack Jr.[2], C. F. Higgs III[2] and Joseph McCarthy[3], (1)Department of Chemical and Petroleum Engineering, University of Pittsburgh, Pittsburgh, PA, (2)Mechanical Engineering, Carnegie Mellon University, Pittsburgh, PA, (3)Chemical and Petroleum Engineering, University of Pittsburgh, Pittsburgh, PA

Granular flows in which shearing plays a key role are prevalent in natural and industrial flows and understanding their behavior and flow characteristics is of considerable importance. Because of their non-linear characteristics, and the general challenges in characterizing and describing multiphase granular flows, it has been difficult to develop models for and to simulate their flow behavior. Experimental studies have frequently been employed to develop a better understanding of their behavior and to both guide and develop theoretical and computational models for them. Building upon our previous validation studies of collisional granular flows, here we focus on flows where friction plays a more significant role. We quantitatively validate DEM simulations with experimental measurements obtained from an annular shear cell and hopper devices to understand the impact of different frictional force models and material properties used in our validation tests on granular shearing flows. In general, the frictional force models range from pragmatic linear techniques to rigorously more complex (nonlinear) contact mechanics inspired routines. Computationally, we examine both ends of the spectrum to compare with the experimental measurements.

(338g) Extending and Filling the Gaps in Solids Flow Theory

Kerry D. Johanson, Material Flow Solutions Inc., Gainesville, FL

In the early 1960's Andrew Jenike initiated a systematic study of the behavior of bulk materials in silos, bins and hoppers. Later in the 1970's Jerry Johanson added some rudimentary two-phase effects to these studies. The resulting theories have helped us understand the basics of bin and silo design and the behaviors of fine powders in typical silos and hoppers. However, these theories were limited by simplifying assumptions and did not extend to complex geometries, situations were external forces govern flow, or conditions where gas-solid interaction dominate behavior. This paper examines the relationship between limiting behaviors and theory explained by early solids flow practitioners and behavior due to two or three phase systems, as well as behavior in complex geometries. Specifically, the existing theories will be extended to include the effect of gas pressure gradients on mass flow, cohesive hang-ups, and rate limitations. This will help bridge the gap between fluidization and cohesive behavior and extend limit analysis to conditions where bulk solids interact with gas, providing a means of using creative gas injection systems to mitigate flow problems in cohesive materials. It will point a direction forward to extending existing fluid-solid systems to handle a wide range of materials. The paper will extend the limiting theories to geometries other than simple conical hoppers and silos. The limiting theory approach is general, but has only been widely applied to very simple geometries. This paper discusses how to relate this theoretical approach to complex hopper geometries, feed devices, and devices with mechanical interaction. It is hoped that extending this solid flow theory will allow the development of design equations for more complex geometries and extend the flow theories to two and three phase systems where engineers struggle with applying typical fluid laws only to discover that the cohesive nature of the system is not well suited to simple fluid behavior. Such an extension to theory will help produce the next generation of process flow vessels that are designed to handle the new cohesive materials currently being created.

(344a) Onset of Bubbling in Geldart Group A Powders

Clive E. Davies, School of Engineering and Advanced Technology, Massey University, Palmerston North 4442, New Zealand and Horng Yuan Saw, School of Engineering and Advanced Technology, Massey University, Palmerston North, New Zealand

In this work, we have tracked transitions in shallow beds aerated with incrementally increasing and decreasing gas flows, in an 80-mm column at ambient conditions. The test materials were six lactose powders, as well as one sample of glass ballotini and two samples of sand. The size and size distributions of the test powders were adjusted to provide powders that spanned the Group A region and part of the Group B region of a Geldart classification diagram. The ratio of loosely packed powder bed height to column diameter, H_i/D, was 1.35–1.80. Changes in bed height, H, bed surface and sides were observed visually. Three transition velocities were identified, *viz* U_{mf}, minimum fluidizing velocity, $U_{mb,v}$, visually determined minimum bubbling velocity, and U_{bv}, the velocity at which the bed was visually uniform, and vigorous bubbling began. Fluctuations in bed pressure drop, ΔP_b, were analyzed using standard deviation, σ, and von Neumann ratio, T. Composite plots of normalized ΔP_b, H, σ, and T versus normalized superficial velocity, U, were constructed to assess the trends exhibited by each bed parameter. Our experimental results and observations indicate that for both increasing and decreasing superficial velocities, there were break points -abrupt changes in slope- in plots of σ and T versus normalised U, at or close to superficial velocities consistent with the values of U_{mf}, $U_{mb,v}$ and U_{bv}. We discuss the relative merits of the use of each index, which may have utility in fluidized bed diagnostics in cases where direct observation of the bed is not feasible.

Effect of Internals and Solid Loading on Bubble Dynamics in a Slurry Bubble Column

Moses Kagumba[1], Yasser I. Abdulaziz[2] and Muthanna H. Al-Dahhan[*]
[1,]Chemical and Biological Engineering Department, Missouri University of Science and Technology, Rolla, MO 65409-1230. USA.*
[2]Chemical Engineering Department, Al-Nahrian University, Baghdad-Iraq.

Abstract

Slurry bubble columns have been used in wide range of applications in industry such as the Fisher-Tropsch synthesis and many others. Effects of solids loading in the presence of heat exchanging internals on the bubble dynamics are lacking in the open literature. Therefore the focus of this study is to investigate the effects of dense internals and solids loading on the bubble dynamics such as; specific interfacial area, local and overall gas hold-up, bubble velocity, bubble frequency and bubble chord lengths using four-point optical probe. The experimental work is carried out in 0.14 m diameter Plexiglas bubble column with air-water and air-water-glass beads (particle size of 150μm and density of 2500 kg/m^3). The superficial gas velocities based on free area for the flow were in the range of 0.03 to 0.45 m/s and solids loading were varied up to 40 % vol. The internals (0.5 inch diameter) occupied 25 % of column cross- section area.

The experimental work is carried out in 0.14 m diameter Plexiglas bubble column with air-water and air-water-glass beads (particle size of 150 μm and density of 2500 kg/m^3). The superficial gas velocities based on free area for the flow were in the range of 0.03 to 0.45 m/s and solids loading were varied up to 40 % vol. The internals (0.5 inch diameter) occupied 25 % of column cross- section area. The experimental data showed that the bubble frequency, local and overall gas hold-up decreases with increasing solids loading. Similar trend was also observed for gas-liquid interfacial area for all the studied superficial gas velocities. Moreover, there is no significant effect of solids loading on the bubble chord length in the churn turbulent flow regime. With internals and when the four-point optical probe was placed in the center of the column at the fully developed flow regime, it has also been found that, the addition of solids enhances the bubble sizes and velocity but lowers the bubble frequency, local gas holdup and specific interfacial area.

Keywords: *Bubble column; Internals; Heat exchanging tubes; Fisher-Tropsch synthesis.*

[*]*To whom correspondence should be addressed. Current address: Chemical & Biological Engineering Department, Missouri University of Science and Technology, Rolla, MO, 65409. Tel.: +1 5733417518. E-mail: aldahhanm@mst.edu.*

1. Introduction

Bubble column and slurry bubble column reactors are being extensively used in a wide variety of chemical and biochemical processes particularly in hydrogenation, oxidation, fermentation, petroleum refining, coal liquefaction, etc., where such reactors not only provide a significant interfacial mass transfer area but also very simple in design and no mechanical agitator is required (Azzopardi et al., 2011). However, most industrial multiphase reactors use internal heat transfer tubes to maintain the desired process temperature and for controlling the flow structures and back mixing in the column (Hulet et al., 2009). Moreover, the installation of internals are the most suitable solution since they provide both a reasonable ratio of reaction volume to heat transfer area and they preclude the need for either an external heat exchanger or a large and expensive slurry pump (Schlüter et al.,1995; and Dyer, 1985). Therefore, the design, scale-up and safe operation of slurry bubble column reactors require better knowledge and understanding of hydrodynamic characteristics.

To understand the performance of slurry bubble column reactors and the impact of internals, the keys factors of hydrodynamics and transport phenomena should be considered such as the gas holdup and the bubble properties (local bubble holdup, gas-liquid interfacial area, bubble frequency, bubble sizes, and bubble velocity). Many researchers attention to report studies that involved multiple vertical internals since fifties from the previous century, where in 1958, Kölbel and Ackermann patented a slurry reactor design for carrying out the Fischer-Tropsch (FT) process. There result was the elimination of the large vertical recirculation loops usually happened in empty bubble columns and the formation of stable slurry - gas suspension system with uniformly sized gas bubbles and a uniform rate of rise. Kölbel and Langheim, (1958) proposed reducing the cooling surface of the heat exchanger tubes in a step-wise manner from the bottom to the top of the reactor by combining two or more tubes together. The main results of the modifications were a higher CO conversion when performing FT synthesis, less methane formation, and an increase in the catalyst‘s performance.

Also hydrodynamic interaction and mass transfer in bubble column reactors were a focused subject for several researchers such as (Saxena et al., 1992; Pradhan et al., 1993; Saxena and Rao, 1993; Saxena and Chen, 1994; Chen et al., 1999 and De et al., 1999). In the last decade, several well-known researchers have made very useful publication concerning the hydrodynamic properties in slurry bubble column reactors such as (Forret et al., 2003; Larachi et al., 2006; Balamurugan and Subbarao, 2006; Youssef and Al-Dahhan, 2009). Recently, Zhang et al., (2010) studied the gas holdup and bubble dynamics by using the double-sensor conductivity probe technique in external loop downcomer slurry bubble column reactor in which the gas is distributed at the bottom and the slurry is circulated through over the column in which consists of three sections, spray section, sieve plates section, and internal loop section. There results show that gas holdup is greatly affected by the geometrical configuration and the effects of solid loading and external slurry circulation velocity are less significant than that of superficial gas velocity. Both local bubble size and bubble rise velocity vary significantly in different regions. Moreover, Al-Oufi et al., (2011) investigated the radial profiles of the local void fraction using a four-point conductivity probe in open tube and concentric annular gap bubble column. They use different inner tube diameter ratios from 0.25 to 0.69 in a system of air – tap water and aqueous solutions of ethanol and isopropanol with different concentration. They found that presence of alcohol inhibited the coalescence between the bubbles and consequently increased the mean gas void fraction at a given gas superficial velocity in both the open tube and the annular gap bubble columns. Moreover, isopropanol results gave slightly higher mean void fractions compared to those for ethanol at the same mass fraction, due to their increased carbon chain length.

It is also worth mentioned that the void fraction profiles in the annular gap bubble column were far from uniform, leading to lower mean void fractions than were obtained in an open tube for the same gas superficial velocity and liquid composition.

The purpose of the current contribution is to extend the understanding of hydrodynamic characteristics affected by presence of solids and high ratio of internals in bubble column reactors mimics the industrial working conditions.

2. Experimental setup

The experiments are conducted in a Plexiglas column of 0.14 m inside diameter and 1.8 m height. A schematic diagram of the experimental setup is shown in Fig. 1. In this study, compressed dry air introduced continuously from the bottom of the column as the gas phase. Soft filtered tap water was used as liquid phase in batch mode. Glass beads with an average size of 150μm and a density of 2500kg/m^2 was used as the solids/fines phase. The solids loading was based on the wet volume and the concentrations varied between (0 – 40) % vol. The experiments were carried out over a range of superficial gas velocities, from 0.03 to 0.45 cm/s based on the free cross-sectional area (CSA) of the column. The internals used in this study were vertical 30 Plexiglas rods of 0.5 inch diameter and 1.8 m high covering 0 %, (no internals) simulating a jacket, and 25 % of the column cross-sectional area that simulates the Fischer-Tropsch. In all experiments, the dynamic height of the bed was maintained constant at a level of 1.56 m (z/D = 11.16) above the gas distributor by adjusting the amount of liquid-solid loaded in the column. A perforated plate was used as the gas distributor with 121 holes of 1.32 mm diameter, arranged in a triangular row pitch, with a total free area of 1.09 %.

Four points fiber optical probe, originally developed, and utilized by Frijlink, (1987) at the Delft University in the Netherlands and advanced by Xue (2004) was used for the bubble dynamics measurements. The probe was mounted in the fully developed flow region at a height of 0.75 m above the gas distributor. Where, the probe tips are facing downward, yielding bubble property measurements for bubbles moving upward.

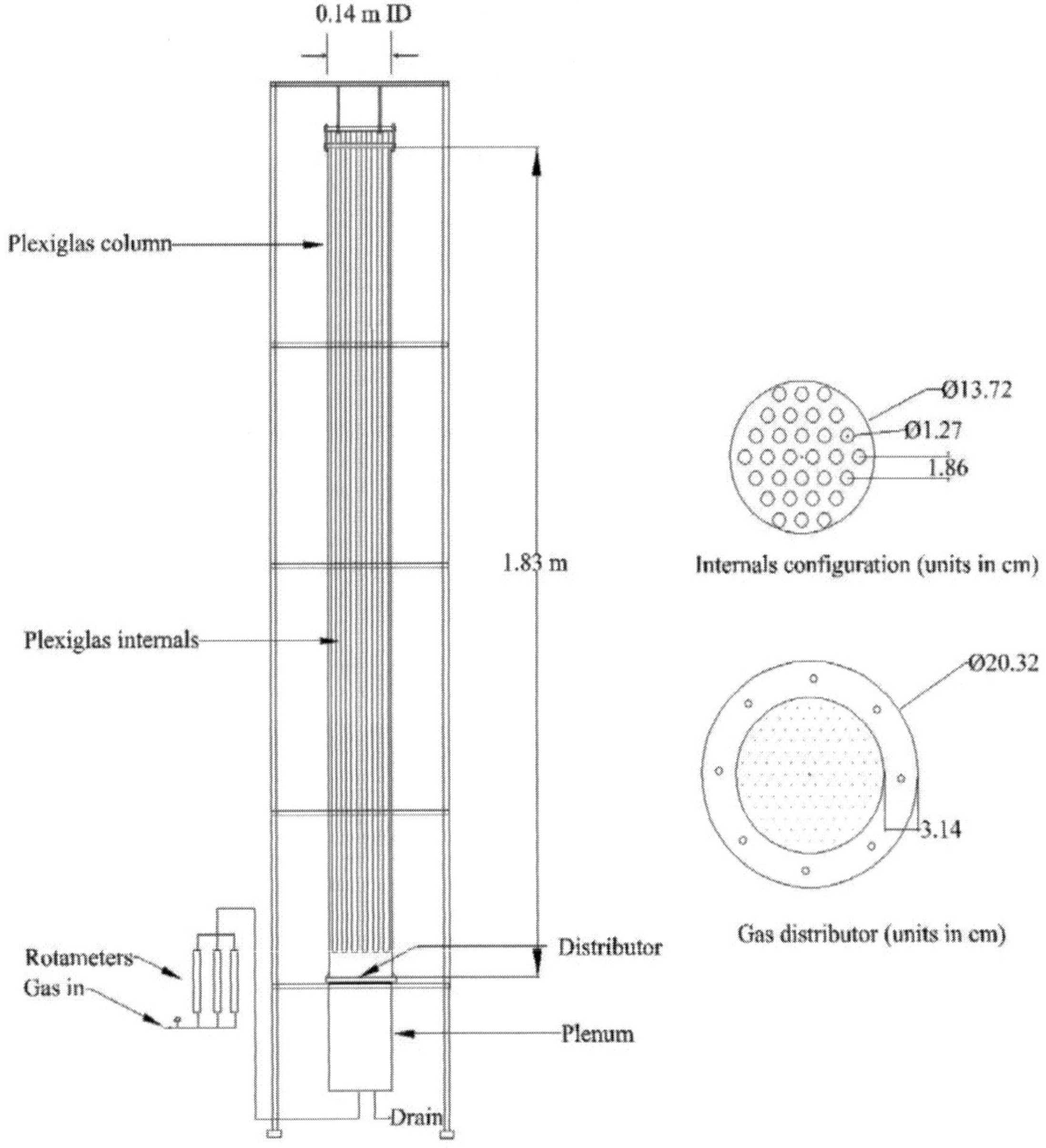

Figure 1. Schematic of the experimental structure and mimicked heat exchanging internals configurations and the gas distributor used.

3. Results and Discussion

3.1. Local gas holdup

Figure (2) illustrates the effect of solids loading on local gas holdup in presence and absence of internals for different gas velocities in bubble column. It is clearly noticed that the gas holdup is significantly increased with increasing superficial gas velocity for all solids concentration. This increase is due to increment in number of bubbles with increasing gas flow rate as determined by Youssef and Al-Dahhan, (2009) and Chen et al. (1999). On the other hand, it was found that the local gas holdup decreases with increase in the solids loading. This can be attributed to increased pseudo-slurry viscosity, which promotes coalescence of large bubbles (Crabtree and Bridgewater, 1971, Li and Prakash 1997), whereas the bubble break-up rate decreases due to dampening of instabilities at bubble-liquid interface. In addition, the possibility of formation of smaller bubbles which lead to the increase in the rise velocity and reduce the residence time of the bubbles as a result the gas holdup would be reduced (Kara et al., 1982; Koide et al., 1984; Li and Prakash, 1987 and Saxena et al., 1989).

The presence of dense internals, that occupy 25 % of the column CSA, leads to general increase in the local gas hold-up regardless of the solids loading. This trend is due to the enhancement of bubble brake up which yields relatively smaller bubbles, with higher bubble passage frequency and lower velocity hence higher residence time in the column and enhanced gas holdup. Similar reasoning was also advanced by Youssef and Al-Dahhan, (2009). However the effect of internals on gas holdup starts to diminish at gas velocities ≥30 cm/s.

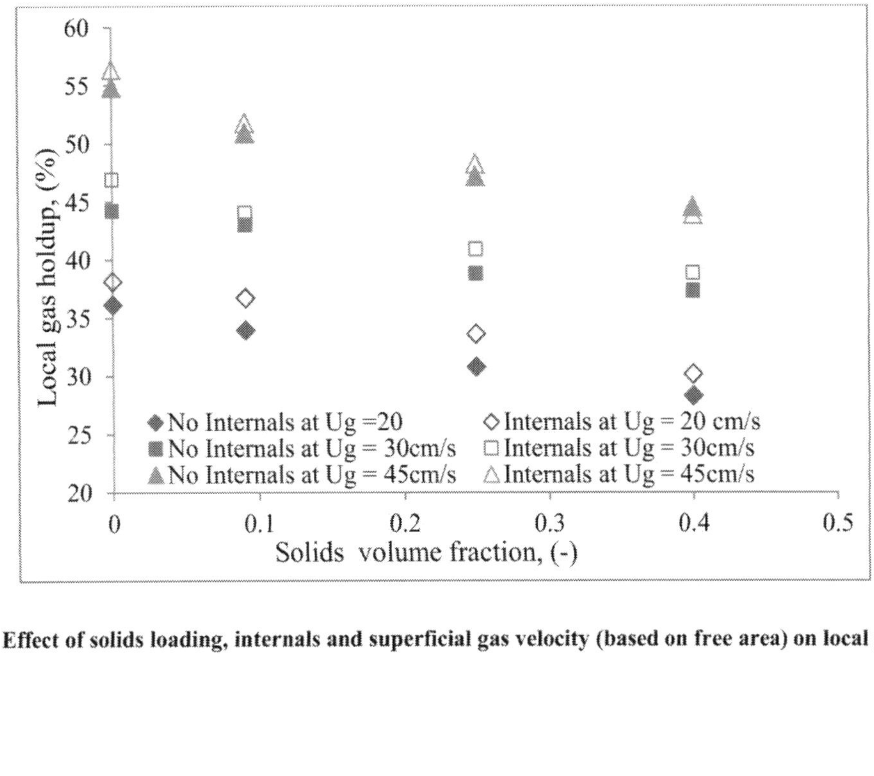

Figure 2. Effect of solids loading, internals and superficial gas velocity (based on free area) on local gas holdup.

3.2. Specific Interfacial area and bubble passage frequency

The influence of solids loading on specific interfacial area at different gas velocities with and without internals is distinctly shown in Figure (3). It is obviously noticed that specific interfacial area decreases with increasing solids concentration. This is due to an increase in bubble coalescence ,where larger bubbles have been formed, which leads to decreased specific interfacial area per unit volume hence the total interfacial area decreases too (Zahradnick et al., 1992). It is noteworthy that increasing the superficial gas velocity leads to dramatic increase in the bubble coalescence rate. But the breakup rate of these are also significantly high thus the population of smaller bubbles increase faster than those of large bubbles. Hence the large coalesced bubbles have a lower interfacial area per unit volume, the significant increase in the population of smaller bubbles which have higher interfacial area resulted in an increase in the average specific interfacial area.

The presence of dense internals was found to increase the specific interfacial area for all the gas velocities and solids loading. This could be attributed to the reduction in bubble sizes, and restricted formation of larger bubbles. Hence, many smaller bubbles result which have a higher specific interfacial area per unit volume.

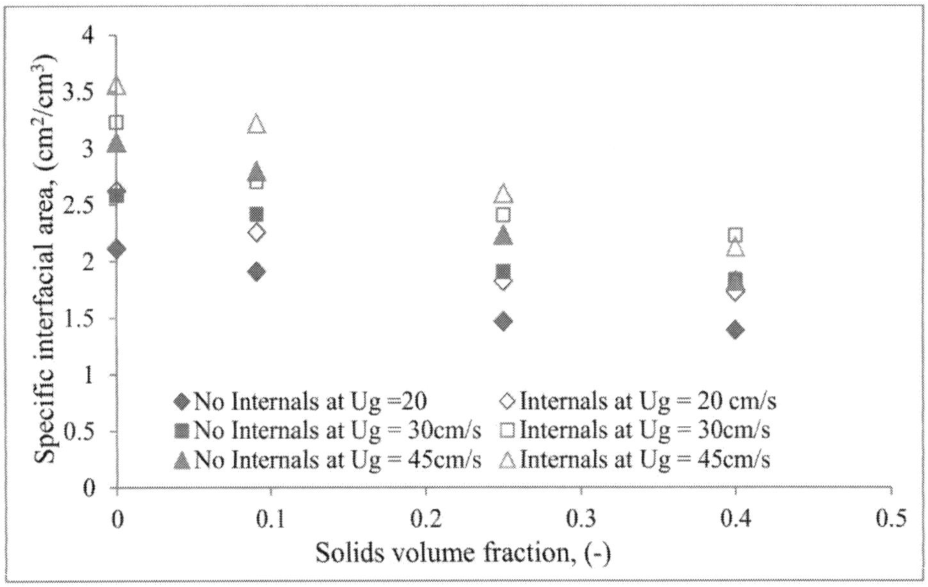

Figure 3. Effect of solids loading, internals and superficial gas velocity (based on free area) on specific interfacial area.

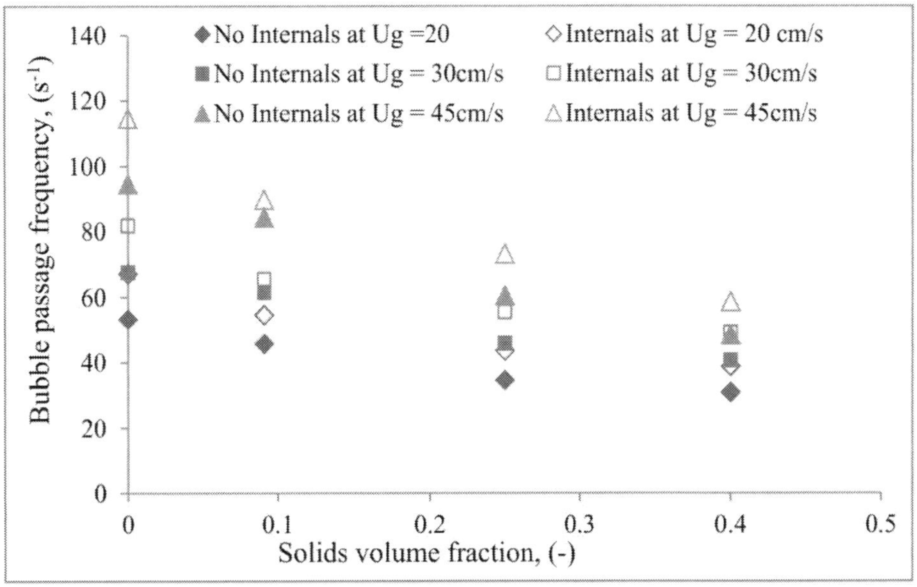

Figure 4. Effect of solids loading, internals and superficial gas velocity (based on free area) on specific interfacial area.

3.3. Axial bubble velocity

The bubble velocity u_b, at any given location depends on mainly two factors: the local liquid velocity, u_l and the local slip velocity, u_s at that location.

$$u_b = u_l + u_s$$

Based on the momentum equation, Gupta (2002) showed that the slip velocity u_s is a function of drag, bubble diameter, axial pressure drop, local gas holdup, and gas and liquid physicochemical properties and hence does the bubble velocity.

As illustrated in figure 5, the increase in solids loading increases the bubble velocity in both bubbly and churn turbulent flow regimes. Similar trend was observed by Li and Prakash, (1997). This increment is attributed to change in the slurry properties, where increased pseudo-slurry viscosity promotes the bubble coalescence (Crabtree and Bridgewater, 1971) and hence increase in the bubble sizes and higher U_b will be achieved. Our results are in agreement with that reported by Luo et al. (1999), where they measured the bubble size distribution in a slurry bubble column by an optic probe under ambient pressure.

It is also, worth to note that the presence of dense internals restrict the formation of larger bubbles and enhances their breakup rate resulting into smaller bubbles with relatively lower axial bubble velocity, particularly in the churn turbulent flow regime. A similar argument was advanced by (Youssef and Al-Dahhan 2009) in their two phase system with 22 % CSA internals.

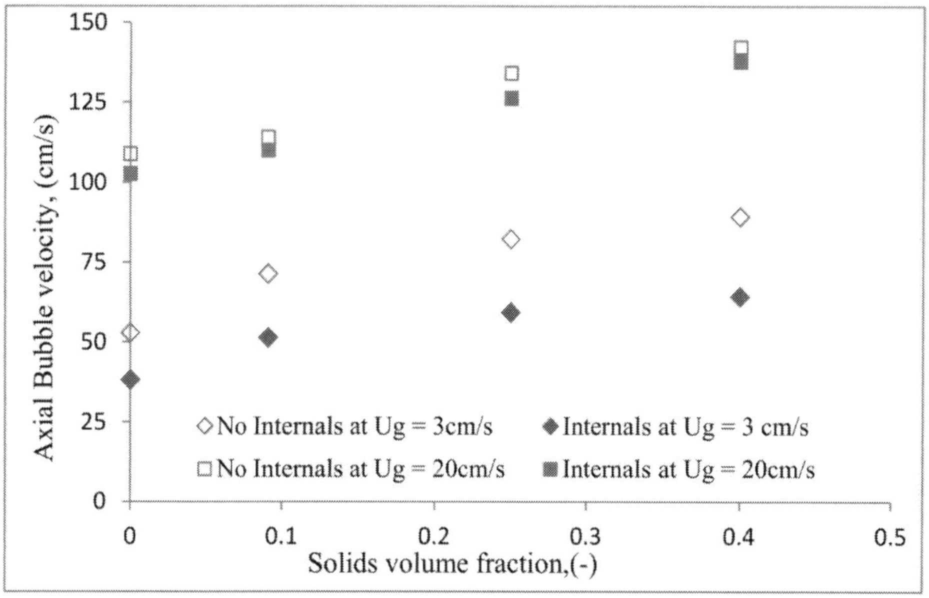

Figure 5. Effect of internals and superficial gas velocity on axial bubble velocity.

3.4. Bubble Chord Length

The bubble chord lengths are characteristic of bubble sizes. Figure 6 shows the distribution of the measured bubble chord lengths. The distribution takes on log-normal for the studied gas velocities and solids loading. Similar trends were reported by Youssef and Al-Dahhan 2009; Poherecki et al 2001 among others. Increase in solids loading increases pseudo-slurry viscosity, which promotes coalescence of large bubbles (Crabtree and Bridgewater, 1971), whereas the bubble break-up rate decreases due to dampening of instabilities at bubble-liquid interface. The addition of solids has greater effect on the chord lengths at lower gas velocity. The inclusion of dense (25 % CSA) internals does not have significant effect on the bubble chord lengths in the bubble flow regime.

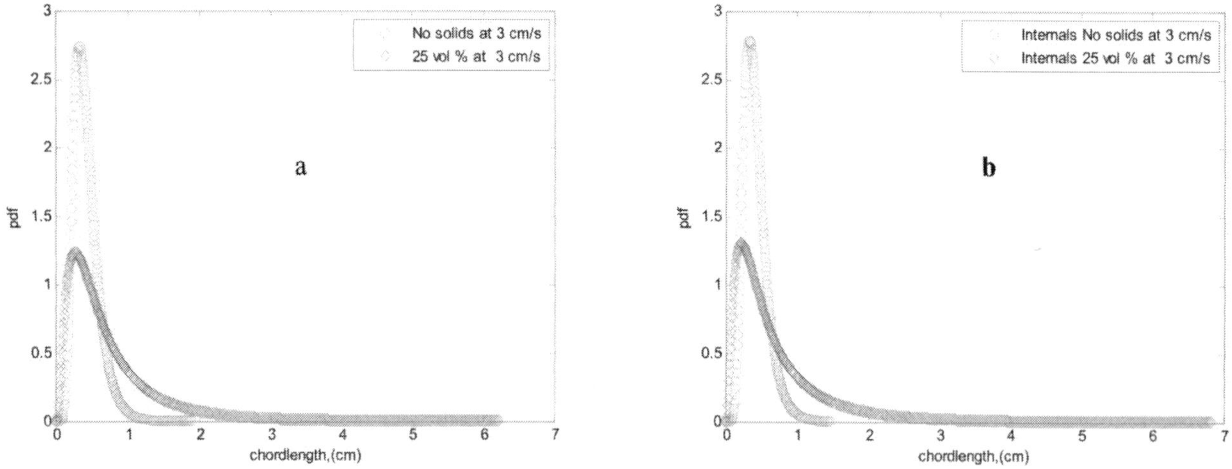

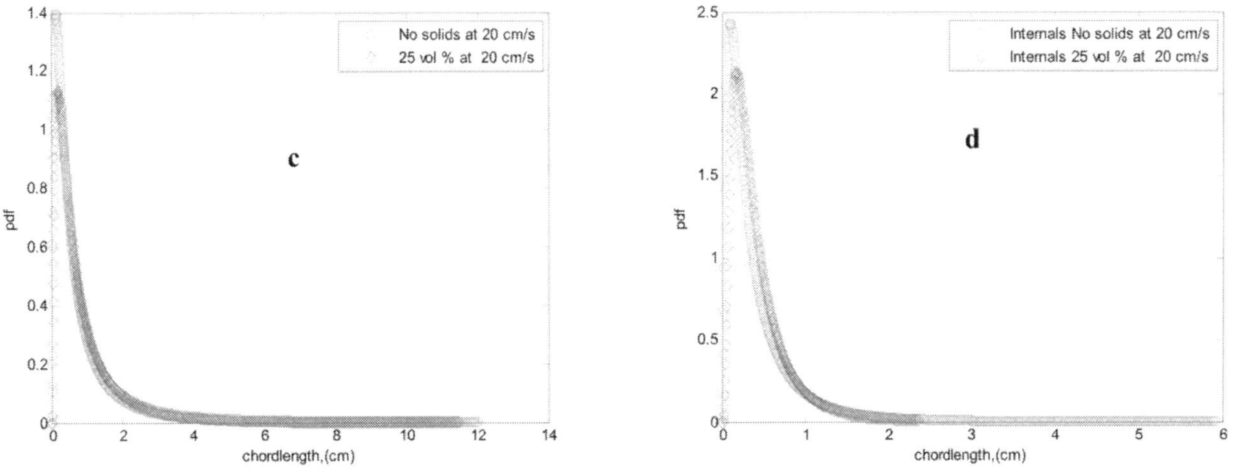

Figure 6. Bubble chord length distribution at column center r/R(-)=0 (a). Empty column at U_g = 3 cm/s . (b) 25 % CSA Internals at U_g = 3 cm/s. (c) Empty column at U_g = 20 cm/s (d)) 25 % CSA Internals at U_g = 20 cm/s.

Remarks:

The effect of internals and solids loading on bubble dynamics were examined for the first time in six inch bubble column using four points optical probe. The solids loading was found to enhance bubble sizes and axial bubble velocity in the bubble column, while the local gas hold-up, bubble passage frequency and specific interfacial area reduced. The effect of solids was more pronounced in the bubble flow regime. The presence of internals enhanced local gas holdup, bubble passage frequency and specific interfacial area in the bubble column with and without solids but lead to a decrease in the bubble velocity and the bubble sizes for all the solids loading used.

References

1. Al-Oufi, F.M.; Rielly, C.D.; Cumming, I.W. An experimental study of gas void fraction in dilute alcohol solutions in annular gap bubble columns using a four-point conductivity probe. Chem. Eng. Sci. 66, 5739–5748 (2011).
2. Azzopardi, B.J.; Mudde, R.F.; Lo, S.; Morvan, H.; Yan, Y.; Zhao, D. Hydrodynamics of Gas – Liquid Reactors. John Wiley & Sons, Ltd. 1st published 2011.
3. Balamurugan, V. and Subbaro, D. bubble size and holdup in bubble columns with vibrating internals, AIChE Spring National Meeting (2006).
4. Chen, J., Li, F., Degaleesan, S., Gupta, P., Al-Dahhan, M. H., Dudukovic, M. P. and Toseland, B. A., fluid Dynamic Parameters in Bubble Columns with Internals, *Chem. Eng. Sci.*, 54 (13-14), 2187-2197 (1999).
5. Chen, J.; Li, F.; Degaleesan, S.; Gupta, P.; Al-Dahhan, M. H.; Dudukovic, M. P.; Toseland, B. A. Fluid Dynamic Parameters in Bubble Columns with Internals. *Chem. Eng. Sci.* 1999, *54* (13-14), 2187–2197.

6. De, S. K., Ghosh, S., Parichha, R. K. and De, P., Gas hold-up in two phase system with internals, Indian Chem. Engr. Section A, 41(2), T54-T58 (1999).

7. Forret, A., Schweitzer, J.M., Gauthier, T., Krishna, R., Schweich, D., Liquid dispersion in large diameter bubble columns, with and without internals, the Canadian Journal of Chemical Engineering 81, 360–366 (2003).

8. Frijlink, J. J. Physical Aspects of Gassed Suspension Reactors, Ph.D. Thesis, Delft University of Technology, the Netherlands, 1987.

9. Hulet,C.; Clement,P.; Tochon, P.; Schweich,D.; Dromard,N.; Anfray, J. Literature Review on Heat Transfer in Two and Three-Phase Bubble Columns. *Int. J. Chem. React. Eng.* 2009, *7*, R1.

10. Kara S.; Kelkar, B.G.; Shah, Y.T. *Hydrodynamics and axial mixing in a three-phase bubble column. Ind. Chem. Eng. Process Des. Dev.* 1982, *21,* 584.

11. Koide, K.; Takazawa, A.; Komura, M.; Matsunga, H. Gas holdup and volumetric liquid-phase mass transfer coefficient in solid – suspended bubble columns. *J. Chem. Eng. Jap.* 1984, *17*, 459.

12. Kölbel, H. and Ackermann, P., US Patent 2,853,369 (Sept. 23, 1958).

13. Kölbel, H. and Langheim, R., US Patent 2,852,350 (Sept. 16, 1958).

14. Larachi, F., Desvigne, D., Donnat, L., Schweich, D., simulating the effects of liquid circulation in bubble columns with internals, Chemical Engineering Science, 61(13), 4195-4206 (2006).

15. Li, H.; Prakash, A. Heat transfer and Hydrodynamics in a three-phase bubble column. *Ind. Eng. Chem. Res.* 1997, *36*, 4688.

16. Maretto, C.; Krishna, R., Design and optimisation of a multi-stage bubble column slurry reactor for Fischer-Tropsch synthesis, Catalysis Today, 66(2-4), 241-248 (2001).

17. Pradhan A., K., Parichia, R., K. and De, P., Gas hold-up in non Newtonian solutions in a bubble column with internals, The Canadian journal of chemical engineering, 71, 468-471 (1993).

18. Saxena, S. C.; Rao, N. S.; and Thimmapuram, P. R., Gas phase holdup in slurry bubble columns for two- and three-phase systems, Chemical Engineering Journal (Amsterdam, Netherlands), 49(3), 151-9 (1992).

19. Saxena, S. C.; Vadivel, R.; Saxena, A. C. Gas-holdup and heat transfer from immersed surfaces in two and three phase system in bubble column. *Chem. Eng. Comm.* 1989, *85*,63.

20. Saxena, S.C., Rao, N. S., Estimation of gas holdup in a slurry bubble column with internals: nitrogen–therminol–magnetite system. Powder Technology 75, 153–158 (1993).

21. Saxena, S.C.; Chen, Z.D., Hydrodynamics and heat transfer of baffled and unbaffled slurry bubble columns, Reviews in Chemical Engineering, 10(3-4), 193-400 (1994).

22. Schlüter, S.; Steiff, A.; Weinspach, P.-M., Heat transfer in two- and three-phase bubble column reactors with internals. *Chemical Engineering and Processing*, 34(3), 157-72 (1995).

23. Xue, J. Bubble velocity, size and interfacial area measurements in bubble columns, Ph.D. Thesis, Washington University, Saint Louis, MO, 2004.

24. Xue, J.; Al-Dahhan, M.; Dudukovic, M. P.; Mudde, R. F. Bubble dynamics measurements using four-point optical probe. *Can. J. Chem. Eng.* 2003, *81*, 1–7.

25. Xue, J.; Al-Dahhan, M.; Dudukovic, M. P.; Mudde, R. F. Bubble velocity, size, and interfacial area measurements in a bubble column by four-point optical probe. *AIChE J.* 2008, *54* (2), 350 363.

26. Xue, J.; Al-Dahhan, M.; Dudukovic, M. P.; Mudde, R., F. Fourpoint optical probe for measurement of bubble dynamics: Validation of the technique. *Flow Measure. Instrument.* 2008, *19* (5), 293–300.

27. Youssef Ahmed A.; Al-Dahhan M.H.; Impact of Internals on the Gas Holdup and Bubble Properties of a Bubble Column. *Ind. Eng. Chem. Res.* 2009, 48, 8007–8013

28. Youssef, A. A. and Al-Dahhan, M. H. Impact of Internals on the Gas Holdup and Bubble Properties of a Bubble Column. Industrial & Engineering Chemistry Research. 48(17), 8007-8013 (2009).

29. Zhang, K.; Qi, N.; Jin J., Lu C.; Zhang H. Gas holdup and bubble dynamics in a three-phase internal loop reactor with external slurry circulation. Fuel. 89 1361–1369 (2010).

(344c) Minimum Fluidization Velocity in a 3D Fluidized Bed Modified with an Acoustic Field

David R. Escudero and **Theodore J. Heindel**, Department of Mechanical Engineering, Iowa State University, Ames, IA

Fluidized beds are used in a variety of process industries because they provide uniform temperature distributions, low pressure drops, and high heat/mass rates. Minimum fluidization velocity is an important factor in understanding the hydrodynamic behavior of fluidized beds, and this characteristic may be modified through high frequency (sound) vibrations. The effects caused by sound wave frequency on the minimum fluidization velocity in a 3D fluidized bed are investigated in this study. Experiments are carried out in a 10.2 cm ID cold flow fluidized bed filled with either glass beads or ground walnut shell, and particle size ranged between 212-600 microns. In this study, four different bed height-to-diameter ratios are examined: H/D = 0.5, 1, 1.5, and 2. Additionally, the sound frequency of the loudspeaker used as the acoustic source ranged between 50-200 Hz, and the sound pressure level ranged 80-130 dB. Results show that the minimum fluidization velocity is influenced by the frequency change. As the frequency increases, the minimum fluidization velocity decreases until a specific frequency is reached, beyond which the minimum fluidization velocity increases. In the same sense, as sound pressure level increases, the minimum fluidization velocity decreases because the additional acoustic energy imparted to the bed helps to loosen the bed and reduces the required energy for the particles to overcome gravity forces. Thus, acoustic fields provide an improvement in the ease of fluidization of these particles. Our next step is to assess the effects of the acoustic field on the bed flow structures using 3D X-ray computed tomography imaging.

PHOSPHORESCENT TRACING TO STUDY SOLIDS MIXING IN FLUIDIZED BEDS

Mike Wormsbecker, Syncrude Canada Ltd., Edmonton, AB, Canada
Craig McKnight, Syncrude Canada Ltd., Edmonton, AB, Canada
Andrew Mezo, Coanda R&D Corporation, Burnaby, BC, Canada
Jeff Mottershead, Coanda R&D Corporation, Burnaby, BC, Canada
Darwin Kiel, Coanda R&D Corporation, Burnaby, BC, Canada
Xiaotao Bi, University of British Columbia, Vancouver, BC, Canada
John Grace, University of British Columbia, Vancouver, BC, Canada

Abstract

A phosphorescent solids tracer technique has been designed to study solids mixing in a pressurized circulating fluidized bed (CFB) model of a FLUID COKING™ reactor. Using pure phosphorescent pigment as a tracer, an external fluidized bed injector activates and then injects a pulse of tracer into the riser. This provided a quick, repeatable, and well-distributed injection of tracer into the top of the fluidized bed. The light emitted from the tracer particles is collected at various locations in the fluidized bed and standpipe using fiber optic probes. Measurements from the tracer tests demonstrate that both reactor solids break-through times and standpipe time-of-flight can be obtained from the solids residence time distributions (RTDs). Leading-edge times from the solids RTD quantify the extent of solids by-passing through the reactor. The by-passing measured corresponds to the bed turnover time predicted by the correlation of Baeyens and Geldart ("Solids Mixing," in *Gas Fluidization Technology*, 1986). This illustrates the accuracy of the tracer technique to measure leading-edge times. Standpipe time-of-flight data are correlated against the riser constriction pressure drop to provide a calibration for external solids circulation.

Introduction

FLUID COKING™ is a process where high molecular weight hydrocarbon is thermally cracked to lower boiling point hydrocarbons, coke, and gases through contact with hot coke solids in a fluidized bed. The spent reactor fluid coke is transferred to a Burner where it reacts with air to generate heat. Hot coke from the burner is transferred back to the reactor to satisfy the heat balance. For a more detailed description of the process, see Matsen (2).

A result of the FLUID COKING™ process is high-velocity fluidization conditions in the reactor. Steam used in the stripping, attrition and feed atomization processes coupled with the generation of cracked hydrocarbon products create turbulent fluidization conditions. Although this is advantageous from a perspective of solids mixing, solids by-pass to the stripper section. This solids by-passing has a detrimental effect on downstream operation in the form of product loss due to incomplete reaction and creates the potential for downstream fouling. Consequently, managing the extent of solids by-passing is critical to reactor performance (3).

The state-of-the-art in FLUID COKING™ reactor hydrodynamics is captured through the works of Matsen (2) and Song et al. (4). Matsen illustrates commercial reactor solids mixing patterns (2), whereas Song et al. (4) designed a fully cylindrical cold flow reactor model to study solids voidage and velocity profiles in the reactor. Similar to Matsen (2), Song et al. (4) found a relatively dense annular region of solids downward flow surrounding a more dilute upward flow of solids in the core region over a time-averaged period. These studies provide detailed information of both local and global hydrodynamics pertinent to FLUID COKING™. The addition of solids residence time distributions (RTDs) will provide further insight into the extent of solids by-passing.

In order to measure the characteristics of a solids RTD, a suitable solids tracer technique is required. Ultimately, phosphorescent solids tracing was chosen because it allows multiple tracer tests to be performed, which is necessary to characterize the by-passing phenomenon. The objective of this paper is to present the methodology used to design the phosphorescent solids tracer technique to study solids mixing in a FLUID COKING™ reactor model. Solids RTD data from the reactor and standpipe are presented in this paper to illustrate the effectiveness of the technique.

Experimental Design

Tracer Technique Selection

Phosphorescent materials absorb long-wave ultraviolet rays (UVA/UVB) and re-emit the photons in the form of visible light. Fluorescence materials exhibit the same photon transformation; however, phosphorescence differs from fluorescence in the duration at which it emits light. Phosphorescent material goes through a unique intersystem crossing phenomenon resulting in singlet and triplet excited energy states. The singlet energy state returns to its ground state in a similar time scale to that of fluorescent material (i.e. nanoseconds). However, its triplet energy state is much more stable. For some materials, it can take up to 24 hours to return to its ground state.

The main advantage of phosphorescence over other tracers is this afterglow phenomenon. Phosphorescent tracers can be left after injection without generating appreciable signal build-up since the intensity of the emitted light decays over time. The same cannot be said when using fluorescent or salt tracers (5-8). These techniques suffer from signal build-up, which manifests into background noise during subsequent tests. As a result, higher and higher doses of tracer or fresh bed inventories are needed to capture RTDs, especially its leading-edge. Secondly, light-based tracers allow for real-time data capture with optical probes. This is conducive to executing a multiple tracer tests to characterize the solids RTD with a high degree of certainty. Off-line tracer techniques, such as salt tracers, require time-consuming sample collection and preparation techniques. As a result, fewer tests are typically executed.

There is a wide array of other solids tracer techniques not discussed here. Techniques including colour, chemical, thermal and radioactive were all considered during the tracer selection process. For a thorough review of these, and other, solids tracer techniques, see Harris et al. (9).

Tracer Material Selection

Afterglow brightness is the critical characteristic of a phosphorescent tracer. There are two main categories of phosphorescent materials - $SrAl_2O_4$ and ZnS based. Typically, $SrAl_2O_4$ pigments are an order of magnitude brighter than their counterpart. As a result, PLC SP-2A, a $SrAl_2O_4$ tracer was sourced (Pete's Luminous Creations, Singapore). Table 1 compares dimensionless scaling parameters for PLC SP-2A and FCC catalyst, the particle type used in the model. The approach taken was to best match the tracer to FCC catalyst through the Archimedes number. Although the Froude number and gas-to-solids density ratio are important for hydrodynamic scaling, it is impossible to scale all dimensionless groups with respect to the tracer. As a result, a reasonable degree of relaxation was accepted. Phosphorescent-doped polyethylene microspheres with a density of 1100 kg/m³ were created to overcome some of these mismatches (Cospheric LLC, USA). Although they did not produce as much light when compared to pure pigments, the results compared well to the pure pigment tracer. This suggests that these tracer density differences were not critical to tracer results for the operating conditions studied.

Table 1. Dimensionless parameters for hydrodynamic scaling

Particle	D_{50} (microns)	Particle density (kg/m^3)	Archimedes number	Particle Froude number	Solids to gas density ratio
FCC catalyst	100	1700	120	528	1538
PLC SP-2A	84	3600	151	629	726

Tracer Injector

A tracer injector is required to get the phosphorescent tracer into the reactor model. It was decided that the injector needs to:

1. Sufficiently activate the phosphorescent tracer.
2. Introduce a pulse injection of solids (or as close to a pulse as possible).
3. Disperse tracer evenly across the top of the reactor dense bed.
4. Operate at elevated pressure to exceed the nominal reactor pressure of 20 psig

It was determined that a pressurized fluid bed injector, similar to the design of Kirbas (10), was required. The main body of the injector used in this study is a 2-in ID borosilicate sight glass with 360º viewing access (Jogler Inc., USA). The sight glass is surrounded by six 4-ft 40W UV fluorescent lights positioned parallel to the sight glass wall. The UV light, which penetrates through the borosilicate glass, energizes the tracer particles near the glass wall. Pressurized air is brought into the system through the freeboard with hard-piped tubing. The tubing extends to the bottom of the fluidized bed (top of the ball valve that isolates the injector from the J-valve). This air is used to fluidize the solids in the injector which in turn keeps them well mixed and therefore exposes all the injector solids to the UV light. The mixing in the fluidized bed was determined to be optimum at a superficial gas velocity 0.5 m/s.

The glass-walled injector section can be pressurized up to 100-psig. This over-pressuring of solids provides the driving force for solids injection. The injector is operated at 50-psig; the system pressure is ~ 20-psig at the bottom of the standpipe. Exhaust gas from the injector passes through an expanded freeboard section (6-in ID) in order to reduce solids carry-over. A schematic of the detailed design of the injector and a P&ID are shown in Figs.1 and 2, respectively.

A single tracer injection consists of loading 200-g of tracer and 300-g of FCC catalyst into the injector. Dilution with FCC catalyst increases the solids volume in the tracer activation zone of the injector. If dilution was not used, the pure pigment would occupy a volume comparable to the volume of the zone between the sight glass flange and the 2-in ball valve. Here, UV light would not activate the particles. With a high fluidization velocity, solids mixing can minimize the impact of this region.

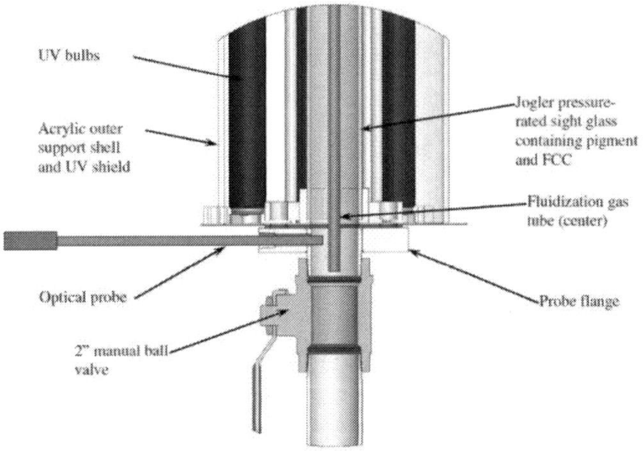

Figure 1. Design of tracer injector

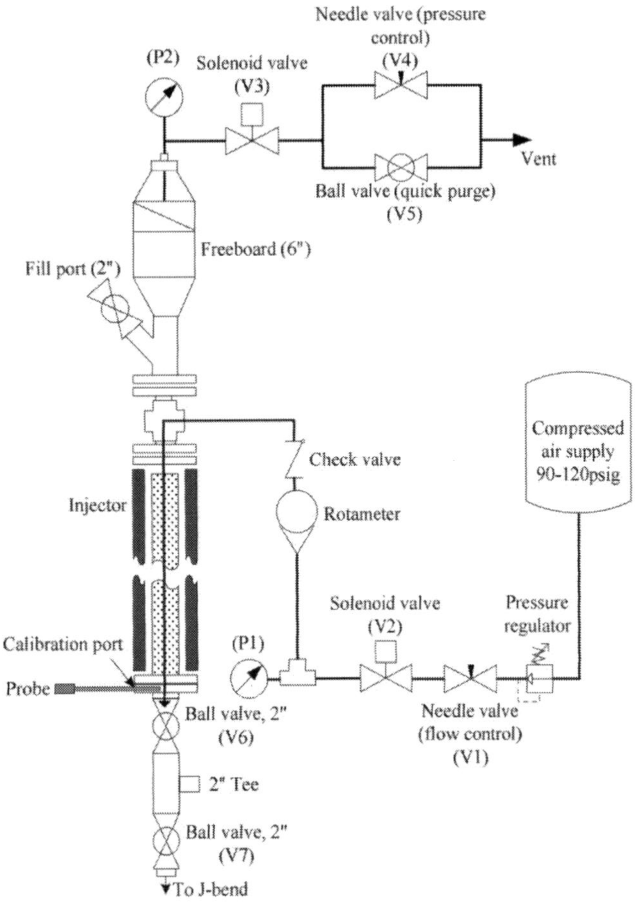

Figure 2. Tracer injector P&ID

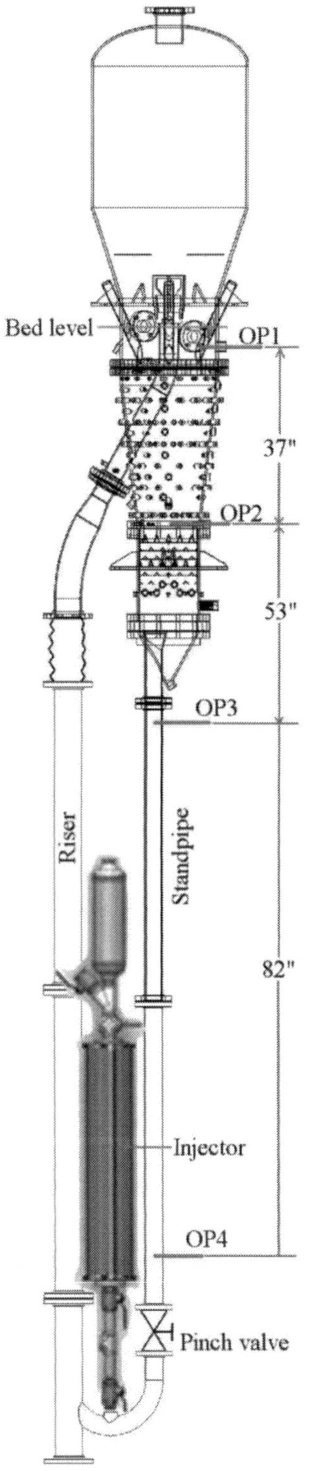

Figure 3. Optic probe location

Fiber optic probes

The fiber optic probes were designed by Coanda R&D Corporation. Each probe consists of a 27-in long glass rod that is optically potted inside a ½-in NPT stainless steel tube. The optical rod is coupled to a photomultiplier, which measures the emitted tracer light. These probes are very sensitive to low levels of visible light due to the large diameter fiber and photomultiplier assembly. As a result, they are ideal for detecting the tracer afterglow at low concentrations. The photomultipliers connect to circuitry boxes that have signal conditioning capabilities.

Four fiber optic probes (OP) were placed strategically around the model: top-of-reactor dense bed (OP1), top-of-stripper (OP2), upper section of standpipe (OP3), and lower section of standpipe (OP4). These locations allow both reactor solids RTD and standpipe time-of-flight data to be determined. The exact location and spacing of the probes are shown in Fig. 3. OP1 and OP2 were inserted 3.0-in and 6.0-in from the reactor wall, respectively, to centre the probes in the flow field. Contrary to this, OP3 and OP4 were inserted ¼-in to minimize their interference with solids down flow. This assumes that the tracer is well mixed in the solids and that solids are in plug flow; both valid assumptions for the standpipe in this model.

Operating conditions

As mentioned above, the FLUD COKING™ reactor model was designed as a scaled replica of commercial units. The reactor model is a 12-in diameter vessel that expands to 19-in due to its tapered vessel design. Song et al. (4) provide a thorough account of typical commercial and scaled operating conditions of the model. In the current study, a similar scaling exercise was conducted; however, a 0.89 m/s top-of-reactor superficial gas velocity condition is used to represent current reactor operation. The rest of the scaling parameters remained unchanged. For a complete description of both the scaling technique and operating conditions used, see Song et al. (4). Equilibrium FCC catalyst is used as the particles in the cold model (see Table 1 for particle properties).

Results

Tracer Activation Data

The activation capabilities of the injector were tested by combining FCC catalyst with 0.1-wt% PLC SP-2A tracer. This concentration represents a 200-g tracer injection into the model, which was the quantity used for this study. A fiber optic probe was placed inside the fluidized bed section of the injector to measure the light emission properties (see Fig. 1). The FCC catalyst and tracer were fluidized for 5 minutes to ensure that the activation process reaches equilibrium. At this point the UV lights were turned off to measure the natural decay of the tracer with time. Fluidization was maintained during the decay period to capture the inherent voltage fluctuations caused by fluidized bed mixing.

The fiber optic probe data in Fig. 4 show that 0.1-wt% tracer in FCC catalyst provides a strong signal response to tracer solids activated using this technique. The tracer lost approximately 75-% of its brightness after 50-s of elapsed time. Although this loss is significant, the tracer is still detectable after 300-s. Besides demonstrating the ability to detect small quantities of tracer, Fig 4 also illustrates the inherent fluctuations in the measurement due to variations in both tracer concentration and bed voidage during fluidization. Using the decay curve in Fig. 4, a one-point calibration for tracer concentration was developed. One-point is sufficient since the probes are designed to operate over a range where the photomultiplier tubes have a linear response to light intensity. The linearity of the optical probes was confirmed experimentally to be within 0.1% over the full output range, at a variety of gain settings, using a custom LED calibration unit. Linearity of the calibration unit is guaranteed through feedback with a calibrated photodetector. With the calibration of one probe, the rest of the probes were calibrated using a custom the LED calibration unit. This unit sets each optical probe to the same zero and span.

Reactor Solids RTDs

Ten tracer tests were executed at baseline operating conditions to determine variations in the reactor break-through time (difference in time between the leading edge times of the top-of-reactor and top-of-stripper RTDs). The RTD data in Fig. 5 were normalized so that the injection time corresponded to $t = 0$. Although not easily seen in Fig. 5, the signals differ with respect to reproducibility in their leading edge responses. A mean break-through time of 1.7 ± 0.4-s was obtained under these conditions, where 0.4-s was the standard deviation for ten runs. The fastest individual break-through time is1.1-s, whereas the slowest is 2.6-s. By running repeat tests, statistical data can be used to describe the effect of hydrodynamics on break-through times. Those data also show the reproducibility of the tracer technique.

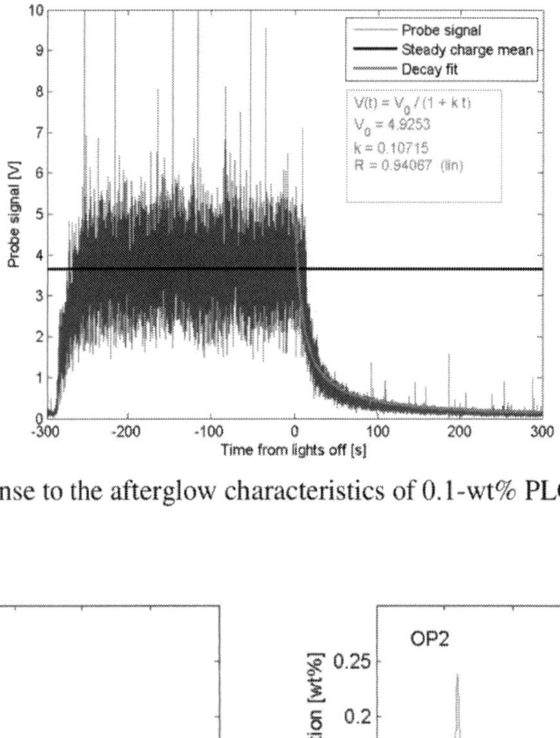

Figure 4. Voltage response to the afterglow characteristics of 0.1-wt% PLC SP-2A in FCC catalyst

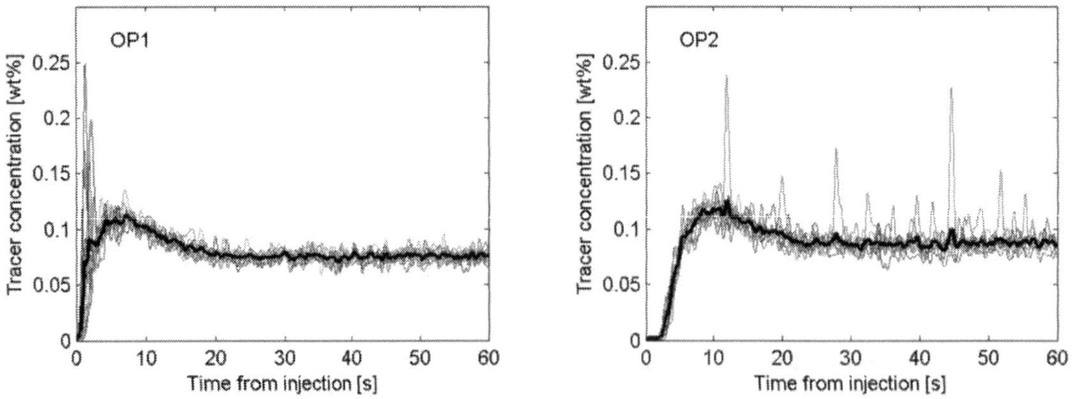

Figure 5. Reactor solids RTDs. Data from ten repeats are shown.

Break-through times were compared with the turnover times from the internal mixing correlations of Baeyens and Geldart (1). They developed a correlation using literature data to estimate the internal mixing in a fluidized bed based on bubble properties. From this mixing model they defined the bed turnover time, τ_T, as the time to displace the entire solids inventory of a fluidized bed through transport in the wake and drift of rising bubbles. Knowing the circulatory nature of solids mixing in fluidized beds, the bed turnover time is analogous to break-through time since downward moving solids inevitably replace the solids carried by upward passing bubbles.

Through a simple derivation, bed turnover time, τ_T, can be equated to

$$\tau_T = \frac{H_{mf}}{Y(U - U_{mf})(\beta_w + 0.38\beta_d)} \tag{1}$$

where H_{mf} is the bed height at minimum fluidization, $U\text{-}U_{mf}$ is the excess superficial gas velocity, Y is the correction factor for deviation from two-phase theory, and , ($\beta_w + 0.38\beta_d$) is a correlation that represents solids transported by the wake and drift of bubbles. Using the Archimedes number representative of the system, values for β_w, β_d and Y can be estimated based on the data provided by Baeyens and Geldart (1). All data required to calculate the bed turnover time are given in Table 2.

There are noteworthy assumptions when applying Eq. 1. First, the derivation of the bed turnover time assumes cylindrical fluidized bed geometry, whereas the model has a tapered geometry. This geometry effect coupled with the non-uniform gas distribution along the height of the dense bed results in a non-uniform superficial gas velocity profile in the reactor. As a result, the superficial gas velocity profile was integrated to generate a characteristic superficial velocity, U, for use in Eq. (1). The integrated superficial velocity was calculated to be 0.72 m/s for the conditions in this study.

Using Eq. 1, the calculated bed turnover time for the model is predicted to be 2.1-s. A sensitivity analysis on β_w, β_d and Y shows this value could vary from 1.6-s to 2.8-s based on reasonable ranges of values for each variable reported in Table 2. In order to compare the bed turnover time to the experimental break-through time, extrapolation of the experimental measurement is required to correct for the probes being located within the bounds of the reactor dense bed. By correcting the break-through times based on mean velocity measured, an overall break-through time of 1.9-s is predicted across the entire dense bed. This matches the calculated turnover time well, adding credibility to the accuracy of the measurement technique.

Table 2. Variables used for turnover time calculations

Variable	FCC catalyst
Minimum fluidization velocity, U_{mf} (m/s)	0.004
Particle diameter, d_p (μm)	100
Particle density, ρ_p (kg/m^3)	1700
Gas density, ρ_g (kg/m^3)	2.34
Gas viscosity, μ_g (Pa s)	1.8 x 10^{-5}
Bed height at minimum fluidization, H_{mf} (m)	0.84†
Archimedes Number $\left(\frac{\rho_g d_p^3(\rho_p-\rho_g)g}{\mu^2}\right)$	120
from Baeyens and Geldart (1):	
Wake fraction, β_w (-)	0.4 (0.3 – 0.5)
Drift fraction, β_d (-)	0.8 (0.7 – 0.9)
Correction factor, Y	0.8 (0.75 – 0.85)

† Estimated from reactor pressure drop data

Standpipe Solids RTDs

The signals from the standpipe shown in Fig. 6 have a more gradual increase in intensity prior to reaching equilibrium. This is attributed to further mixing in the stripper and cone sections of the model. Visual comparison indicates that the shape of the standpipe RTD profiles does not change significantly across its length. This confirms its plug-flow behaviour. At this point of the tracer test the material has had only 10 seconds to decay. Although the tracer brightness has been significantly reduced (recall Fig. 4), there is still sufficient signal to obtain solids RTDs in the standpipe.

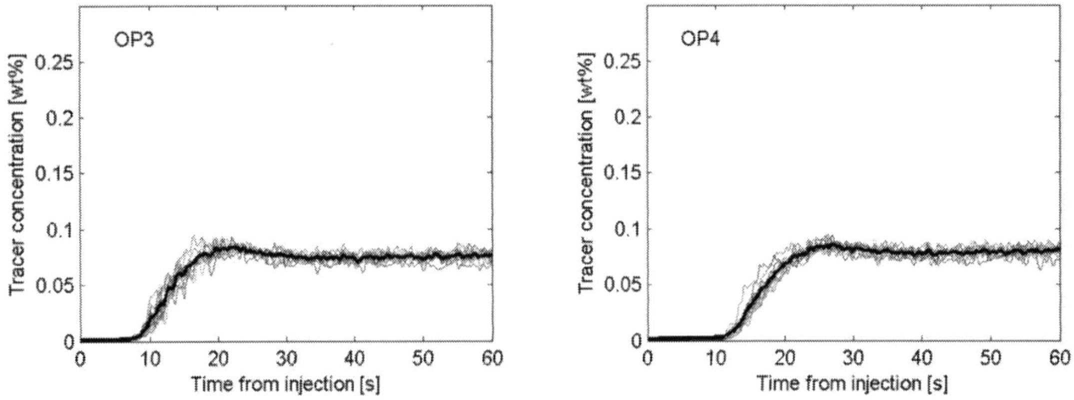

Figure 6. Standpipe solids RTDs. Data from ten repeats are shown.

The tracer time-of-flight in the standpipe is used to measure circulation rates in the model. Using the probe spacing, Δx_{SP}, and elapsed time, Δt_{SP}, between the leading edge times of the upper (OP3) and lower (OP4) standpipe detectors, the standpipe particle velocity, U_{SP}, is calculated as

$$U_{SP} = \frac{\Delta x_{SP}}{\Delta t_{SP}} \qquad (2)$$

The standpipe mass flow rate, $\dot{m}_{SP}$, is then

$$\dot{m}_{SP} = U_{SP} A_{SP} (1 - \varepsilon_{SP}) \rho_{FCC} \qquad (3)$$

where the standpipe cross-sectional area, A_{SP}, is $\frac{1}{4}\pi(3.07\text{-in})^2$, and the standpipe voidage, ε_{SP}, is estimated from standpipe pressure drop as

$$\varepsilon_{SP} = 1 - \frac{\left(\Delta P / L\right)_{SP}}{\rho_{FCC}\, g} \qquad (4)$$

Using those data, the solids circulation rate is correlated against the pressure drop across the riser contraction (see in Fig. 7). The linearity of the fit is attributed to maintaining upward dilute phase flow through the riser contraction. This technique demonstrated accuracy within ± 20%.

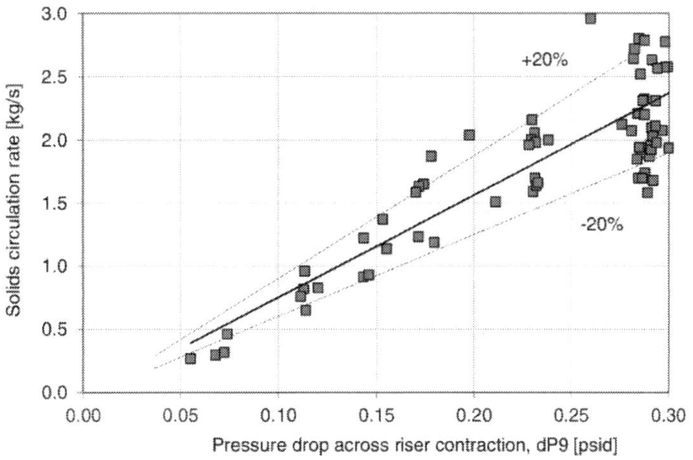

Figure 7. Calibration data for solids circulation

Conclusions

A phosphorescent solids tracer technique has been successfully deployed in a cold flow FLUID COKING™ reactor model. Using a pressurized fluidized bed injector, 200-g of pure phosphorescent pigment were externally activated, then injected at the J-valve. Ten solids RTDs were collected to characterize the variability in solids break-through time in the reactor. RTD measurements indicate a mean reactor break-through time of 1.7 ± 0.4-s. Individual RTDs show that break-through times can vary between 1.1-s to 2.6-s. This suggests local gas-solids hydrodynamics affect the rate that the fastest particles move through the system. The bed turnover time is calculated using Baeyens and Geldart (1) to quantify the accuracy of the tracer technique to measure break-through times. The turnover time compares well to the experimental results. The tracer technique also successfully captured standpipe time-of-flight data, which can be used to quantify solids circulation rates.

References

1. Baeyens, J. and D. Geldart (1986), "Solids Mixing," in *Gas Fluidization Technology*, D. Geldart, ed. Chichester, UK, John Wiley & Sons Ltd.
2. Matsen, J.M. (1996), "Scale-up of fluidized bed processes; principle and practice," *Powder Technology*, 88, pp. 237-244.
3. Wyatt, JR., J.T., E.N. Jones, A.U. Chen, C.R. Sutton, T.M. Healy, R. Suryo, L. Lampert, J. Miller (2011), "Circulating Fluid Bed Reactor with Improved Circulation," United States Patent Application Publication, US2011/0206563.
4. Song, X., X. Bi, C.J. Lim, J.R. Grace, E. Chan, B. Knapper and C. McKnight (2004), "Hydrodynamics of the reactor section in fluid cokers," *Powder Technology*, 147, pp.126-136.
5. Muramoto, T., M. Nakajima, and S. Morooka (1985), "Measurement of flow of solid particles in gas fluidized bed," *Chemical Engineering Communications*, 35 (1-6), pp. 193-201.
6. Baba, T., M. Nakajima, S. Morooka, H. Marsuyama (1984), "A New Measuring System for flow Patterns of Solid Particles in Gas-solid Fluidized Bed," *Journal of Chemical Engineering of Japan*, 17, pp. 275-279.
7. Bader, R., J. Findlay and T. M. Knowlton (1988), "Gas/Solids Flow Patterns in a 30.5-cm Diameter Circulating Fluidized Bed," in *Circulating Fluidized Bed Technology II*, P. Basu and J. F. Large, ed. New York, NY, Oxford, UK, Pergamon Press.
8. Rhodes, M. J., S. Zhou, T. Hirama and H. Cheng (1991), "Effects of operating-conditions on longitudinal solids mixing in a circulating fluidized-bed riser," *AIChE Journal*, 37(10), pp. 1450-1458.
9. Harris, A.T., J.F. Davidson and R.B. Thorpe (2002), "A novel method for measuring the residence time distribution in short time scale particulate systems," *Chemical Engineering Journal*, 89, pp. 127–142.
10. Kirbas, G. (2004), "Solids motion and mixing in high-density circulating fluidized beds," Ph.D. Thesis, University of British Columbia.

(344e) Magnetic Resonance Imaging (MRI) of Jets in Gas-Solid Systems

Maximilian Koehl[1], Guang Lu[1], James Third[1], Klaas Pruessmann[2] and **Christoph R. Müller**[1], (1)Department of Mechanical and Process Engineering, ETH Zurich, Zurich, Switzerland, (2)Department of Information Technology and Electrical Engineering, ETH Zurich, Zurich, Switzerland

Owing to their excellent heat and mass transfer characteristics, fluidized beds have various applications in industry ranging from the cracking of hydrocarbons over the gasification of carbonaceous fuels, to the coating of tablets. However, since gas-fluidized beds are visually opaque it is intrinsically difficult to perform measurements in these systems. Only a few measurement techniques exist that can probe non-intrusively the dynamics of gas-fluidized beds, such as positron emission particle tracking (Stein et al. 1997), electron capacitance tomography (Du, Waristo and Fan, 2006), X-ray attenuation (Rowe and Yocono, 1976) and magnetic resonance imaging (MRI) (Müller et al. 2008).

In this study we apply magnetic resonance imaging to image the formation of jets in gas-solid systems. MRI is an imaging technique that is able to measure both the solids' distribution and the velocity of the solids in granular systems.

In this study beds, constructed of PMMA, of square cross section (L = 46.8 mm and L = 72 mm) were used. The orifice was also of square geometry with Lo=3.6 mm. The following MRI detectable seeds were used as particles: (i) Iceland poppy seeds dp = 0.74 ± 0.08 mm,(ii) opium poppy seeds dp = 0.94 ± 0.04 mm, and (iii) mustard seeds dp = 1.34 ± 0.18 mm).

MRI measurements were performed in a Philips 3T Achieva system equipped with dual Quasar gradients. The maximal gradient strength and slew rate were 80mT/m and 200 mT/m/ms, respectively. A spin echo sequence of repetition time 1500 ms, echo time 12.4 ms and flip angle 60° was used. The voxel size was 1 mm x 1 mm x 1 mm.

Müller et al. (2009) found that the start-up procedure crucially influenced jet heights. In this study two different start up procedures were evaluated: (I) Filling of the bed with particles at a low orifice velocity followed by a stepwise increase of the orifice velocity and (II) filling of the bed at a high orifice velocity followed by a stepwise reduction of the orifice velocity. The MRI measurements revealed that the jet height differed by a factor of two between the two start -up procedures investigated. Subsequently all experiments were performed using procedure (II) since this procedure was determined as more relevant for practical applications and showed an excellent reproducibility.

Further measurements were performed to determine the jet length as a function of (i) fill level, (ii) particle diameter, (iii) bed dimensions and (iiii) orifice velocity. It was observed that the reduction of the orifice velocity led to a rapid transition from a spouting bed to a stable jet in a packed bed, of jet height approximately half the height of the bed. Subsequently, the jet height decreased roughly linearly with decreasing orifice velocity. Interestingly, the fill level did not influence the jet height. However, with increasing fill level the transition to a spouting bed shifted to higher orifice velocities. As expected the jet height increased with increasing orifice velocity and decreased with increasing particle diameter. Additional, in particular for the larger particles, a substantial difference in the jet height (at a given orifice velocity) was observed for beds with different cross sectional areas. Based on the MRI measurements we postulate that once a critical ratio of bed dimension to particle diameter of approximately 70 is obtained, the jet length becomes independent of the bed dimensions, i.e. above that critical ratio the walls of the bed do not affect the jet height.

References

Kunii, Daizo, and Levenspiel, Octave: Fluidization Engineering, second edition, Butterworth-Heinemann, Eds: Brenner, Howard, 1991

Stein, M., Martin, T. W., and Parker, D.J.: Positron emission Particle tracking: Particle velocities in gas Fluidised beds in gas Fluidised beds, mixers and other applications, Non-Invasive Monitoring of Multiphase Flows, Elsevier, Eds: Chaouki, J., F. Larachi, and M.P. Dudukovic, 1997

Du, Bing, Warsito, W., and Fan, Liang-Shih: Flow dynamics of gas-solid fluidized beds with evaporative liquid injection, Particuology 4(1), volume 4, 1–8, 2006

Rowe, P. N., and Yacono, C. X. R.: The bubbling behaviour of fine powders when fluidised, Chemical Engineering Sience 31(12), volume 31, 1179–1192, 1976

Müller, C.R., Holland, D.J., and A.J. Sederman, M.D. Mantle L.F. Gladden J.F. Davidson: Magnetic Resonance Imaging of fluidized beds, Powder Technology 183, volume 183, 53 – 62, 2008

Müller, C.R. et. al Geometrical and Hydrodynamical Study of Gas Jets in Packed and Fluidized Beds Using Magnetic Resonance, Canadian Journal of Chemical Engineering 87 volume 87, 517 – 525, 2009

(344f) Magnetic Particle Tracking in Spouting and Bubbling Fluidized Beds

Jack Halow, Separation Design, Garards Fort, PA and Stuart C. Daw, FEERC, Oak Ridge National Laboratories, Knoxville, TN

Magnetic Particle Tracking in Spouting and Bubbling Fluidized Beds

J.S. Halow* (Separation Design Group, Waynesburg, PA 15370),

Stuart Daw (Oak Ridge National Laboratory, Knoxville, TN 37932)

*halow@windstream.net Introduction

Fluidized beds are extensively used for many types of fluid-solid contacting and reaction applications. Spouted beds are used where uniform inter-phase contacting is important in applications such as drying and coating. Bubbling beds are often used where uniform temperature and good contacting are important. Fluidized beds have been extensively studied for applications which use a single type of bed material, but the basic hydrodynamics and mixing patterns in systems with disparate types of bed particles are not well characterized. Some biomass gasification processes, for example, use bubbling beds with a fine sand bed into which larger biomass particles are fed. Biomass material usually comprises a small fraction of the bed so the dynamics involve a few coarser lighter particles circulating in a finer denser fluidized bed. Spouted bed applications can sometimes involve mixed particle sizes and densities. Both types of systems can lead to segregation and sometimes defluidization.. We have used a magnetic particle tracking to measure particle motion and segregation behavior in both spouting and bubbling beds.

Experimental Materials and Methods

We previously reported (1,2) on a new magnetic particle tracking method in 39 and 55 mm three-dimensional bed fluidized with air at ambient conditions. Safe inexpensive magnetic tracers and detectors are used thus avoiding the issues associated with past particle tracking methods. In this study, we used neodymium magnets embedded in plastic and wooden spherical particles as flow tracers and externally positioned magnetic field detectors to continuously locate the position a tracer particle. The method takes advantage of the tendency of the tracer particles to align their magnetic axes with the earth's magnetic field like a compass needle. Proper positioning of the probes can take advantage of this tendency, which simplifies the data analysis. We also collected high-definition slow-motion videos of the bed surface to observe top segregated tracer motion.

In the spouting bed size studies, binary mixtures of 0.7 to 3 mm glass beads were used. For the density difference tests we used mixtures of glass bead and zirconium oxide (density ratio of ~2). In both studies, various ratios of materials were studied over a range of velocities. A bead with an imbedded magnet having a composite density equivalent to the glass bed material was used to track bed motion along with visual observation of the bed surface. In the bubbling bed tests 200 micron glass beads were studied over a range of velocities with magnetic tracers with densities ranging from 0.53 to 1.2 g/cc. Single tracer particles were dropped into the beds and their positions determined over a 5 minute run time. Tracer positions were calculated using an algorithm we previously reported on. Statistical analysis on the time series position data was used to characterize the tracer behavior. Autocorrelation analysis was particularly useful to determine periodic patterns of solids flow.

Results and Discussion

In the spouted bed tests, segregation occurred when the size ratio was 3/1 or greater. This segregation exhibited a hysteresis pattern with velocity. With a density ratio of 2.2 (zircon and glass) little segregation occur but segregation did occur with higher density ratio mixtures.

We previously reported on the position and segregation tendencies of light and heaver tracer particles in the fine glass bead . In this report we present our recent times series analysis of the bubbling bed tests data. Autocorrelation analysis of vertical position provides characteristic time scales for solids circulation and bubble growth rated. Auto correlation of radial and angular positions reveal other characteristic times and point out the lack of angular symmetry in bubbling beds and the need to use 3-dimensional modeling to achieve realistic results.

1. Patterson et.al., An Innovative Method Using Magnetic Particle Tracking to measure Solids Circulation in a Fluidized Bed, Ind. Eng. Chem. Res. 2010, 49, 5037-5043.
2. Halow et. Al. Observed Mixing Behavior of Single Particles in a Bubbling Fluidized bed, submitted

Abstract submitted for presentation at the 2012 Annual A.I.Ch.E. meeting in Pittsburgh, PA

(344g) Effect of Fluidized Bed Scale and Bed Height On Flow Regimes Transition Using Optical Probe and Pressure Transducer Measurements

Faraj M. Zaid, Department of Chemical and Biological Engineering, Missouri University of Science and Technology, Rolla, MO and Muthanna H. Al Dahhan, Chemical and Biological Engineering, Missouri University of Science and Technology, Rolla, MO

Effect of Fluidized Bed Scale and Bed Height on Flow Regimes Transition Using Optical probe and Pressure Transducer Measurements

Faraj Zaid, and Muthanna H. Al Dahhan, Department of Chemical and Biological Engineering, Missouri University of Science and Technology, Rolla, MO

Different flow regimes are encountered in fluidized bed reactors. The reactor hydrodynamics and hence, performance change significantly as a result of flow regime change. It is very important to know what type flow regime the reactor is operating at and how to identify the boundaries of the flow regimes. Therefore, the reactor volume productivity, mixing, mass and heat transfer are affected by the prevailing flow regime. In this work pressure transducer and optical probe were used to measure and identify bubbly and churn turbulent flow regimes and their transitions in two different plexiglas fluidized beds (small scale 0.14 m and pilot plant scale of 0.44 m in ID) at different axial and radial positions. Pressure transducer was measured at the wall while optical probe was pleased at different radial position from the wall to the center of the bed. Statistical analyses have been performed on the time series measurement of these techniques for identifying the the flow regimes and their transitions.

Notable differences were observed in flow regimes transition along the bed axial and radial locations particularly among the sparger (Z/D=0.286) the fully developed flow (Z/D=1.7) and free band regions. There are some variations between the flow regime transitions obtained using pressure transducer at the wall and at local points inside the bed using optical probe at the studied bed heights. Detailed data and findings will be discussed and presented in this talk.

Keywords: Fluidized beds, optical probe, Pressure fluctuations, Flow regimes identification, Various transition velocities

(366a) Coupling of Crystallizers for Efficient Enantioseparation - Comparison of Two Different Process Strategies

Matthias Johannes Eicke, Physical and Chemical Foundations of Process Engineering, Max Planck Institute for Dynamics of Complex Technical Systems, Magdeburg, Germany, Guillaume Levilain, Max Planck Institute for Dynamics of Complex Technical Systems, Magdeburg, Germany, Martin Peter Elsner, Georg-Simon-Ohm University of Applied Sciences, Nuremberg, Germany and Andreas Seidel-Morgenstern, Physical and Chemical Foundations of Process Engineering, Max-Planck Institute for Dynamics of Complex Technical Systems, Magdeburg, Germany

Chiral molecules can be a particular challenge when used in drugs because of the existence of their stereoisomers. Although both enantiomers are merely mirror images, they can show drastically different effects in living organisms, which are chiral themselves and thus able to distinguish between the two molecules. Owing to this, it is essential to obtain enantiomers in their pure form. A wide array of methods has been studied and brought to application which on the one hand is concerned with the direct synthesis and on the other hand focuses on a more engineering biased approach that is, the separation of the racemate using well-known as well as novel techniques.

This contribution takes on the engineering route and puts the application of preferential crystallization (PC) as a separation technique at its center. It is well known that a racemic solution containing both enantiomers at equal proportions can be subcooled and still remain in a state where only the liquid phase exists. This phenomenon, called metastability, applies to other solutions as well but it is the key for applying PC. Within this region of kinetic inhibition it is possible to grow seed crystals of only one enantiomer. At some point however, the unseeded species will undergo nucleation and contaminates the product. In order to prevent this, the process must be stopped and the product has to be removed from the mother liquor. Certainly, this leads to great limitations with respect to yield because a considerable amount of product is still left in the dissolved state.

An improvement is possible for systems that crystallize as conglomerates such as DL-threonine/H_2O. It has been shown that two crystallizers coupled via a constant exchange of crystal free solution and seeded with opposite enantiomers can overcome these limitations [1, 2].

Recently more work has been done by the authors to further improve coupled PC. Different polythermal modes of operation have been investigated theoretically indicating the possibility for significant enhancements with respect to yield and productivity. Furthermore, a variant of coupled PC originally described in [3] has been investigated in detail both theoretically and experimentally for the model system DL-threonine/H_2O. The scheme of the process is depicted in Figure 1.

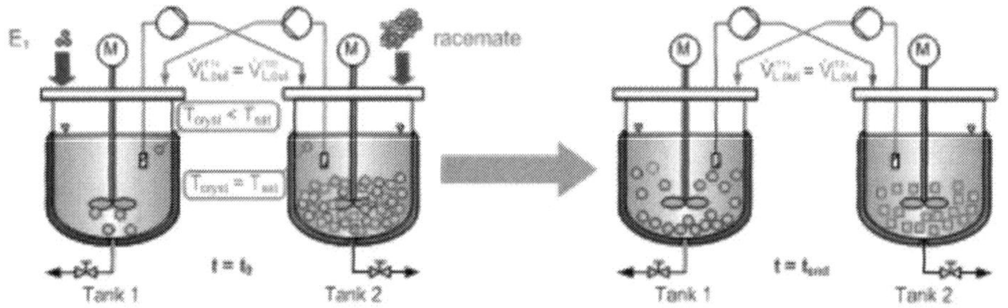

Figure 1: Operating principle for the connection of selective crystallization and dissolution.

Contrary to the first strategy, where both tanks are seeded with the respective pure enantiomers, this variant requires only one type of homochiral seeds. One crystallizer (Tank 1) holds a supersaturated racemic solution to which the preferred enantiomer E_1 is added in solid form. The other crystallizer (Tank 2) contains a saturated racemic solution with the same concentration to which solid racemate is added. Since the temperature in Tank 1 is below saturation level the seeds will grow reducing the level of supersaturation. Due to the exchange of clear solution a transient undersaturation with respect to E_1 will appear in Tank 2, leading to a selective dissolution of this enantiomer from the solid racemate. In this case preferential crystallization is happening in one tank while a selective dissolution in the other leads to a purification of the supplied racemate with respect to the second enantiomer E_2. The process thus yields two pure molecules while only one is needed as an investment.

This contribution will focus on how coupling crystallizers can help to make enantioseparation more efficient. The core will be a comparison of both process strategies with respect to their productivity but also their strengths and weaknesses. Further insight will be given by complementing experimental results with simulation studies directed at the influence of different process parameters.

1. Elsner, M.P., G. Ziomek, and A. Seidel-Morgenstern, *Efficient Separation of Enantiomers by Preferential Crystallization in Two Coupled Vessels*. AIChE Journal, 2009. **55**(3): p. 640-649.

2. Elsner, M.P., G. Ziomek, and A. Seidel-Morgenstern, *Simultaneous preferential crystallization in a coupled batch operation mode. Part II: Experimental study and model refinement*. Chemical Engineering Science, 2011. **66**(6): p. 1269-1284.

3. Merck & Co. Inc., *Resolution of Racemic Mixtures of Optically Active Enantiomorphs*. Patent DE1543238, 1965.

(366b) PBE-Aided Design of Slurry Milling Process to Reduce Particle Size of Intermediate and Active Pharmaceutical Ingredients

Carla Luciani, CPR&D, Eli Lilly & Co, Indianapolis, IN, Kevin D. Seibert, Chermical Product Research and Development, Eli Lilly & Co., Indianapolis, IN, Daniel Jarmer, CPRD, Eli Lilly & Co., Indianapolis, IN and Eric Moher, CPR&D, Eli Lilly & Co., Indianapolis, IN

A variety of milling processes are used in pharmaceutical industry to reduce the particle size of intermediate and active pharmaceutical ingredients. Slurry milling processes are of particular interest because they significantly reduce the transfer of heat to the product compared to dry mill equipment and they can eliminate a unit operation when crystallization-milling sequence are combined prior to the isolation. In spite of its extensive use in the pharmaceutical industry, the design of slurry milling processes are not trivial due to the multivariate nature of the process. One of the most intuitive ways to describe milling processes is the use of population balances equations (PBEs). However, relating PBE parameters (i.e., breakage rate and breakage kernel) to actual milling conditions (i.e., rotor tip speed or shear rate, slurry density, mechanical, morphological, and rheological properties of the milled material milled and the suspension medium, etc.) is challenging.

In this work, a PBE approach is applied to describe the slurry milling process of a pharmaceutical intermediate. Correlations between milling conditions and model parameters are proposed. Finally, the model is inverted to determine set of operating conditions that produce desired final particle sizes.

(366c) Development of a High Shear Crystallization Process for a Pharmaceutical Intermediate

Brenda Remy, Amit Joshi, Junying Fan, Jason Sweeney, Kenneth Natalie, Aghogho Pedro and Shawn Pack, Chemical Development, Bristol-Myers Squibb Co., New Brunswick, NJ

Batch crystallization processes for pharmaceutical intermediates are generally design to maximize the removal of impurities and to minimize yield losses. Crystallization performance is dependent on the ability to control several mass transport mechanisms such as growth, nucleation, impurity deposition etc. These mechanisms, in turn, can be impacted by controlling the amount of shear during the crystallization. Wet milling via rotor-stator devices is commonly employed in batch crystallization to control shear. While wet milling is mostly used as a particle size reduction technique, it can be used to control the purity of the isolated solids and to improve yield.

This talk will highlight the use of wet milling in a pharmaceutical intermediate crystallization to improve quality, purity and cycle time. The crystallization involves an initial mass nucleation followed by a solid-solid form conversion. Upon initial scale-up the process suffered from inconsistent yields and impurity rejection. These inconsistencies appeared to correlate with final product particle size. Wet milling achieved consistently smaller crystals and enabled faster desaturation leading to consistently higher yields. Additionally, wet milling improved the purity of the crystallized material by mitigating impurity inclusion. The use of a DoE study to determine the main parameters affecting desaturation and impurity removal will be highlighted. Scale up results from a pliot plant campaign will also be presented.

(366d) Developing a Selective Crystallization Process for a Complex Reaction System

Chenchi Wang, Truc Vu and Lucius Rossano, Research and Development, Bristol-Myers Squibb Company, New Brunswick, NJ

Developing a Selective Crystallization Process for a Complex Reaction System

Authors: Chenchi Wang, Truc Vu, Lucius Rossano.

Research and Development, Bristol-Myers Squibb Company, New Brunswick, New Jersey

A reactive crystallization process often involves multiple complex steps, such as chemical reaction, solid precipitation/crystallization, and particle size control if required. In this case study, an active pharmaceutical ingredient (API) is prepared in a telescoped process integrating deprotection, free basing and salt formation. Apart from the chemistry issues encountered that are not the focus of this presentation, the challenges of the crystallization lie in (1) maintaining chemical stability of the free base prior to crystallization, (2) quantitating the free base available for salt formation, (3) maximizing yield of the free base under process conditions, (4) enhancing purity by trace metal removal, and (v) achieving desired particle size for drug product development. In order to address these important process issues and improve process robustness, a two-step, selective crystallization process was developed. As a result of this development, a robust API synthesis step, incorporating this well-controlled reactive crystallization, was demonstrated at various scales and consistently generated material with desired purity, yield, and particle size. The designing and outcomes of this process will be discussed in details.

(366e) Crystal Nucleation Control Using Microscopic Online Imaging

Tod Canty, J.M. Canty Inc., Buffalo, NY

An innovative method of inline microscopic imaging has been developed to allow crystal viewing and measurement of size, shape, color and concentration in batch or continuous lab reactors ,pilot plant crystallizers and full production vessel at pressure(10,000psi) and temperature (-450 to 800 deg F). The high resolution images presented will show how this technology succesfully over comes the limitations of other systems where aprticle resoltions where limit to 30 micron and above due to front lighting even though the optical lenses where capable of higher pixel resoltuion. The particle imaging system captures high resolution 2 megapixel images down to .7 micron at speeds up to 1/100,000 of a second using back lighting. The system has a fiber optic lighting system that allows viewing through a wide variety of fluids (crude oil, polymer ,inks ,paint etc.). It does this at speeds through 10ft/sec in liquids. The software then analyzes the size and shape of particles compiling the data to determine particle distribution by length, width or volume along with ppm concentration. The Lab scale reactors then allow the user to optimize the seeding process to provide control of Nucleation to optimize yield and filterability.

(366f) Precipitation Kinetics of Aluminum Solids Formation During the Caustic Side Solvent Extraction (CSSX) Process

Rebecca K. Toghiani[1], Punith P. Naik[2], Jeffrey S. Lindner[2], Laura T. Smith[2] and Larry Pearson[2], (1)Dave C. Swalm School of Chemical Engineering, Mississippi State University, MS State, MS, (2)Institute for Clean Energy Technology, Mississippi State University, Starkville, MS

The Caustic Side Solvent Extraction Process (CSSX) process is used for cesium separation at the Savannah River Site (SRS). The CSSX process is comprised of 4 steps: 1) extracting cesium into the solvent; 2) scrubbed of the loaded solvent to remove co-extracted sodium and potassium; 3) stripping to transfer the cesium out of the solvent; and 4) washing of the stripped solvent to remove solvent degradation products if present. During the scrubbing process, aluminum precipitation has been observed. This is a concern as solids formation may have undesirable consequences during operation. The work reported was undertaken to identify conditions under which solids formation may occur and to identify an operating region where solids formation is avoided or minimized.

Experiments were conducted at room temperature using simulants based on different dissolution fractions of SRS Tank 25F contents. Combination of a given amount of simulant with a 0.05 M nitric acid solution mimicked the carryover during the CSSX process. Aluminum concentration and solution pH were measured over time for the prepared samples. Simulation of these experiments were also completed using the Environmental Simulation Program (ESP, OLI Systems, Inc.) and comparisons made. Solids formation was observed for a select subset of samples both experimentally and through ESP predictions. For these samples, a second set of samples was prepared. The temporal variation of aluminum concentration was measured, until solids were visually observed in each sample. The resulting data were analyzed using a first order reversible rate expression, and forward and backward reaction rate constants determined. Solids formed were isolated, dried and identified using polarized light microscopy.

(388a) Percolation Segregation in a Sheared Bed of Particles (Poster)

Kevin J. Brennan and Carl R. Wassgren, Mechanical Engineering, Purdue University, West Lafayette, IN

Segregation negatively impacts many powder processing operations, but few tools exist to quantitatively evaluate and predict this phenomenon. In order to assess segregation due to percolation quantitatively, a discrete element method (DEM) model has been developed to simulate a shear cell containing a bimodal size distribution of particles. The steady state flux at which fine particles percolate downward through the bed is recorded for a wide range of shear rates, normal loads, fines concentrations, and particle size ratios. The simulation results are compared to the analytical random, fluctuating sieve model developed by Savage and Lun (1988).

(388aa) Oxidation Mechanism of Metal Particles Derived From Thermal Analysis

Shasha Zhang, Department of Chemical, Biological & Pharmaceutical Engineering, New Jersey Institute of Technology, Harrison, NJ and Edward L. Dreizin, Otto H. York Department of Chemical, Biological, and Pharmaceutical Engineering, New Jersey Institute of Technology, Newark, NJ

Thermal analysis is widely used to establish rates of oxidation of metal powders, which can be correlated with their ignition mechanisms. Measurements are performed with polydisperse powders and are commonly processed considering a representative average particle size. However, particles of different sizes may react with different rates; understanding and quantifying these differences is of interest for developing a comprehensive oxidation mechanism. As a simple example, reaction between metal and oxidizer may occur at the surface of the growing oxide layer, or inside this layer, at the metal-oxide interface, depending on whether metal or oxidizer diffuse faster through the growing oxide. As the oxide layers grow, the changes in reactive surface to volume ratio will change differently depending on the location of the reactive surface. This work describes a methodology of processing thermo-analytical measurements aimed to distinguish such differences in oxidation rates of particles of different sizes. Thermo-gravimetric (TG) measurements describing oxidation of different aluminum powders are processed. For each powder, particle size distribution is measured and used to process the experimental data. For each time step, the measured weight gain is distributed among particles proportionally to the available surface of the reactive interface. This distribution depends on the selected reaction model, and thus, the interface location. The surface area of the reactive interface is recalculated for each experimental data point. Note that the selected reaction model can include effects of the growing oxide thickness and changes in diffusion rates as a function of temperature and other parameters. As a result of this processing, for each TG run, individual weight gain curves are obtained for particles of different sizes. The appropriate reaction model is validated comparing weight gain curves obtained for particles of the same size for powders with different, but overlapping particle size distributions. These weight gain curves must be identical, assuming that the oxidized particles do not interact with one another. Results and conclusions for oxidation of spherical aluminum powders will be presented and discussed.

(388ab) Lattice Boltzmann Method (LBM) Simulations of Gas-Solid Flows Containing Non-Spherical Particles

Yong Chen, James Third and **Christoph R. Müller**, Department of Mechanical and Process Engineering, ETH Zurich, Zurich, Switzerland

A good understanding of particle-fluid interactions is crucial to predict better the dynamics of various industrial processes and natural phenomena, e.g. chemical reactors, the transport of pollutants in the atmosphere or the transport of sediments in rivers. In those systems, the dynamics of particles are highly complex due to (i) particle-particle, (ii) particle-fluid and (iii) particle-wall interactions.

Currently the drag force correlations of Ergun or Wen &Yu are commonly used. However, these correlations are often applied for conditions in which their validity is questionable. Direct numerical simulations using e.g. the lattice Boltzmann method would allow us to measure fluid-particle interactions and, thus, formulate accurately drag force correlations.

Koch (Koch and Sangani 1999; Koch and Hill 2001; Koch and Ladd 2001a, b) was the first to develop a drag force correlation based on lattice Boltzmann simulations. Van der Hoef and Beetstra (van der Hoef 2004; Beetstra 2007) extended the work of Koch et. al. establishing a new drag force correlation for mono- and bidisperse arrays of spheres. In these simulations, the flow field around a random array of spheres was computed using a lattice Boltzmann method with a grid sufficiently small to model the detailed flow around the spheres. The drag force correlations derived from lattice Boltzmann simulations showed a better agreement with experimental measurements than the traditional Ergun and Wen &Yu correlations.

Numerical simulations of gas-fluidized beds have been restricted mostly to spherical particles due to a lack of drag force correlations for non-spherical particle assemblies. Indeed, even for a single non-spherical particle only limited work has been reported. For example, Holzer (2007) using lattice Boltzmann simulations proposed a drag force correlation for a single non-spherical particle as a function of particle orientation and Reynolds number. However, these correlations cannot be applied directly to simulate gas-fluidized beds. Here, we develop a drag force correlation for non-spherical particles. The effect of particle shape on the drag force was studied in detailed. The new drag force correlation derived here is suitable for Euler-Euler and Euler-Lagrangian simulations of gas-fluidized beds.

References

Koch, D. L., Sangani, A. S. 1999 Particle pressure and marginal stability limits for homogeneous monodisperse gas fluidized bed: kinetic theory and numerical simulations. J. Fluid Mech. 400, 229

Koch, D. L., Hill, R. j. 2001 Inertial effects in suspension and porous media flow. Annu. Rev. Fluid Mech. 33, 619

Hill, R. J., Koch, D. L., Ladd, A. J. C. 2001a The first effects of fluid inertia on flows in orderd and random arrays of spheres. J. Fluid Mech. 448, 213

Hill, R. J., Koch, D. L., Ladd, A. J. C. 2001b Moderate-Reynolds –number flows in order and random arrays of spheres. J. Fluid Mech. 448, 243

Van der Hoef, M. A., Beetstra, R., Kuipers, J. A. M. 2004 Lattice-Boltzmann simulations of low Reynolds number flow past mono- and bidisperse arrays of spheres: results for the permeability and drag force. J. Fluid Mech. 528, 233

Beetstra, R., van der Hoef, M. A., Kuipers, J. A. M. 2007 Drag force of intermediate Reynolds number flow past mono- and bidisperse arrays of spheres. J. AIChE. 53,489

Holzer, A., Sommerfeld, M. 2007 New simple correlation formula for the drag coefficient of non-spherical particles. Powder Technology, 184, 361

(388ac) Photocatalytic Activity of the Liquid Flame Spray Generated Titanium Dioxide Nanoparticle Coating On a Lacquered Metal Surface

Miia Luhtalampi[1], Mikko Aromaa[2], Kimmo Lahtonen[3], Oskari Kangasniemi[2], Janne Haapanen[2], Mika Valden[3] and Jyrki M. Mäkelä[2], (1)Explosives and CBRN Protection Technology, Defence Forces Technical Research Centre, Lakiala, Finland, (2)Aerosol Physics Laboratory, Tampere University of Technology, Tampere, Finland, (3)Surface Science Laboratory, Tampere University of Technology, Tampere, Finland

In order to achieve new properties, like photocatalytic activity or self-cleaning, for lacquered surfaces, nanosized liquid flame spray [1] generated nanoparticles may be deposited on such surfaces. Here we demonstrate the functionality of lacquered surfaces with titanium dioxide nanocoating. The morphology of the surfaces is similar to lotus surface having both micro- and nanostructure. The microstructure is due to the roughness of the lacquered surface and titanium dioxide particles form the nano-sized morphology. Titanium dioxide nanoparticles work as photocatalyst in a process where organic material degrades when the surface is exposed to UV-radiation. We used a common indicator methylene blue to test the coatings. The purpose was to degrade methylene blue in water solution where lacquered sample coated with titanium dioxide nanoparticles was placed. The tests showed that methylene blue degraded in solution which was exposed to UV-radiation and when there were nanoparticles deposited on to the sample surface. Degradation did not happen without UV-light or nanoparticles. The chemical composition and uniformity of selected surfaces with flame spray generated coatings were investigated utilizing X-ray photoelectron spectroscopy. In addition, the morphology of the coating was studied using scanning electron microscopy [2].

[1] J. Tikkanen, M. Eerola, V. Pitkänen, M. Rajala, Patent No. FI98832, 16.3., 1997.

[2] M. Luhtalampi, M. Aromaa, K. Lahtonen, O. Kangasniemi, J. Haapanen, M. Valden, J.M. Mäkelä. Superhydrophobic properties of the liquid flame spray generated titanium dioxide nanoparticle coating on a lacquered metal surface. *Thin Solid Films*, submitted 2012.

(388ad) Fundamental Aspects of the Carbonation and Calcination Reactions and the Synthesis of Stable, Calcium-Based CO2 Sorbents

Marcin Broda, Agnieszka Kierzkowska and **Christoph R. Müller**, Department of Mechanical and Process Engineering, ETH Zurich, Zurich, Switzerland

The increasing concentration of greenhouse gases in the atmosphere is most likely linked to climate change. One of the most promising options to mitigate climate change is CO_2 capture and storage (CCS). Here, CO_2 is captured from power plants, compressed, transported and stored in geological formations rather than emitted into the atmosphere. However, the costs of CO_2 capture using amine based scrubbing technology are estimated to be around 39-96$ per ton of CO_2 captured.[1] Thus, there is an urgent need to develop more efficient and less costly CO_2 capture technologies. One of the emerging CO_2 capture technologies is based on the carbonation and calcination of CaO and $CaCO_3$, respectively, i.e. $CaO + CO_2 \leftrightarrow CaCO_3$. This CO_2 capture technology is commonly referred to as calcium looping. Economic analyses showed that such a process is economically attractive, in particular due to the relatively inexpensive, naturally occurring sorbents, i.e. limestone or dolomite.[1] However, most natural sorbents show a rapid decrease in CO_2 uptake with cycle number. It has been argued that the drop in CO_2 capture capacity is mainly caused by thermal sintering resulting in a reduction of surface area and the volume of pores with diameter < 100 nm.[2] It has been hypothesized that the carbonation reaction can be divided into two regimes.[3] The first, fast reaction stage finishes when the volume available in small pores is filled by the newly-formed $CaCO_3$. Once the internal pores in the grain are filled, a layer of product ($CaCO_3$) is deposited on the surface of a grain. Owing to the typically short contacting times in industrial fluidized bed reactors, only the CO_2 uptake achieved during the fast reaction stage will be of practical relevance. However, it is likely, that this simple carbonation model does not describe the dynamics of the carbonation reaction in full detail. In this work focused ion beam – high resolution scanning electron microscopy (FIB-HRSEM) was used to probe the structural changes occurring on a nanometer scale during the carbonation and calcination reactions. Using FIB-HRSEM we are able to determine whether $CaCO_3$ is formed on the CaO skeleton homogeneously layer by layer or via the merging of heterogeneous $CaCO_3$ islands. However, to derive robust conclusions a Ca-based material with regular pore structure and a high stability over repeated cycles is required. So far, mostly "simple" techniques, such as wet-impregnation, mechanical mixing or co-precipitation[4,5] were applied to develop synthetic Ca-based sorbents. However, these methods do not allow to tailor key structural properties, such as pore size distribution, that have shown to influence strongly the overall CO_2 uptake. In this work we report the development of a Ca-based sorbent using a sol-gel technique. An important facet was to determine relationships between (i) the preparation conditions, (ii) the structure and (iii) the performance of the material. The calcium precursor and the ratio of Ca^{2+} to Al^{3+} were identified as the sol-gel parameters which influenced the CO_2 uptake characteristics most. Calcium precursors containing a weak organic acid, i.e. $Ca(C_5H_7O_2)_2$ or $Ca(CH_3COO)_2$ yielded sorbents which possessed a homogeneous, nano-structured morphology as well as a high surface area and pore volume. In turn, these sorbents also showed a high and very stable CO_2 uptake over many cycles. After 30 cycles of calcination and carbonation the CO_2 uptake was 0.51 g CO_2 /g sorbent which is about 250 % higher than for natural Havelock limestone.

References

(1) A. MacKenzie, D. L. Granatstein, E. J. Anthony, J. C. Abanades, *Energy Fuels* **2007**, 21, 920-926.

(2) J.C. Abanades, E.J. Anthony, D.Y. Lu, C. Salvador, D. Alvarez, *AIChE J.* **2004**, *50*, 1614-1622

(3) J. S. Dennis, R. Pacciani, *Chem. Eng. Sci.* **2009**, *64*, 2147-2157.

(4) M. Broda, A. M. Kierzkowska, C. R. Müller, *ChemSusChem* **2012**, *5*, 411-418.

(5) A. M. Kierzkowska, C. R. Müller, *Energy Environ. Sci.* **2012, *5***, 6061-6065.

Effect of ignition energy on dust explosion behavior:

Phenomena of under-driving and over-driving (ID# 250682)

Niansheng Kuai, Weixing Huang*, Bing Du, Jingjie Yuan, Zongshan Li, Yuan Gan, Jingyi Tan

School of Chemical Engineering, Sichuan University, 29 Wangjiang Road, Chengdu 610065, PR China

Abstract

To improve the accuracy of dust explosion tests, systematic experiments were conducted by using a Siwek 20 L vessel. Explosion behavior of typical light metal and carbonaceous dusts induced by different ignition energies was explored. Both the under-driving phenomenon in the determination of thermodynamics and the over-driving in kinetics were investigated. Results show that weak ignition source causes incomplete deflagration of carbonaceous dust, referred to as under-driving, which results in under-estimation in thermodynamics. Nevertheless, under-driving is not notable for light metal dust. Overly strong energy will cause over-driving so that the measured kinetic parameter will be over-estimated. Also it is found that light metal dust is easier to be over-driven than carbonaceous. A dimensionless parameter was introduced to evaluate the occurrence of over-driving. Authors suggest that the most appropriate ignition energy to gain realistic data in both thermodynamics and kinetics may be the energy at which the minimum dimensionless flame thickness is attained.

Keywords: Dust explosion; Ignition energy; Under-driving; Over-driving; Dimensionless flame thickness

Introduction

The dust explosion hazard continues to represent a recognized threat to the process industries which manufacture, use or handle powders and dusts of combustible materials [1], which presents urgent needs for the knowledge and technology in explosion prevention and mitigation. In assessing venting requirement and for specifying automatic explosion suppression equipment, an accurate knowledge of explosion thermodynamics and kinetics is quite essential. It is well-established that dust clouds are inherently more difficult to ignite than gases, and there needs to be an ignition source with adequate energy, more energetic than for gases, for igniting a dust cloud to result in an explosion [2]. Consequently, powerful ignition source such as pyrotechnical ignitors are usually employed in the determination of explosion characteristics, but such level of initialing energy may cause a pronounced impact on explosion behavior.

Energy of pyrotechnic ignition sources used to determinate the crucial explosion characteristics is always prescribed as a fixed value according to the specifications of current testing standards. Although several studies indicate that higher explosion severity characteristics are found to be somewhat related to stronger ignition sources [3-5], little attention has been given to evaluate whether the parameter measured under weak/strong ignition energy may be under-/over- estimated to some extent. The present work aims neither to repeat tests made already by others nor to prove their results, but to attempt to gain an insight into the effects of ignition energy on the amount of energy liberated from explosions, the initiation of explosions and the progress of flame propagation, and above all, to propose practical methods to distinguish the under-/over- estimated experimental data induced by inadequate ignition energy level from the realistic results.

Experimental

Experiments were performed in the well-known 20 L vessel developed by Siwek [6]. The vessel is an explosion resistance hollow sphere made of stainless with a volume of 20 L according to the specifications of European standard EN 14034 [7-9], ASTM standard E1226 [10] and Chinese standard GB/T 16425 [11] (Fig.1).

The evolution of explosion overpressure during the explosion was measured with a piezoresistive pressure

* Corresponding author. Tel.: +86 28 85408126; fax: +86 28 85403397.
E-mail address: hwx@scu.edu.cn (Prof. W. Huang).

transducer, for gauge pressures in the range of 0-17 bar. The Dytran transducer, model 2300V1, has a dynamic response of 250 kHz. The data acquisition is composed by a PC of 2.4 GHz clock speed inside of which is installed a data acquisition board. This Aavantec PCL-818L board can provide 16 single-ended or 8 differential analog inputs. Sixteen output channels for A/D conversion with 12-bits resolutions at 100 kHz are also provided. The processes of pressurization, dust dispersion and ignition are controlled by Panasonic FPX-L14 PLC.

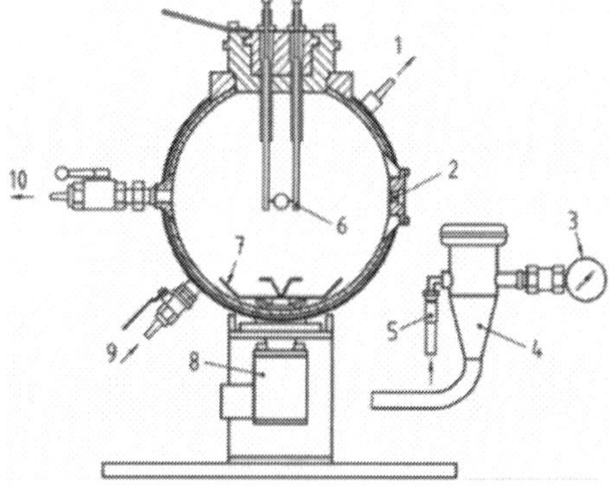

Fig.1. Schematic diagram of the 20 liter sphere: 1-circulating water inlet 2-piezoelectric pressure sensor 3-air pressure gage 4-dust container vacuum gage 5-compressed air inlet 6-ignition source 7-dispersion nozzle 8-solenoid valve 9-circulating water outlet 10-to vacuum pump

Prior to testing, the explosion vessel was sealed and partly evacuated. After ignition source and dust sample were placed, the computer control program was initialized and the auxiliary dust deposit was pressurized. To create a dust cloud, the solenoid valve was opened and the jet flow from the deposit dispersed the dust sample into the vessel, raising its pressure to the desired initial absolute pressure p_0. Finally, PLC energized the ignitor after a given delay 60 ms. Dust clouds in the testing were ignited by electrically activated pyrotechnical ignitors. The ignition energy depends on the mass of pyrotechnical composition composed of 40 % zirconium, 30 % barium nitrate and 30 % barium dioxide.

The present experimental study employed magnesium, bituminous coal and sweet potato fine powders as the typical combustible dusts. All the samples were systematically dried in a vacuum before handling. Their particle size distributions were gained by Leitz metallographic microscope. The heat of combustion was determined by DDS Cal 2koxygen bomb calorimeter. All the analytical data above-mentioned were listed in Table 1. Moreover, SEM photomicrographs of unburned and post-explosion particles were gained by FEI Inspect F50 scanning electron microscope.

Table 1 Analytical data for dust samples studied

Samples	Granulometry			Heat of combustion (kJ/g)
	d_{10} (μm)	d_{50} (μm)	d_{90} (μm)	
Bituminous coal	6.2	29.4	60.0	20.06
Sweet potato	8.5	29.8	74.9	16.27
Magnesium	12.3	22.4	88.8	23.21

Fig. 2 indicates a typical profile of pressure-time trace for sweet potato dust at the dust concentration of 400 g/m^3, by using an ignitor of 5 kJ. As can be seen, the essential explosion severity parameters measured are maximum explosion overpressure p_{max} and maximum rate of pressure rise $(dp/dt)_{max}$. The value of p_{max} is typically related to the thermodynamics concerned with the amount of energy liberated during deflagration, whereas $(dp/dt)_{max}$ is concerned with the rate at which the explosion energy is liberated. In addition to these results, some parameters can be derived from pressure-time traces. At the early stage of explosions, it is essential to point out the delay in the pressure development. The apparent burning velocity is slow because the flame front wave needs developing and particles need preheating prior to combustion, which could be the reason why the flame propagation undergoes an ineffective period I, i.e. the induction duration. After the self-induced flame acceleration is fully developed within the induction period, the fast pressure development in period II gives the idea that the dust flame propagation wave becomes quite efficient. At the end of pressure development, the top pattern of the trace (period III) means that the slowdown of burning velocity is verified actually. This behavior is related to the existence of flame thickness by reasoning that in the final stage of explosions the flame front has already reached the vessel wall, while the remaining part is still propagating through the wall. The significant heat loss, when the thick combustion reaction zone contracts with the chamber wall, results in that $(dp/dt)_{max}$ usually occurs before the end of explosions. The inflection point between periods II and III corresponds to the moment that flame front contacts with the chamber wall, as well as the transition from the initial adiabatic stage to the subsequent non-adiabatic stage. An original dimensionless parameter, flame thickness defined as δ = Period III / (Period I+Period II) , was introduced to characterize the behavior of flame propagation.

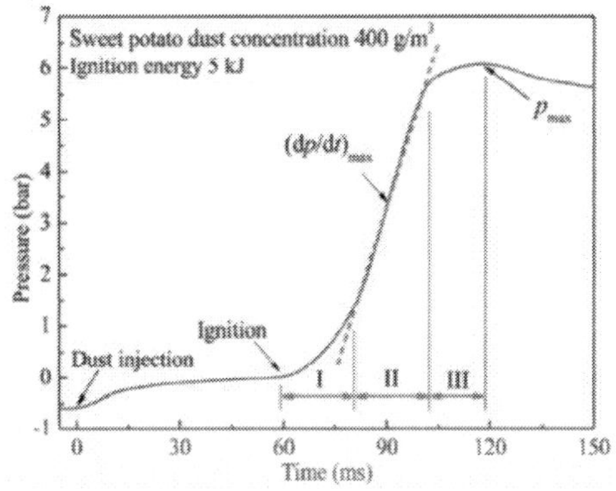

Fig.2. Typical pressure-time trace at dust concentration of 400 g/m³, for sweet potato dust induced by ignitor of 5 kJ

Result and discussion

Ignition behavior of ignitors

Before discussing the effect of ignition energy on dust explosion behaviors, it is worthwhile having an insight into the acting behavior of pyrotechnical ignitors. In the ignition process, pyrotechnical ignitor explodes rapidly with a very strong flame spread. The convection of flame products with high temperature acting like a jet not only heats the unburned particles but induces a certain degree of turbulence in the test vessel [5]. This kind of volumetric ignition is obviously much stronger than an electrical spark. Heat liberated by pyrotechnical ignitor would directly lead to a pressure rise p_{ign}. Nevertheless, the presence of p_{ign} is always under the mask of p_{max}. In the present work, p_{ign} was picked out by determining pressure rise due to ignitors themselves without adding dust samples. Fig. 3 shows that p_{ign} is proportional to ignition energy. Values of p_{ign} in the present work are similar to those reported by Cashdollar and Chatrathi [12], Mintz et al. [13] and Pilao et al. [4].

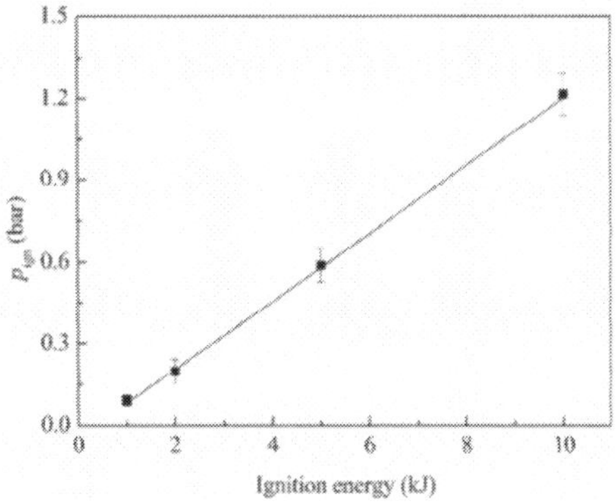

Fig.3. Evolution of p_{ign} with ignition energy

Explosion thermodynamics

The influence of ignition energy on the explosion thermodynamics was evaluated for bituminous coal, sweet potato and magnesium dusts, using ignition energies of 1, 2, 5 and 10 kJ. Figs. 4a and c represent the results of p_{max} at dust concentrations of 125 and 400 g/m³, initiated with the four tested ignition sources. As can be seen, it is well-established that p_{max} increases with ignition energy [4, 5]. Nevertheless, it should be pointed out that the conclusion that higher ignition energy always promotes the amount of the net energy release of dust explosion cannot be drawn, because values of p_{max} are always measured by the total energy release including p_{ign}.

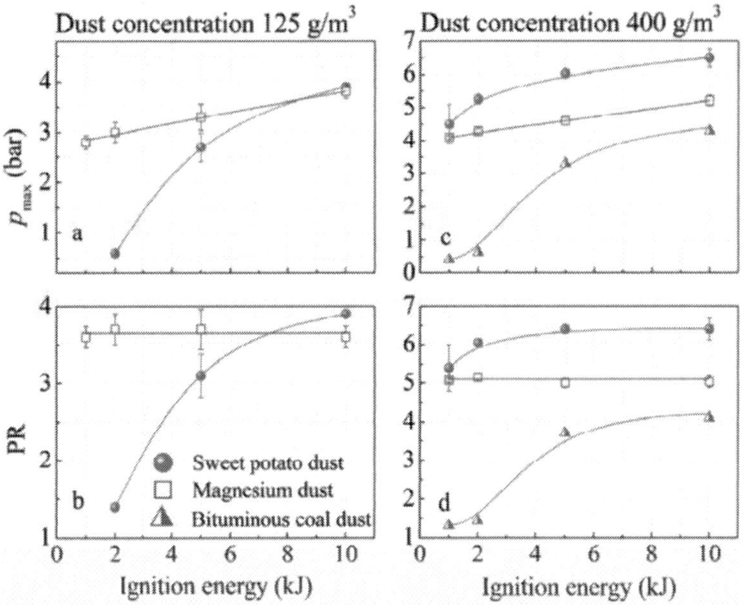

Fig.4. Evolutions of p_{max} and PR with ignition energy, for sweet potato, magnesium and bituminous coal dusts with concentrations of 125 and 400 g/m^3

A dimensionless parameter pressure ratio was introduced to subtract the impact of ignitiors. Pressure ratio is defined as PR = (p_{max} - p_{ign} + p_0)/ p_0, where p_0 is the initial absolute pressure (1.0 bar). Different characteristics of PR under various ignition energies were observed, indicating that influence of ignition energy on the explosion thermodynamics depends on the type of dusts (Figs. 4b and d). The increase of PR for bituminous coal and sweet potato dusts is the result of the increase of volatile content in the atmosphere. This behavior is ascribed to the dependence of volatilization effectiveness on ignition energy, that is, the volatile matter yields of carbonaceous dusts are related to the strength of ignition sources [14]. Moreover, the influence of ignition energy is more pronounced for bituminous coal dust than sweet potato dust, which is ascribed to lower volatility of bituminous coal. It should be pointed out that the incomplete deflagration of carbonaceous dusts under low ignition energy results in the under-driving phenomenon, which yields under-estimated thermodynamic data in dust explosion testing.

Nevertheless, the independence of PR on ignition energy for magnesium dust is the most distinguishing feature in comparison with carbonaceous dusts. The good stabilization of PR reveals that the amount of energy liberated during the explosion process is relatively independent of ignition energy. Hence the under-driving phenomenon, which usually takes place for carbonaceous materials, is not notable for magnesium dust. This is because the magnesium particle does not undergo a volatilization process dominated by the activation of ignition energy but tends to represent an extremely efficient surface heterogeneous oxidation [15, 16] only dominated by the kinetics of oxygen dissolution in the molten particles [17-19].

Under-driving phenomenon

To gain an insight into the under-driving phenomenon, systematic experiments were performed for sweet potato powder with concentration ranging from LEL to 1000 g/m^3, using ignitors of 1, 2, 5 and 10 kJ. Fig. 5 shows that PR is somewhat independent of ignition energy under fuel-rich conditions. Nevertheless, below the concentration of 500 g/m^3, a significant decrease of PR with the decay of ignition energy is observed, which means that the under-driving phenomenon becomes conspicuous as the dust concentration decreases. Due to better inter-particle heat transfer and less heat loss, the highly loaded dust cloud usually has a much better flame propagation in comparison with the fuel-lean mixture [3, 20], which means that the feeble activation of weak ignition source is somewhat partially balanced by the sufficiently efficient particle preheat process under fuel-rich conditions. This may be the reason why highly loaded carbonaceous dust cloud hardly suffers under-driving.

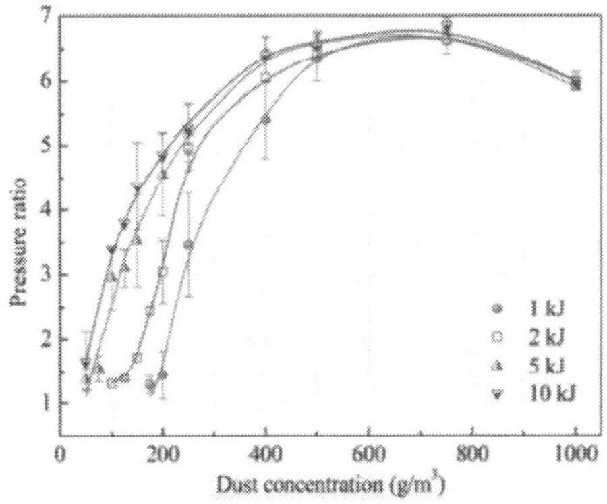

Fig.5. Evolutions of PR with sweet potato dust concentration for various ignition energies

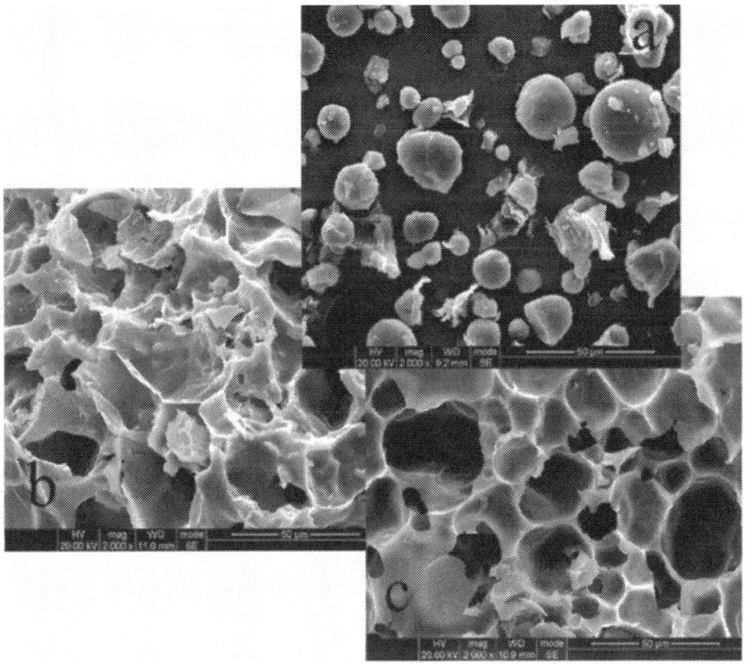

Fig.6. Photomicrographs of original unburned (a) and post-explosion particles resulting from 1 kJ (b) and 10 kJ (c)

Date obtained with dust concentration of 250 g/m^3 were picked out for a typical discussion. Results show a clear ignition energy dependency and give a PR of 3.47 at 1 kJ, 4.97 at 2 kJ, 5.20 at 5 kJ and 5.27 at 10 kJ. The rapid increase of PR means that the volatile matter yield is significantly promoted by the increasing ignition energy. However, the stabilization of PR is identified when ignition energy ranges from 5 to 10 kJ, which gives the idea that the dust explosion system is no longer under-driven due to the sufficiency of ignition energy. Fig. 6 shows SEM photomicrographs of sweet potato particles before (a) and after explosion at ignition energy of 1 kJ (b) and at 10 kJ (c). Remarkable morphological changes in the residue surface are verified by comparing the original unburned particles on the top side of the figure with the post-explosion particles on the bottom of the figure, indicating that the burned particles form cenospheres [21] and volatile matter emits via the 'blow holes' [22] porous structure. Note that the residue resulting from 10 kJ has better porosity than 1 kJ, which confirms that stronger ignitor can promote the pyrolysis process and contributes to more complete devolatilization.

Explosion kinetics

The influence of ignition energy on the apparent explosion kinetics was evaluated for sweet potato and magnesium dusts, at the dust concentration of 500 g/m^3, where the incomplete deflagration caused by under-driving does not occur. In addition to using ignition sources of 1, 2, 5 and 10 kJ, tests employs an overly strong ignitor with energy of 20 kJ as an additional testing. The activation behavior of 20 kJ ignitors can be found in author's previous study [3]. Evolutions of $(dp/dt)_{max}$ were plotted as a function of ignition energy (Fig.7). As can be seen, stronger pyrotechnical ignitor promotes the acceleration of combustion rate. By considering that the overall reaction rate of magnesium dust combustion is mainly controlled by the transport of oxygen to the molten particle surface according to the heterogeneous mechanism [17-19], the above-mentioned evolution can be related to the behavior that the increasing temperature rise resulting from stronger ignition source can promote the melting of particles and thereby enhances the surface heterogeneous oxidation. Alternatively, the turbulence induced by strong ignitors also promotes the burning velocity because it gives rise to mixing of the hot burned and burning parts of the dust cloud with the unburned parts [23]. For sweet potato dust, nevertheless, the evolution of $(dp/dt)_{max}$ is primarily associated with the acceleration of devolatilization rate caused by the stretching of effective flame front area [5], rather than faster vapor-phase combustion reaction rate due to the promoted temperature and the violent turbulence induced by the activation of ignition charge. It is worthwhile emphasizing that $(dp/dt)_{max}$ varies more significantly with the rise of ignition energy for carbonaceous particles, which gives the idea that there is less of an ignition effect on burning velocity for the surface-heterogeneous-oxidation-prone dust than for dusts undergoing devolatilization followed by vapor-phase combustion.

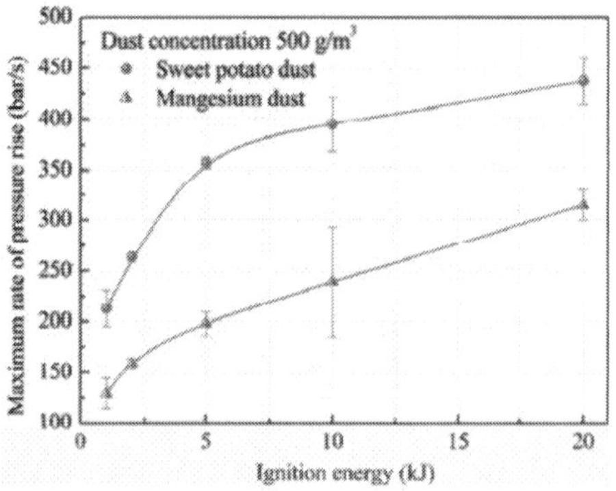

Fig.7 Evolutions of $(dp/dt)_{max}$ with ignition energy, for sweet potato and magnesium dusts with a concentration of 500 g/m^3

Evaluating over-driving by using dimensionless flame thickness

Either sweet potato or magnesium dust represents a rapid rise of $(dp/dt)_{max}$ with energy ranging from 1 kJ to a critical value, followed by a gentle linear increase for higher energies, but fails to represent a horizontal asymptote. The critical value, that is, the inflection point between the rapid rise and the gentle linear increase, is set at 5 kJ for sweet potato and at 2 kJ for magnesium dust respectively. This behavior to some extent gives the idea that values of $(dp/dt)_{max}$ may be over-estimated by using strong ignition sources.

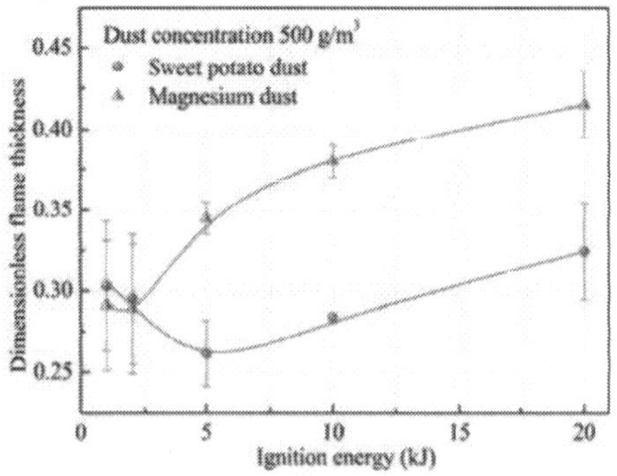

Fig.8. Evolutions of δ with ignition energy, for sweet potato and magnesium dusts with a concentration of 500 g/m^3

To gain an insight into the dust flame propagation under different activations of ignition charge, evolutions of δ for the set of measurements in Fig. 7 were plotted as a function of ignition energy in Fig. 8. Unlike the monotonic increase of $(dp/dt)_{max}$ with ignition strength, δ-values of both sweet potato and magnesium dusts decrease and then increase with the rise of ignition energy. Note that the peak points are identical with the corresponding inflection points illustrated in Fig. 7. The decrease of δ could be ascribed to the acceleration of combustion rate, by considering that the trend of thinner flame thickness is found to be associated with faster flame propagation [15, 24]. Nevertheless, the subsequent increment of δ could be related to wider combustion zone. Strong ignitor explodes with more violent energy release, which will excessively enhance the flame front propagation. In this situation, the reaction rate in combustion reaction zone cannot match the overly accelerated propagation wave of flame front, and thereby the reaction zone becomes wider again. It follows that the overall system is over-driven and even the self-sustained dust flame propagation could be somewhat superimposed by the violent activation of ignition charge, which means that measured kinetic parameter $(dp/dt)_{max}$ as well as the burning velocity is over-estimated.

Consequently, the occurrence of over-driving can be verified when the decreasing δ tends to increase. Moreover, it is found that the surface-heterogeneous-oxidation-prone dust is much easier to be over-driven than dusts undergoing devolatilization followed by vapor-phase combustion. Note that there is a pronounced conflict between the thermodynamics and kinetics measurements. It is advocated to employ strong ignition energy like 10 kJ ignitor in thermodynamic parameter determination to prevent tested sample from under-driving, whereas such powerful activation of ignition charge would lead to over-driving and thereby causes the over-estimation in kinetics testing. Hence, these two facts taken together, it is suggested that the most appropriate ignition energy in both thermodynamics and kinetics determinations may be the energy at which the minimum of δ is attained. From this point of view, the procedure of EN 14034-2, which demands 10 kJ ignition sources in kinetics testing, would result in over-driving, and it therefore is somewhat inappropriate.

Conclusions

Impacts of ignition energy are much more pronounced for carbonaceous dust than light metal dust, because the devolatilization of carbonaceous dust is strongly dependent on the activation of ignition sources, especially for powder with less volatile. Weak ignition sources will cause the incomplete deflagration of carbonaceous dust, referred to as under-driving. The occurrence of under-driving in determining thermodynamic characteristics will yield under-estimated data. The under-driving is found to happen mostly for fuel-lean conditions but seems to be inconspicuous as the dust concentration is raised. For light metal dust, nevertheless, under-driving is not notable due to the efficient surface heterogeneous reaction.

Also it is found that apparent combustion rate increases with ignition energy. However, over-driving will occur when the energy level is excessive for initialing the explosions, so that the measured explosion kinetic parameter will be over-estimated. Moreover, light metal dust is easier to be over-driven than carbonaceous. To verify the occurrence of over-driving, a dimensionless parameter δ was introduced based on the behavior of flame propagation. The analysis shows that δ presents a minimum with the rise of ignition energy, and the minimum value is found to correspond to a critical energy above which the system will be over-driven. It is advocated that the determination of explosion severity should be performed at the critical energy to avoid both under- and over-driving. As a result, the procedure of EN 14034-2 seems to be inadequate somewhat.

References

[1] R.K. Eckhoff, Current status and expected future trends in dust explosion research, J Loss Prevent Proc, 18 (2005) 225-237.

[2] J.E. Going, K. Chatrathi, K.L. Cashdollar, Flammability limit measurements for dusts in 20 L and 1 m^3 vessels, J Loss Prevent Proc, 13 (2000) 209-219.

[3] N. Kuai, J. Li, Z. Chen, W. Huang, J. Yuan, W. Xu, Experiment-based investigations of magnesium dust explosion characteristics, J Loss Prevent Proc, 24 (2011) 302-313.

[4] R. Pilao, E. Ramalho, C. Pinho, Overall characterization of cork dust explosion, J Hazard Mater, 133 (2006) 183-195.

[5] G.P. Zhen, W. Leuckel, Effects of ignitors and turbulence on dust explosions, J Loss Prevent Proc, 10 (1997)

317-324.
[6] R. Siwek, Determination of technical safety indices and factors influencing hazard evaluation of dusts, J Loss Prevent Proc, 9 (1996) 21-31.
[7] CEN/TC305, Determination of explosion characteristics of dust clouds, Part1: Determination of the maximum explosion pressure p_{max} of dust clouds, European Committee for Standardization, Brussels, 2004.
[8] CEN/TC305, Determination of explosion characteristics of dust clouds, Part2: Determination of the maximum rate of explosion pressure rise $(dp/dt)_{max}$ of dust clouds, European Committee for Standardization, Brussels, 2004.
[9] CEN/TC305, Determination of explosion characteristics of dust clouds, Part3: Determination of the lower explosion limit LEL of dust clouds, European Committee for Standardization, Brussels, 2004.
[10] ASTM, Standard test method for pressure and rate of pressure rise for combustible dusts, ASTM international, West Conshohocken, 2007.
[11] MCI, Determination for minimum explosive concentration of dust cloud, State Administration for Quality Supervision and Inspection and Quarantine of China, Beijing, 1996.
[12] K.L. Cashdollar, K. Chatrathi, Minimum explosible dust concentrations measured in 20 L and 1 m^3 chambers, Combustion Science and Technology, 87 (1993) 157-171.
[13] K.J. Mintz, M.J. Bray, D.J. Zuliani, P.R. Amyotte, M.J. Pegg, Inerting of fine metallic powders, J Loss Prevent Proc, 9 (1996) 77-80.
[14] N. Chawla, P.R. Amyotte, M.J. Pegg, A comparision of experimental methods to determine the minimum explosible concentration of dusts, J Loss Prevent Proc, 75 (1996) 654-658.
[15] E.L. Dreizin, V.K. Hoffmann, Constant pressure combustion of aerosol of coarse magnesium particles in microgravity, Combustion and Flame, 118 (1999) 262-280.
[16] E.L. Dreizin, V.K. Hoffmann, Experiments on magnesium aerosol combustion in microgravity, Combustion and Flame, 122 (2000) 20-29.
[17] E.L. Dreizin, Phase changes in metal combustion, Prog Energ Combust, 26 (2000) 57-78.
[18] E.L. Dreizin, Effect of phase changes on metal-particle combustion processes, Combust Explo Shock+, 39 (2003) 681-693.
[19] E.L. Dreizin, C.H. Berman, E.P. Vicenzi, Condensed-phase modifications in magnesium particle combustion in air, Combustion and Flame, 122 (2000) 30-42.
[20] S. Goroshin, I. Fomenko, J.H.S. Lee, Buring velocities in fuel-rich aluminum dust clouds, 26th. Symposium (International) on Combustion, The Combustion Institute, 1996, pp. 1961-1967.
[21] C.F. Lai, Z.H. Duan, Y.F. Zhang, Explosion mechanism of carbon powder, Explosion and Shock Waves, 30 (2010) 325-328.
[22] K.L. Cashdollar, Coal dust explosibility, J Loss Prevent Proc, 9 (1996) 65-76.
[23] R.K. Eckhoff, Dust explosions in the process industries, 3rd ed., Gulf Professional Publishing, Amsterdam, 2003.
[24] A.E. Dahoe, J.F. Zevenbergen, S.M. Lemkowitz, B. Scarlett, Dust explosions in spherical vessels: The role of flame thickness in the validity of the 'cube-root law', J Loss Prevent Proc, 9 (1996) 33-44.

(388af) Effect of the Flow Field of An Internally Spout-Fluid Bed On Dispersion of Glass Fibre for Thermoplastic Composites Manufacture

Yuan Zong[1], Xiaogang Yang[2], Gance Dai[3] and Ling Zhao[3], (1)Chemcial Engineering, East China University of Science and Technology, Shanghai, China, (2)Glyndŵr University , Wrexham, United Kingdom, (3)East China University of Science and Technology, Shanghai, China

Effect of the Flow Field of an Internally Spout-fluid Bed on Dispersion of Glass Fiber for Thermoplastic Composites Manufacture

Yuan Zong[a], Xiaogang Yang[b], Gance Dai[a], Ling Zhao[a*]

[a] Department of Chemical Engineering

East China University of Science and Technology, Shanghai 200239 China

[b] Institute for Arts, Science and Technology

Glyndŵr University, Wrexham, LL11 2AW UK

*Corresponding Author's E-mails: zhaoling@ecust.edu.cn

Keywords: Fibre flocs; Dispersion; Spout-fluid bed; Numerical simulation

Abstract

The mechanical and thermal properties of glass fibre-reinforced thermoplastics have close relationship with the impregnation of glass fiber by high viscosity resin. A novel methodology of dispersion and pre-mixing for components integratedly by applying a modified spout-fluid bed has been pioneered by State Key Laboratory of Chemical Engineering (China). The modified spout-fluid bed was mounted draft tube and disk-baffle to endow it with special flow field. The mechanics of fiber dispersion in the internally spout-fluid bed was investigated using Large Eddy Simulation (LES) modeling and reported in a previous paper (Zong, Yang and Dai, 2011). This study attempts to reveal correlations between fibre dispersion and the flow field by numerical simulation. The motion of fibers was modeled by species transport equation, coupled with the turbulent shear flow in the spout-fluid bed. The correlation between the fibre concentration and local vorticity was obtained from the simulation. The results showed existence of a strong correlation between the glass fiber concentration and local vorticity. A local high vorticity corresponds a relatively high glass fiber concentration. Due to the addition of the internals (draft tube and disk-baffle) into the spout-fluid bed, the flow field were significantly altered and has a characteristics of relatively uniquely distribution of high turbulence, which was proved to be benefit to the manufacture of fiber reinforced thermoplastic.

References

1. Zong, Y., Yang, X. and Dai, G. (2011) Design simulation of glass-fiber-loaded flow in an internally spout-fluidized bed for processing of thermoplastic composites. I. Flow characterization. Ind. Eng. Chem. Res. 2011, 50(15), pp. 9181-9196 .

Figures

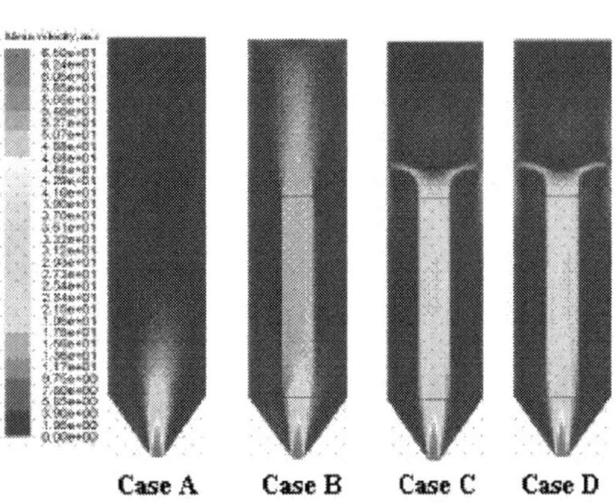

Fig.1. Time-averaged LES velocity distributions

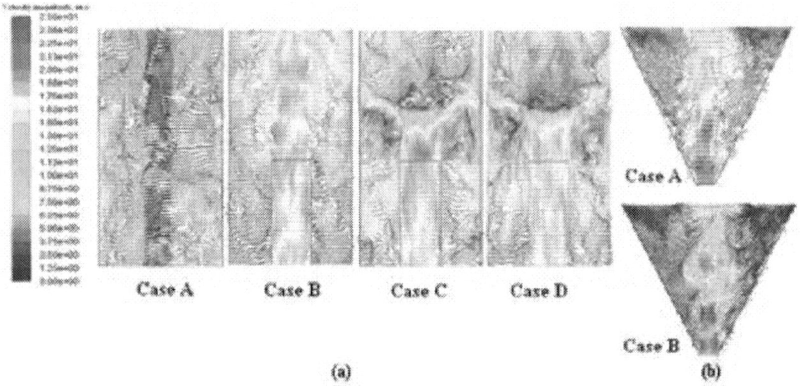

Fig. 2 Velocity vector in entry region and impinging region. (a) impinging region; (b) entry region.

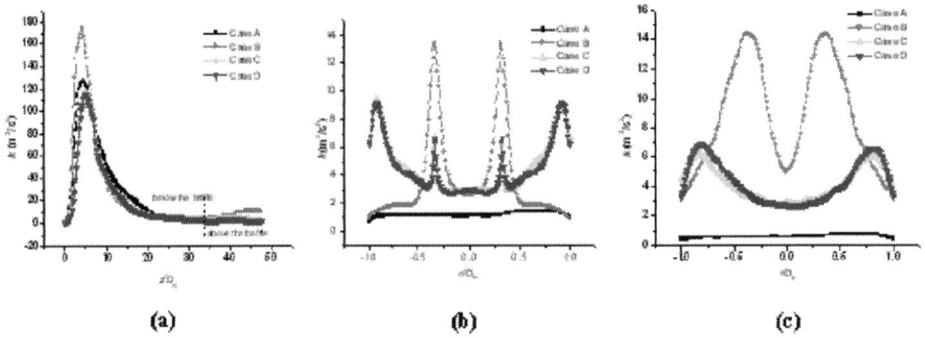

Fig. 3 Turbulent kinetic energy distribution at different position(D_N is the diameter of the spouting nozzle). (a) Along the centerline; (b) impinging region z/DN=32.2;(c) outlet region z/DN=38.9.

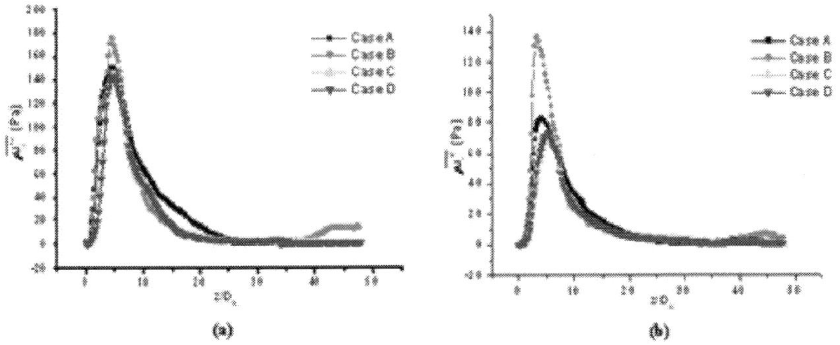

Fig. 4 Axial Reynolds stress along the centerline. (a) ¦$\tilde{N}u_z'^2$;(b) ¦$\tilde{N}u_r'^2$

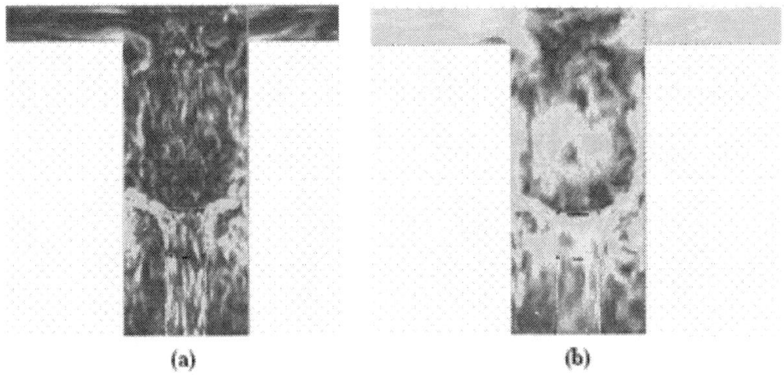

Fig. 5 Vorticity distribution and corresponding fiber distribution inside the reactor. (a) Vorticity distribution ; (b) Fiber concentration distribution

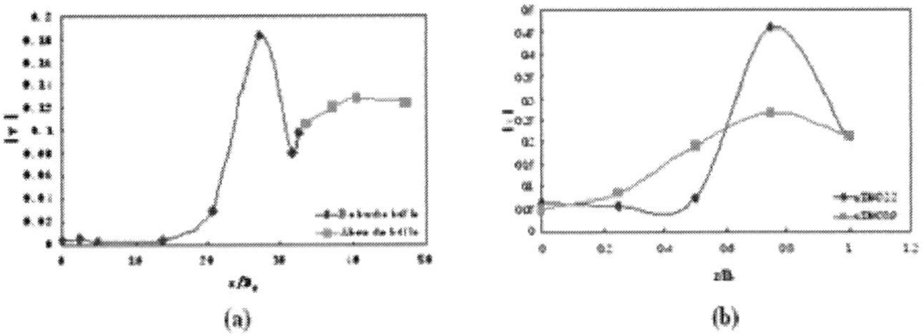

Fig.6 Distribution of correlation coefficient |¦Ã| at different position.(a) axial distribution along the centerline;(b) radial distributions at z/D_b=32.2 and z/D_b=38.9 .

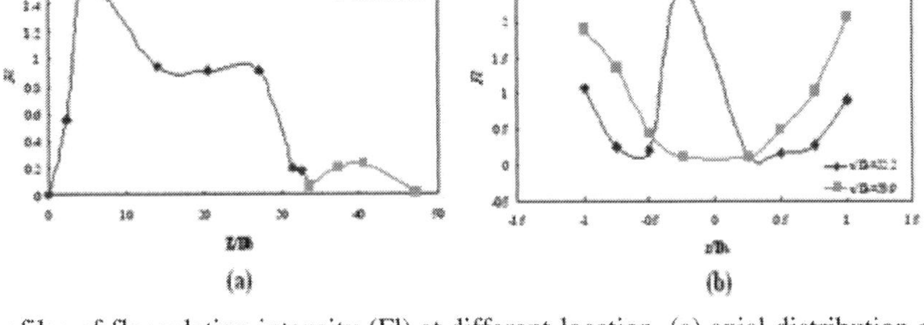

Fig.7 Profiles of flocculation intensity (Fl) at different location. (a) axial distribution along the centerline£»(b) radial distributions at z/D_b=32.2 and z/D_b=38.9 .

(388ag) Increasing the Clay Content and Size of Iron Ore Pellets Reduces Their 'Dustiness'

Joe Halt, Michigan Tech University, Houghton, MI

Potentially hazardous dust is generated when making, handling, and shipping iron ore pellets. Fugitive dust is considered a surrogate for hazardous air pollutants such as Pb, Hg, and incomplete combustion products, and may be a health concern in its own right. Dust generation may depend on many factors, such as the size of iron ore pellets and their microstructure (which is influenced by binders). To test this hypothesis, iron ore pellets were made in the laboratory from an un-fluxed, magnetite concentrate at various sizes and using bentonite clay as a binder. Pellets were made at various sizes ranging from 6.3 – 12.7 mm in diameter, with bentonite clay doses ranging from 0 – 8.9 kg/t, and tested on MTU's dust tower. Pellet dustiness ranged from 10 to 100 mg dust per kg pellets, similar to levels observed in industry-made pellets. Pellet dustiness decreased with increasing pellet size and bentonite clay content.

Selective Collection of Fine Particles by Water Drops

K.R. Liyanaarachchi, G.B. Webber, K.P. Galvin

Centre for Advanced Particle Processing and Transport, Newcastle Institute for Energy and Resources, University of Newcastle, Callaghan, NSW 2308, Australia

ABSTRACT

Froth flotation is normally used to separate the valuable particles from the worthless gangue by taking advantage of the differences in physical properties such as hydrophobicity and hydrophilicity. Although it is a remarkably effective technology for recovering and concentrating hydrophobic particles, there remains a considerable scope for significant improvement. The recovery of fine hydrophobic particles in conventional flotation is very low compared to coarser particle recovery, though due to gravitational forces recovery decreases when the particles become too large. The exceedingly fine hydrophobic particles do not have sufficient momentum to collide with the air bubbles. These particles tend to follow the streamlines, avoiding collision (Figure 1(a)). Even when these particles deviate from the streamlines and collide, they may still have insufficient kinetic energy to rupture the air-liquid interface around the air bubble. These problems arise because of the viscous drag of water. Furthermore, fine hydrophilic particles, which contaminate the product, are difficult to remove because the liquid drains slowly from the froth. The hydrophobic product also retains significant moisture even after filtration.

Recently two research groups examined the potential of an alternative method of fine particle beneficiation. Tran et al. (2009) initiated the preliminary experimental studies on selective separation of ultra-fine particles less than 10 microns using film flotation in a planar trough. Galvin et al. (2010) separated fine coal less than 90 microns using individual drops of water fixed in position under a gas dispersion of fine coal and glass ballotini.

This paper, which builds upon the previous work, is concerned with a new approach for selectively separating particles on the basis of their wetting properties. Here, a dry particle dispersion interacts with falling water drops (Figure 1(b)). The paper also outlines the basic laboratory systems used to test this concept. Hydrophilic particles pass directly through the gas-liquid interface of the falling drops, while hydrophobic particles choose to remain with the gas phase or adhere to the outer surface of the water drops. The underflow collection efficiency achieved for a range of particles interacting with a given flux of falling drops was determined experimentally.

The central process of selective collection is the wetting and particle adhesion. Thomas Young (1804) and Gibbs (1873) provided the foundation used today to describe the interfacial phenomena involved in wetting and adhesion. The basic thermodynamic criterion for the drop-particle attachment involves a finite contact angle at the line formed by the intersection of three phases. The contact angle is a quantitative measure of the wetting of a solid by a liquid (Chau, 2009) and indicates the affinity between the water and the solid (Nicol, 1992). If the contact angle has a higher value (measured through the water), the solid is described as hydrophobic and if a relatively small angle, is described as hydrophilic.

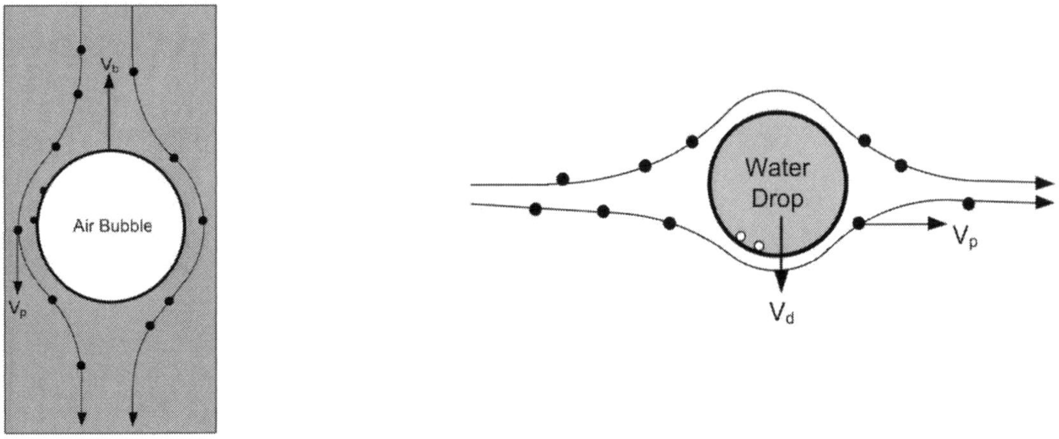

(a) Conventional Flotation (b) Proposed Selective Collection

Figure 1: Comparison between particles interacting with bubbles in water and drops in air.

Two experimental approaches have been used to test this concept. Preliminary tests have been conducted using single water drops to test the validity of the phenomena at the drop level. Secondly, a laboratory scale experimental system has been designed and developed for detailed investigations of the concept at the multiple drop level.

Dry particles of glass ballotini and coal (75 to 90 microns size range) were used in single drop experiments. The first set of single drop experiments was conducted using a vibratory table under a hanging water drop to energise the system of particles. For glass ballotini (Figure 2(a)), most of the particles exhibited engulfment into the water drop. The second set of single drop experiments was conducted by sprinkling the particles above the hanging water drops. Video microscopy was used to monitor the phenomena. When a mixture of ballotini and coal was used, a considerable amount of ballotini particles was observed (Figure 2(b)) to again penetrate the drop, while there was some evidence of fine coal at the drop surface. These single drop experiments clearly demonstrate selectivity in the process; that is, the ballotini is engulfed into the bulk of the drop while the few coal particles that do interact with the drop remain at the interface.

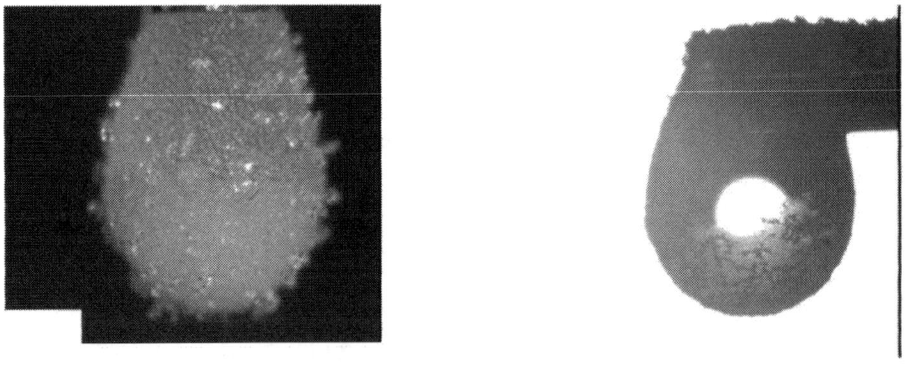

Figure 2: (a) Pendent drop of water saturated with glass ballotini. (b) Pendent drop covered with coal while ballotini filled the inside of the drop

An experimental system was then designed and fabricated to test this novel concept at the multi-drop level in the laboratory. A schematic diagram of the initial prototype is shown in Figure 3. The system consists of a feed column, water distribution chamber mounted on a collision chamber and an air suction pump. Falling drops are surrounded by gas with a viscosity 100 times lower than the viscosity of water. The capture efficiency of the hydrophilic particles by the water drops should increase significantly due to the low viscosity of the air. In other words, inertial forces should dominate, hence ultrafine particles should engage with the gas-liquid interface. Hydrophilic particles covering a broad size range should pass through the interface with little or no resistance. Hydrophobic particles should fail to pass through the interface, and hence should remain with the gas phase or attach to the outside of the drops. Thus the proposed approach should result in improved separation kinetics.

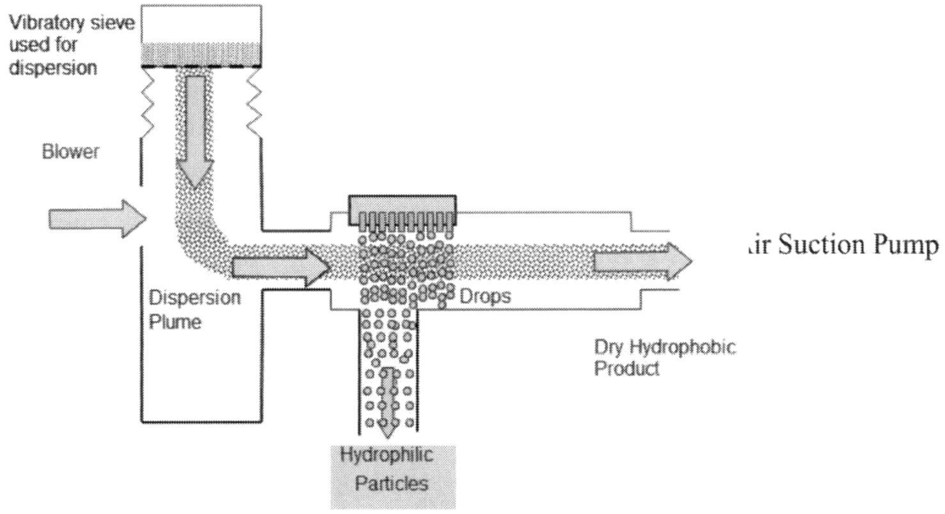

Figure 3: Schematic illustration of modified experimental system.

The feed was projected into the collision chamber horizontally while water drops were released from the water distribution chamber vertically. The particles were fed to the system in multiple 10g portions within a time period of 30 seconds between each portion, while maintaining a constant water drop size and flux.

The dispersion then collided with the water drops, with the hydrophilic particles (silica) engulfed. Water collected at the base of the chamber, and then drained into the underflow collection tank. The recovery of the particles in the underflow was calculated relative to the total feed that passed through the experimental system. In all of these experiments the velocity of the air stream and hence the velocity of the particles was kept constant.

The results from these experiments produced a linear correlation between the recovery of the particles in the underflow and the water flux. The underflow recovery for two types of particles provided a measure of the selectivity of the process. The selectivity of different samples of glass ballotini (glass ballotini as received from the commercial pack and glass ballotini treated with Extran® to clean the surface) were tested initially. The results were found to be reproducible, hence the experimental system was found to be reliable.

Interestingly the experimental data indicated some selectivity between the ballotini samples, as received, versus those that were cleaned, though the difference was slight (Figure 4).

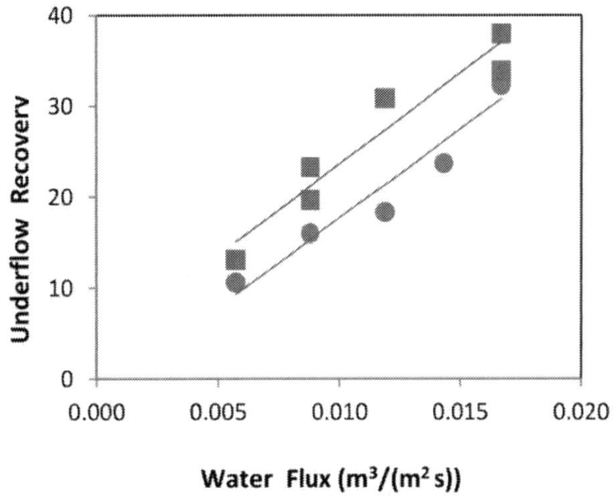

Figure 4: Underflow recoveries of two different ballotini samples, as received, and cleaned.

The underflow recovery of hydrophilic silica particles were observed to be in the same range as that of cleaned ballotini. For hydrophobic coal particles relatively low recoveries were obtained. This work proved that there is selectivity between hydrophilic silica and hydrophobic coal. These results are supported by the experimental data shown in Figure 5.

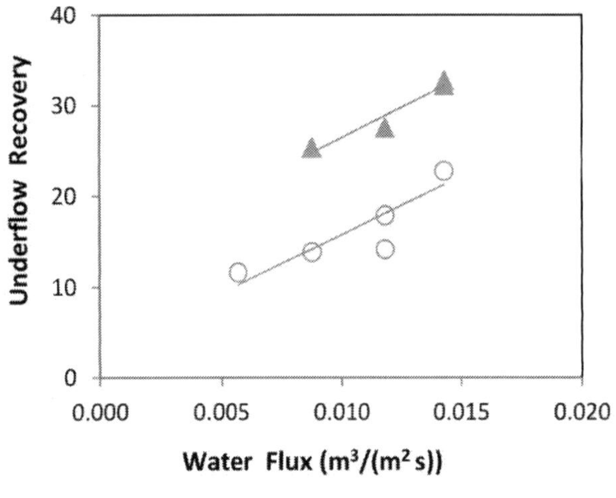

Figure 5: Underflow recoveries of silica and coal showing clear, but limited, selectivity.

One key advantage of the technique described in this paper is that the hydrophobic product obtained is dry. Overall, there was clear selectivity measured, dependent on the particle size and other surface properties of the material. Further work is needed to study other variables such as the air flow velocity, and water chemistry, as well as mixtures of hydrophobic and hydrophilic particles. Given the limited selectivity observed, however, it is unlikely that this concept will develop into a new technology competitive with conventional froth flotation.

Acknowledgements

The authors thank the Australian Coal Association Research Program (ACARP) for financial support for this work.

References:

Adamson, A.W., Gast, A.P., 1997. Physical chemistry of surfaces, sixth ed. John Wiley & Sons, New York.

Tran, D.N.H, Whitby, C.P., Fornasiero, D., Ralston, J. 2010. Selective separation of very fine particles at a planar air-water interface, International Journal of Mineral Processing 94, 35-42

Galvin, K.P., Webber, G.B., Mason, M., Liyanaarachchi, Inverse Flotation-A New Method of Fine Particle Beneficiation, Chemeca 2010, Adelaide.

Young, T.1805. An essay on the cohesion of fluid, Philosophical Transactions, Royal Society Publish.

Chau, T.T., 2008. A review of techniques for measurement of contact angles and their applicability on mineral surfaces, Minerals Engineering 22 (2009) 213-219

Nicol, S.,1992. The Principles of Coal Preparation, Australian Coal Preparation Society, Newcastle, NSW, p. 232

(388c) Deposition of Non-Spherical Particles in a Bifurcating Airway

Poom Bunchatheeravate, Chemical Engineering, University of Florida, Gainesville, FL and Jennifer Sinclair Curtis, University of Florida, Gainesville, FL

Title: Deposition of Non-Spherical Particles in a Bifurcating Airway

Poom Bunchatheeravate and Jennifer Curtis

Department of Chemical Engineering, University of Florida, Gainesville, FL 32611. USA

Particles between the sizes of 1-10um are more susceptible to deposition in the human lung. Previous investigations of particle deposition in the human airways focus on the deposition of spherical particles. Only a handful of studies focus on the effect of particle morphology. These studies include the deposition of fibrous particles in response to pollutants such as asbestos, deposition of porous particles for higher drug loadings, as well as deposition of nano-size particles. The current study seeks to quantify the deposition of different shaped particles and establish a relationship between particle shape and deposition. The airway model used in this study is that of a triple bifurcating airway that was glass blown with the dimensions similar to that of Weibel's Model A. The simplicity of the airway model helps to isolate the effect of particle morphology on deposition patterns. The particles used in this study were classified into three general shapes - spherical, rod-like, and flake-like. The effect of particle density was also studied for spherical particles and flake particles. All of the particles studied had a mean aerodynamic diameter between 1-10 microns. Particles were dyed using a near-infrared dye and sent into the airway model with a fixed volumetric flowrate. After the deposition, the particle were visualized using IVIS Lumina fluorescence imaging machine. The IVIS Lumina provides two essential pieces of information - the fluorescence image showing the position of particle and the photon counts emitted in each region of the airway model relating to the local concentration. The deposition in each individual airway was quantified for each type of particle. The deposition associated with each type of particle is compared and an empirical model was developed relating the deposition efficiency in each branching airway with the particle properties. It was found that the particle properties that influence the deposition are the particle Stokes number, the diffusion coefficient, the density and the sphericity.

(388d) Biomass Size Reduction of Microalgae

Rajasekhar Anumolu, Chemical Engineering, University of Utah, Salt Lake City, UT and Leonard F. Pease III, Chemical Engineering and Pharmaceutics & Pharmaceutical Chemistry, University of Utah, Salt Lake City, UT

Microalgae derived fuels have great potential to replace fossil fuels with environmental friendly alternatives. Microalgae are advantageous because they grow quickly, offer higher photosynthetic efficiencies, better accumulation of lipids, may offer hydrogen for fuel cells, and synthesis of energy storing starch relative to most biofuel feedstocks. However, the high water content of microalgae restricts the optimization of biomass energy conversion. Here, we report on a new technique to remove the water from microalgae leaving dry biomass. This technique atomizes the micron sized algae within the droplets. The nebulized droplets dry in a diffusion drier prior to entering a differential mobility analyzer (DMA), in which they are classified by their size-to-charge ratio. The dry algal biomass was observed to have a dry size ranging from 50-300 nm and was approximately spherical in shape. The same algae were sized between 10-15 microns using asymmetric field-field flow fractionation (AFFFF) in liquid form. DMA classified algae were electrostatically deposited and compared with the unaltered algae. Electron microscopy reveals that DMA classified algae preserved organelles yet consistently separated the biomass from water. This technique may improve the efficiency of biomass energy conversion.

ID#: 254227

Experimental investigation on the influences of the moisture content on coal dust explosion

Jingjie Yuan, Weixing Huang, Niansheng Kuai, Bing Du, Jingyi Tan
School of Chemical Engineering, Sichuan University, Chengdu, Sichuan 610065, P. R. China

Abstract

The dust explosion severity and sensitivity are related to dust moisture content to a large extent. To investigate the effects of moisture on dust explosion behavior is therefore of practical needs for the prevention and mitigation of dust explosion. In this work, systematic experiments were performed on pre-humidification coal dust explosions by using the Siwek 20-L vessel. The influences of moisture content on coal dust explosion behavior had been studied by testing the maximum explosion pressure, the maximum rate of pressure rise and the explosive limit concentrations. Inhibition phenomena of moisture have been observed for various dust concentrations and particle size distributions, and there always exists a critical moisture content to fully suppress the explosion for the coal dust with a given size distribution. The suppressing effectiveness of moisture increases with the reduction of dust heat value, but weakens with the decrease of particle size. For the explosions of large size coal dust, the moisture mainly acts as the thermal sick to absorb heat and forms water vapor to dilute oxygen. Consequently, with the rise of moisture content, the maximum explosion pressure and the maximum rate of pressure rise decrease linearly and the dust explosible concentration range narrows significantly until the coal dust explosion is fully suppressed. For the explosions of small size coal dust, a suddenly enhanced inhibition on the explosion severity and the explosive limits can be observed when the coal dust has higher moisture content. It has been verified that this enhanced inhibition is caused by the notable dust agglomeration emerged in the small size coal dust with higher moisture content. Based on the specific role of moisture in affecting dust explosion behavior, to prevent coal dust explosion and, responsibly, determine the explosion characteristics, the influences of moisture content should be taken into account.

Keywords: *Moisture content; Coal dust explosion; Explosion characteristic*

Introduction

As an important physical property, the moisture content (or humidity) plays a crucial role in the combustion and dispersion of dust particles. The difference of moisture content will give rise to the variations of particle ignitability and dispersibility so as to affect the dust explosion behavior[1]. Thus, in the process of assessing dust explosion risk, the effects of moisture content should be taken into account.

However, the majority of test standards related to dust explosion characteristics neglect the effect of moisture content and only specify to write down the moisture content of test sample if necessary[2-5]. To study and clarify the influences of moisture content is of great importance for the reliable determination and full knowledge of dust explosion characteristics.

In addition, based on the moderation principle of inherent safety, to study the explosion behaviors of dust with different moisture content is also essential for the inerting or suppression of dust explosion if considering the moisture or water as a special inertant. The inerting or suppression of dust explosion is to prevent the occurrence or mitigate the consequences of a dust explosion by adding some inert material[6]. Generally speaking, most of inert materials used in inerting or suppression are mainly the solid (e.g. rock dust) and gas (e.g. inert gas). If the liquid can be used as an inertant or suppressant, the inerting or suppression technology of dust explosion will be greatly improved and enriched due to the unique advantages of liquid in extinguishing the flame and altering the particle properties. Therefore, it is necessary to investigate the influences of moisture content on dust explosion behavior.

Based on the above-mentioned facts, the purpose of this paper is to present the effects of moisture on dust explosion behavior by means of systematic tests in the 20-L Siwek vessel using coal dusts. Abundant data about the maximum explosion pressure, the maximum rate of pressure rise and the explosive limits were given, and the experimental results were analyzed and explained in the view of suppression.

Experimental

As shown in Fig. 1, the experimental set-up used in this paper is the 20-L explosion sphere, which was initially developed by Siwek[7] in order to characterize the ability of powders to explode and had become a standard laboratory test chamber used extensively for dust explosion studies[8-11].

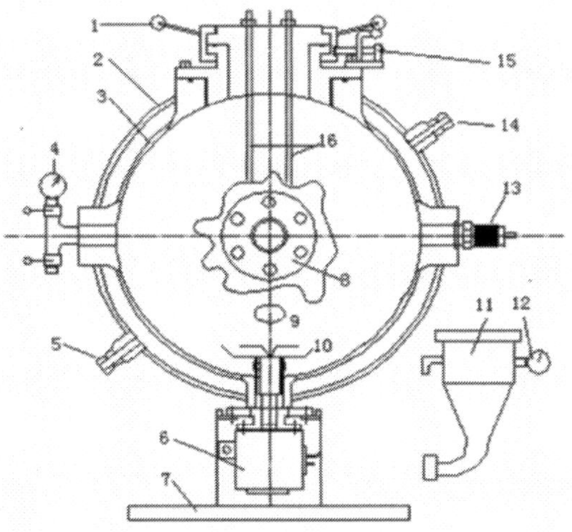

Fig. 1 Schematic diagram of the 20-L sphere
1-End Closure 2-Jacket 3-Sphere Shell 4-Vacuum Gauge 5-Circulating Water Inlet 6-Two-way Mechanical Valve 7-Base 8-View Port 9-To Vacuum Pump 10-Rebound Nozzle 11-Dust Container 12-Air Pressure Gauge 13-Piezoelectric Pressure Sensor 14-Circulating Water Outlet 15-Safety Limit Switch 16-Ignition Electrode

The test chamber is made of stainless steel with a double jacket, and there is a 0.6 L dust container connected to the 20-L sphere. At the bottom of this sphere, the Rebound Nozzle is used as the dust dispersion device, and above this dispersion device, a chemical ignitor is centrally-mounted to trigger the dust explosion. During an explosion, the pressure inside the sphere is measured by the piezoelectric sensors mounted in the vessel wall. The results are recorded by a computer and the expressed as the pressure-time curve. It should be pointed out that, in accordance with the recommendations of IEC standard[12] and Chinese standard[13], the explosion criterion used in this work is that explosion overpressure excluding ignitor effects is not lower than 0.4 bar.

The ignition energy used in the tests mainly involves 5 kJ and 10 kJ in this paper. The 5 kJ ignitors were used to measure the explosion severity, and the 10 kJ ignitors were used to measure the minimum explosible concentration. In addition, the test sample of this work is pr-humidification coal dust, and at room temperature, the compositions, size distributions and heat value of test sample are given in Table 1 and 2, respectively.

Table 1. Composition analysis of test sample

Composition	Fixed carbon	Volatile	Ash	Moisture
Coal sample A (wt%)	37.01	20.12	41.76	1.11

Table 2. Size distributions and heat values of coal samples

Sample	Coal sample A			Coal sample B
	No. 1	No. 2	No. 3	
Size distributions (μm)	125-550	75-125	43-75	75-125
Heat value (kJ/g)		21.698		30.939

Results and discussion

Typical pressure-time curves of coal dust explosion

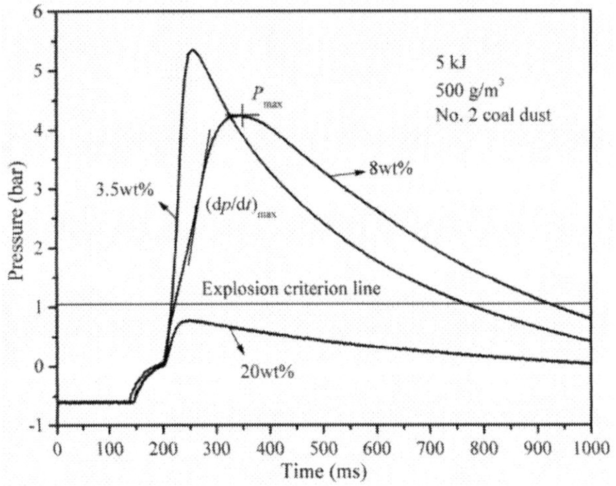

Fig. 2 Representative pressure-time curves of dust explosion

Under different moisture content, Fig. 2 gives the representative pressure-time curves of coal dust explosion. As shown in Fig. 2, the maximum explosion pressure and the maximum rate of pressure rise can be directly obtained from the pressure-time curve, where the maximum explosion pressure is a value including ignitor effects. To obtain the pure pressure rise of an explosion, the pressure rise caused by the ignitor should be excluded. Furthermore, due to the increase of moisture content, an obvious decrease on the maximum explosion pressure and rate of pressure rise also can be found from Fig. 2, and especially, when the moisture content is high enough, the dust explosion even can not happen. These phenomena indicate that the variation of moisture content will significantly change dust explosion behavior.

Influences of moisture content on coal dust explosion severity

With respect to dust explosion severity, the maximum explosion pressure and the maximum rate of pressure rise are both typical severity characteristics[14], where the maximum explosion pressure is concerned with the heat release amount of an explosion and the maximum rate of pressure rise is concerned with the heat release rate of an explosion. To investigate the influences of moisture content on dust explosion severity, a series of tests were preformed using Coal sample A at an extensive moisture content range with the concentration of 500 g/m^3. Evolutions of explosion severity parameters are plotted as the functions of moisture content shown in Fig. 3 (a) and Fig. 3 (b).

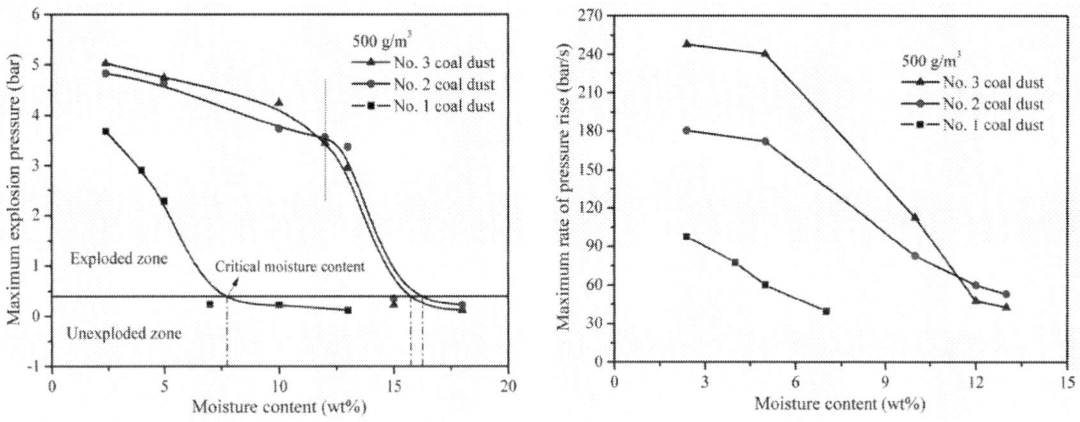

(a) Maximum explosion pressure **(b)** Maximum rate of pressure rise

Fig. 3 Evolutions of explosion severity parameters with moisture content

As shown in Fig. 3 (a), a remarkable inhibition on the maximum explosion pressure can be observed, and there always exists a critical moisture content above which the coal dust explosion will be fully suppressed. For No. 1 coal dust, the decrease of maximum explosion pressure is linear approximately until the explosion is fully suppressed. Due to the worse ignitability and flammability of large size particle, a much stronger moisture suppression effect exists in the explosions of No. 1 coal dust. For No. 2 and No. 3 coal dusts, a suddenly enhanced decline on the maximum explosion pressure can be found when the moisture content is high enough, especially for the No. 3 coal dust which has a smaller particle size, although the maximum explosion pressure of No. 1, No. 2 and No. 3 coal dusts has a similar variation trend under lower moisture content.

Compared with Fig. 3 (a), it can be seen from Fig. 3(b) that the variation of maximum rate of pressure rise is similar to that of maximum explosion pressure with the increase of moisture content. For No. 1 coal dust, a stronger suppression effect can be observed, and the maximum rate of pressure rise decreases linearly with the rise of moisture content. For No. 2 and No. 3 coal dusts with high moisture content, the maximum rate of pressure rise also has a suddenly enhanced decline, especially for the No. 3 coal dust.

In fact, the specific role of moisture in affecting dust explosion mainly involves two aspects of meanings. On the one hand, the moisture may hinder the ignition and combustion processes of dust cloud by absorbing heat and diluting oxygen. On the other hand, the moisture may impact the dispersion behavior of dust cloud by varying the particle dispersion property.

For the large size dust (No. 1 coal dust), the dust dispersion property is almost not affected by the moisture in an extensive moisture content range due to the very weak inter-particle cohesion of large size dust, and the moisture mainly acts as the thermal sick to absorb heat and forms the water vapor to dilute oxygen in the ignition and combustion processes of dust cloud. On the contrary, for the small size dust (No. 2 and No. 3 coal dusts), when the moisture content is high enough, an obvious dust agglomeration will appear, and then further enhance with the increase of moisture content or decrease of particle size, even though these particles can still maintain the good dispersibility under lower moisture content. Fig. 4 gives the photographs of No. 1, No. 2 and No. 3 coal dusts with higher moisture content. It can be confirmed that the obvious dust agglomeration just exists in the small size dust (No. 2 and No. 3 coal dusts).

No. 1 coal dust No. 2 coal dust No. 3 coal dust

Fig. 4 Photographs of Coal sample A with the moisture content of 13 wt%

Due to this suddenly appeared agglomeration, the dust dispersibility will be weakened significantly so as to affect the dispersion behavior of dust cloud. With the further increase of moisture content, the dust agglomeration will become the main controlling factor of dust explosion, especially for the dust with smaller particle size. Therefore, a suddenly enhanced decline on the maximum explosion pressure and the maximum rate of pressure rise can be found for No. 2 and No. 3 coal dusts with higher moisture content, and because of the smaller particle size, this enhanced decline of No. 3 coal dust is more notable compared with that of No. 2 coal dust.

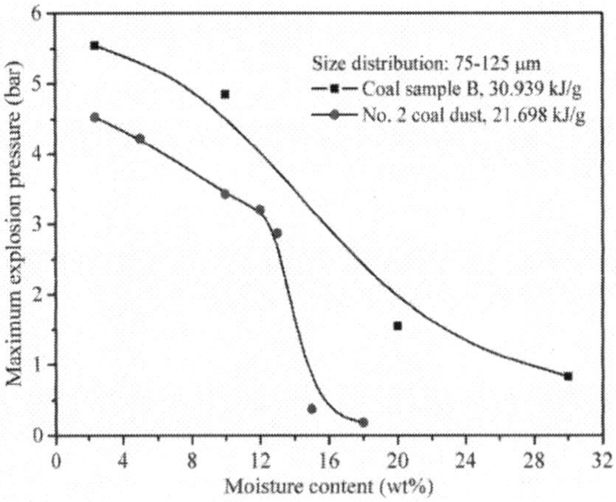

Fig. 5 Maximum explosion pressures of coal dusts with different heat values

In order to further reveal the moisture inhibition effects, Fig. 5 presents the maximum explosion pressures of the coal dusts with different heat values at the concentration of 300 g/m^3.

Compared with Fig. 3(a), the similar moisture inhibition effects can be observed, although the test concentration is totally different. Furthermore, in contrast to the two dusts in Fig. 5, it can be found that, at a given dust size distribution, the Coal sample B has a much higher critical moisture content to fully suppress the explosion. This is because the Coal sample B has a much higher heat value than the No. 2 dust of Coal sample A, and at a given dust size distribution, the higher dust heat value means a better ignitability and flammability. Thus, the suppressing effectiveness of moisture will reduce with the increase of dust heat value.

As a conclusion, the moisture can affect the ignition, combustion and dispersion of dust particles so as to give rise to a significant inhibition on dust explosion severity, especially for the coal dust with smaller particle size or lower heat value. To reliably determine dust explosion severity, the influences of moisture content should not be ignored.

Influences of moisture content on coal dust explosive limits

To form a dust explosion, dust concentration should be within the explosible range. The boundaries of this explosible range are referred to as the explosive limits which contain the lower limit and the upper limit.

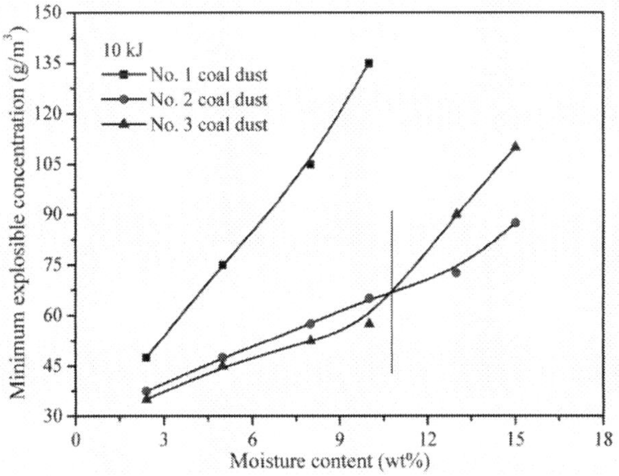

Fig. 6 Development of minimum explosible concentration with moisture content

To study the influences of moisture content on the lower limit, a number of tests were carried out using No. 1, No. 2 and No. 3 dusts of Coal sample A, and development of minimum explosible concentration was plotted as a function of moisture content given in Fig. 6. As expected, the minimum explosible dust concentration significantly enhances with the rise of moisture content, especially for No. 1 coal dust. Due to the weak inter-particle cohesion of large size dust, the particle dispersibility of No. 1 coal dust basically will not be changed with the increase of moisture content in an extensive moisture content range, so that the minimum explosible concentration has a stationary linear variation. To the opposite, because of the existence of notable dust agglomeration, the minimum explosible concentration has an obviously enhanced uptrend for No. 2 and No. 3 coal dusts with high moisture content. Moreover, if the moisture content continues to increase, the minimum explosible concentration of No. 3 coal dust will become higher than that of No. 2 coal dust due to the more notable dust agglomeration existed in No. 3 coal dust.

Besides the lower limit, the upper limit is also significantly affected by the moisture content. Fig. 7 gives the maximum explosion pressure of No. 2 coal dust with different moisture content in an extensive concentration range.

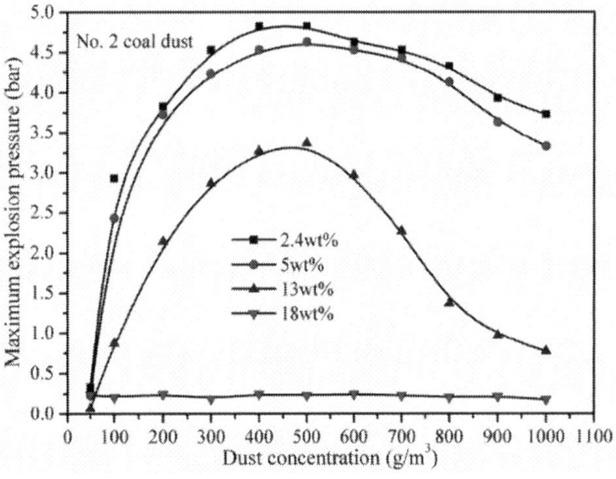

Fig. 7 Evolution of maximum explosion pressure with dust concentration

As presented in Fig. 7, with the rise of dust concentration, evolutions of maximum explosion pressure are different for the coal dust with different moisture content. For the coal dust with lower moisture content (not exceeds 5 wt%), the maximum explosion pressure levels off at the concentration of 300 to 500 g/m^3, and at even dust concentration of 1 kg/m^3, the maximum explosion pressure nevertheless has no evidence of a decline. This fact implies that there is a very high upper limit concentration for the coal dust with lower moisture content. However, for the coal dust with higher moisture content (e.g. 13 wt%), the maximum explosion pressure becomes more sensitive to dust concentration, especially that a sharp decline on the maximum explosion pressure can be observed at higher dust concentration. Moreover, if the moisture content is overly high (e.g. 18 wt%), the coal dust explosion will be fully suppressed. It means that the upper limit concentration of coal dust explosion is significantly weakened with the increase of moisture content.

Consequently, the moisture content of coal dust has a significant influence on the explosion limits. With the rise of moisture content, the lower limit will increase and the upper limit will weaken so as to narrow the explosible concentration range of coal dust.

Conclusion

Systematic study on the explosion behaviors of the pre-humidification coal dusts has been conducted by using the Siwek 20-L sphere. Abundant severity and sensitivity data of coal dust explosion were reported. The influences of moisture content on coal dust explosion were investigated by measuring the explosion characteristics at different dust moisture content conditions.

As a consequence, a remarkable inhibition on coal dust explosion severity and explosive limits can be observed, and there always exists a critical moisture content to fully suppress the explosion for the coal dust with a given size distribution. Because there is a worse ignitability and flammability for the coal dust with lower heat value or larger size distribution, the suppressing effectiveness of moisture enhances with the reduction of dust heat value, but weakens with the decrease of particle size.

For the large size coal dust, due to the very weak inter-particle cohesion, the particle dispersion property is basically not affected by the moisture content, and the moisture mainly acts as the thermal sick to absorb heat and forms the water vapor to dilute oxygen in the ignition and combustion processes of dust cloud. Consequently, with the rise of moisture content, the maximum explosion pressure and the maximum rate of pressure rise have a linear decrease and the explosible concentration range notably narrows until the dust explosion is fully suppressed.

However, for the small size coal dust, once the moisture content is high enough, a notable dust agglomeration will appear, and further enhance with the increase of moisture content or decrease of particle size. Therefore, a suddenly enhanced inhibition on the explosion severity and the explosive limits can be observed for the small size coal dust with higher moisture content, even though the inhibition effects of moisture are similar under lower moisture content for both the large size coal dust and the small size coal dust.

In conclusion, to reliably measure the coal dust explosion severity and explosive limits, it is very necessary to take the influences of moisture content into account.

References

1. Eckhoff, Rolf K. (2003), "Dust Explosions in the Process Industries," third ed., Boston, Gulf Professional Publishing.
2. IEC standard 1241-2-1 (1994), "Methods for Determining the Minimum Ignition Temperature of Dust," International Electrotechnical Commission.
3. IEC standard 1241-2-3 (1994), "Methods for Determining the Minimum Ignition Energy of Dust/Air Mixtures," International Electrotechnical Commission.
4. Chinese standard GB/T 16428 (1996), "Determination of the Minimum Ignition Energy of Dust Cloud," State Administration for Quality Supervision and Inspection and Quarantine of China.
5. Chinese standard GB/T 16429 (1996), "Determination of the Minimum Ignition Temperature of Dust Cloud," State Administration for Quality Supervision and Inspection and Quarantine of China.
6. Amyotte, Paul R. (2006), "Solid Inertants and Their Use in Dust Explosion Prevention and Mitigation," *Journal of Loss Prevention in the Process Industries*, 19(2-3), pp. 161-173.
7. Siwek, R. (1996), "Determination of Technical Safety Indices and Factors Influencing Hazard Evaluation of Dusts," *Journal of Loss Prevention in the Process Industries*, 9(1), pp. 21-31.
8. Calle, S., Klaba, L., Thomas, D., Perrin, L. and Dufaud, O. (2005), "Influence of the Size Distribution and Concentration on Wood Dust Explosion: Experiments and Reaction Modelling," *Powder Technology*, 157(1-3), pp. 144-148.
9. Chawla, N., Amyotte, Paul R. and Pegg, Michael J. (1996), "A Comparison of Experimental Methods to Determine the Minimum Explosible Concentration of Dusts," *Fuel*, 75(6), pp. 654-658.
10. Yuan Jingjie, Huang Weixing, Ji Hua, Kuai Niansheng and Wu Yi (2012), "Experimental Investigation of Dust MEC Measurement," *Powder Technology*, 217, pp. 245-251.
11. Garcia-Agreda, A., Di Benedetto, A., Russo, P., Salzano, E. and Sanchirlco, R. (2011), "Dust/Gas Mixtures Explosion Regimes," *Powder Technology*, 205(1-3), pp. 81-86.
12. IEC SC3H sheet 2-5 (1990), "Methods for Determining the Minimum Explosible Concentration of Dust/Air Mixtures," International Electrotechnical Commission.
13. Chinese standard GB/T 16425 (1996), "Determination for Minimum Explosive Concentration of Dust Cloud," State Administration for Quality Supervision and Inspection and Quarantine of China.
14. Dufaud, O., Traore, M., Perrin, L., Chazelet, S. and Thomas, D. (2010), "Experimental Investigation and Modeling of Aluminum Dusts Explosions in the 20 L Sphere," *Journal of Loss Prevention in the Process Industries*, 23(2), pp. 226-236.

(388f) Impact of Operating Conditions On Physical Properties in an Agitated Filter Dryer

Ekneet Sahni, Department of Pharmaceutical Sciences, University of Connecticut, Storrs, CT and Bodhisattwa Chaudhuri, Pharmaceutics, University of Connecticut, Storrs, CT

Drying is often a bottleneck and in an effort to minimize drying times, there is often fiscal pressure to select aggressive drying conditions. Nonetheless, the extreme conditions can have a major impact on the properties, and quality of the material thereby affecting further downstream operations. The study investigates the effect of critical process parameters (wall temperature, impeller speed, pressure, and bed depth) on the drying performance in an agitated filter-dryer with effect on physical properties (particle size, shape, specific surface area, roughness) as the output attributes. The variation of the process parameters in the range investigated revealed that the rate of solvent drying is enhanced with increasing wall temperatures (40 – 80 °C) and decreasing bed depth (28 – 74 mm) due to increased driving force and reduced resistance to heat transfer respectively. The effect of agitation speed (5 – 25 rpm) was variable for both glassbeads and lactose monohydrate suggesting that the influence of operating parameter is dependent on the material properties of the compound under study. This change in the effect of speed was mainly attributed to frictional behavior which was demonstrated from flow dynamics using the velocity profiles. The velocity profiles lactose showed the heap formation leading to formation of a recirculation zone due to high friction coefficients. Because of low friction as in the case of glassbeads, recirculation zones were not created. Although mechanical agitation aids in improving heat transfer, consequences on particle size can be profound. Decrease in the particle size D[4,3] and specific surface area was evident with an increase in impeller speed and decrease in the bed depth due to increased collision frequency and reduction in the fill volume. It was concluded from the particle size distribution studies that attrition dominates the drying process when the loss of solvent drying rate was low and/or the shear rate was high. Under these conditions, the drying time was sufficiently long for particles to encounter countless particle-impeller and particle–particle collisions that can result in small fragments. Particle attrition was much higher towards the end of drying where solvent acting as a lubricant was reduced further. For higher dryer loads, the number of collisions decreased, stronger agglomerates were formed and thus agglomeration becomes more dominant. Particle morphology was greatly affected with fines being generated at higher impeller speeds. Thermograms obtained from conventional DSC scans showed endothermic peaks at 205 – 210 °C demonstrating the melting of α-lactose monohydrate. This was further confirmed by PXRD with specific diffraction peaks at 12.0–12.5°, 16.3–16.6° and 19.10–20.05° two-theta values which were consistent with those of α-lactose monohydrate. The knowledge of how particle properties affect the macroscopic properties is important for an effective operation, scale-up and design.

(388g) A New Type of Unimolecular Polymeric Nanoparticle: Janus Double-Brush Copolymers

Yukun Li, Efrosyni Themistou, Jiong Zou, Biswa P. Das, Marina Tsianou and Chong Cheng, Department of Chemical and Biological Engineering, University at Buffalo, The State University of New York (SUNY), Buffalo, NY

Representing a new type of unimolecular nanoparticle, double brush copolymers (DBCs) with each graft site quantitatively carrying two different grafts have been studied in our laboratory. Two types of DBCs, one with polystyrene and polylactide grafts and the other with polylactide and poly(ethylene oxide) grafts, were synthesized through the preparation of well-defined diblock macromonomers with pendent norbornene groups at block junctions, followed by "grafting through" these macromonomers via ring-opening metathesis polymerization. Well-defined chemical structures of these DBCs were proven through MALDI and NMR analysis of the macromonomers, as well as GPC analysis of both macromonomers and DBCs. The nanostructures of DBCs were characterized by DLS, AFM, and TEM. Specifically, the Janus morphologies of the DBCs with both polystyrene and polylactide grafts were visualized by TEM imaging, providing critical evidence to verify DBCs as Janus nanomaterials. With amphiphilic nanostructures, the DBCs with both polylactide and poly(ethylene oxide) grafts showed different bulk solution properties as compared with their macromonomer precursors. Moreover, these amphiphilic DBCs can effectively stabilize biphastic systems, and miniemulsions using these DBCs as surfactants exhibited remarkably high stability.

Three-dimensional simulation of a solid-liquid flow by a Lagrangian-Lagrangian approach

Xiaosong Sun, Mikio Sakai, Yoshinori Yamada
Graduate School of Engineering, the University of Tokyo, Japan

Introduction

Solid-liquid flows are often encountered in various industrial processes. Accurate numerical simulations would be helpful for both the design and investigation of the operational conditions. For these simulations, modeling of the free surface and the solid-liquid coupling is important, where a Lagrangian approach becomes a valuable candidate in the numerical simulations. In comparison with some traditional Eulerian methods such as the two-fluid model [1], Lagrangian schemes, e.g. the smoothed particle hydrodynamics (SPH) [2], the moving particle semi-implicit (or moving particle simulation, MPS) method [3] and the discrete element method (DEM) [4], are able to capture characteristic aspects in the two-phase flow accurately. Besides, the Lagrangian schemes have advantage that complicated processes such as mesh generation or adaptive mesh are not required in the calculations. Accordingly, they can simulate large deformation and fragmentation of free surface flows relatively easily. Hence, a Lagrangian-Lagrangian approach was developed by coupling the DEM with MPS method, where the effects of hydrodynamics forces and physical properties on macroscopic behavior of solid particles were investigated in a two dimensional system [5]. It is illustrated that the other hydrodynamics forces except for drag force was negligible and softer spring could be applied from the viewpoint of engineering. On the other hand, application of two-way Lagrangian approaches was not investigated in three-dimensional systems. Furthermore, modeling of the boundary forces within arbitrarily shaped boundaries has remained to be an arguable aspect requiring further research in the SPH method. Consequently, adequacy of the modeling of solid-liquid flows was not proved by using the DEM and SPH method.

In the present study, we developed a new Lagrangian-Lagrangian approach, namely the DEM-SPH method, where DEM and SPH were coupled by the local averaging technique [1]. We designed the numerical scheme with caution to balance the interphase action, and brought in a wall boundary formulation suitable for three-dimensional simulations. We performed three-dimensional simulations of a solid-liquid flow in a cylindrical tank and the validation test. With good agreement between the simulation and the validation, the DEM-SPH method is shown to be adequate.

Local average modeling of solid-liquid flows

As a place of departure, we start from the mechanical description for solid-liquid flows [1], where a local mean voidage ε is used to indicate volume ratio of different phases and empirical relations with respect to ε are employed to estimate the interphase momentum exchange term, i.e. coefficient related to the drag force. Besides, a mean density $\bar{\rho}_f$ is related to the actual material density ρ_f through

$$\bar{\rho}_f = \varepsilon\rho_f \ . \tag{1}$$

According to the local averaging technique, the continuity and momentum equations for fluids are transformed into the following form:

$$\frac{D\bar{\rho}_f}{Dt} = -\bar{\rho}_f \nabla \cdot \boldsymbol{u}_f \ , \tag{2}$$

$$\bar{\rho}_f \frac{D\boldsymbol{u}_f}{Dt} = -\varepsilon\nabla p - \boldsymbol{f}^f + \nabla \cdot \tau + \bar{\rho}_f \boldsymbol{g} \ , \tag{3}$$

where $\boldsymbol{u}_f, p, \boldsymbol{f}^f, \tau$ and $\boldsymbol{g}$ are fluid velocity, pressure, drag force per unit volume, deviatoric stress tensor and gravity, respectively. Note that the drag force is a function of the local mean voidage ε and its concrete form is left blank for empirical relations.

The SPH kernel approximation

In the SPH method, the domain of interest is divided into particles carrying physical quantities such as mass, velocity and pressure. Unlike fixed grids in Eulerian approaches, these particles move with their own velocities. To distinguish between SPH particles and solid particles, in this work, we use subscripts i and j for fluid particles, and k and l for solid particles.

The mathematical basis for SPH frames lies in the integral interpolant theory. Consider some generic field $A(\boldsymbol{x})$ of which the values are scattered among arbitrarily placed points. In SPH the following spatial convolution is used to approximate (or to "smooth") its value,

$$\langle A \rangle(\boldsymbol{x}) = \sum_j A_j V_j W_j(\boldsymbol{x}_i; h) \ , \tag{4}$$

where V_j is the tributary volume of particle j, $W_j(\boldsymbol{x}; h) = W(\boldsymbol{x} - \boldsymbol{x}_j; h)$ is the smoothing kernel and h is called the smoothing length defining the influence domain of the kernel. In this work a cubic spline function [2] is adopted with a compact support of $2h$. With the SPH notation, the spatial derivative of $A(\boldsymbol{x})$ could be calculated by

$$\langle \nabla A \rangle(\boldsymbol{x}) = \sum_j \left(A_j - A(\boldsymbol{x})\right) V_j \nabla W_j(\boldsymbol{x}) \ , \tag{5}$$

with $\nabla W_j(\boldsymbol{x}) = \nabla W\left(\boldsymbol{x} - \boldsymbol{x}_j\right)$.

However, it is known that using the SPH summation Eq. (4) will lead to inaccurate results near boundaries, i.e. if $\operatorname{supp} W \cap \partial V \neq \emptyset$. To fix this, we choose to define the kernel approximation through a least-square process, as to minimize the following error functional,

$$\langle A \rangle(\boldsymbol{x}) = \arg\min_{\hat{A}} \int_{\Omega} \left(A(\boldsymbol{r}) - \hat{A}\right) W(\boldsymbol{r} - \boldsymbol{x}) d\boldsymbol{r} \ . \tag{6}$$

This in turn yields a corrected form of kernel approximation as

$$\langle A \rangle(\boldsymbol{x}) = \frac{\sum_j A_j V_j W_j(\boldsymbol{x})}{\Gamma(\boldsymbol{x})} \ , \tag{7}$$

$$\Gamma(\boldsymbol{x}) = \int_{\Omega} W(\boldsymbol{r}-\boldsymbol{x})d\boldsymbol{r} \ , \tag{8}$$

where $\Gamma(\boldsymbol{x})$ is referred to as a correction factor. This approach is equivalent to that proposed by [6].

The SPH local averages

Using the corrected kernel approximation, the average density of fluid phase is calculated as

$$\bar{\rho}_i = \frac{\sum_j m_j W_j(\boldsymbol{x}_i)}{\Gamma_i} \ , \tag{9}$$

where m_j is the mass associated with the particle. In fact, this equation also readily describes the mass conservation in SPH dynamics. In a similar way, we give the local mean voidage as the complement of volume fraction of nearby solid particles,

$$\varepsilon_i = 1 - \frac{\sum_k V_k W_k(\boldsymbol{x}_i)}{\Gamma_i} \ , \tag{10}$$

where V_k is the volume of a solid particle. Subsequently the reference fluid density is estimated by $\rho_i = \bar{\rho}_i/\varepsilon_i$. Generally for single-phase flows, the number of neighbor particles roughly keeps constant and thus the SPH density Eq. (9) varies little from that of the real fluids. In contrast, in solid-liquid flows, the number of local fluid particles shall decrease as solid particles intruding into the influence domain. As a result, both the local SPH density $\bar{\rho}$ and the voidage ε decrease on the existence of solid particles in the vicinity, while the corresponding reference density ρ is not changed.

Numerical methodology

The DEM-SPH method is developed for three-dimensional simulation of a solid-liquid flow involving free surfaces. Details of the numerical methodology are shown in the following sections.

Solid phase: the discrete element modeling

In this study, the forces exerted on a solid particle are taken into consideration as follows:

$$m_k \boldsymbol{a}_k = \sum_l \boldsymbol{F}_{kl}^C + m_k \boldsymbol{g} + \sum_l \boldsymbol{F}_{kl}^L + \boldsymbol{F}_k^D + \boldsymbol{F}_k^B \ , \tag{11}$$

where the right-hand side terms are, respectively, the contact force, gravitational force, lubrication force, drag force and the buoyancy. In this momentum equation, the drag force $\boldsymbol{F}_k^D$ and buoyancy $\boldsymbol{F}_k^B$ are modeled as solid-liquid interaction terms.

The forces $\boldsymbol{F}_{kl}^C$ due to particle-particle or particle-wall contacts are evaluated using the Discrete Element modeling [4]. The DEM model combines springs, dashpots and a friction slider to represent the compacting and friction between granular matters. The forces $\boldsymbol{F}_{kl}^L$ are introduced for the lubricant effect in contact between wet particles. Moreover, the drag force, calculated for each solid particle immersed in the fluid region, has the general form

$$\boldsymbol{F}_k^D = \frac{\beta_k}{1-\varepsilon_k}(\overline{\boldsymbol{u}}_k - \boldsymbol{v}_k)V_k \ , \tag{12}$$

where β_k is an interphase momentum transfer coefficient, and $\bar{\boldsymbol{u}}_k - \boldsymbol{v}_k$ is the relative flow velocity. A proper drag model for the description of β is vital in solid-liquid interaction problems. In the current study, a combination of the equations of Ergun [7] and Wen-Yu [8] has been employed to determine its value. A voidage of 0.8 is adopted as a threshold to toggle between these two regimes. For details of these circumstances, reader is referred to literature [5]. The buoyancy, arising from the gradient of fluid pressure at the particle position, is modeled by

$$\boldsymbol{F}_k^B = -V_k \langle \nabla p \rangle_k = -V_k \sum_j \frac{m_i p_i}{\Gamma_i \bar{\rho}_i} \nabla W_i(\boldsymbol{x}_k) \ . \tag{13}$$

Note that the form of buoyancy is found from the coupled Lagrangian dynamics of the solid-liquid system, which is to be further explained in next section.

Liquid phase: Lagrangian-based SPH equations

The SPH equations are derived starting with the Lagrangian of the particle system,

$$\mathcal{L} = \sum_j m_j \left(\frac{1}{2} \boldsymbol{v}_j^2 + \boldsymbol{x}_j \cdot \boldsymbol{g} - u_j \right) \ , \tag{14}$$

where u_j is the internal energy density assumed to be related with the fluid pressure. Subsequently, the Euler-Lagrangian equations with non-conservative generic forces read

$$\frac{d}{dt}\left(\frac{\partial \mathcal{L}}{\partial \boldsymbol{v}_i} \right) - \frac{\partial \mathcal{L}}{\partial \boldsymbol{x}_i} = \boldsymbol{\Pi}_i + \boldsymbol{\Lambda}_i \ , \tag{15}$$

where $\boldsymbol{\Pi}_i$ is the viscous force and $\boldsymbol{\Lambda}_i$ is reaction of drag forces. Expanding this equation gives

$$m_i \boldsymbol{a}_i = \boldsymbol{T}_i + m_i \boldsymbol{g} + \boldsymbol{\Pi}_i + \boldsymbol{\Lambda}_i \ , \tag{16}$$

$$\boldsymbol{T}_i = \sum_j m_j \frac{\partial u_j}{\partial \boldsymbol{x}_i} \ . \tag{17}$$

This gives the momentum equation of the SPH phase and herein we call $\boldsymbol{T}_i$ the internal force of the Lagrangian system, which incidentally corresponds to the pressure gradient term in Navier-Stokes equations.

The internal force is derived via the variation of the internal energy with the SPH density attached as a constraint on the Lagrangian system. Finally, it proves to take the following form,

$$\boldsymbol{T}_i = \boldsymbol{T}_i^f + \boldsymbol{T}_i^s + \boldsymbol{T}_i^w \ , \tag{18}$$

where

$$\boldsymbol{T}_i^f = -\sum_j m_i m_j \left(\frac{\varepsilon_j p_j}{\Gamma_j \bar{\rho}_j^2} + \frac{\varepsilon_i p_i}{\Gamma_i \bar{\rho}_i^2} \right) \nabla W_j(\boldsymbol{x}_i) \ , \tag{19}$$

$$\boldsymbol{T}_i^s = -m_i \frac{p_i}{\Gamma_i \bar{\rho}_i} \sum_k V_k \nabla W_k(\boldsymbol{x}_i) \ , \tag{20}$$

$$\boldsymbol{T}_i^w = m_i \frac{p_i}{\Gamma_i \bar{\rho}_i} \frac{\Gamma_i'}{h} \boldsymbol{n}_i^B \ , \tag{21}$$

with $\boldsymbol{n}_i^B$ standing for the inward normal vector from the rigid boundary. Incidentally, this form of internal force is divided into three different actions: $\boldsymbol{T}_i^f$ is in fact a locally averaged version of the pressure gradient term in conventional SPH formulations; $\boldsymbol{T}_i^s$ involves the information of immersed solids and defines some kind of fluid lift force, namely the buoyancy (13); $\boldsymbol{T}_i^W$ gives the boundary force, similar to that proposed by [6]. Specifically, we remark on the boundary force (21) that it is to be evaluated explicitly using boundary information and thus no other special SPH boundary particles are needed. This model improves computational efficiency as auxiliary SPH particles are removed, and what's more, it allows direct deployment of boundary shapes and sharing of boundary conditions between DEM and SPH. These features make it an extraordinary tool for practical three-dimensional simulations. In this study, we introduce the distance function (DF) [9] for both definition of boundary shape and fast evaluation of boundary distance vectors.

In original SPH method, the fluid is assumed to be weakly compressible. Following this statement, the equation of state is used to calculate the fluid pressure directly.

$$p_i = \begin{cases} p_i^{EoS}, & \bar{\rho}_i > \varepsilon_i \rho_{i0} \\ 0, & \bar{\rho}_i \leq \varepsilon_i \rho_{i0} \end{cases} \tag{22}$$

$$p_i^{EoS} = \frac{\rho_{i0} c_i^2}{\gamma} \left(\left(\frac{\bar{\rho}_i}{\varepsilon_i \rho_{i0}} \right)^{\gamma} - 1 \right) , \tag{23}$$

where c_i and ρ_{i0} are the speed of sound and the physical density of the fluid, respectively. For simple fluids, γ is often set to be 7.0. This pressure calculation is a convention in SPH, and it relates the change in fluid pressure to the variation of fluid density. In the context of solid-liquid flows, it works as well when solid particles are submerging in the fluid region: local voidage ε decreases and pressure increases, causing the fluid particle pushed out of the way. Accordingly as a macroscopic phenomenon, it can be observed that solid occupy the place and fluids are displaced.

The dissipative force is modeled by the laminar viscosity of Newtonian fluids. The viscous force on a fluid particle has been proposed by [10] as

$$\boldsymbol{\Pi}_i = \sum_j m_j \left(\frac{4\nu(\boldsymbol{x}_i - \boldsymbol{x}_j) \cdot \nabla W_i(\boldsymbol{x}_j)}{(\bar{\rho}_i + \bar{\rho}_j)|\boldsymbol{x}_i - \boldsymbol{x}_j|^2} \right) (\boldsymbol{v}_i - \boldsymbol{v}_j) , \tag{24}$$

where ν is the kinematic viscosity of the fluid. The reaction force $\boldsymbol{\Lambda}_i$ is defined according to the Newton's third law of motion as follows,

$$\boldsymbol{\Lambda}_i = \sum_k \boldsymbol{\Lambda}_{ik} , \tag{25}$$

$$\boldsymbol{\Lambda}_{ik} = -\frac{V_i W_i(\boldsymbol{x}_k)}{\sum_j V_j W_j(\boldsymbol{x}_k)} \boldsymbol{F}_k^D . \tag{26}$$

Additionally, both the artificial viscosity and the empiric pressure are attached to the momentum equation to improve the numerical stability of SPH simulation.

Numerical example

A three-dimensional simulation was performed in this study applying the DEM-SPH method to the solid-liquid flow in a cylindrical tank. The results were compared with those obtained from validation experiments.

Calculation conditions

Figure 1 shows a schematic diagram of the compute domain, which is a cylindrical tank with inner diameter and depth of both 100 mm. The cylindrical tank was rotated at 104 rpm in the simulation. The behavior of the solid-liquid mixture in the tank was studied. For the solid phase, the stiffness, restitution and friction coefficient are set to be 1000 N/m, 0.8 and 0.3, respectively. The particle diameter is 2.7 mm and density is 2500 kg/m^3. For the liquid phase, the average particle distance is set to be 3.0 mm, and the smoothing length h is 1.25 times larger than that. Totally 7755 DEM particles and 11007 SPH particles are used in the simulation. The total mass of solids counts 199.80 g and the number of fluid particles is adjusted to fill half the tank. The calculation is updated with a time step of 1.0e-5 sec.

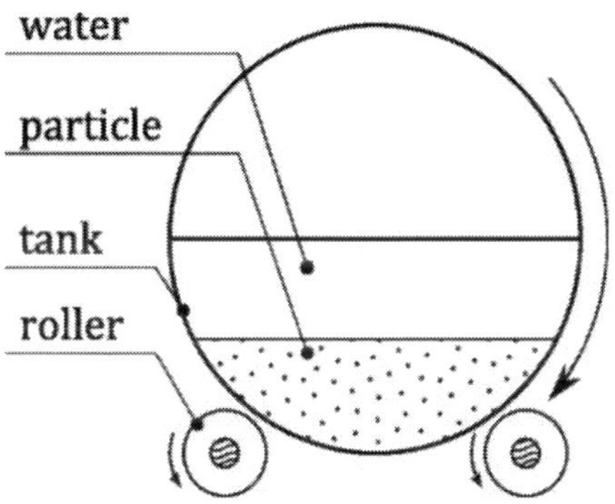

Figure 1. Schematic diagram of the cylindrical tank.

Validation test

The experimental device is composed of a platform and a cylindrical jar made of glass. The cylinder's inner diameter and depth were both 100 mm. The platform was equipped with two rollers by which the tank was driven to rotate at 104 rpm. The rotation speed was measured using a tachometer. Glass beads 199 g were placed in the vessel. They had a mean diameter of 2.7 mm, and their density was 2500 kg/m^3. The fluid was water. Photographs and videos of the rotating cylinder were recorded by a high-speed camera. From these visual data, the macroscopic behaviors of the solid particle bed, viz. the bed shape, width and height, were measured for validation. Also, these pictures were processed using a DANTEC Particle Image Velocimetry (PIV) to obtain the velocity distribution of the circulating particle bed.

Results and discussions

In this study, we give attention to the behavior at a quasi-steady state. The results are described as below. A typical experimental photograph is shown in Figure 2 (left). The bed slope has become bilinear in shape, as indicated by the lines in the figure. From a series of photos picked randomly, we measured the bed widths and heights, of which the mean values read 73.41 mm and 62.18 mm, respectively. Correspondingly, a representative snapshot of the simulation results is shown in Figure 2 (right). In that scene, the shape of solid bed seemed to be bilinear as well. Similarly, by choosing snapshots with time intervals sufficiently large, the dimensions of solid bed were averaged to be 71.45 mm in width and 59.80 mm in height, respectively. The resultant values were quantitatively comparable to those of the validation test, as shown in Table 1.

Figure 2. Macroscopic behaviors of the solid bed. (Left: experiment; right: simulation.)

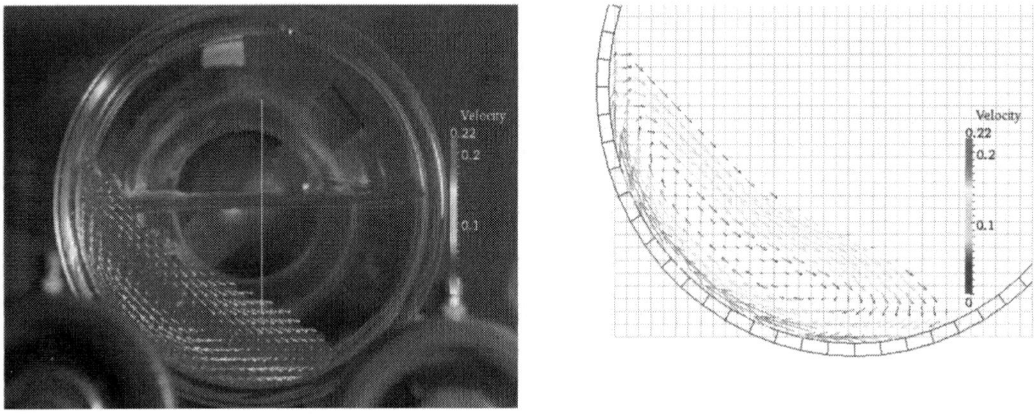

Figure 3. Velocity profiles of solid bed. (Left: experiment PIV; right: simulation.)

Table 1. Comparison of shape and dimensions of the solid bed.

	Bed shape	Bed width (standard deviation)	Bed height (standard deviation)
Experiment	Bilinear	73.41 (1.24) mm	62.18 (1.71) mm
Simulation	Bilinear	71.45 (0.79) mm	59.80 (0.61) mm
Error	N/A	2.67 %	3.83 %

The velocity profiles obtained from the experiment and the simulation were compared for validation as well. To be compared with those PIV results, the compute domain was divided into grids and the velocities of DEM particles contained in grids were filtered using a Cubic spline kernel. As shown in Figure 3, similar patterns and phenomena were observed for both the experiment and simulation: solid particles in the bottom were carried upward and settled rolling down the slope so that the solid bed appeared to be circulating, the maximum velocity magnitude was around 0.2 m/s, and the central part of the solid bed was moving much slower than other parts.

Hence, it has been shown that the simulation results are reasonably close to those obtained in the validation test. The close correspondence between them implies that the DEM-SPH method can simulate solid-liquid free-surface flows accurately.

Summary and conclusions

In this study, the DEM-SPH method was developed to model the dynamics of solid-liquid flows involving free surfaces. It was validated with the two phase flow in a rotating cylindrical tank. Hence, we have shown that the DEM-SPH method is adequate to the simulation of solid-liquid free-surface flows.

Acknowledgement

This study was financially supported by a Grant (22760579) from the Ministry of Education, Culture, Sports, Science and Technology (MEXT), Japan.

References

[1] T.B. Anderson, R. Jackson, Fluid Mechanical Description of Fluidized Beds. Equations of Motion, Ind. Eng. Chem. Fund. 6 (1967) 527–539.

[2] J.J. Monaghan, An introduction to SPH, Computer Physics Communications. 48 (1988) 89–96.

[3] S. Koshizuka, Y. Oka, Moving-particle semi-implicit method for fragmentation of incompressible fluid, Nuclear Science and Engineering. 123 (1996) 421–434.

[4] P.A. Cundall, O.D.L. Strack, A discrete numerical model for granular assemblies, Géotechnique. 29 (1979) 47–65.

[5] M. Sakai, Y. Shigeto, X. Sun, T. Aoki, T. Saito, J. Xiong, et al., Lagrangian–Lagrangian modeling for a solid–liquid flow in a cylindrical tank, Chemical Engineering Journal. 200-202 (2012) 663–672.

[6] S. Kulasegaram, J. Bonet, R.W. Lewis, M. Profit, A variational formulation based contact algorithm for rigid boundaries in two-dimensional SPH applications, Computational Mechanics. 33 (2004) 316–325.

[7] S. Ergun, Fluid flow through packed columns, Chem. Eng. Prog. 48 (1952) 89–94.

[8] C. Wen, Y. Yu, Mechanics of fluidization, Chem. Eng. Prog. Symp. Ser. 62 (1966) 100.

[9] S. Osher, R. Fedkiw, Level Set Methods and Dynamic Implicit Surfaces (Google eBook), Springer, 2002.

[10] E. Y.M. Lo, S. Shao, Simulation of near-shore solitary wave mechanics by an incompressible SPH method, Applied Ocean Research. 24 (2002) 275–286.

(388i) Computational Investigation of Nanoparticle Synthesis by Flame Spray Pyrolysis

Arto J. Gröhn, Department of Mechanical and Process Engineering, ETH Zurich, Zurich, Switzerland, Sotiris E. Pratsinis, Department of Mechanical and Process Engineering, ETH Zürich, Zurich, Switzerland and Karsten Wegner, Particle Technology Laboratory, ETH Zurich, Zurich, Switzerland

Computational Investigation of Nanoparticle Synthesis by Flame Spray Pyrolysis

Arto J. Gröhn, Sotiris E. Pratsinis and Karsten Wegner

Particle Technology Laboratory, Swiss Federal Institute of Technology (ETH) Zurich, Sonneggstrasse 3, CH-8092 Zurich, Switzerland.

Flame Spray Pyrolysis (FSP) is a promising method for production of ceramic nanoparticles at the industrial level (Strobel et al., 2007). In order to assist scale-up from the current laboratory process and optimize reactor design, a computational fluid dynamics (CFD) model has been constructed (Gröhn et al., 2012). Taking zirconia nanoparticle synthesis as a case study, the model accounts for spray atomization of the liquid raw material (precursor), combustion and fluid dynamics as well as nanoparticle growth by the aerosol dynamics.

Using a commercial CFD solver, product nanoparticle characteristics are predicted without adjustable parameters or need of experimental input data. By assuming immediate particle formation on precursor oxidation and a self-preserving particle size distribution a robust and computationally feasible description of the process could be employed. The model has been validated with measurements of fuel spray characteristics, flow fields, temperature profiles and primary particle sizes. To obtain insight into the FSP process for scale-up the model has been extended to include both laboratory- and pilot-scale reactor geometries.

References:

Strobel, R. and Pratsinis, S.E. Flame aerosol synthesis of smart nanostructured materials. *Journal of Materials Chemistry*, 2007. 17, 4743-4756.

Gröhn, A.J. Pratsinis, S.E. and Wegner, K. Fluid-particle dynamics during combustion spray aerosol synthesis of ZrO_2. *Chemical Engineering Journal*, 2012. 10.1016/j.cej.2012.02.093.

(388j) Numerical Parametric Investigation of Breakage in a Ball Mill

Matthew J. Metzger, Chemical and Biochemical Engineering, Rutgers University, Piscataway, NJ and Benjamin J. Glasser, Chemical and Biochemical Engineering, Rutgers, the State University of New Jersey, Piscataway, NJ

Size reduction is an essential operation across many industries. Despite its prevalence, the efficiency of the conversion of applied force to the creation of new surfaces is extremely poor, none worse than the commonly encountered ball mill. However, tuning of operational parameters may be a means to increase the low efficiency levels, if the proper connection between operating conditions and breakage conditions can be determined. Described here is the implementation of the Bonded Particle Model (BPM) within the Discrete Element Method (DEM) framework to analyze the breakage of bonded agglomerates within a batch ball mill. It is found that agglomerate strength and grinding media diameter affect breakage significantly; whereas grinding media fill level has a minor affect. The majority of breakage occurs near the mill shell, rather than at the point of impact between the media and material and the extent of material breakage has a significant influence on the material flow within the mill.

(388k) Liquid Distribution in Wet Fluidized Beds

Matthew Girardi, Department of Chemical and Biological Engineering, Princeton University, Princeton, NJ, Stefan Radl, Institute for Process and Particle Engineering, Graz University of Technology, Graz, Austria and Sankaran Sundaresan, Chemical Engineering, Princeton University, Princeton, NJ

Gas-fluidized beds of particles are widely used in industries to carry out catalytic and non-catalytic reactions, drying, and granulation. In many applications, liquid is injected at some locations within the bed, creating a system of wet particles with a potential to agglomerate. In the event that particles agglomerate, liquid is transferred from particle surfaces into capillary bridges.

Many questions remain unanswered regarding the fundamental behavior of liquid spreading in wet fluidized beds, including the steady state liquid distribution and how this distribution changes with cohesive force. To begin answering these questions, we consider a small (on the order of 1 cm^3 three dimensional periodic domain in which discretely represented solid particles are coated in a thin film of liquid and fluidized by a continuous gas phase. We perform a time-driven Discrete Element Method based simulation (coupled with the average equation of motion for the gas), and consider liquid transfer[1]depletion of liquid on the surface of particles into pendular bridges, and the viscous and capillary forces between individual particles[2,3,4].

Figure 1 shows the structure of agglomerates in a wet system (a) compared to the clustering of a dry system (b). The domain-averaged slip velocity for the wet system is approximately 70% that of the dry system. Initially all particles are assigned a random position and identical liquid contents. However, after some time, particles reach a statistical steady state in which nearly half of the liquid is on the particles and half is in bridges. The standard deviation of liquid content on the particles is 10% of the initial liquid present on a particle while the standard deviation of liquid content in bridges is 3% of the initial liquid present on a particle. We perform a sensitivity analysis with respect to the initial configuration of the particles before they are wetted and provide statistics on domain-averaged slip velocities, liquid distribution, and filtered quantities for domain-averaged volume fractions ranging from 0.05 to 0.40. Analysis of data gathered from these simulations provides quantitative insight into liquid spreading and fluidization of agglomerates within wet fluidized beds.

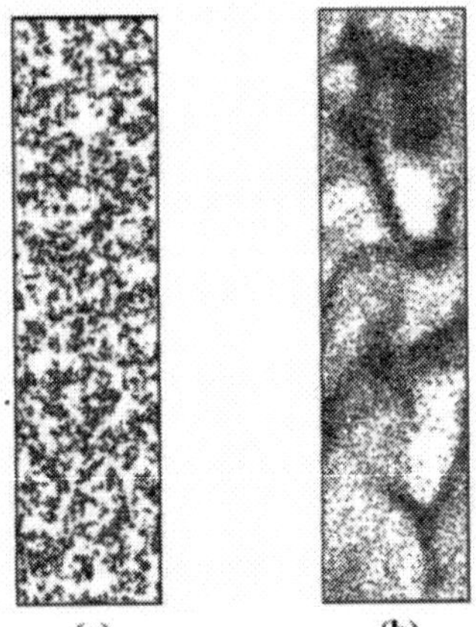

Figure 1: Cross sectional view of particle positions in the 3D domain. (a) the wet system and (b) the corresponding dry system. The system is 2 cm^3, with 256,600 particles. The domain-averaged solid volume fraction is 0.2 and the liquid content in the system is 0.25 % of the solid particle mass.

References

[1] D. Shi and J. McCarthy, "Numerical simulation of liquid transfer between particles," Powder Technology, vol. 184, pp. 64-75, May 2008.

[2] T. Mikami, H. Kamiya, and M. Horio, "Numerical simulation of cohesive powder behavior in a fluidized bed," Chemical Engineering Science, vol. 53, no. 10, pp. 1927-1940, 1998.

[3] O. Pitois, P. Moucheront, and X. Chateau, "Rupture energy of a pendular liquid bridge," Physical Journal B, vol. 86, pp. 79-86, 2001.

[4] P. Darabi, T. Li, K. Pougatch, M. Salcudean, and D. Grecov, "Modeling the evolution and rupture of stretching pendular liquid bridges," Chemical Engineering Science, vol. 65, no. 15, pp. 4472-4483, 2010.

(388l) Growth of Metal Nanoparticles by Molecular Dynamics

Beat Buesser, Department of Mechanical and Process Engineering, ETH Zurich, Zurich, Switzerland and Sotiris E. Pratsinis, Particle Technology Laboratory, Institute of Process Engineering, Department of Mechanical and Process Engineering, ETH Zurich, Zurich, Switzerland

Growth of Metal Nanoparticles by Molecular Dynamics

B. Buesser[1,2] and S.E. Pratsinis[1]

[1]Particle Technology Laboratory, Institute of Process Engineering, Department of Mechanical and Process Engineering, ETH Zürich, 8092 Zürich, Switzerland

[2]Department of Chemical Engineering, Massachusetts Institute of Technology, Cambridge, MA 02139, USA

Metal nanoparticles are attractive in catalysis, magnetic separations[1] or sensors[2], to name only a few applications. The performance of these particles, however, depends considerably on their primary particle size and structure. Gas-phase processes allow making economically such particles in large quantities with close control of their size and extent of aggregation[3]. In gas-phase reactors these two characteristics are determined mostly by particle sintering. The detailed understanding of sintering is crucial for the development and scale-up of such reactors to target product particle size and morphology at maximal yield especially when precious metals are involved.

Silver nanoparticles are one of the most studied noble nanomaterials. For example Shimada *et al.*[4] proposed a sintering rate for silver nanoparticles in gas-phase. They found their particle dynamics model in good agreement with the measured particle size evolution in their hot wall reactor, where primary particle sizes bigger than d_p = 8 nm were observed. The evolution of smaller primary particles (d_p < 8 nm) is increasingly difficult to determine experimentally although this would be the key size range where nanoparticle exhibit their extraordinary performance and exciting new properties.

Here, sintering of silver nanoparticles is investigated using MD simulations accelerated by graphical processing units (GPU) in the range of d_p = 2 – 5 nm. The sintering rate is determined by calculating the surface area evolution comparable to BET surface area measurements. First observations of the atom trajectories reveal that surface atoms exhibit a much higher mobility than bulk ones indicating that sintering by surface diffusion dominates[5] at these particle sizes and temperatures (Figure 1). The dependence of the sintering rate on particle morphology has been investigated during sintering of straight chains, triangles and stars of three and four particles.

An expression for the sintering rate as function of primary particle size and temperature has been extracted from MD, filling the gap of knowledge between clusters of a few atoms up to particles of several nanometers. This sintering rate will facilitate the design of large scale manufacture and processing of these small nanoparticles based on phenomenological models[6] or allow engineering estimations of the particle morphology by comparing it to the coagulation rate[3].

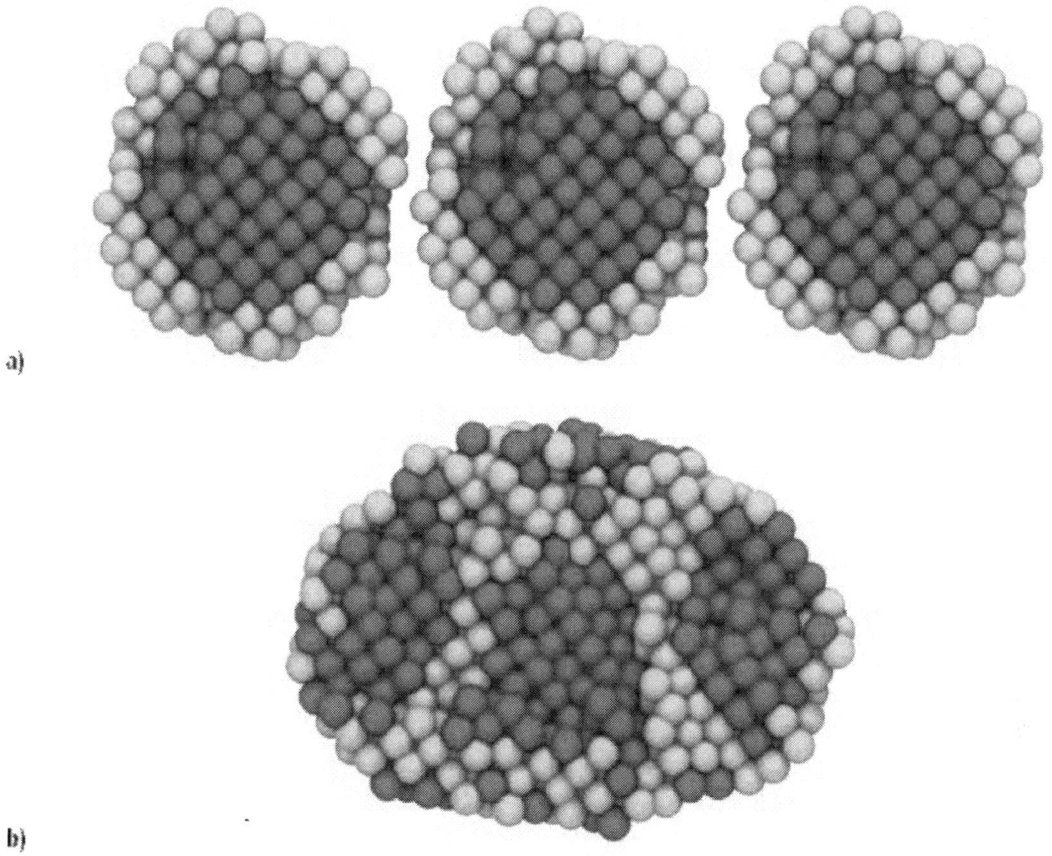

Figure 1 Snapshots of 3 silver nanoparticles with a diameter of d_p = 3 nm sintering at T = 800 K and time t = a) 0 ns and b) 100 ns. The atoms are colored green at the surface and red in the bulk at t = 0 ns. It is fascinating to see that these surface atoms largely move to the concave areas between these particles at t = 100 ns, revealing the dominance of surface diffusion during their sintering or coalescence.

Financial support from the European Research Council is gratefully acknowledged.

1. Grass RN, Athanassiou EK, Stark WJ. Covalently Functionalized Cobalt Nanoparticles as a Platform for Magnetic Separations in Organic Synthesis. *Angewandte Chemie International Edition.* 2007;46(26):4909-4912.

2. Sotiriou GA, Sannomiya T, Teleki A, Krumeich F, Vörös J, Pratsinis SE. Non-Toxic Dry-Coated Nanosilver for Plasmonic Biosensors. *Advanced Functional Materials.* 2010;20(24):4250-4257.

3. Buesser B, Pratsinis SE. Design of Nanomaterial Synthesis by Aerosol Processes. *Annual Review of Chemical and Biomolecular Engineering.* 2012;3:103-127.

4. Shimada M, Seto T, Okuyama K. Size change of very fine silver agglomerates by sintering in a heated flow. *Journal of Chemical Engineering of Japan.* 1994;27(6):795-802.

5. Buesser B, Gröhn AJ, Pratsinis SE. Sintering Rate and Mechanism of TiO_2 Nanoparticles by Molecular Dynamics. *J. Phys. Chem. C.* 2011;115:11030-11035.

6. Koch W, Friedlander SK. The effect of particle coalescence on the surface area of a coagulating aerosol. *Journal of Colloid and Interface Science.* 1990;140(2):419-427.

(388m) Sphero-Cylindrical Particle Dynamics and Stresses in a Vertical Axis Mixer

Xia Hua, School of Mechanical Engineering, Purdue University, West Lafayette, IN, Jennifer Sinclair Curtis, University of Florida, Gainesville, FL, Bruno C. Hancock, Pfizer Inc., Groton, CT, William R. Ketterhagen, Pfizer Global Research and Development, Pfizer Inc., Groton, CT and Carl Wassgren, Mechanical Engineering, Purdue University, West Lafayette, IN

A discrete element method model is used to examine the stresses in sphero-cylindrical particles agitated in a vertical axis mixer. The model is validated against experimental measurements of the rotating shaft torque. Using this model, the effects of particle aspect ratio from one to approximately nine, on blade torque, flow patterns, and average particle internal stress are investigated. Increasing particle aspect ratio generally decreases the flow rate of particles over the blade and increases the height of the heap in front of the blade. The mean particle hydrostatic and Mises stresses are found to increase with increasing bed depths ranging from one to two times the blade height and impeller speeds from 0 to 90 rpm. The stress distributions also appear to follow two different exponential distributions, depending on the stress magnitude. Increasing the particle aspect ratio increases the mean Mises stress due to an increase in the bed height in front of the mixer blade, as well as due to the interference of particles as they flow past one another. Increasing either the inter-particle friction coefficient from 0.11 up to 0.66 or the particle elastic modulus from 1 MPa up to 100 MPa results in larger internal stresses. However, increasing the particle-drum friction coefficient from 0.17 up to 0.68 has a much weaker influence.

(388n) Thermite Reaction Based On Nano-Al and Iodine Pentoxide: Ignition and Combustion Characterization

Guoqiang Jian[1], Snehaunshu Chowdhury[2], Jingyu Feng[1] and Michael Zachariah[3], (1)Department of Chemistry and Biochemistry, University of Maryland, College Park, MD, (2)Department of Mechanical Engineering, University of Maryland, College Park, MD, (3)Department of Mechanical Engineering and Chemistry, University of Maryland, College Park, MD

Thermite reactions with high energy release and biocidal agent production are of interest for their potential applications in bio-agent defeat. In this study, ignition and combustion behavior of nano-Al and iodine pentoxide were investigated. Cyro-milling was used to produce fine particles, but found only to be effective for preparation of micro-sized iodine pentoxide. The ignition and reaction of the prepared nano-Al/micro-I_2O_5 thermite samples were systematically studied by a rapid wire heating technique using a temperature-jump/time-of-flight mass spectrometer and a high speed digital camera. Time-resolved mass spectra results showed that nano-Al/micro-I_2O_5 reactions can produce a lot of iodine species which were suggested to be good biocidal agents. The ignition temperature of nano-Al/micro-I_2O_5 thermite in air at atmospheric pressure was found to be ~810 K, lower than ignition temperature ~940 K in vacuum. Constant-volume pressure cell tests were run to characterize the pressure and optical signals during the combustion, and the performance of the iodine containing thermites was compared with other thermite systems. The results show that Al/I_2O_5 thermite composite outperform the traditional Al/CuO and Al/Fe_2O_3 thermite systems in terms of peak pressure, pressurization rate and burning time. Transmission electron microscope with X-ray compositional microanalysis characterization were performed to analyze the post-combustion products and showed that the production of iodine species and aluminum oxides. Our study of nano-Al and iodine pentoxide reactions shows that these iodine containing thermite reactions are very reactive and suggested to effectively produce iodine gas which can be used in microbial agent defeat applications.

Development of a boundary model for complex systems in the discrete element method

Yusuke Shigeto*1
Mikio Sakai[2]
[1]Graduate School of Engineering, the University of Tokyo, Tokyo, Japan
[2] School of Engineering, the University of Tokyo, Tokyo, Japan
*Presenting Author, shigeto@mps.q.t.u-tokyo.ac.jp

Introduction

In the industrial powder handling processes, large amount of powder is treated in complicated shaped devices. Application of computational simulations is desired to reduce the cost in designing or planning operational conditions of these devices. The discrete element method(DEM)[1] is widely used in the powder simulations. The location of each particle of the powder is calculated with its movement in the DEM, and these movements are calculated in results of particle-particle interactions and particle-boundary interactions. When the wall boundary conditions of the complicated shape is modeled, there are some methods, namely, meshes and signed distance functions(SDF). The meshes are widely used in the computer aided design (CAD) systems, and they are also well used in the DEM, such as our previous study[2] or an open source code[3]. On the other hand, the SDF is volume based method using a scalar field based on the level set function. The SDF was originally used in the level set method[4], and recently it was applied in two dimensional DEM simulation[5]. The SDF is reported as a robust method, and it might be useful in powder-fluid coupling simulations. However, the SDF was applied only in two dimensions, and it is not applied in the three-dimensional simulations.

In the current study, we show that the SDF is able to be applied to the three-dimensional simulations. Besides it, we compare the accuracy and the computation time between the mesh and the SDF in the usage of the wall boundary. Consequently we illustrate the SDF is effective for DEM simulations with complicated shapes.

Conventional wall boundary

In designing of these complicated shaped devices, the CAD systems are used with meshes which are widely used to represent shapes of objects; these meshes are composed of multiple triangles or squares in some cases. These meshes generated by CAD systems usually represent only the surfaces of objects, unlike the meshes used in the finite element method. The mesh can express almost all shapes; even smooth curves can be expressed by the mesh if the resolution is high enough. In our previous study we performed DEM simulations using CAD generated meshes as the complicated boundary walls[4]. In the calculation, particle-boundary interaction is performed by detecting collisions between each triangle element in the boundary mesh and a particle. The collision detection cost may grow with the number of particles and with the number of triangular elements of the mesh. Therefore DEM simulations in a complicated shaped system might require high calculation cost.

Signed distance function

The signed distance function (SDF) is the mathematical expression of surfaces by using the scalar field based on distance. In the previous studies, SDF is used in the level set method[2] and distance scalar field was used in DEM simulation in two dimensions[3]. The SDF is a kind of implicit function and it is defined as

$$\phi(\boldsymbol{x}) = d(\boldsymbol{x})s(\boldsymbol{x}),$$

where $d(\boldsymbol{x})$ is the minimal distance from the point $\boldsymbol{x}$ to the surface of the shape, and $s(\boldsymbol{x})$ is the sign positive if $\boldsymbol{x}$ is outside of the shape, and negative if $\boldsymbol{x}$ is inside. The function gives the distance from the point $\boldsymbol{x}$, and the sign of the function gives the information of if the point is inside or outside of the shape. The normalized normal vector of a SDF expressed shape is given as a normalized gradient of $\phi(\boldsymbol{x})$. The SDF can be defined on general form of a CAD generated mesh. In the simulations, the values of SDF are discretized, i.e., they are calculated in advance on each sampling point before the simulation starts. This pre-calculation is equivalent to the pre-calculation of the distance between the sampling point and the mesh.

Accuracy comparison between shapes by the mesh and the SDF

In the present study, first we compared the results of DEM simulation between using shapes expressed by the mesh(Case 1-M) and the SDF(Case 1-S). In the comparison, DEM simulations were performed with only one particle in a very simple sloped box (slope angle is 45 degrees) as shown in Fig. 1. At the initial condition, the particle was placed at the same location among two cases. The calculation conditions are shown in Table 1. In the simulation, the movements of the particles were measured as shown in Fig. 2. It was found that the movements of two cases matched very well and they were almost the same.

Comparison of computational time between two boundary expressions

In the current study, we compare calculation times between the mesh and the SDF with a screw conveyer, which is an example of complicated systems. Some numerical analyses were performed with the shape varying the number of particles, and measured the particle-boundary interaction time of each case.

The screw conveyer used in this study has a pair of helical fins fixed inside a cylinder as shown in Fig. 3. It can contain some powder inside the cylinder, and it can rotate along the height axis of the cylinder to convey powder. It is an example of a powder conveyer system generated by a CAD system. The mesh consists of 448 triangles. In this study, the cylinder is placed with the rotation axis parallel to the ground. In the simulation, the particles were placed on one bottom of the cylinder with the gravity along the axis of the cylinder, and when the calculation begins, the gravity is set to cross the axis of the cylinder and the screw conveyer rotates at 30 rpm. In the simulation, the number of particles and the particle radius was changed for three cases as shown in Table 2. The total mass of the particles were kept to be the same among all cases.

In the calculation results, the particles were transported with the time in all cases. Typical snapshots are shown in Fig. 4. The calculation times consumed for particle-boundary interactions in one step was measured as the average of first 200 steps. The results are shown in Fig. 5. The interaction time grows longer with the number of the particles becomes much in all cases, meanwhile times spent in cases using the SDF was always shorter than the times spent in cases using the mesh.

Conclusion

We performed three-dimensional DEM simulations using the mesh and the SDF for boundary expressions. It was shown that the SDF can be applied in three-dimensional simulations. We shows that the calculation time can be reduced drastically by using the SDF, though the accuracy were regarded to be equivalent between them. Eventually, the SDF is effective for DEM simulation with complicated shapes.

References

[1]: P.A.Cundall and O.D.L. Strack: "A Discrete Numerical Model for Granular Assembles", Geotechnique, 29, 47-65 (1979)

[2]: Y. Shigeto, M. Sakai, "Parallel Computing of the Discrete Element Method on Multi-Core Processors", Particuology, 9(4), pp.398-405 (2011)

[3]: CFDEM Project, "LIGGGHTS Open Source Discrete Element Method Particle Simulation Code", http://www.liggghts.com/

[4]: S.Osher and R.Fedkiw: "Level Set Methods and Dynamic Implicit Surfaces", Applied Mathematical Sciences, Vol.153, Springer, Berlin, 2003

[5]: K.Yokoi: "Numerical method for interaction between multiparticle and complex structures", Physical Review E, 72, 046713 (2005)

Table 1 Calculation Conditions for Case 1

Parameter	Unit	Case 1-M	Case 1-S
Particle radius	m	0.5×10^{-2}	0.5×10^{-2}
Particle density	kg/m^3	2.5×10^{3}	2.5×10^{3}
Spring constant	N/m	1.0×10^{3}	1.0×10^{3}
Coefficient of restitution	-	0.9	0.9
Coefficient of friction	-	0.3	0.3
Boundary Type	-	Mesh	SDF
Resolution of signed distance field	m	-	2.5×10^{-3}

Table 2 Calculation conditions for cases 2, 3 and 4

Name	Unit	Case 2-M Case 2-S	Case 3-M Case 3-S	Case 4-M Case 4-S
Number of particles	-	50,000	200,000	400,000
Particle radius	m	2.0×10^{-3}	1.3×10^{-3}	1.0×10^{-3}
Particle density	kg/m^3	2.5×10^{3}	2.5×10^{3}	2.5×10^{3}
Total mass of particles	kg	1.68×10^{-3}	1.68×10^{-3}	1.68×10^{-3}
Spring constant	N/m	1.0×10^{3}	1.0×10^{3}	1.0×10^{3}
Coefficient of restitution	-	0.9	0.9	0.9
Coefficient of friction	-	0.3	0.3	0.3
Resolution of signed distance field	m	1.0×10^{-3}	0.63×10^{-3}	0.5×10^{-3}

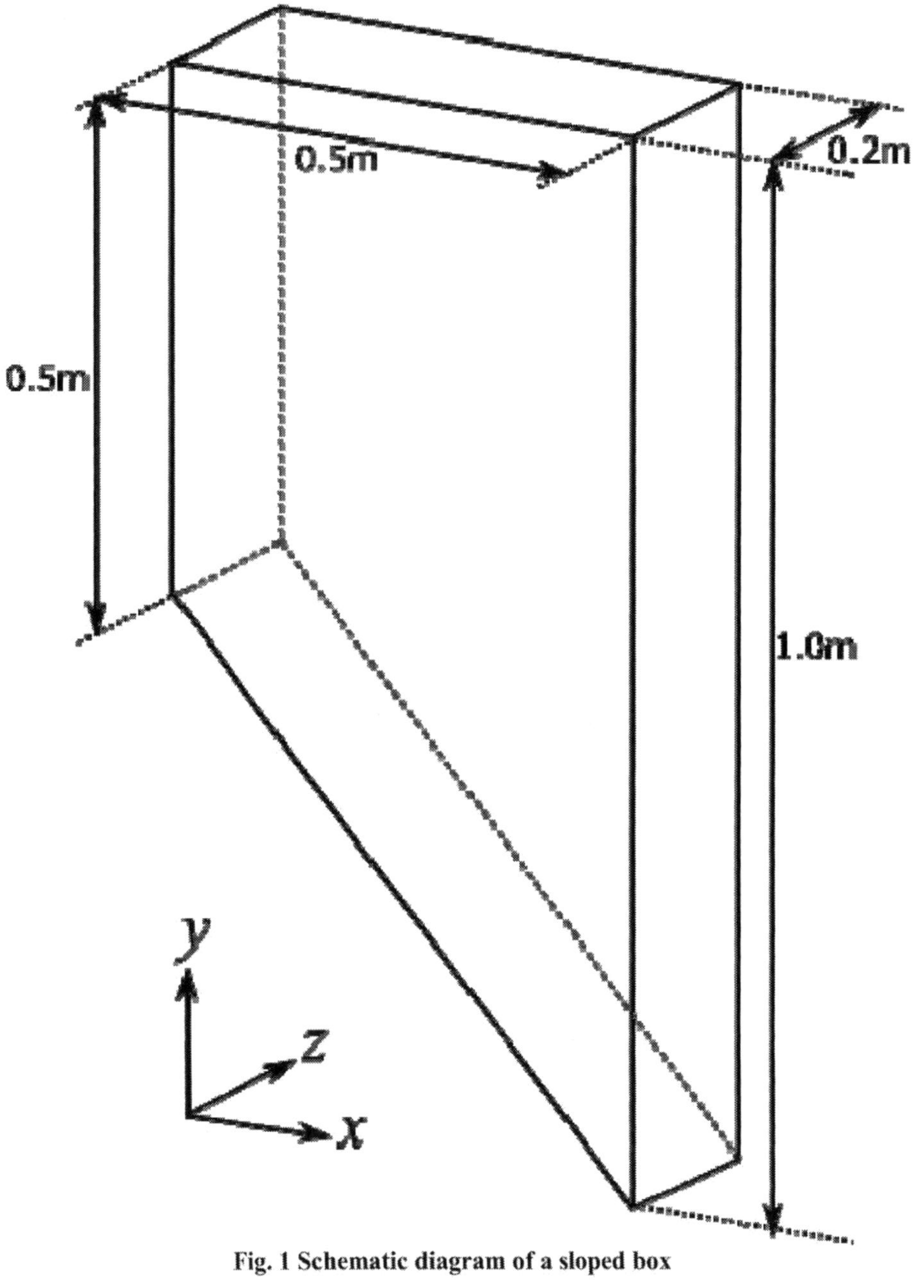

Fig. 1 Schematic diagram of a sloped box

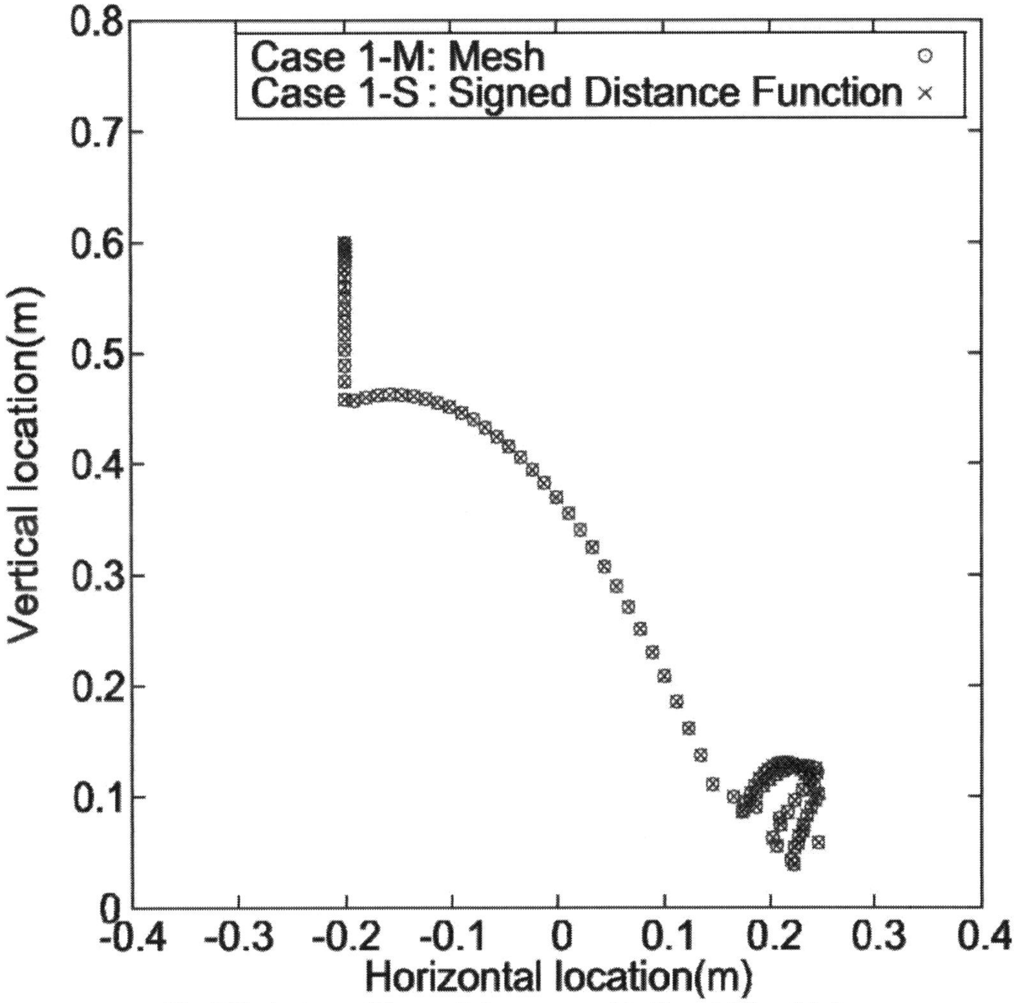

Fig. 2 Trajectory of the particle movement in Case 1-M and 1-S

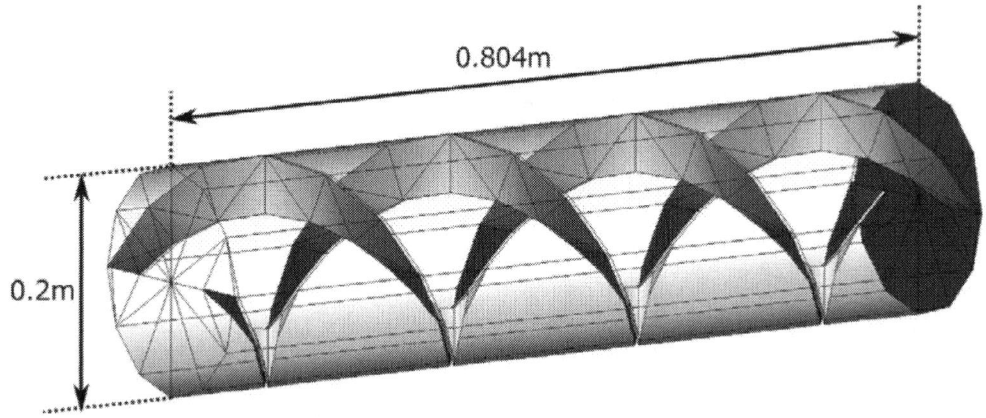

Fig. 3 Schematic diagram of the screw conveyer

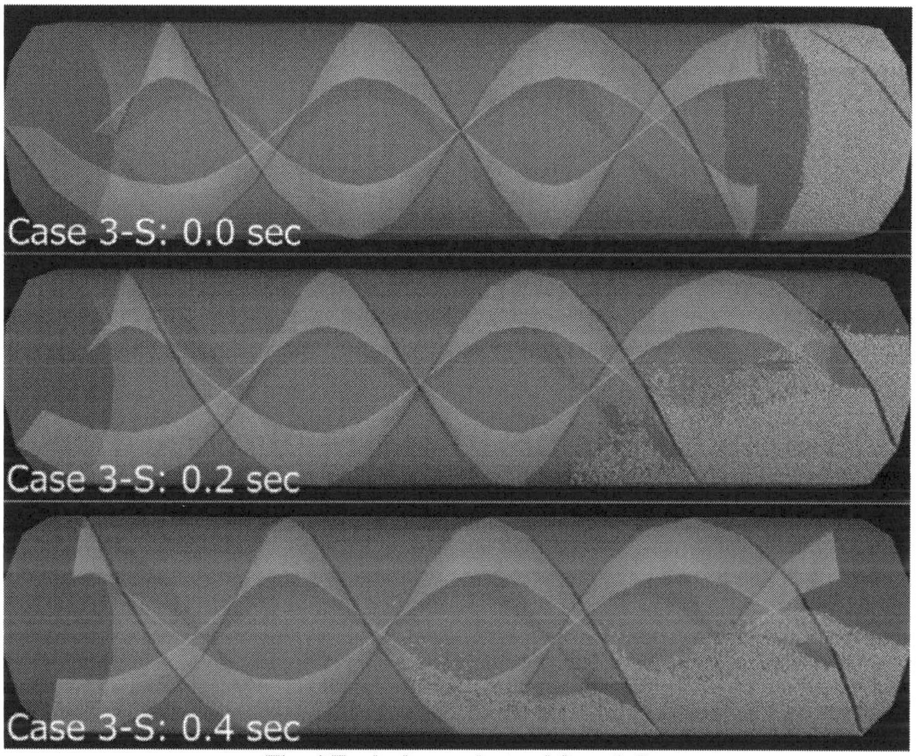

Fig. 4 Typical snapshots of Case 3-S

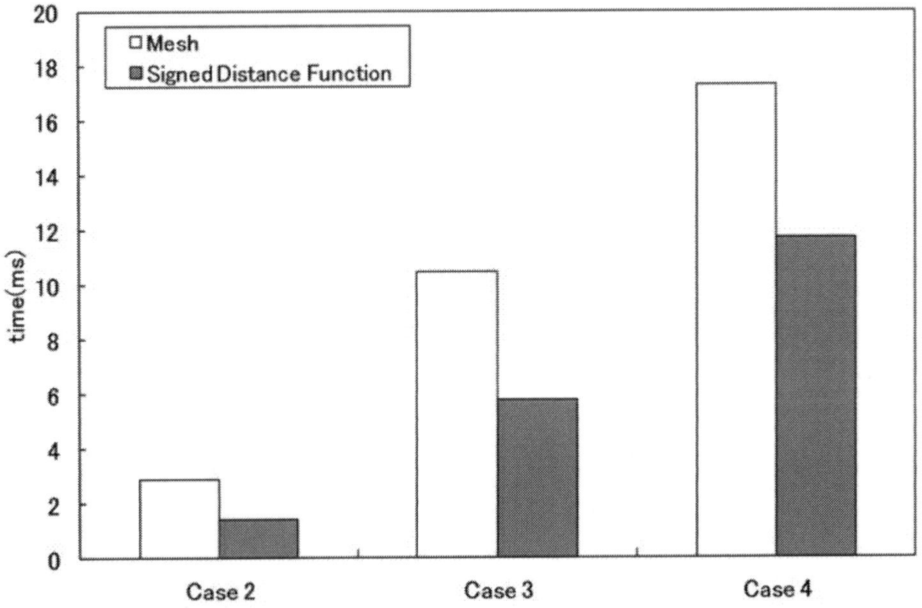

Fig. 5 Time consumed for particle-wall collision in cases 2, 3 and 4

Biosensing with plasmonic silver nanostructures

Georgios A. Sotiriou, Christoph O. Blattmann and Sotiris E. Pratsinis

Particle Technology Laboratory, Institute of Process Engineering
Department of Mechanical and Process Engineering
Swiss Federal Institute of Technology
Zurich (ETH Zurich)

ABSTRACT

Silver (Ag) nanoparticles dispersed in an amorphous silica (SiO_2) matrix or coated by a SiO_2 layer were synthesized by flame spray pyrolysis (FSP). The coated nanoparticles were produced by using a modified enclosed FSP setup, in which the SiO_2 precursor was injected through a ring above the FSP nozzle at various burner-ring-distances (BRDs), after the core Ag nanoparticles had been formed. The produced nanoparticles were characterized by XRD, BET, TEM and UV/vis analysis. The Ag particle size was possible to be controlled by tuning the FSP parameters. For the SiO_2 coated nanoparticles, larger Ag core sizes were obtained for higher BRDs. All the produced nanoparticles exhibited the characteristic plasmon resonance frequency of Ag nanoparticles.

INTRODUCTION

Silver (Ag) nanoparticles play an important role in photography, catalysis, biological labeling, photonics, optoelectronics and surface enhanced Raman scattering *(1)*. This is attributed to the unique optical properties of Ag that originate from the collective oscillations of conduction electrons, termed as surface plasmon polariton resonances *(2)*. This intrinsic property of Ag nanoparticles is mainly determined by their size, shape, composition, crystallinity, and structure. In principle, one could control any one of these parameters to fine-tune the corresponding optical properties. An attractive way to do this is by tuning the dielectric medium which surrounds them *(3)*. Such a dielectric medium is silica (SiO_2), with which the unique shape- and size-dependent optical properties of Ag nanoparticles can be tailored. Nanosilver structures show great potential as plasmonic biosensors because of the strong dependency of their plasmonic properties on their environment *(4-6)*.

Coating Ag with SiO_2 can also enhance the stability of Ag nanoparticles and provide tunable solubility in various solvents *(7)*. Monodispersed SiO_2-coated Ag particles readily self-assemble into photonic crystals with an optical band gap in the visible region *(8)*. Moreover, SiO_2 shells can be easily functionalized, allowing the nanoparticles to be conjugated to other substrates or molecules, a property extremely useful in bio-labeling *(9)*.

There have been reported various wet techniques to encapsulate fine metal nanoparticles, such as Ag, in a SiO_2 shell. These methods are effective but time-consuming and difficult to control *(7)*. Additionally, most of these methods can only be used for coating large metallic particles (~50 nm in diameter) because small particles are unstable and tend to aggregate in the alcoholic solutions typically used for wet methods *(7)*. Therefore, it is attractive to make Ag nanoparticles coated with a layer of SiO_2, by flame spray pyrolysis (FSP), a dry, faster, more versatile and with a proven scalability process *(10, 11)*.

Recently, Teleki *et al.* *(12)* reported the in-situ coating of TiO_2 nanoparticles with an ultrathin SiO_2 layer, using a modified FSP unit, in which the SiO_2 precursor was injected above the FSP burner after the formation of core TiO_2 particles had been completed. However, unlike metal oxide nanoparticles which can be easily coated with SiO_2, Ag nanoparticles are considered not to be suitable for direct coating due to their low chemical affinity to SiO_2 *(3, 7)*.

Here, Ag nanoparticles dispersed within an amorphous SiO_2 matrix or coated with a nanothin layer of SiO_2 are made by FSP, of solutions of silver nitrate in ethanol and diethylene glycolmonobutyl followed by in-situ SiO_2 coating from hexamethylsiloxane (HMDSO) vapor injected downstream of the FSP burner. For the SiO_2 coated nanoparticles, the size of the Ag core particles was controlled by varying the distance above the FSP burner. The as produced particles were characterized by TEM, X-ray diffraction (XRD), nitrogen adsorption (BET) and UV/vis spectroscopy.

EXPERIMENTAL SECTION

Silver nitrate (Fluka, purity >99%) and hexamethyldisiloxane (HMDSO, Aldrich, purity >97%) were used as silver and silicon precursors, respectively. Appropriate amount of the silver nitrate was dissolved in ethanol (Alcosuisse) and diethylene glycolmonobutyl ether (Fluka, purity >98%) with a volume ratio 1:1 and stirred at room temperature for 3 hours. Immediately before the precursor solution was fed into the FSP setup, appropriate amount of HMDSO was added and stirred for a few minutes. The total (Ag and Si) metal concentration was 0.5 M. The Ag content was 70 and 85 wt % (*70Ag/SiO$_2$* and *85Ag/SiO$_2$*). The particle synthesis was carried out using a flame spray pyrolysis (FSP) reactor which is described in detail elsewhere *(10)*. The synthesis of the SiO_2-coated particles was carried out using a modified enclosed FSP reactor *(12)*. In brief, the FSP reactor was enclosed with a metallic tube (internal diameter, ID = 40 mm) on top of which a metallic ring with 12 equidistant openings was placed, at two (15 and 25 cm) burner-ring-distances (BRDs). The SiO_2 coating was applied after the core particle formation had been completed, by the injection of vapor HMDSO from the metallic ring. Through these openings, nitrogen gas (purity >99 %), carrying HMDSO vapor, the precursor source for the SiO_2 coating, was injected along with 15 L/min nitrogen (N_2 mixing) at room temperature. The HMDSO vapor was supplied by bubbling 0.8 L/min nitrogen through approximately 350 mL liquid HMDSO in a 500 mL glass flask which was kept at 10 °C with cooling water, adding approximately 5.7 g/h of SiO_2, at saturated conditions, in the product *(13)*. The total Ag content of the coated particles was kept constant at 70 wt %.

The produced particles were characterized by TEM (FEI; LaB6 cathode, operated at 300 kV, point resolution ~2 Å) and EDAX analysis (EDXS; detector attached to the Tecnai F30 microscope), X-ray diffraction (Bruker AXS D8 Advance spectrometer (Cu Kα, 40 kV, 40 mA), specific surface area (*SSA*) of the samples was measured by five-point nitrogen adsorption at 77 K (BET: Micrometrics Tristar 3000) after degassing the samples at 150 °C for 1.5 hours in nitrogen. The optical properties were examined by UV/vis spectroscopy (Cary Varian 500), by dispersing in deionized water 50 μg/ml of the as-prepared nanoparticles.

RESULTS AND DISCUSSION

Ag nanoparticles dispersed within an amorphous SiO_2 matrix (uncoated), or coated by a thin SiO_2 layer were synthesized by FSP. The uncoated particles were produced by an open FSP setup unit, while the SiO_2 coated by a modified enclosed unit. Figure 1 shows the XRD patterns of the as-prepared nanoparticles. The spectra have been normalized for the (111) peak. Starting from the bottom, the patterns of the uncoated particles are presented (*70Ag/SiO*$_2$ and *85Ag/SiO*$_2$), followed by the patterns of the SiO_2 coated ones, for two different BRDs (*70Ag/SiO*$_2$*-15cm* and *70Ag/SiO*$_2$*-25 cm*). The peak positions on top indicated by the black lines in this figure correspond to the (111), (200) and (220) crystal planes of the Ag metal (ICSD nr. 064995). It is observable that the obtained peaks correspond to this metallic structure, for all the products. Additionally, the particle sizes, calculated by using the Rietveld method for the (111) peak, are also shown.

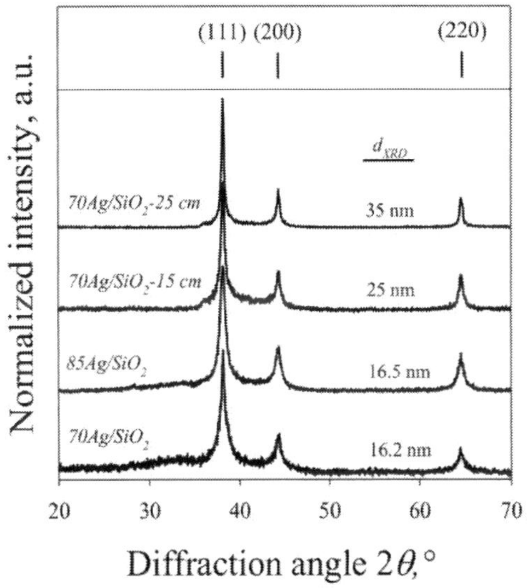

Figure 1: Normalized spectra of the as-prepared nanoparticles. The black lines on top represent the peak positions of the Ag metal structure. The d_{XRD} calculated by the Rietveld method is also shown.

In flame synthesis, the particle size depends on several parameters, such as the flame temperature, the particle residence time and the concentration of the precursor which is used *(10)*. In respect to the precursor concentration which is used, larger sizes are obtained for higher concentrations. However, for the uncoated particles, in respect to the d_{XRD}, it can be observed that the Ag average particle sizes are equal for the two different Ag contents, indicating that the presence of SiO_2 inhibits the further growth of the Ag nanoparticles. On the other hand, when the d_{XRD} sizes of the coated samples are examined, it can be observed that for a higher BRD, larger particle size is obtained. This could be attributed to the larger residence time that the Ag nanoparticles have for the higher BRD, before their entrapment by the SiO_2, and due to the fact that an open flame has much higher cooling rates than an enclosed one. Therefore, it can be concluded that by using the modified enclosed setup, the size of the Ag nanoparticles can be precisely controlled, while when the open FSP setup is used, the size of the Ag nanoparticles is influenced mostly by the SiO_2.

Table 1: The values of the *SSA* for the as-produced nanoparticles.

Sample	*SSA*, m^2/g
70Ag/SiO$_2$	107
85Ag/SiO$_2$	70
70Ag/SiO$_2$ – 15 cm	53
70Ag/SiO$_2$ – 25 cm	35

The values of the specific surface area (*SSA*) of the produced samples can be observed in Table 1. The *SSA* includes the area from both silver and silica. It is noticeable that the uncoated samples have higher values, while the coated, lower ones. In case the enclosed setup is used, the particles have larger residence time in higher temperatures than when the open setup is used and thus, the particles grow bigger, resulting in lower *SSA*. When the two coated samples are compared, the one with the higher BRD has lower *SSA*, which is line with the Ag core particle sizes obtained from the XRD analysis, having the higher *SSA* values for the smaller particle sizes.

The morphology of the produced particles can also be observed from the obtained TEM images. In Figure 2, two TEM images of the uncoated particles *70Ag/SiO$_2$* (a) and the SiO_2 coated *70Ag/SiO$_2$-15cm* (b) are shown. It can be observed that the uncoated Ag particles are dispersed within an amorphous SiO_2 matrix, while the coated ones have a SiO_2 layer surrounding them.

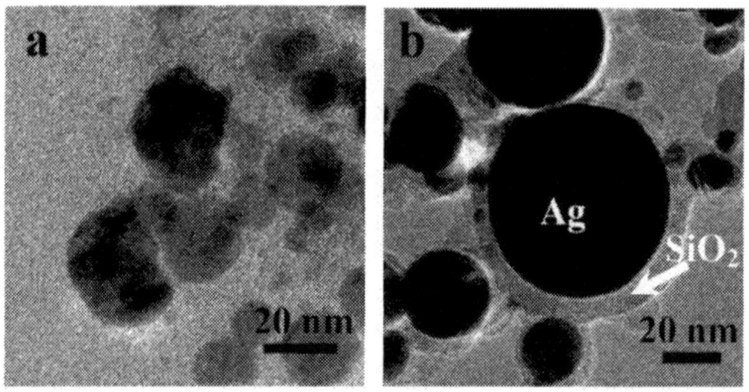

Figure 2: Two TEM images of the uncoated particles *70Ag/SiO$_2$* (a) and the SiO_2 coated *70Ag/SiO$_2$-15cm* (b) are shown.

The optical properties of the produced nanoparticles are presented in Figure 3. The UV/vis spectra of the produced nanoparticles are shown. The spectra have been normalized in respect to the strong absorption band at ~400 nm, which corresponds to the plasmon resonance frequency. For the uncoated *70Ag/SiO$_2$*, there is the plasmon resonance frequency peak at ~400 nm present *(14)*. However, as the Ag particle size is increased, in the spectrum corresponding to the *85Ag/SiO$_2$*, apart from the sharp peak at ~400 nm, there is also a broad hump appearing at ~520 nm, due to the increased state of aggregation and the larger particle size, as predicted by the literature *(14)*. For the SiO_2 coated samples, it is also observable that for the lower BRD, there is only the peak at ~400 nm, however, as the Ag core particle size is increased, there is a second peak appearing at higher wavelengths. For the higher BRD, the peak at higher wavelengths is more intense, in respect to the 400 nm peak, than the one resulting from the open flame *85Ag/SiO$_2$*. It is observable that even though the d_{XRD} sizes of the two samples resulting from the open flame are similar, there are differences in the corresponding UV/vis spectra. Therefore, it seems that the band at higher wavelengths is not only dependent on the primary particle size of the Ag nanoparticles, but also other factors affect it, such as the aggregation state and the size of the aggregate structures, as reported by the literature *(14)*. The secondary peak in the extinction spectra could also be due to too large particle size and non-spherical particle shape such as elongated rod like structures, besides coupled mode of multiple spheres.

Therefore, with both FSP setups, it is possible to control the optical properties of the produced Ag nanoparticles, by controlling the Ag particle sizes. The advantage of the coated particles lies in the possibility of fine tailoring the Ag nanoparticles sizes, and thus, tuning their optical properties depending on the desired application. Moreover, the SiO_2 layer surrounding the Ag core particles is easier to be functionalized and thus, organic molecules can be conjugated on its surface, a property extremely important in applications such as bio-labeling *(9)*.

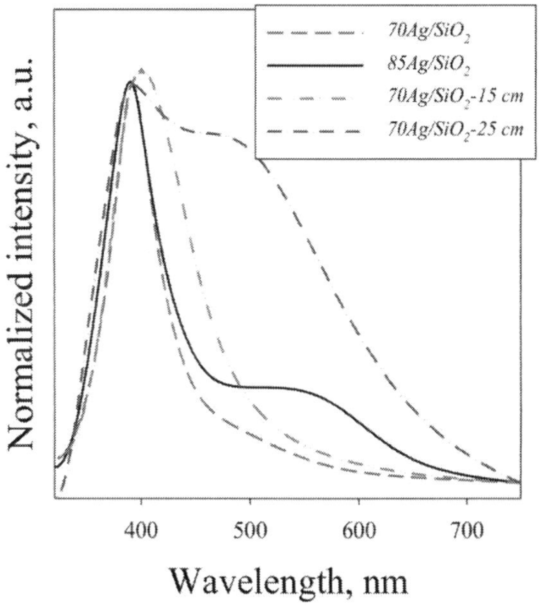

Figure 3: Normalized UV/vis spectra of the as-prepared samples.

CONCLUSIONS

Silver nanoparticles dispersed within an amorphous SiO_2 matrix or coated by a SiO_2 layer were produced by FSP. The uncoated Ag/SiO_2 nanoparticles were produced by an open FSP setup, while the synthesis of the SiO_2 coated by using a modified enclosed setup. In case the open setup was used, the Ag particle sizes depend mostly on the Ag content, however, limitation in the controllability due to the SiO_2 do exist. However, when the enclosed setup is used, precise control over the Ag core particle size can be achieved. Depending on the BRD, larger Ag particles are obtained for higher BRDs. All the produced Ag nanoparticles exhibited a peak at ~400 nm in their UV/vis spectra, characteristic peak corresponding to the plasmon resonance frequency of Ag nanoparticles. The precise Ag core size controllability, as well as the SiO_2 coating layer, are properties extremely important in several applications, such as bio-labeling.

REFERENCES

1. Sun, Y. G.; Xia, Y. N. Shape-controlled synthesis of gold and silver nanoparticles *Science* **2002**, *298*, 2176-2179.

2. Evanoff, D. D.; Chumanov, G. Synthesis and optical properties of silver nanoparticles and arrays *ChemPhysChem* **2005**, *6*, 1221-1231.
3. LizMarzan, L. M.; Giersig, M.; Mulvaney, P. Synthesis of nanosized gold-silica core-shell particles *Langmuir* **1996**, *12*, 4329-4335.
4. Sotiriou, G. A.; Hirt, A. M.; Lozach, P. Y.; Teleki, A.; Krumeich, F.; Pratsinis, S. E. Hybrid, silica-coated, Janus-like plasmonic-magnetic nanoparticles *Chem. Mater.* **2011**, *23*, 1985-1992.
5. Sotiriou, G. A.; Pratsinis, S. E. Engineering nanosilver as an antibacterial, biosensor and bioimaging material *Curr. Opin. Chem. Eng.* **2011**, *1*, 3-10.
6. Sotiriou, G. A.; Sannomiya, T.; Teleki, A.; Krumeich, F.; Vörös, J.; Pratsinis, S. E. Non-toxic dry-coated nanosilver for plasmonic biosensors *Adv. Funct. Mater.* **2010**, *20*, 4250-4257.
7. Han, Y.; Jiang, J.; Lee, S. S.; Ying, J. Y. Reverse microemulsion-mediated synthesis of silica-coated gold and silver nanoparticles *Langmuir* **2008**, *24*, 5842-5848.
8. Lu, Y.; Yin, Y. D.; Li, Z. Y.; Xia, Y. A. Synthesis and self-assembly of Au@SiO_2 core-shell colloids *Nano Lett.* **2002**, *2*, 785-788.
9. Liu, S. H.; Han, M. Y. Synthesis, functionalization, and bioconjugation of monodisperse, silica-coated gold nanoparticles: Robust bioprobes *Adv. Funct. Mater.* **2005**, *15*, 961-967.
10. Madler, L.; Kammler, H. K.; Mueller, R.; Pratsinis, S. E. Controlled synthesis of nanostructured particles by flame spray pyrolysis *J. Aerosol. Sci.* **2002**, *33*, 369-389.
11. Strobel, R.; Pratsinis, S. E. Flame aerosol synthesis of smart nanostructured materials *J. Mater. Chem.* **2007**, *17*, 4743-4756.
12. Teleki, A.; Heine, M. C.; Krumeich, F.; Akhtar, M. K.; Pratsinis, S. E. *In-situ* coating of flame-made TiO_2 particles by nanothin SiO_2 films *Langmuir* **2008**, *24*, 12553-12558.
13. Yaws, C. L.; Narasimhan, P. K.; Gabbula, C. Yaws' Handbook of Antoine coefficients for vapor pressure (Electronic edition) *Knovel* **2005**,
14. Quinten, M. The color of finely dispersed nanoparticles *Appl. Phys. B-Lasers Opt.* **2001**, *73*, 317-326.

(388q) The Polydispersity of Primary Particles in Aggregates Undergoing Sintering

Max Eggersdorfer and Sotiris E. Pratsinis, Particle Technology Laboratory, Swiss Federal Institute of Technology (ETH Zurich), Zurich, Switzerland

The Polydispersity of Primary Particles in Aggregates undergoing Sintering

Fractal-like aggregates consist of multiple particles that are connected by chemical (e.g. sintering) bonds. Such aggregates form by natural and man-made processes, typically at high temperatures, like fly ash from coal combustion as well as aerosol synthesis of ceramics (titania, fumed silica, alumina) and metals (Ni, Fe, Ag etc.). The morphology of such particles has critical implications in their performance. In nanoparticle synthesis by aerosol processes the sintering of aggregates during particle formation narrows their primary particle size distribution affecting mechanical and optical properties. The understanding of the evolution of primary particle size distributions during sintering is also important for the design of aerosol reactors by computational modelling. Furthermore, the fractal dimension, D_f, of agglomerates decreases steadily with increasing polydispersity of constituent primary particles [1].

Here a recently developed model for viscous sintering of amorphous [2] and crystalline (e.g. TiO_2 or Ag) [3] aggregates is used to elucidate the detailed evolution of their primary particle distribution during sintering. This model reproduces nicely the initial neck growth and evolution of particles center-to-center distance for equally sized pairs of particles and is compared to the classic models [4]. So the evolution of the radius of gyration and D_f of ensembles of irregular particles is presented as they asymptotically approach full compactness by sintering. It is shown that initially monodisperse primary particles become slightly polydisperse during sintering exhibiting even a small maximum in geometric standard deviation before converging to a fully coalesced sphere. Initially polydisperse primary particles monotonically reduce the breadth of their size distribution by sintering.

[1] Eggersdorfer, M.L. and Pratsinis, S.E., *Aerosol Sci. Technol.* **46** (2012) 437-353.

[2] Eggersdorfer, M.L., et al., *Langmuir* **27** (2011) 6358-6367.

[3] Eggersdorfer, M.L., et al., *J. Aerosol Sci.* **46** (2012) 7-19.

[4] Coblenz, W.S., et al., *Mat. Sci. Res.* **13** (1980) 141-157.

Simulation of Flow Behavior of Particles in a Magnetic Liquid-Solid Fluidized Bed with Uniform Magnetic Field

Shuyan Wang [1, 2], Lixin Wei [2], Jinsen Gao [3], Xingying Lan[3]

1 Department of Chemical and Biological Engineering, Princeton University, Princeton, NJ, United States

2 School of Petroleum Engineering, Northeast Petroleum University, Daqing 163318, China

3 State Key Laboratory of Heavy Oil Processing, China University of Petroleum, Beijing, 102249, China

ABSTRACT

Magnetic fluidized beds (MFB) of magnetically susceptible particles are considered as one of the technologies developed to eliminate the drawbacks of fluidized beds. The MFB results in an improved fluid-solid contactor that combines the most desirable characteristics of both fluidized and packed beds. The MFB deals with two basic magnetization modes: Magnetization FIRST (preliminary magnetization of fixed bed and consequent fluidization) and Magnetization LAST (magnetization of preliminarily fluidized beds). In present study, the flow behavior of solid phases is simulated by means of DEM-CFD in a liquid-solid fluidized bed with LAST mode along an axial uniform magnetic field. The magnetic force on the particle due to the non-uniform magnetic field resulting from the interaction of the external uniform field and non-uniform fields of the magnetized particles is predicted. By changing the magnetic field strength, the distribution of particles is studied within the bed. The distributions of velocity and concentration of ferromagnetic particles are analyzed at the different magnetic field intensities. The simulations show a significant effect on the motion of particles with axial magnetic fields applied. When the magnetic field strength is increased to a value at which the fluidization of strings starts, the particles are found to form straight-chain aggregates in the direction of the magnetic field. At very high magnetic field strengths, the defluidization is observed at which particles are fixed in the bed. Simulations indicate the granular temperature of particles increases, reaches a maximum, and then decreases with the increase of magnetic-flux density. Through the analysis of the motion of particles, it is concluded that the moderate strength magnetic field gives a high fluctuation of particles.

(388s) Template-Directed Synthesis of Porous CoFe2O4/Al2O3 Reactive Structure with Atomic Layer Deposition for Solar Thermochemical Fuel Production

Darwin Arifin[1], Victoria J. Aston[2], Xinhua Liang[2], Anthony H. McDaniel[3] and Alan W. Weimer[2], (1)Chemical and Biological Engineering, University of Colorado Boulder, Boulder, CO, (2)Chemical and Biological Engineering, University of Colorado at Boulder, Boulder, CO, (3)Sandia National Laboratories, Livermore, CA

In this work, we examine a novel chemistry for a two-step, non-volatile metal oxide CO_2 splitting cycle that utilizes the change in iron oxidation states ($Fe^{2+/3+}$) between $CoFe_2O_4$ and $FeAl_2O_4$ spinel compounds within a novel nano-engineered reactive structure. The nano-engineered material was synthesized by template directed atomic layer deposition (ALD) of Al_2O_3, Fe_2O_3 and CoO. The structure maintained structural integrity over 10 heating cycles under conditions that mimic a concentrated solar power application, namely an oxidation temperature of 1000°C, reduction at 1460°C, and a heating rate of 16°C/s from low to high temperature.

The nano-engineered structure is ideal for solar thermochemical cycling due its highly porous nature that allows fast radiative heat transfer and high gas/solid reaction interface. Perhaps more important than the fast heat transfer is the ability to precisely control the thickness of the porous structure via ALD; this allows for the minimization of large excess Al_2O_3. A large excess of alumina would lead to diffusion of Co and Fe deeply into the bulk and undoubtedly have a detrimental impact on the redox kinetics. Oxygen uptake and release behavior of the nano-structure material is similar to that of ceria, which is well known to have high oxygen ion conductivity and rapid exchange kinetics. Raman spectroscopy was used to verify cycle chemistry.

(388t) Gas Phase Synthesis of Nickel Catalyst Nanoparticles Via Atomic Layer Deposition

Troy D. Gould, Alia M. Lubers, J.Will Medlin, Alan W. Weimer and John L. Falconer, Chemical and Biological Engineering, University of Colorado at Boulder, Boulder, CO

Efforts to design new catalysts for renewable energy applications have revolved around increasing selectivity & activity by exploring advantages of catalysis at the nanoscale, mainly due to geometric & electronic effects created by nanostructures or bimetallic structures. This work investigates the synthesis of Ni nanoparticle catalysts through the scalable gas phase synthesis route of Atomic Layer Deposition (ALD). Typically ALD deposits Ni as a NiO film when using oxygen as the secondary precursor. However; by using hydrogen as the secondary precursor, metallic Ni nanoparticles were synthesized with particle diameters of 2 nm with 1 ALD cycle. The size of the nanoparticles can be increased by performing more ALD cycles to give the ALD synthesis an element of particle size control. The catalysts were characterized via chemisorption, temperature programmed desorption (TPD), transmission electron microscopy (TEM), inductively coupled plasmon mass spectroscopy (ICP-MS), energy dispersive x-ray spectroscopy (EDS) & compared to traditional incipient wetness Ni catalysts. The incipient wetness Ni catalysts were also used as a base comparison of catalytic activity for alkene hydrogenolysis reactions using propylene as a probe reactant molecule. Hydrogenation turnover frequencies were at least one order of magnitude higher for the nanoscale ALD catalysts than the incipient wetness catalysts. In tuning the size of the nanoparticles, certain sized nanoparticles were found to have hydrogenolysis activity in addition to the typical hydrogenation. To further explore the benefits of Ni ALD, bimetallic ALD catalyst nanoparticles were investigated for a NiPt bimetallic system. Pt ALD has been an established technique for creating Pt nanparticles. However, by combining the novel Ni ALD with Pt ALD, nanoscale bimetallic catalysts were created. To address issues of specific Pt deposition on Ni particles, temperature programmed reduction (TPR) and chemisorptions were used to characterize the Ni & Pt depositions separately and together to determine the extent of bimetallic interaction.

(388u) Dissipative Particle Dynamics Simulation of Sonication-Mediated Particle Interactions

Shu Li[1], Melissa H. Lash[1], Steven R. Little[2] and Joseph McCarthy[1], (1)Chemical and Petroleum Engineering, University of Pittsburgh, Pittsburgh, PA, (2)Departments of Chemical and Petroleum Engineering, Bioengineering, Immunology, and The McGowan Institute for Regenerative Medicine, University of Pittsburgh, Pittsburgh, PA

Sonication is a powerful dispersion technique that has been used successfully to disperse agglomerates and break partially formed bridges of nano-scaled particles. For micro-scaled particles, recent work in our lab suggests that it may also serve as a viable "thermalization" technique in order to promote close-packed, highly ordered colloidal crystal formation of large (non-Brownian) particles on a surface. In this study, we computationally explore the impact of sonication on particle cluster formation using the Dissipative Particle Dynamics (DPD) technique. DPD is a particle-based (i.e. Lagrangian) simulation technique that captures meso-scopic dynamics of multiphase flows by coarse-graining intermolecular forces. At its most basic, this coarse-graining results in a conservative, dissipative, and random force. Here, the effect of sonication is incorporated into the DPD method as an oscillatory external force with high frequency and low amplitude. By adding the external force from the sonication, we can inspect the microscopic details of the relative importance of the particle interactions as a function of driving parameters (frequency and amplitude) and compare our results to experimental measurements.

(388v) Mixing of Solid Particles in Coarse Particle Fluidized Beds

Haihong Fan[1], **Yanxin Chen**[2], Hui Li[2], Jinhe Fan[2] and Yong Min[2], (1)Xi'an University of Architecture &Technology, Institute of Powder Engineering, XI'AN, China, (2)Xi'an University of Architecture &Technology, Institute of Powder Engineering, Xi'an, China

The mixing of solid particles in fluidized bed reactions has been recognized as an important factor determining the heat and mass transfer rates as well as the overall reaction rates. Many studies have been made in the past in order to understand the underlying mechanisms and predict the behaviour of mixing and segregation, including the investigation of factors affecting the mixing/segregation. Almost all the studies in this area are concerned with monosized particles,and Few studies on mixing of single-size particles in Coarse Particle were reported. This paper extended the work of these work and aimed to the mixing behaviors of solid single-size particles and influencing factors by means of a discrete particle simulation.

In this simulation, the gas phase was taken as continuum medium and the particle phase was discrete phase. The distinct element method (DEM) based on Lagrangian approach was applied to calculate the movement of particles. By this approach, the particle trajectories, particle-particle and particle-wall interactions were considered distinctly for each particle existing within the studied system. The continuity and momentum equations was used for the fluid phase. The motion of the particles was calculated by solving Newton's equation of motion for each individual particle by taking into account the gravity force, the drag force, the fluid pressure and the contact force.

The value of Ashton index was calculated to elucidate the dynamic mixing process.

The solid flow patterns were examined first to generate some visual understanding of the mixing process. The minimal mixing time was defined as the time to reach a macroscopically stable state.

In this simulation, the tracer particles were monitored momentarily to acquire their locations and the mixing index. This paper focused on five factors affecting the mixing extent, the gas velocities, the bed height, the gas temperature, the particle density and the particle size. It was showed that there is a maximum value of mixing index and a relevant minimum mixing time for a given operation condition. These five factors have different effects on particles mixing.

(388w) Recovery of Iron From Gold Mine Tailings by Decentralized Magnetization Roasting-Magnetic Separation

Yanxin Chen[1], Yanfei Yao[2], Bo Zhao[2] and Yong Min[1], (1)Xi'an University of Architecture &Technology, Institute of Powder Engineering, Xi'an, China, (2)Xi'an University of Architecture& Technology

Recovery of iron from Gold mine tailings by decentralized magnetization roasting-magnetic separation

CHEN Yan-xin YAO Yan-fei JIU Shao-wu ZHAO Bo MIN Yong

Xi'an University of Architecture& Technology

Abstract:

Decentralized magnetization roasting-magnetic separation was used to process a sort of gold mine tailings, with the purpose of recycling the iron resources. Tailings was magnetization roasting under the conditions of 850 , CO volume percentage was 2%, then determined the right roasting time to get roasted ore. Magnetic separation was proceed with orthogonal test, focused on the effect of Magnetic field intensity and the dosage of dispersant and flocculent to concentrate grade and iron recovery. The experimental results indicate that after magnetization roasting under the conditions of 850 ,CO volume percentage was 2%, roasting time was 5min,then proceed magnetic separation at Magnetic field intensity was 900Oe, the dosage of dispersant and flocculent was 0.05g/20g and 0.1128g/20g, a concentrate assaying 57.15% Fe with 81.43% recovery was achieved.

Keywords: Gold mine tailings, Magnetization roasting-magnetic separation, Concentrate grade, Iron recovery

(388x) Theoretical and Experimental Study On the Improvement of Powder Bulk Density

Yanxin Chen[1], Feng Wu[1], Yaru Hu[1], Haihong Fan[2] and Yong Min[1], (1)Xi'an University of Architecture &Technology, Institute of Powder Engineering, Xi'an, China, (2)Xi'an University of Architecture &Technology, Institute of Powder Engineering, XI'AN, China

Theoretical and Experimental Study on the Improvement of Powder bulk density

CHEN Yan-xin, WU Feng, HU Ya-ru, FAN Hai-hong, MIN Yong

Xi'an University of Architecture &Technology, Institute of Powder Engineering, Xi'an, China

Abstract

Based on the theory of Andreason and compressible packing model, the calculations of modifying and packing density of the coal particles have been carried out, and the calculation results have been also verified by experiments. Meanwhile the influences of the size range and add volume on packing density has been discussed.

The result indicated that: 1) The particle size distribution can be optimized by modifying, and it is close to the one of close packing; 2) The packing density of coal particles was well predicted with compressible packing model and it is increased by 5.977% after modifying; 3) Under the condition of close packing, the large size range is helpful for improving the packing density of the system; 4) With the increasing of modifying material, the packing density is also increasing and it is close to the packing density under close packing; 5) Just adding power, the packing efficiency is relevant with particle size composition of raw materials.

Key words: coal water slurry size distribution compressible model

(388y) Thin Film, Big Difference – Atomic Layer Deposition Functionalized Oxide and Polymer Particles

Staci A. Van Norman[1], Joseph W. Tringe[2], John L. Falconer[1] and Alan W. Weimer[1], (1)Chemical Engineering, University of Colorado at Boulder, Boulder, CO, (2)Lawrence Livermore National Lab, Livermore, CA

Functionalization of insulating particles by Atomic Layer Deposition (ALD) could provide a unique range of electrical conductivity and outperform metal doped polymers for applications as replacements for metal electrodes in devices. The characterization and method of applying conductive metal ALD coatings on insulating materials to selectively produce precisely controlled properties is discussed.

In this study, functionalization of four types of oxide and polymer particles, 18 μm diameter or larger, was achieved by ALD of conformal metallic W thin films. Hollow and solid oxide particles as well as high and low surface area porous polymers were examined for conductivity range with increasing W film thickness. Particles were characterized for electrical and thermal conductivity using a four point probe planar van der Pauw test geometry. Measurable decreases in resistance over bare oxide support particles have been seen for film thicknesses of 25 nm or greater, additional decreases observed by increasing W thickness. Combining resistance characterization with ALD of conductive films on non-conductive particles allows for the development of a novel technology to selectively control electrical and thermal properties for defined applications.

(388z) Stabilization of Pt Catalyst Via Deposition of Ultrathin Porous Films

Alan Izar[1], Sandro Hutter[2], John L. Falconer[3] and Alan W. Weimer[3], (1)Chemical & Biological Engineering, University of Colorado - Boulder, Boulder, CO, (2)ETH, Switzerland, (3)Chemical Engineering, University of Colorado at Boulder, Boulder, CO

Porous nanolayers of alumina can be fabricated by the deposition of alucone, an aluminum alkoxide polymer, onto zirconia nanoparticles, followed by the calcination of the layers in order to remove the organic portion of the molecule. Two monomer structures of alucone were used: AB, which consists of trimethylaluminum (TMA) and ethylene glycol (EG); and ABC, composed of trimethylaluminum (TMA), ethanolamine (EA) and maleic anhydride (MA). Using molecular layer deposition, self-assembled monolayers of these polymers were coated on zirconia nanoparticles, with varying film thickness. The particles were subsequently calcinated at different temperatures in order to produce different porous layer structures. These porous layers were characterized with TEM, EDS, and other techniques in order to study their structure and porosity. The same procedure was repeated for platinum particles supported on zirconia, which were used in an oxidation reactor. A possible future application of these layers is in prevention of catalyst sintering, as well as selectivity control.

(402a) Continuous Pharmaceutical Hot-Melt Extrusion and Hot-Die Face Pelletizing

Daniel Treffer[1], Andreas Eitzlmayr[1], Clemens-Michael Smola[2], Gerold Koscher[2], Eva Roblegg[3] and Johannes G. Khinast[1,2], (1)Institute for Process and Particle Engineering, Graz University of Technology, Graz, Austria, (2)Research Center Pharmaceutical Engineering GmbH, Graz, Austria, (3)Institute of Pharmaceutical Sciences, University of Graz, Graz, Austria

The pharmaceutical industry goes through a paradigm change. So far batch manufacturing was nearly the only way to produce pharmaceuticals. A new approach, however, is continuous manufacturing. Hot melt extrusion (HME) is such a continuous process with significant potential for the manufacturing of different dosage forms with an improved control of the drug release profile[1,2]. HME produces homogeneous strands of molten material, which are further processed in a downstream process. A hot die-face pelletizer is one possible downstream option. It cuts the strands directly at the die plate into pellets. Some authors reported that spherical pellets can be obtained directly with a hot die-face pelletizer[3]. The benefit of spherical pellets is better flowability compared to non-spherical pellets, which is an important requirement for further particle processes. In this work the application of a hot die-face pelletizer as a continuous downstream process is demonstrated.

The experimental setup consists of a co-rotating twin screw extruder ZSK 18 (Coperion, Germany) and a hot die-face pelletizer (Automatik Plastics Machinery GmbH, Germany). Powders or pellets consisting of active pharmaceutical ingredient (API) and excipients are fed continuously to the intake zone of the extruder with loss-in-weight feeders (K-Tron, Switzerland). The extruder itself combines several unit operations, such as conveying, mixing, melting, and degassing within one apparatus. At the extruder outlet the molten material is pushed through an extrusion die and homogeneous strands emerge from the die plate. The strands are cut continuously into pellets with rotating knifes directly at the die plate. An airstream cools the strands before the cutting as wells as the particles after the cutting process. Furthermore, it conveys the particles to the next processing step. In order to ensure good particle qualities the particle size and shape is monitored in-line with the Eyecon (Innopharma Labs, Ireland). Most process parameters are collected with SIPAT (Siemens, Belgium) which enables an efficient process development and monitoring.

Several pharmaceutical polymers and formulations have been investigated for hot die face pelletizing. Not all formulations are suitable for this downstream process. Rheological investigations of the formulations have been performed and approaches to predict applicability of the hot die-face pelletizer are derived.

References:

1. Breitenbach J.: Melt extrusion: from process to drug delivery technology, *European Journal of Pharmaceutics and Biopharmaceutics* 2002, 54, 107-117.
2. Roblegg E, et. al.: Development of sustained-release lipophilic calcium stearate pellets via hot melt extrusion, *European Journal of Pharmaceutics and Biopharmaceutics 2011, 79, 635-645.*
3. Bailleck S; Rein H, Preparation of starch-based pellets by hot-melt extrusion, *European Journal of Pharmaceutics and Biopharmaceutics, 2011, 79, 440-448*

(402b) Predictive Modeling of Hot Melt Extruders

Andreas Eitzlmayr[1], Daniel Treffer[1], Gudrun Hoerl[2], Sarah Windhab[2], Gerold Koscher[2] and Johannes G. Khinast[1], (1)Institute for Process and Particle Engineering, Graz University of Technology, Graz, Austria, (2)Research Center Pharmaceutical Engineering GmbH, Graz, Austria

Introduction

During recent years, extrusion processes increasingly attracted attention in the area of pharmaceutical manufacturing. As a continuous manufacturing operation, extrusion provides the potential of increased efficiency and reduced operation costs together with a steady product quality. Depending on the used materials and conditions, different types of pharmaceutical extrusion processes can be distinguished, as hot melt extrusion, wet extrusion and solid lipid extrusion. In contrast to wet extrusion and solid lipid extrusion, hot melt extrusion is characterized by a temperature level above the melting point of the material. Since the production of solid dispersions and solid solutions can be achieved, hot melt extrusion is a potential pathway to increase the bioavailability of poorly soluble drugs.

The most common extrusion devices are co-rotating twin screws. Compared to different types as counter-rotating twin screws and single screws the co-rotating twin screws are often preferred due to their mixing performance and self cleaning screw profiles.

Although melt extrusion processes are known from plastics industry for many years and spatially resolved CFD (Computational Fluid Dynamics) simulations of extruders screws are already achieved today [1], there is still a lack of predictive tools supporting design, scale-up and optimization of extrusion processes and a high amount of empirical methods and experimental effort is required. Finite element and finite volume CFD methods can be used to solve the hydrodynamics of completely filled screws. Partially filled screw sections cannot be addressed by conventional methods, since their facilities to recover free surface flows are very limited. The interactions of the involved physical phenomena are highly complex and first-principle design methods are not available for extrusion processes.

Methodology

In our work, we are following two different approaches. The first is a one-dimensional (1D) screw model, which describes physical process variables as pressure, material temperature and screw filling ratio along the screw, and follows the idea of a continuous stirred tank reactor cascade with backmixing, as shown by Choulak et al. [2]. This model targets a global, approximate description of the material along the twin-screw. Since a 1D model cannot resolve the three-dimensional (3D) geometry of the screws, it depends on empirical screw parameters which close the gap between 3D geometry and 1D flow description. These parameters can be determined either by experiments or by 3D simulations. The latter is a target of our second approach, which is a spatially resolved screw model based on a relatively new CFD method providing increased potential for problems with complex, moving boundaries as extruder screws.

Results

Results of the 1D model are profiles of filling ratio, pressure, material temperature and power input along the screw. A residence time distribution gives a measure for back-mixing in screw direction. The calculated profiles are tightly related to the geometrical configuration of the screw. Experimental work to validate the results is ongoing.

References

[1] K. Kohlgrüber, *Der gleichläufige Doppelschneckenextruder*, Carl Hanser Verlag, München, 1st edn., 2007, ch. 7, pp. 140 – 142.

[2] S. Choulak, F. Couenne, Y. Le Gorrec, C. Jallut, P. Cassagnau, A. Michel, Generic Dynamic Model for Simulation and Control of Reactive Extrusion, *Ind. Eng. Chem. Res.*, 2004, 43, 7373-7382.

(402c) Foam Granulation Extrusion: A Novel Method to Continuous Wet Granulation of Powder Drug Formulations

Michael R. Thompson[1], Sharleen Weatherley[1], Rohan Pukadyil[1] and Paul Sheskey[2], (1)Dept. of Chemical Engineering, McMaster University, Hamilton, ON, Canada, (2)The Dow Chemical Company, Midland, MI

The present work seeks to demonstrate a new continuous method for wet granulation using an aqueous foam binder to produce stable continuous operations in a twin screw extruder. Furthermore, we endeavor to understand the influence that foam properties have on the performance of a binder during foam granulation.

EXPERIMENTAL: The study was conducted using α-lactose monohydrate (Pharma FlowLac® 100, Meggle) as a model excipient to be granulated in a co-rotating twin screw extruder (ZSE 27HP, Leistritz) with side stuffer at 10-40 kg/h and 220-320 RPM. The extruder barrel was controlled to 25°C using a cooling system. An aqueous binder solution, composed of 4% (w/w) hypromellose (either METHOCEL™ A15PLV, E3PLV, E6PLV or E15PLV; Dow Chemical, Midland, MI), was foamed to 75-95% foam quality (FQ) using a mechanical foam generator and added at 6% or 11% (w/w) to the extruder. Foam properties were evaluated using two shear stability tests. Evaluation of granule properties was done by particle size analysis and determination of the characteristic fracture strength.

RESULTS: The initial studies to realize the method were done with A15 binder. Using a side stuffer (auxiliary feeder affixed to the extruder) it was found that the foamed binder could be accurately metered into the extruder at all operating conditions. A visualization window showed spreading of the foam over the surface of the lactose bed along an axial distance more than 4 L/D. To determine the value of the new method, wet granulation was compared by introducing the same aqueous binder solution at 6% or 11% (w/w) either as a foam (FA) or by direct addition using a pulseless metering pump (LI). The FA process demonstrated excellent stability over different flow rates (up to 40 kg/h) and screw speeds with no surging, attributed to uniform wetting of the powder; no segregated dry regions were noted, unlike by the LI process. Exiting temperatures of granules varied from 23-40°C (FA) versus 23-80°C (LI). Final granule properties were shown to be comparable between the two methods of binder addition though not at similar operating conditions and fewer operating conditions produced good yields with the LI approach. Processing was only found possible for the drier formulation with 6% binder solution using the FA approach.

A second set of studies looked more directly at the influence of the foamed binder on the granulation process to better understand the mechanism. In this case, granulation was done using binders of comparable surface tension but different molecular weights selected from the E-series of METHOCEL products. The half-life drainage time for the low MW binder, E3, was 9±3 min at 75% FQ (70 Pa-s), increasing to 23±2 min at 95% FQ (109 Pa-s); the value in brackets was the apparent viscosity of the foam to indicate its stiffness. For the highest MW binder, E15, its foam half-life was 82±6 min at 75% FQ (103 Pa-s) and 165±12 min at 95% FQ (232 Pa-s). Under shear in a cylinder viscometer, the average collapse rate of foam (based on foam stiffness) was 60% greater using E15 than E3. In comparison, the average collapse rate was 168% higher with foams of 95% FQ versus 75% FQ. The total fraction of particles larger than 1180μm was a good differentiating response variable to determine effectiveness of granulation. This fraction was nominally 54% (75% FQ) to 52% (95% FQ) for E3, and 42% (75% FQ) to 48% (95% FQ) for E15. Flow rate and screw speed were determined to be the most significant factors ($P<0.05$) affecting both particle size and granule fracture strength whereas binder MW had a minor influence on granule properties and foam quality showed no significant influence.

The negligible influence of foam quality on granule properties was reconciled with the earlier mentioned attributes of foam stability and visual observations of the process to present a proposed two-stage nucleation regime inside the extruder. The talk will present this nucleation mechanism and how foam properties can control the spreading length of the binder inside the extruder.

CONCLUSIONS: A robust continuous process for wet granulation using a twin screw extruder was presented using a foamed binder. Lower material temperatures resulted and a broader set of operating conditions were possible using the foamed binder in comparison to direct liquid addition to the extruder. The negligible influence of foam quality to granule properties points to a consistent wetting mechanism where the foam collapses at a similar spreading length regardless of foam stiffness as it is introduced to the extruder. This regulation of wetting rate is related to the environment of the extruder where foams of lower stiffness favor collapse over stagnant solids and foams of higher stiffness, near the high shear planes of the barrel. In comparison to foam properties, operating conditions (flow rate and screw speed) most strongly affected granule properties suggesting a mechanical dispersion dominated nucleation mechanism.

(402d) Powder Feeder Pairing with Downsteam Continuous Mixer

William E. Engisch Jr. and Fernando J. Muzzio, Chemical and Biochemical Engineering, Rutgers, The State University of New Jersey, Piscataway, NJ

In pharmaceutical powder handling, as well as other continuous powder processing industries, the challenge to consistently feed powder can be one of the most important parts of the overall processing. Inconsistent flow from a powder feeder can lead to inconsistencies in subsequent unit operations, such as mixing. There are methods to improve feeding consistency, but due to the intrinsic nature of powder, feedrate variability cannot be entirely removed. The dilemma then arises of what is an acceptable level of variability for the process.

The focus of this work is developing a method to pair a subsequent unit operations with a powder feeder or vice versa. By using pulse experiments to determine the characteristic residence time distributions (RTD) of a mixer, it is possible to develop a transfer function model that can be used to predict the effect of fluctuations and/or setpoint deviations in an inlet feed stream. By changing the configuration or design of either the mixer or feeders, it is possible to greatly improve overall performance of the feeders-mixer system.

(402e) Powder Lubrication in a Continuous Mixing Process

Daniel O. Blackwood[1], Bruce C. MacDonald[1], Joseph Kushner IV[1], Bruno C. Hancock[1] and Yang A. Liu[2], (1)Drug Product Design, Pfizer Worldwide Research and Development, Groton, CT, (2)Analytical Development, Pfizer Worldwide Research and Development, Groton, CT

In traditional batch processes, a lubricant, such as magnesium stearate, is incorporated into the blend during the final processing step before tablet compression. Recent models, from Kushner and Moore, have proposed a process scaling parameter for a number of different formulations to facilitate scale up of these batch-based processes. In the current work, a series of powder lubrication experimental trials have been completed using an in-line powder mixer. The influence of mass throughput rate, hold up mass of powder in the mixer, and the impeller rotational speed is studied. A new process scaling term, based on the tip speed of the impeller and the mean residence time is proposed for this continuous mixing process and compared to the model proposed by Kushner and Moore for batch processes. Additionally, two new sensors (NIR and torque) are evaluated as part of these experimental trials and introduced as potential on-line PAT techniques for monitoring powder lubrication in a continuous mixer.

(402f) Effect of Paddle Shape On Flow of Powder Through Feed Frames On Rotary Tablet Presses

G. Singh[1], J. Givand[2], P. Rajniak[2], R. Chern[2], S.D. Reynolds[2] and I.C. Sinka[1], (1)University of Leicester, Leicester, United Kingdom, (2)Merck, Sharp & Dohme, West Point, PA

Effect of paddle shape on flow of powder through feed frames on rotary tablet presses

G. Singh*, J. Givand**, P. Rajniak**, R. Chern**, S.D. Reynolds** and I.C. Sinka*

*Department of Engineering, University of Leicester, University Road, Leicester, LE1 7RH, UK
**Merck, Sharp & Dohme, Sumneytown Pike, West Point, PA19486, USA

A commercial scale high speed rotary tablet press can produce in excess of 0.5 million pharmaceutical tablets per hour. This involves the same number of die fill operations, which must be executed consistently to ensure weight and content uniformity from tablet to tablet. High productivity can only be achieved if the formulation of the powder and the selection of process parameters are optimised.

Die fill is a complex and dynamic process and consists of a number of distinct stages that present unique features: delivery of the powder to the press (in batches or continuously), powder flow from hopper to feed frame, powder flow through feed frame to die opening, and die fill. In this paper we examine the flow of powders through a lab scale rotary feeder which was developed at the University of Leicester to explore the mechanisms present in the feed frames of rotary tablet presses. The feeder pictured below consists of an inlet hopper, a disk shaped frame housing a paddle wheel and a powder outlet.

The flow regimes were examined for a poorly flowing placebo formulation and the following parameters were considered:

· Paddle wheel speed: 25, 50, …, 400 rpm

· Outlet diameter: 5, 10, 15, 20 mm

· Shape of paddle wheel spoke cross section: round, square and inclined

It is shown that the powder delivery is not always as continuous as the figure below implies and that poor flowing materials can perform surprisingly well under the conditions imparted by the rotary feeder. The results are quantified in terms of mass flow rate through the feeder, which increases and saturates as the freed frame speed is increased. It was interesting to note, however, that the mass flow rate per unit area of the outlet can increase or decrease as the outlet diameter is reduced, depending on powder. For all cases a significant effect of the shape of the paddle cross section was observed.

Experimental set-up of lab scale rotary feeder

(402g) Supply Chain Analysis of Continuous Pharmaceutical Manufacturing

Arul Sundaramoorthy, James M.B. Evans and P.I. Barton, Department of Chemical Engineering, Massachusetts Institute of Technology, Cambridge, MA

In traditional pharmaceutical supply chains, production takes place in two major steps: 1) primary production, and 2) secondary production. In primary production, the raw materials are converted into active pharmaceutical ingredients (APIs) through multiple intermediate production stages. It is common in practice to carry out the first few stages of production in one location, and the other stages in a different location. In secondary production, the APIs are formulated into different drug products (DPs) through various unit operations such as granulation, molding, coating, and packaging. The secondary production facilities are usually located closer to the major markets. Finally, the DPs are supplied to warehouses or retailers for distributions. The above traditional supply chain design has several disadvantages such as high costs of inventory and logistics, long supply-chain cycle times, high degree of supply-chain disruptions, and less resilience. Furthermore, the pharmaceutical industry has been dominated by the batch mode of manufacturing for decades, which greatly contributes to the above inefficiencies in the supply chain. Recently, the Novartis-MIT Center for Continuous Manufacturing has embarked on an innovative project to shift from the batch mode of manufacturing to the continuous mode, where raw materials through the APIs to the finished products are processed seamlessly in an integrated facility. Such an integrated end-to-end production scheme is more promising for several reasons: a) low costs of inventory, logistics, and production, b) short supply-chain cycle times, c) less exposure to supply-chain disruptions, and d) resilient supply chain. Thus, the option of continuous manufacturing brings in additional challenges and opportunities in the context of pharmaceutical supply chains.

In this work, we develop an optimization framework to analyze various supply chain configurations based on continuous production facilities. Since the option of storing APIs contributes to increased degrees of freedom, we also analyze supply chain configurations based on the decoupled production scheme, where primary and secondary production are separated by storage units. However, the mode of production in the decoupled scheme is still continuous. We consider three different supply chain configurations: 1) centralized configuration, where APIs and DPs of all products are produced in a single location, 2) semi-centralized configuration, where APIs of all products are produced in one location, and DPs are produced in different locations closer to the markets, 3) localized configuration, where both APIs and DPs of various products are produced in different locations closer to the markets. We present a mixed-integer linear programming (MILP) model for the analysis of the various supply chain configurations. The key decisions of the proposed framework are which production scheme is economical, which supply chain configuration is efficient, and what should be the sizes of facilities given the demands of products over the planning horizon. We illustrate the application of the proposed framework using three different drug molecules of different dosage levels and different demand volumes.

(403a) A Smart Oil Spill Dispersant Formulation for Reduced Environmental Impact and Consumption

Courtney A. Ober and Ram B. Gupta, Chemical Engineering, Auburn University, Auburn, AL

The application of chemical dispersants is one of the few feasible response measures for minimizing the impact of a large oil spill at sea. Effective use of dispersants can accelerate oil biodegradation by naturally occurring microorganisms and prevent the oil spill from coming ashore and damaging sensitive coastal ecosystems. However, indiscriminate application of dispersant to subsea and surface oil spills results in a significant amount of dispersant not coming into contact with the oil. The aqueous solubility and miscibility of oil spill dispersants and inevitable sea currents cause dispersant to be washed away and wasted. To compensate for inefficiency, large amounts of dispersants are used resulting in high costs and a large environmental load. For example, an estimated 2.2 million gallons of dispersant were applied to the Gulf of Mexico during the Deepwater Horizon oil spill of 2010. A smart dispersant formulation that microencapsulates the active ingredient of chemical dispersants, a surfactant, in a water-insoluble shell has been developed. In this formulation the surfactant payload is only released when in contact with oil, at which time the shell dissolves in the aromatic compounds present in crude oil and refined petroleum products. The components used in this study were the surfactant, dioctyl sodium sulfosuccinate, and aromatic-soluble polymer, polystyrene. A hardening agent, poly(methyl methacrylate), was added to the surfactant to facilitate mechanical grinding into microparticles. The surfactant microparticles were then fluidized while a polystyrene solution was sprayed onto their surface. Following optimization of spray-coating conditions for maximum encapsulation efficiency, the kinetics of surfactant release from the coated microparticles were studied using representative oil components and simulated sea water. This study demonstrates that novel core-shell surfactant-polymer microparticles, prepared by simplistic and scalable technologies including hardening, grinding, and spray-coating, have significant potential for improving oil spill dispersion. Selective delivery of surfactant directly to an oil spill reduces the amount of surfactant needed for effective dispersion, thus reducing the environmental impact of the surfactant.

(403b) A New Approach to Prepare Polystyrene(Pst)/CaCO3 Core-Shell Composite Nanoparticles

Yang Hu, Guanzhi Cheng, Wenjiao Ji, Yiming Jiang and Chengyou Kan, Key Laboratory of Advanced Materials, Department of Chemical Engineering, Tsinghua University, Beijing, China

Yang Hu, Guanzhi Cheng, WenJiao Ji, Yiming Jiang, Chengyou Kan, Key Laboratory of Advanced Materials, Department of Chemical Engineering, Tsinghua University, Beijing, China

Polystyrene (Pst)/$CaCO_3$ core-shell composite nanoparticles have been attracted great attentions of many researchers. Such core/shell structure composite particles are utilized in broad areas for its improved compression strength and anti-scratch resistant. Here a new approach was developed to prepare polystyrene(Pst)/$CaCO_3$ core-shell composite nanoparticles. Sodium poly(4-styrenesulfonate) was utilized as a stabilizing agent, which was grafted on the Pst particles to form the hair structure as the core particles. The Pst core particles with hairy structure was prepared by modifying the particle surface with sodium poly(4-styrenesulfonate). A certain amount of $Ca(OH)_2$ was added into the core latex, followed by feeding CO_2, and the $CaCO_3$ was synthesized through the reaction of $Ca(OH)_2$ and CO_2, and the $CaCO_3$ amorphous primary particles deposited on Pst core particles with hairy structure. The thickness of hair structure of core particles was 10 nm, which was determined by Dp^{DLS} 350 nm and Dp^{TEM} 330 nm. The influences of key parameters including particle size, shell thickness, the amount of initiator, the density of sulfonic acid groups, temperature and pH on deposition process was investigated, and the optimized condition was obtained.

(403c) Fabrication and Characterization of Porous Alumina Nanolayers Via Molecular Layer Deposition

Alan Izar, Chemical & Biological Engineering, University of Colorado - Boulder, Boulder, CO, Sandro Hutter, ETH, Switzerland, John L. Falconer, Chemical & Biological Engineering, University of Colorado at Boulder, Boulder, CO and Alan W. Weimer, Chemical and Biological Engineering, University of Colorado at Boulder, Boulder, CO

Porous nanolayers of alumina can be fabricated by the deposition of alucone, an aluminum alkoxide polymer, onto zirconia nanoparticles, followed by the calcination of the layers in order to remove the organic portion of the molecule. Two monomer structures of alucone were used: AB, which consists of trimethylaluminum (TMA) and ethylene glycol (EG); and ABC, composed of trimethylaluminum (TMA), ethanolamine (EA) and maleic anhydride (MA). Using molecular layer deposition, self-assembled monolayers of these polymers were coated on zirconia nanoparticles, with varying film thickness. The particles were subsequently calcinated at different temperatures in order to produce different porous layer structures. These porous layers were characterized with TEM, EDS, and other techniques in order to study their structure and porosity. The same procedure was repeated for platinum particles supported on zirconia, which were used in an oxidation reactor. A possible future application of these layers is in prevention of catalyst sintering, as well as selectivity control.

(403d) Inkjet Printing and Spraying of Composite Calcium Alginate Particles with Solid Inner Inclusions

Jiri Dohnal, Department of Chemical Engineering, Institute of Chemical Technology, Prague, Prague, Czech Republic

InkJet printing and spraying of composite calcium alginate particles with solid inner inclusions

Dohnal J., Štìpánek F.

Institute of Chemical Technology, Department of Chemical Engineering

Laboratory of Chemical Robotics

Tel: +420 220 443 046, email: dohnalj@vscht.cz

Calcium alginate is very often used for the encapsulation of living cells or enzymes for bioapplications [1, 2, 3]. An active substance (living cells, enzymes or catalyst particles) can be dispersed or dissolved in the starting sodium alginate solution and subsequently encapsulated in the beads by calcium ions.

Alginate beads can be prepared by different techniques from simple dropping of alginate solution to a calcium pool to precipitation in a microfliudic device. We use inkjet printing in this work. This technique can generate small microdroplets of a liquid with viscosity of up to 25 mPas with a diameter from 40 to 80 µm. Stable dispersions are possible to printing and this work uses it for their aims.

The aim of this work is preparation of composite microparticles from calcium alginate with inner inclusion for artificially started control release of active substances or for fabrication of alginate beads with inner solid inclusions or catalysts which can be used as microreactors for artificially started reaction. 1% (w/w) of sodium alginate solutions filled by TiO_2 particles for photocatalysis respectively Fe_3O_4 particles which are able to concentrate by magnetic field in current location. For massive production of these composite particles spray drying nozzle was used and this work compare process limitation and differences between InkJet and Spray nozzle in respect to quality of final product.

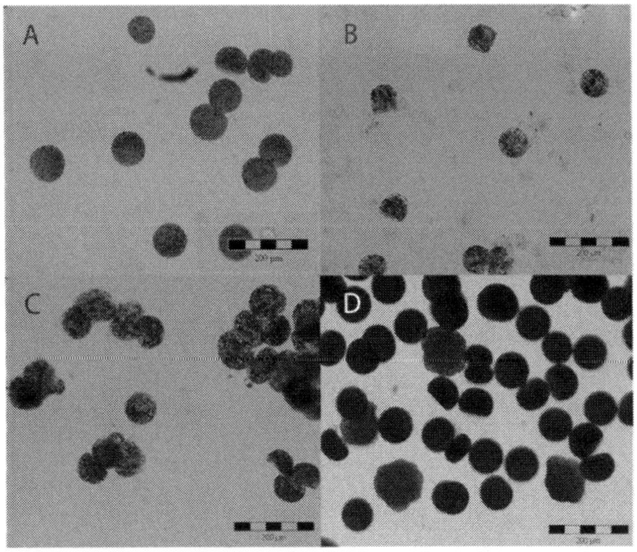

Figure SEQ Figure * ARABIC 1: Calcium alginate beads from dispersion of TiO_2 in alginate solution with

concentration 1.95 (A), 3.9 (B) 5.85 (C) and 7.8 (D) vol. % of inner solids.

1) Anal, A. K.; Stevens, W. F. Chitosan-alginate multilayer beads for controlled release of ampicillin. *Int. J. Pharm.* **2005**, *290*, 45–54.

2) Murua, A.; et al. Cell microencapsilation technology: Towards clinical application. *J. Controlled Release* **2008**, *132*, 76–83.

3) Morch, Ý. A; et al. Effect of Ca^{2+}, Ba^{2+} and Sr^{2+} on Alginate Microbeads. *Biomacromolecules* **2006**, *7*, 1471–1480.

INVESTIGATION OF SPECIFIC PARTICLE ADHESION TOWARDS MODEL AND BIOLOGICAL SUBSTRATES

Viola Tokárová, František Štěpánek

Institute of Chemical Technology Prague, Department of Chemical Engineering, Prague, Czech Republic, e-mail: frantisek.stepanek@vscht.cz

Introduction

Composite microparticles or microcapsules for targeted delivery of active substances such as antibacterial agents or pharmaceutical compounds are of both practical and fundamental interest [1]. The purpose of such delivery vehicles is to store the active compound at a high concentration, transport it through an environment in which the presence of the active compound is not required (or not desirable), and eventually anchor to the target site and release the encapsulated payload either spontaneously or upon an external stimulus [2, 3]. For the design a targeted delivery system, not only the release kinetics of the encapsulated compound from the microcapsule [4] but also the interactions of the microcapsules with various substrates and bacterial strain in the presence of a flowing fluid – i.e. their adhesion properties – are crucial [5, 6]. Moreover, it would be desirable to have the possibility to control the adhesion/detachment of a microcapsule by some external stimulus.
The adhesion force between a particle and a given substrate depends on the nature of their interaction (electrostatic interaction, hydrophilic/hydrophobic interaction, specific biological interaction such as ligand-receptor binding, etc.). The present work describes two kind of composite microparticles and their different adhesion character. It will be shown that either adhesion force of microparticle inside microfluidic cell or bioadhesive property of chitosan particle with bacterial strain have a considerable effect on a target delivery mechanism.

The first kind of particle studied inside the microfluidic cell was prepared via Pickering emulsion polymerization [7] and consisted of a poly-N-isopropylacrylamide (PNIPAM) hydrogel core and a monolayer of SiO_2 nanoparticles on the surface. PNIPAM is known as a temperature-responsive polymer that changes its hydrophobic/ hydrophilic character with temperature [2]. At temperatures below the lower critical solution temperature (LCST) of 32°C, the amide groups interact strongly with water molecules through hydrogen bonding and the hydrogel swells. At temperatures above the LCST, the hydrogen bonds between water and the amide groups are disrupted, water molecules are expelled from PNIPAM and the hydrogel shrinks. This shrinking/swelling behavior can be used not only for controlling the release rates of active substance incorporated inside PNIPAM particles [1], but also for controlling adhesion.

The second type of composite microparticles consists of polymer chitosan, cross-linker tripolyphosphate (TPP) and embedded silver nanoparticles. Chitosan particles improve bioavailability of drug molecules and protect them from degradation. It is also known that the antibacterial action of silver nanoparticles occurs not only through the release mechanism but also via to so-called contact mechanism [8].Therefore, it can be expected that the bioadhesive character of chitosan would further enhance the antibacterial action of silver nanoparticles. It has been shown that chitosan microparticles not only adhere onto cell surfaces [9] but also that chitosan can act as a flocculant thanks to its strong adhesion to a range of bacteria [10]. It will be shown in the present work that by the combination of bioadhesive properties of chitosan and controlled diffusion of nanoparticles from the polymer matrix, effective delivery of nanoparticles to the target can be achieved and their functionality enhanced.

Experimental

Synthesis of PNIPAM/SPION/SiO$_2$ microcapsules

PNIPAM/SPION/SiO$_2$ microcapsules were prepared by the Pickering emulsion polymerization Fig. 1 as follows [3]. Dried silica nanoparticles (50 mg) prepared according to the to the Stöber method [11] was first dispersed in 20 ml of toluene. Separately, the monomer NIPAM (400 mg), crosslinker BIS (15 mg) and initiator of the polymerization reaction APS (20 mg) were dissolved by means of a homogenizer in 2 ml of deoxygenated water containing citrate-stabilized SPIONs [3]. The water solution was mixed with the toluene phase and the mixture was stirred at 1400 rpm under nitrogen bubbling for 20 min to form the Pickering emulsion stabilized by the SiO$_2$ nanoparticles. Then the mixture was heated to 70°C in order to initiate the polymerization reaction and stirred under nitrogen atmosphere for the next four hours. The final PNIPAM/SPION/SiO$_2$ microcapsules were washed with pure ethanol and re-suspended in water.

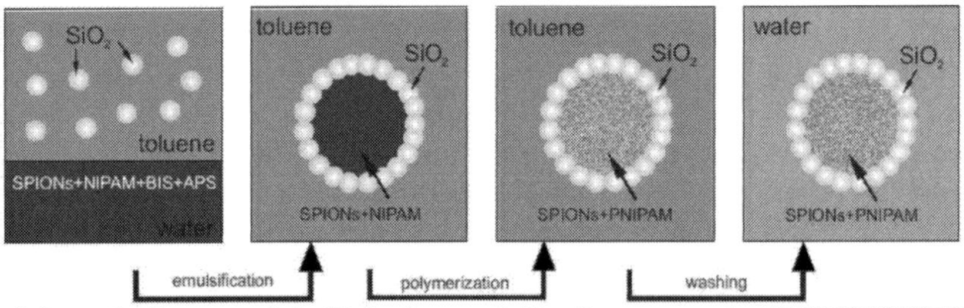

Fig. 1: Schematic diagram of the Pickering emulsification process for PNIPAM/SPION/SiO$_2$ microcapsules.

Synthesis of silver nanoparticles

The silver nanoparticles were synthesized by the reduction of silver nitrate by sodium citrate [3]. Briefly, 48 ml of water was added to a 100 ml double-neck flask equipped with a reflux condenser and heated to its boiling point in an oil bath; 39.4 mg of silver nitrate dissolved in 1 ml of water was added to the boiling solution and stirred at 500 rpm. Then, 22.7 mg of sodium citrate dissolved in 1 ml water was added to the solution dropwise. The mixture was kept at its boiling point for the next 15 min with continuous stirring at 500 rpm. The final concentration of silver nanoparticles was established to be 0.5 g/l.

Preparation of chitosan microparticles by spray drying

Chitosan solutions were prepared by dissolving the required amount of chitosan in 25 ml of aqueous solutions of acetic acid at room temperature. The chitosan solution was stirred at 1000 rpm for 30 min in each case to ensure complete dissolution of the polymer. Cross-linking was achieved by adding of 2 ml of a 1% (w/v) TPP solution dropwise with a syringe (type 25Gx1") to the chitosan solution under constant stirring and then further stirred at 650 rpm for 60 min to ensure complete cross-linking of chitosan. After addition of TPP the solution changes from transparent to opalescent. Opalescence occurs due to the formation of chitosan-TPP nanoparticles. The colloidal suspension of cross-linked chitosan nanoparticles was then spray-dried by a 2-fluid nozzle to form microspheres with uniformly dispersed nanoparticles. The mass fraction of Ag in the final composite microparticles was 0.35 or 3.5% depending on the used volume of Ag nanoparticle suspensions.

The solutions were spray dried using the Buchi B-290 laboratory spray drier with a high performance cyclone in open mode configuration. The values of operation parameters that ensured stable and robust process for the entire range of compositions were determined experimentally. They were: inlet air temperature 160°C, aspirator rate 90%, the total liquid flowrate was 6.35 ml/min and the atomizing gas flow rate was 414 normlitres/hr. The outlet temperature was typically 70-75°C. The typical mass of dry powder produced in a single batch was around 0.1-0.3 g.

Adhesion studies PNIPAM/SPION/SiO_2 composite microparticles

The adhesion properties of the composite microcapsules were studied in a custom-designed adhesion cell where it is possible to measure the ratio of adhered microcapsules with temperature changes and with controlled volumetric flow rate, and which enables rapid exchange of the tested substrates. The adhesion cell consists of three parts (Fig. 2). The upper and bottom parts are made from plexiglass and the middle section is a rubber spacer which determines the flow field character. The lower section contained a milled groove for placing microscope slide glass, which is coated by the different studied substrates. The layer of the substrate material was uniformly distributed on the glass slide by spin coating and then drying its solution in a suitable solvent.

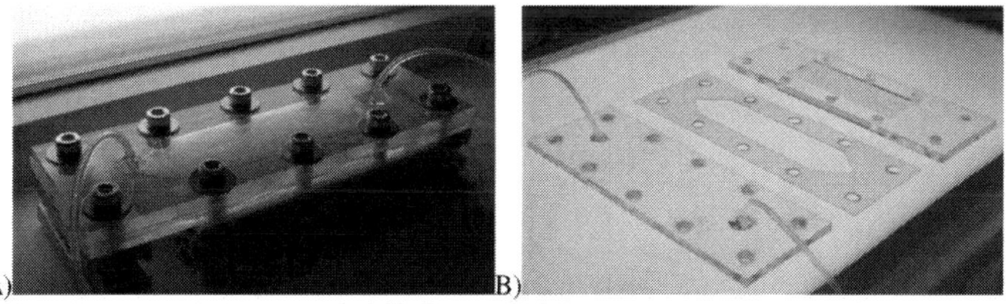

Fig. 2: Adhesion cell: A) assembled B) disassembled (upper part, spacer and bottom part).

The experimental set-up included the adhesion cell and a syringe pump (Lambda Vit-Fit) to control the volumetric flow rate. A camera attached to a stereo microscope (Olympus SZ-CTV) was used to record the movement of particles inside the adhesion cell and the images are evaluated by the software NIS Elements. During the experiment, the adhesion cell was placed on a heating plate which was used to control the temperature. The experiment with increased temperature in the cell requires the fluid inside the syringe to be heated to the required temperature too. The temperature in the adhesion cell as well as at the fluid outlet was monitored.

Bioadhesion and antibacterial character of composite chitosan/Ag microparticles

The antibacterial properties of silver nanoparticles were tested on different strains of bacteria: gram negative *Escherichia coli* and from gram positive *Micrococcus luteus* and *Staphylococcus epidermis*. The tests using composite chitosan/Ag microparticles had the following arrangement: a chosen amount of the composite microparticles (10 or 20 mg) and 1 ml of bacterial broth with a known OD_{600} were added to a flask with 10 ml of fluid LB medium. This solution was put on a shaker and incubated for one day at 37°C. The solution was then inoculated on a Petri dish with a solid LB medium and left overnight at 37°C. As a control, bacterial culture without any added microparticles and a culture with blank (pure chitosan without Ag) microparticles were always compared with the measured sample.

Results and discussion

Adhesion properties – effect of substrate

Adhesion experiments were carried out with the three-component composite PNIPAM/SPION/SiO_2 microcapsules at room temperature using a range of substrate materials. Their selection was based so as to simulate various surfaces of both natural (cholesterol, palm oil, gelatin) and man-made (teflon, paraffin, polystyrene, rubber, glass) origin that could potentially be encountered during targeted delivery applications. The investigated materials also vary from hydrophilic to hydrophobic, which was characterized by the measuring the equilibrium contact angle of water at room temperature on each substrate. The values of contact angles are given in Tab. 1.

Tab. 1: Contact angles of deionized water on the investigated substrate materials at room temperature.

material	contact angle (°)
cholesterol	6.0 ± 1.0
palm oil	14.1 ± 2.0
gelatin	23.0 ± 2.8
glass	28.0 ± 2.8
teflon	83.3 ± 1.5
rubber	87.4 ± 1.7
paraffin	93.0 ± 1.7
polystyren	93.3 ± 1.9

The adhesion experiments proceeded as follows. An aqueous dispersion of studied micorcapsules was spread on the glass coated by the investigated material and inserted into the machined groove in the bottom part of the adhesion cell. The rubber spacer and the upper part were then put in position and the cell was closed. The inlet Tygon hose was connected to a syringe in the linear pump. A stereo microscope was used for recording the number of adhered particles as function of time at each flow rate. The volumetric flow rate was increased in a step-wise manner and maintained at the set level until the number of adhered microcapsules stabilized. At the values of volumetric flow rate studied in this work, we observed that the number of adhered particles became stable typically after 3-4 minutes. Three values of flow rates were used, namely 0.058, 0.086 and 0.130 ml/s which corresponds to a linear flow velocity of 1.12, 1.65 and 2.50 mm/s in the flow cell. The Reynolds number was lower than 10^{-5} in all cases, i.e. the flow was laminar.

The fraction of adhered microcapsules decreased with increasing flow rate for all the investigated model surfaces (Fig. 3). However, it can be seen that both the strength of the decrease and the actual fraction of adhered microcapsules for a given flow rate varied widely among the investigated materials. As the adhesion strength should depend on the nature of the surface interaction between the microcapsules (whose surface is formed by the silanized SiO_2 nanoparticles) and the substrate. A correlation between the substrate wettability and the fraction of adhered microcapsules can be seen in Fig. 3B. The fraction of adhered microcapsules on materials with a hydrophilic character (contact angle between 0° and approximately 30°) decreased sharply with increased flow rate. On the other hand, although a similar trend was observed on hydrophobic substances with contact angles between 80° and 100°, the dependence of the fraction of adhered microcapsules on flow rate was much weaker and in some cases there was nearly no decrease at all even at the highest flow rate.

This means that the composite PNIPAM/SPION/SiO_2 microcapsules adhered more strongly towards hydrophobic surfaces, which in turn implies that microcapsules themselves have a hydrophobic character even at temperatures below the LCST of PNIPAM, at which PNIPAM itself is hydrophilic. This finding provides interesting information about the structure of the microcapsule surface – essentially, it means that no PNIPAM chains are grafted on the external surface of the SiO_2 nanoparticles and no PNIPAM hydrogel protrudes through the monolayer of silica nanoparticles that cover the surface of the microcapsules. To verify this affirmation a simple test of wettability of the composite PNIPAM/SPION/SiO_2 microcapsules was made. Their water dispersion was mixed with oleic acid, and the microcapsules went across the interface to the oleic phase by little shaking, thereby proving their hydrophobic surface. It is interesting to note that despite their preference for a non-polar solvent, the composite microcapsules are dispersible in water. To the best of our knowledge, the composite PNIPAM/SiO_2 microcapsules below the LCST of PNIPAM have always been assumed to be hydrophilic in the literature.

A) 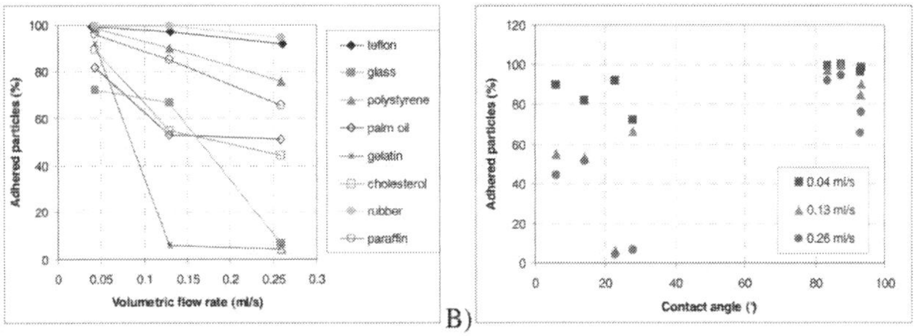 B)

Fig. 3: Adhesion experiments on 9 different substrates A) dependence of adhesion on volumetric flow rate (0.04, 0.13 and 0.26 ml/s); B) dependence of the contact angles of substrates on adhesion test.

Bioadhesion and antibacterial performance of composite chitosan/Ag

The antibacterial activity tests of composite chitosan/Ag microparticles were carried out using particles with different nanoparticle loading, i.e. samples Ag1 and Ag10, each in two concentrations (10 and 20 mg per 11 ml of broth). Fig. 4 shows an example of the results of antibacterial tests against *E. coli*. The negative controls (Fig. 4A) show that pure chitosan at these concentrations does not inhibit bacterial growth while each particle type containing Ag (Ag1 and Ag10) is active at the high concentration (Fig. 4B and Fig. 4C) while the lower concentration (10 mg) is insufficient to inhibit bacterial growth. Note that the antibacterial action is not directly proportional to the amount of Ag present in the system: by increasing the Ag1 particle dose from 10 to 20 mg, the quantity of silver nanoparticles in the system is doubled, but by changing the particles from Ag1 to Ag10 at 10 mg, the quantity of silver increases by a factor of 10 (cf. Fig. 4B and Tab. 2). Despite the higher concentration of silver nanoparticles, bacterial growth was not inhibited by the combination of Ag10 particles and 10 mg. From this it can be concluded that not only the amount of free Ag particles but also the way in which they are introduced may play a role in the overall antibacterial performance.

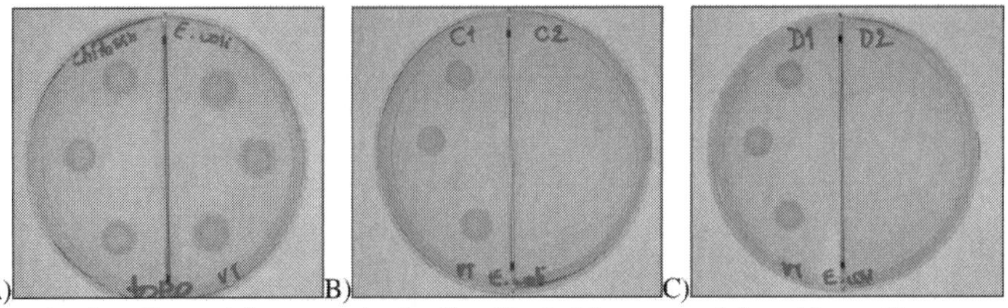

A) B) C)

Fig. 4: Example of antibacterial test results for E. coli: a) chitosan particles without silver nanoparticles against a control of bacterial culture with no particle addition; b) composite chitosan/Ag particles type Ag1_TPP2 with variable quantity (10 and 20 mg); c) composite chitosan/Ag particles type Ag10_TPP2 with variable quantity (10 and 20 mg).

Results for all used bacterial strains are summarized in Tab. 2 where a plus mark (+) indicates successful antibacterial effect and a minus mark (-) indicates no antibacterial effect. It can be seen that in the case of *M. luteus*, the particles have antibacterial effects at all four Ag concentrations, whereas in the case of *S. epidermis*, the increase of Ag concentration from the lowest value (Ag1 at 10 mg) leads to antibacterial effect regardless of whether the increase of silver concentration was achieved by using 10 mg of highly loaded chitosan carriers (Ag10) or 20 mg of the lower-loading variety (Ag1). Let us now return to the interesting case of *E. coli* where increasing the chitosan particle concentration was more effective than increasing the silver loading. It has been reported that Ag nanoparticles can be active by two mechanisms – the so called contact and release mechanism. In our case, the contact mechanism would mean physical interaction (collision followed by adhesion between the bacteria and the chitosan carriers), whereas the release mechanism would mean liberation of Ag nanoparticles from the chitosan carriers and their subsequent interaction with the bacteria.

Tab. 2: Summary of the results of antibacterial tests.

	m (mg)	Ag0	Ag1	Ag10
E. coli (OD_{600} 0.10)	10	-	-	-
	20	-	+	+
M. luteus (OD_{600} 0.09)	10	-	+	+
	20	-	+	+
S. epidermis (OD_{600} 0.04)	10	-	-	+
	20	-	+	+

To confirm whether the contact mechanism contributes to the observed antibacterial activity in this case and to quantify the extent of possible adhesion between bacteria and chitosan microparticles, the following experiment was carried out. The size distributions of bacteria and of the chitosan microspheres (composition Ag0) were measured by light scattering first independently and then after mutual mixing. If there were no adhesion, the particle size distribution (PSD) of the mixture should not change and be just a superposition of the bacteria and chitosan PSD. On the other hand, if adhesion between bacteria and chitosan particles occurred, the size distribution should shift to the right (agglomeration) and the original peaks should diminish. The results of the light scattering experiments are summarized in Fig. 5. As can be seen, all three bacterial strains have a strong adhesion towards chitosan microparticles and form stable agglomerates. While individual chitosan particles have a mean diameter of approximately 8.7 μm and the mean particle size of *E. coli*, *M. luteus* and *S. epidermis* is 1.1, 6.9 and 1.2 μm, respectively, the mode particle size of their binary mixtures was 70, 120 and 90 μm, respectively. Fig. 5 also shows that the peaks corresponding to the bacteria have nearly disappeared once the chitosan carriers were added, meaning that most bacteria initially present in the broth participated in the formation of larger clusters with chitosan particles. On the other hand, the PSD after chitosan addition is bi-modal with the lower mode coinciding with the original chitosan particle size. This means that chitosan particles were probably in excess in this case and not all of them participated in the cluster formation.

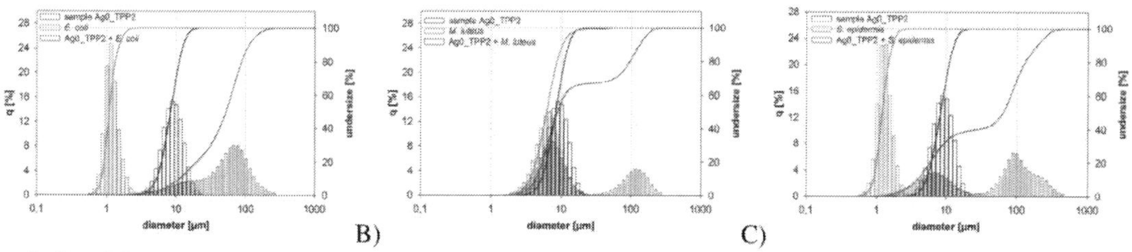

A) B) C)

Fig. 5: Particle size distributions of bacteria and chitosan particles before mixing and once agglomerates have formed (as indicated by the legends): a) E. coli, b) M. luteus, c) S. epidermis.

Another indication of the fact that bacteria have been collected by the added chitosan microparticles can be provided is a reduction of OD_{600} of the bacterial broth after chitosan addition. Specifically, the OD_{600} has decreased from 0.27, 1.89, 0.62 to 0.20, 1.52, 0.49 in the case of *E. coli*, *M. luteus* and *S. epidermis*, respectively.

Conclusions

This work described synthesis and adhesion character of two composite microparticles. The first type of the microparticle is formed from thermo-responsive hydrogel PNIPAM and SiO_2 compact shell. The phase transition of the underlying PNIPAM, which is hydrophilic below its LCST and hydrophobic above its LCST, does not affect the adhesion force of the structured microcapsules whose surface character is dictated by the SiO_2 nanoparticles, which are hydrophobic as a result of their previous silanization. This affirmation was proved by the test of wettability of studied particles. Even though PNIPAM is hydrophilic below its LCST, the composite microcapsules partitioned from water to an oil phase. The fact that the external surface of the silica nanoparticles remained ''protected'' by their immersion in the toluene phase during the Pickering emulsion polymerization provides opportunities for their further functionalization to tailor the adhesion towards specific substrates, including biological ones. The second type of composite microparticle was prepared by the spray-drying process from biodegradable polymer chitosan and silver nanoparticles uniformly distributed in final particle volume. The antibacterial and bioadhesion tests on three types of bacteria have shown that the chitosan carriers have a high affinity towards the bacteria, forming stable aggregates in which the contact mechanism of silver nanoparticle action can be enhanced. This functionality enhancement was especially profound in the case of *E. coli* where a two-fold increase in the concentration of silver nanoparticles achieved by increasing the number of chitosan microcarriers proved to be more effective than a ten-fold increase in nanoparticle concentration realized by their higher loading in an otherwise constant number of chitosan microcarriers. The ability of the chitosan microcarriers to form stable aggregates with the bacteria was proved by light scattering experiments and provides opportunities for the design of targeted nanoparticle delivery carriers.

Acknowledgments

This work has been supported by the European Research Council (200580-Chobotix), by the Specific University Research (MSMT 21/2012) and by the Specific Science Cooperation Program AKTION (MEB061108).

References

1. Champion, J.A., Y.K. Katare, and S. Mitragotri, *Particle shape: A new design parameter for micro- and nanoscale drug delivery carriers.* Journal of Controlled Release, 2007. 121(1-2): p. 3-9.

2. Kumar, A., et al., *Smart polymers: Physical forms and bioengineering applications.* Progress in Polymer Science, 2007. 32(10): p. 1205-1237.

3. Tokarova, V., et al., *Thermo-responsive adhesion properties of composite hydrogel microcapsules.* Soft Matter, 2012. 8(4): p. 1087-1095.

4. de Vos, P., et al., *Multiscale requirements for bioencapsulation in medicine and biotechnology.* Biomaterials, 2009. 30(13): p. 2559-2570.

5. Podczeck, F., J.M. Newton, and M.B. James, *Variations in the adhesion force between a drug and carrier particles as a result of changes in the relative humidity of the air.* International Journal of Pharmaceutics, 1997. 149(2): p. 151-160.

6. Felicetti, M.A., et al., *Influence of removal time and particle size on the particle substrate adhesion force.* Brazilian Journal of Chemical Engineering, 2008. 25(1): p. 71-82.

7. Cejkova, J., J. Hanus, and F. Stepanek, *Investigation of internal microstructure and thermo-responsive properties of composite PNIPAM/silica microcapsules.* Journal of Colloid and Interface Science, 2010. 346(2): p. 352-360.

8. Li, Z., et al., *Two-level antibacterial coating with both release-killing and contact-killing capabilities.* Langmuir, 2006. 22(24): p. 9820-9823.

9. Goldberg, S., R.J. Doyle, and M. Rosenberg, *Mechanism of Enhancement of Microbial Cell Hydrophobicity by Cationic Polymers.* Journal of Bacteriology, 1990. 172(10): p. 5650-5654.

10. Strand, S.P., T. Nordengen, and K. Ostgaard, *Efficiency of chitosans applied for flocculation of different bacteria.* Water Research, 2002. 36(19): p. 4745-4752.

11. Stober, W., A. Fink, and E. Bohn, *Controlled Growth of Monodisperse Silica Spheres in Micron Size Range.* Journal of Colloid and Interface Science, 1968. 26(1): p. 62-&.

(403f) Dynamic Encapsulation of Hydrophilic Nanomaterials in Hydrophobic Polymer-Based Particles by Simple Emulsion Technique

Shaowen Ji, Devesh Srivastava, Nathan J. Parker, Ankush A. Gokhale and Ilsoon Lee, Department of Chemical Engineering and Materials Science, Michigan State University, East Lansing, MI

Various strategies of encapsulating hydrophilic materials, such as magnetite nanoparticles and drug molecules, have been developed and performed for applications in biological and medical fields. Polymer-based particles are attractive as carriers because of their high stability, process feasibility, and low cost. Even though emulsion polymerization or double emulsion has been reported for successful entrapment of hydrophilic nanomaterials within the hydrophobic polymer matrix, simple and efficient encapsulation is still being investigated. Here we introduce the novel dynamic encapsulation of hydrophilic magnetic nanoparticles or small drug molecules within hollow polylactic acid (PLA) particles by one-step oil in water (O/W) emulsion diffusion method. The formation of hollow polymer particles was based on the control of emulsification in the turbulent fluid flow and emulsification temperature. The fabricated PLA particles were found size and shape controllable by manipulating emulsification conditions. PLA-containing ethyl acetate solution was emulsified in the hydrophilic materials-containing aqueous medium and followed by the solidification of PLA particles in the diffusion step. The emulsion droplets dynamically formed multiple emulsion structure which entrapped hydrophilic materials within hydrophobic polymer matrix. Studies on optimizing the formulation, encapsulation efficiency, and emulsification process were performed. This method shows advantages of simple operation and easy control on fabricating the promising carrier of different hydrophilic materials for widespread use.

(403g) Influence of [Alpha]-Zrp Particle Size and Crystalline Level On Dust Explosion Suppression Effectiveness

Diana Castellanos[1], Andres F. Mejia[2], Agustin Diaz[3], Chad Mashuga[4], Victor H. Carreto[5], Zhengdong Cheng[2] and M.Sam Mannan[6], (1)Materials Science and Engineering, Texas A&M University, College Station, TX, (2)Chemical Engineering, Texas A&M University, College Station, TX, (3)Chemistry, Texas A&M University, College Station, TX, (4)Dow Chemical Company, (5)Mary Kay O'Connor Process Safety Center, Artie McFerrin Department of Chemical Engineering, College Station, TX, (6)Mary Kay O'Connor Process Safety Center, Department of Chemical Engineering,Texas A&M University, College Station, TX

We introduce the use of zirconium phosphate ($Zr(HPO_4)_2 \cdot H_2O$, α-ZrP) crystals as a dust explosion suppressant. The influence of size and crystallinity of the crystals during a dust explosion was systematically studied. During α-ZrP synthesis, the size and crystallinity of the pristine platelets were tailored by adjusting the phosphoric acid (H_3PO_4) concentration and the reaction time [1, 2]. The thermal stability of the mixtures containing corn starch and ZrP crystals at different size and crystallinity was evaluated using thermo gravimetric analysis (TGA) and differential scanning calorimeter (DSC). In addition, the maximum pressure (P_{max}) and the maximum pressure rise ($(dP/dt)_{max}$) of these mixtures were obtained by performing dust explosion tests using a 36 L vessel. As a reference, monoammonium phosphate ($NH_4H_2PO_4$, MAP) was utilized. MAP is a widely studied compound commonly used as an explosion suppressant [3, 4]. From experiments, TGA revealed a higher thermal stability of mixtures containing corn starch and α-ZrP, in comparison with mixtures of corn and MAP. The DSC results confirmed that α-ZrP crystals of small size and low crystalline level possess the ability to absorb a great amount of energy during the corn starch decomposition. Finally, dust explosion tests of the mixtures confirm a reduction of P_{max} and ($(dP/dt)_{max}$). In general, a significant improvement of the suppression effectiveness was observed by reducing the α-ZrP size and crystallinity. We expect that the results from this study can be used as guidance on the synthesis of more efficient suppressant materials for dust handling industries with dust explosions hazards.

References

1. Clearfield, A., R.H. Blessing, and J.A. Stynes, *New crystalline phases of zirconium phosphate possessing ion-exchange properties.* Journal of Inorganic and Nuclear Chemistry, 1968. 30(8): p. 2249-2258.

2. Sun, L., et al., *Preparation of [small alpha]-zirconium phosphate nanoplatelets with wide variations in aspect ratios.* New Journal of Chemistry, 2007. 31(1): p. 39-43.

3. Chatrathi, K. and J. Going, *Dust deflagration extinction.* Process Safety Progress, 2000. 19(3): p. 146-153.

4. Liu, Q., et al., *Methane/coal dust/air explosions and their suppression by solid particle suppressing agents in a large-scale experimental tube.* Journal of Loss Prevention in the Process Industries, 2011.

(427a) Efficient Parallelization of Large Scale Fluidized Bed Flow Computations

Amit Amritkar, **Surya Deb** and Danesh Tafti, Mechanical Engineering, Virginia Tech, Blacksburg, VA

In a gasifier, the carbonaceous material undergoes several processes at high temperatures, including pyrolysis, combustion and gasification reactions. Fluidized bed technology is used in gasifiers because the process provides good mixing and promotes uniform heat and mass transfer between the gases and solid particles (e.g., granular material). There is usually a notable difference in the fluidization behavior between various solid fuel particles due to the particle characteristics, gas-particle interactions and especially the reaction kinetics. Research has been conducted on some aspects of the gasification process, but the physics and chemistry of reacting particles in a fluidized bed are less well understood. Therefore, it is important and timely to increase the fundamental understanding of particle mixing in fluidized bed gasifiers to help maximize the benefit of this technology. Current work aims to increase the fundamental understanding of large scale fluidized beds for solid fuel conversion using computational fluid dynamics (CFD). This is achieved using in-house CFD code GenIDLEST which has the capability of Discrete Element Method (DEM) for getting accurate predictions for gasifier flows. Traditionally the CFD simulation of the fluidized beds has been limited to two dimensional (2D) flow simulations. The current work deals with the three-dimensional (3D) flow simulations of an industrial gasifier with $O(10^6)$ particles. With large number of particles the coupled CFD-DEM computations become very expensive in terms of computational resources. Using the domain decomposition strategy, two different bed configurations are evaluated for communication and computational costs. To study the computational efficiency of the system, strong scaling study of the multiphase flow problem has been performed on the Nautilus system at NICS. Computational costs exceed communication costs for 1.3 million particle case with 256 cores, clearly indicating the limits on the minimum number of particles per core. The scaling of the DEM solver is much better as compared to the fluid solver since the fluid computational workload per core is much smaller as compared to particulate phase.

(427b) Efficient Euler-Lagrangian Models for Large Scale Fluidized Beds

Shitanshu Ratilal Gohel[1], **Shailesh Ozarkar**[2], Ulrich Becker[3], Markus Braun[3] and L. Srinivasa Mohan[4], (1)Support and Services, Ansys Fluent India Pvt Ltd., Pune, India, (2)ANSYS, Inc., Pittsburgh, PA, (3)ANSYS Germany GmbH, Darmstadt, Germany, (4)ANSYS Fluent India Pvt. Ltd., Pune, India

Mathematical models of fluidized beds enable engineers to gain insight, and understanding of the various physical processes that occur in these reactors. Detailed engineering models of fluidized beds differ in how the particulate phase is described. In the Eulerian description, partial differential equations are derived by averaging the motion of particles over various configurations and time. In the Lagrangian description, particles or parcels are tracked individually using Newton's laws of motion through the domain. In this talk, we will discuss an efficient Euler-Lagrange model for flow and chemical reactions for large scale fluidized beds.

The Dense Discrete Phase Model (DDPM) with a soft sphere based model for handling contacts and collisions between particles (the Discrete Element Method, or DEM) is proving to be an efficient model to handle large fluidized beds in an efficient manner. In this model, the concept of parcels (which represent a collection of particles of the same size) is used in lieu of individual particles generally associated with DEM models. This, coupled with an efficient algorithm to account for fluid-particle interactions, as well as the parallelization of the models allows for an efficient and accurate description of large fluidized beds.

In this talk we will highlight the application of the DDPM model to various gas-solid fluidized beds, including the bubbling and circulating fluidized beds of the NETL Fluidization challenge 2010, the segregation experiments of Goldshmidt et al. (2003). In addition to cold flow, experiments involving homogeneous and heterogeneous reactions in fluidized beds are also simulated using the DDPM model and compared with experimental measurements, for example the ozone decomposition reaction of Fryer and Potter (1976).

References.

1. Goldschmidt, M.J., Link, J.M., Mellema, S., Kuipers, J.A., 2003. Digital image analysis measurements of bed expansion and segregation dynamics in dense gas-fluidised beds. Powder Technology 138, 135–159.
2. Colin Fryer, and Owen E. Potter, 1976, Experimental investigation of models for fluidized bed catalytic reactors, AIChE J, 22, 38--47.

(427c) Erosion of Granular Beds - Particle-Resolved Simulations

Jos Derksen, Department of Chemical & Materials Engineering, University of Alberta, Edmonton, AB, Canada

Erosion of granular beds - particle-resolved simulations

Erosion of granular beds by a shear flow of liquid over its surface has many natural and engineered manifestations. In terms of industrial applications one can think of flow in slurry pipelines, dredging, and solids suspension in mixing tanks. The extent to which solids are entrained by the flow is a result of a competition between hydrodynamic stress and net gravity. The dimensionless Shields number embodies this competition and a critical Shields number demarcates between static and mobilized solids in the (top layers of the) granular bed.

Simulations of erosion processes are also interesting for fundamental and numerical reasons. The suspension flow closely above the bed is very inhomogeneous with a solids loading that can change from random close packing to dilute over a distance of only a few particle diameters. This poses challenges to continuum-based and point-particle-based numerical approaches; e.g. not much is known about the drag force in the presence of a strong solids volume fraction gradient; not every method can deal with suspensions having such wide ranges of solids volume fractions. Simulations in which we resolve the flow around the particles ('particle-resolved simulations') should be able to overcome the challenges but at the same time face issues of themselves. In addition to their stiff computational demand, these issues mainly relate to interactions between the particles when their surfaces are very close to or actually in (dry) contact. Then these 'direct' simulations need to revert to modelling (lubrication forces for near-contact and friction and restitution coefficients upon contact).

In this contribution we present different types of erosion-related flow systems that we have approached through resolved-particle simulations (with the particles being solid spheres). (1) Erosion of a granular bed by a laminar simple shear flow; (2) erosion of a granular bed by a mildly turbulent channel flow; (3) erosion of a solids bottom layer by the (again mildly turbulent) flow driven by an impeller in a mixing tank. Impressions of these flow systems are given in the figure.

The aim of this work is to enhance our understanding of the way solids are entrained by the liquid flow, and to assess how critical the overall results depend on numerical setings and on modelling choices regarding particle-particle (near) contacts.

Multiphase Simulation of a Commercial Fluidized Catalytic Cracking Regenerator

Samuel M. Clark, CPFD Software, LLC, Albuquerque, NM, USA
Dale M. Snider, CPFD Software, LLC, Albuquerque, NM, USA
Ray P. Fletcher, INTERCAT$_{JM}$, Utrecht, NL

Abstract

A multiphase, thermal, and chemically reacting simulation of a commercial Fluidized Catalytic Cracking (FCC) regenerator was conducted. The simulation was full-scale, 3-dimensional, and included complex internal structures. The regenerator is 70 ft tall, with a diameter of 47 ft. The particle inventory in the system is about 400 tons. The computational method used was an Eulerian-Lagrangian approach called MP-PIC (multi-phase particle-in-cell). Barracuda® VR™, a commercially available CFD software package, was used to perform the simulation. Discrete particle-level chemistry was applied for the chemical reactions of coke burn-off from the catalyst particles.

The simulation had several goals, including: (1) gain insight into the performance of the unit regarding effectiveness of coke combustion from the spent catalyst particles; (2) predict the severity of afterburn in the freeboard, and understand its causes; (3) predict the production rates of pollutants such as SOx from the regenerator; and (4) identify any maldistribution of particles or gas from the highly complex spent catalyst distributor.

Results of the simulation are presented, including particle-fluid flow behavior, thermal profiles throughout the vessel, and gas compositions exiting the system. A comparison of actual operations with the simulation results will be presented to demonstrate the predictive capability of the computational model.

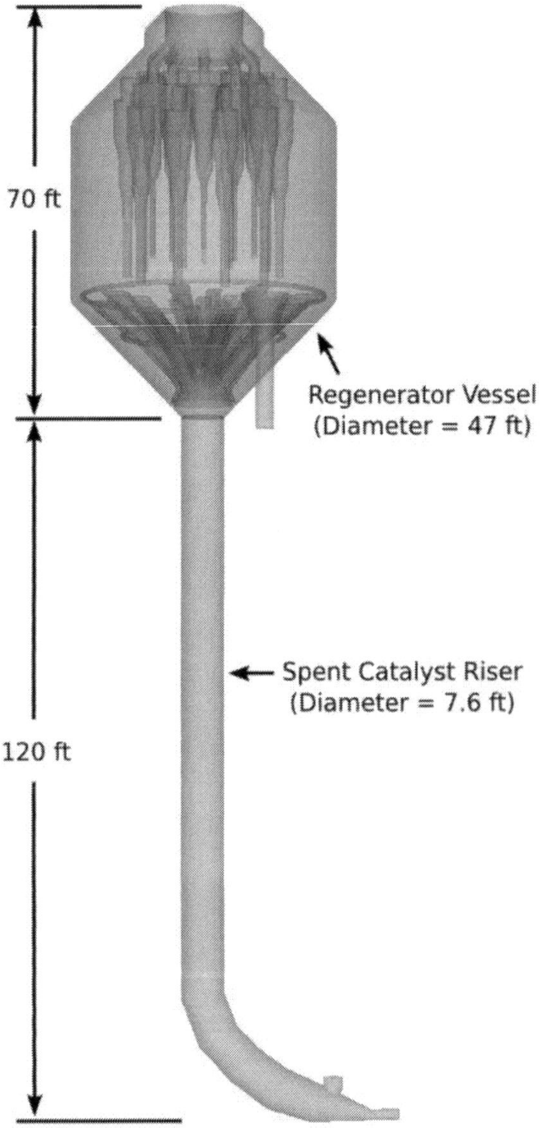

Figure 1: FCC regenerator system geometry

FCC Regenerator As-Built System Geometry

The portions of the FCC regenerator which were modeled included the spent catalyst return riser, the spent catalyst distributor, and the inside of the regenerator vessel. Figure 1 shows a CAD model of the regenerator system considered. The spent catalyst riser was modeled starting at the entrance of lift air, and a short portion of the spent catalyst feed pipe was included. The interior flow region of the regenerator vessel was modeled, with internal structures such as the spent catalyst distributor, combustion air rings, and 12 pairs of internal cyclones. A portion of the regenerated catalyst standpipe was included, allowing particles and gas to exit the regenerator.

Computational Method

Barracuda VR, a commercially available CFD software package, was used to perform the simulation. Barracuda VR uses a proprietary implementation of the MP-PIC method, known as the Computational Particle Fluid Dynamics (CPFD®) method. The MP-PIC method has been described previously (Andrews and O'Rourke, 1996), and several specific aspects of the method as applied in Barracuda VR have also been published (O'Rourke, Zhao, and Snider, 2009; Snider, 2001; Snider, Clark, and O'Rourke, 2011). The CPFD method solves the fluid and particle momentum equations in three dimensions, treating the fluid field(s) as Eulerian and the particles as Lagrangian. There is strong coupling between the fluid and the particles, i.e. the fluid drags the particles, and the particles displace the fluid. Thermal and chemistry calculations are also available, with the fluid and particle phases coupled for energy and reaction purposes. The reader is referred to the papers listed above for details of the CPFD method.

Additional studies of applications of Barracuda VR, with comparisons against operational experience and/or experimental data, have also been conducted (Parker, 2011; Snider, 1998; Snider, 2007; Snider and Banerjee, 2009; Snider, Guenther, Dalton, and Williams, 2010; Zhao, Snider, and Williams, 2006).

Figure 2: Spent catalyst riser simulation boundary conditions

Multiple-Simulation Strategy for the Regenerator System

For a system as complex as the commercial FCC regenerator considered in this paper, a simulation of the entire geometry, and also including thermal and chemical reaction calculations, would result in a long calculation time. Rather than simulating the entire system with a single calculation, three distinct computational models were constructed and run. The first model was an isothermal, non-reacting simulation of the spent catalyst return line. The second model was a thermal, chemically reacting simulation of the spent catalyst distributor. The third model was a thermal, chemically reacting simulation of the inside of the regenerator vessel. Each of these models is discussed in turn in the following sections.

Spent Catalyst Riser Simulation

Computational Grid and Particles

The computational grid used for the spent catalyst riser simulation was specified with simulation speed in mind. Since the results from this simulation were primarily intended to provide inlet conditions to the subsequent spent catalyst distributor simulation, it was important that the simulation run quickly so as not to delay the overall schedule of the three simulations. The grid consisted of 30,360 real cells, with a typical cell side-length of about

6 inches (0.15 m).

At steady operating conditions, from 100 s to 150 s of simulation time, the spent catalyst riser had an average of about 550,000 computational particles. The average particle mass in the system during this time-period was about 46 tons (42,000 kg).

Boundary Conditions

The spent catalyst riser had three boundary conditions, as shown in Figure 2: (1) a flow boundary condition to bring spent catalyst and carrier gas into the system; (2) a flow boundary condition to bring lift air into the riser; and (3) a pressure boundary condition at the top exit of the riser. The flow rates and pressures were specified to match typical operating conditions of the unit.

Simulation Results

The main purpose of the spent catalyst riser simulation was to determine the distribution of particles and gas exiting the top of the riser. This information would then be used to define boundary conditions in the spent catalyst distributor simulation. Flux planes, as shown in Figure 3, were used to measure the particle and gas flow rates across several regions defined near the outlet. The measured fluxes are given in Table 1.

Table 1: Spent catalyst riser simulation mass fluxes from flux planes

Flux Plane	Fluid Mass Flux (kg/s/m^2)	Particle Mass Flux (kg/s/m^2)
A	9.40	254
B	9.80	275
C	12.52	184
D	9.80	275
E	10.34	214
F	16.75	165
G	10.34	214
H	11.13	283
I	17.03	170
J	11.13	283

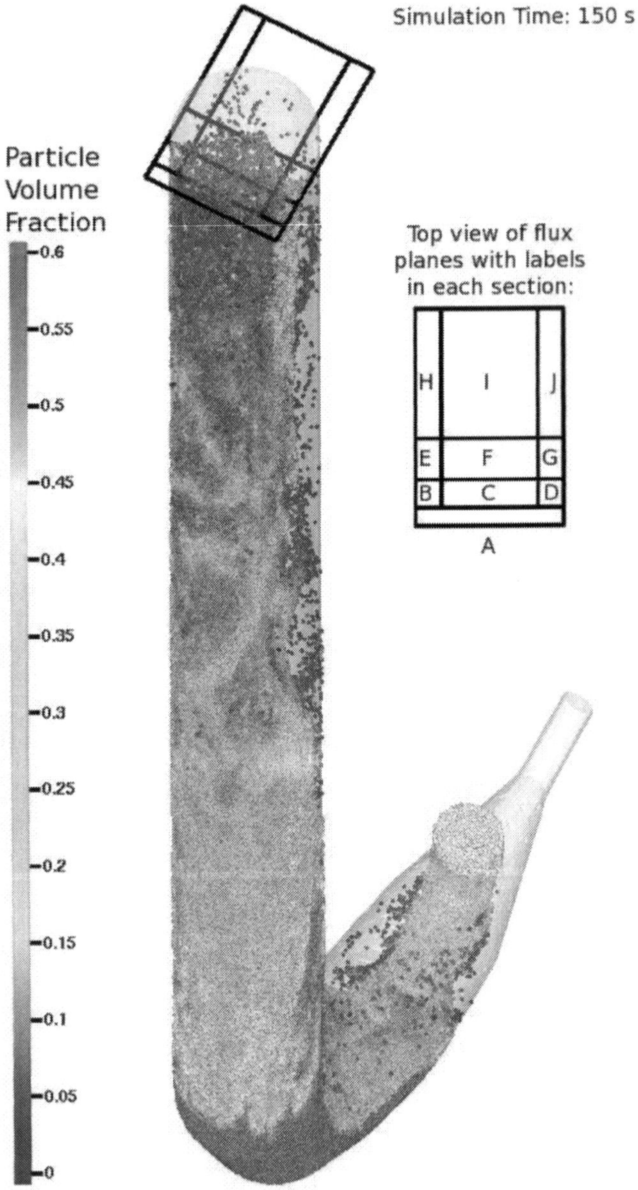

Figure 3: Spent catalyst riser simulation results with flux planes shown and labeled

Potential Improvements

The spent catalyst riser simulation was successful in providing a good estimate of the distribution of particles and gas as they exited the riser. In hindsight, it might have been useful to run the simulation with thermal and chemical reaction calculations. This could have provided better gas and particle composition information for use in the subsequent regenerator vessel simulations, especially if reactions were occurring as particles travelled through the riser.

Spent Catalyst Distributor Simulation

Computational Grid and Particles

The computational grid used for the spent catalyst distributor simulation was designed to capture individual nozzles in the spent catalyst distributor arms. There were 192 nozzles, and particles and gas passed through the nozzles from the insides of the arms into the regenerator vessel. The computational grid consisted of 407,200 real cells.

The computational particles used in the spent catalyst distributor simulation were composed of multiple materials, including catalyst, carbon, hydrogen, sulfur, and nitrogen. The multi-material particle specification was chosen because chemistry and thermal calculations were included in the simulation. Particle composition and size distribution information was based on operating conditions at the refinery.

Boundary Conditions

The results of the spent catalyst riser simulation were used to specify the flow rates of gas and particles at the bottom flow boundary condition in the spent catalyst distributor simulation. Flow and

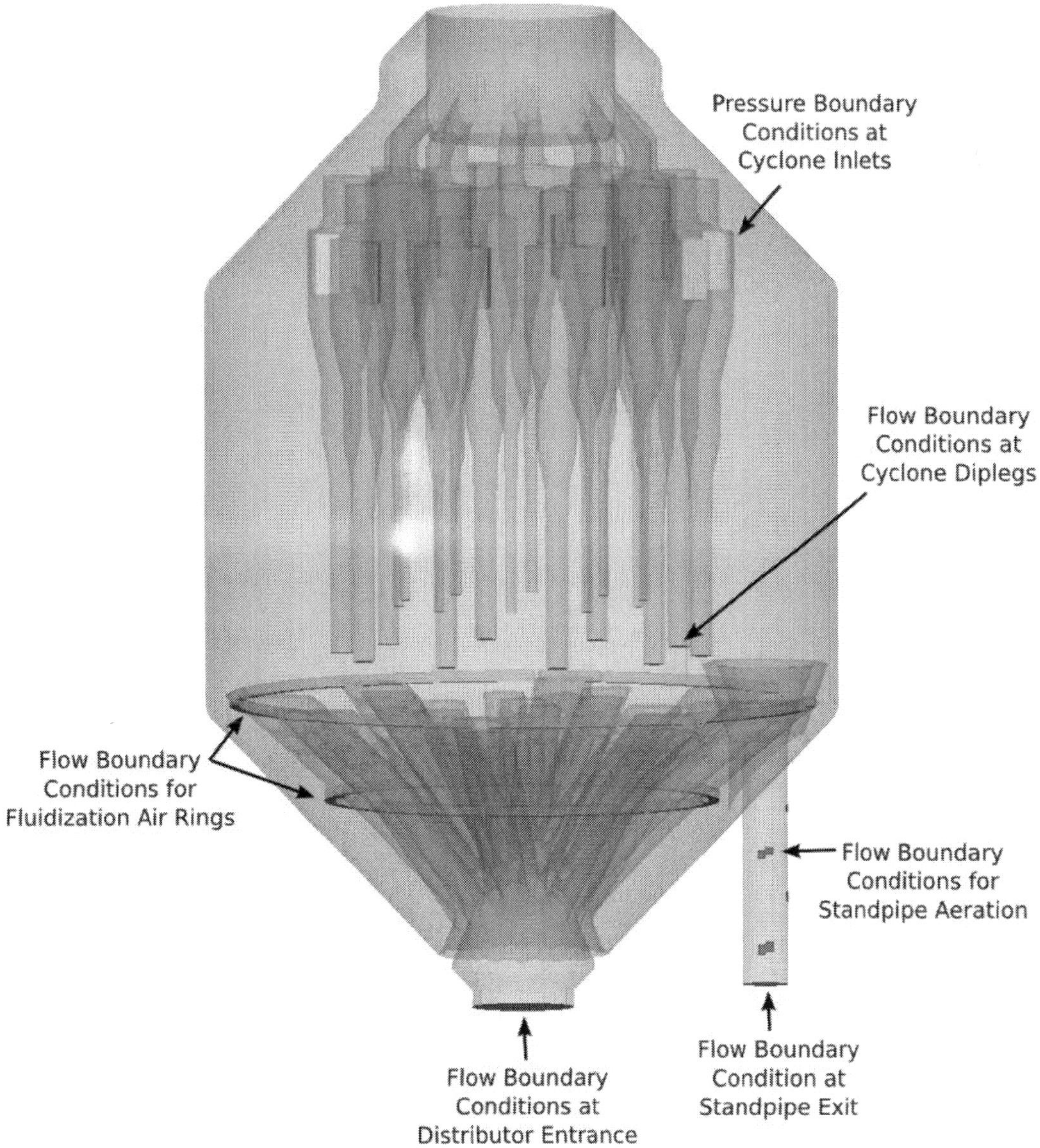

Figure 4: Spent catalyst distributor simulation boundary conditions

pressure boundary conditions were defined as shown in Figure 4, with parameters set to match actual regenerator operating conditions as closely as possible.

Simulation Results

The main purpose of the spent catalyst distributor simulation was to provide particle and gas flow rate information for the individual distributor arms. The flux planes use to measure the particle and gas flow rates are shown in Figure 5, and the numerical flux plane results are presented in Table 2. This information was ultimately used to define flow boundary conditions on the distributor arm surfaces in the final model of the entire regenerator vessel.

Table 2: Spent catalyst distributor simulation gas and particle flow rates.

Arm	Gas (kg/s)	Particles (kg/s)	Arm	Gas (kg/s)	Particles (kg/s)	Arm	Gas (kg/s)	Particles (kg/s)
1	3.78	69.1	7	5.87	57.8	13	1.26	23.6
2	5.14	77.6	8	6.34	65.7	14	1.31	22.2
3	3.86	75.2	9	3.60	40.1	15	1.36	23.6
4	5.06	78.2	10	3.45	40.6	16	1.35	22.9
5	3.56	70.0	11	3.71	37.8	17	0.53	17.4
6	6.34	66.4	12	3.50	42.2			

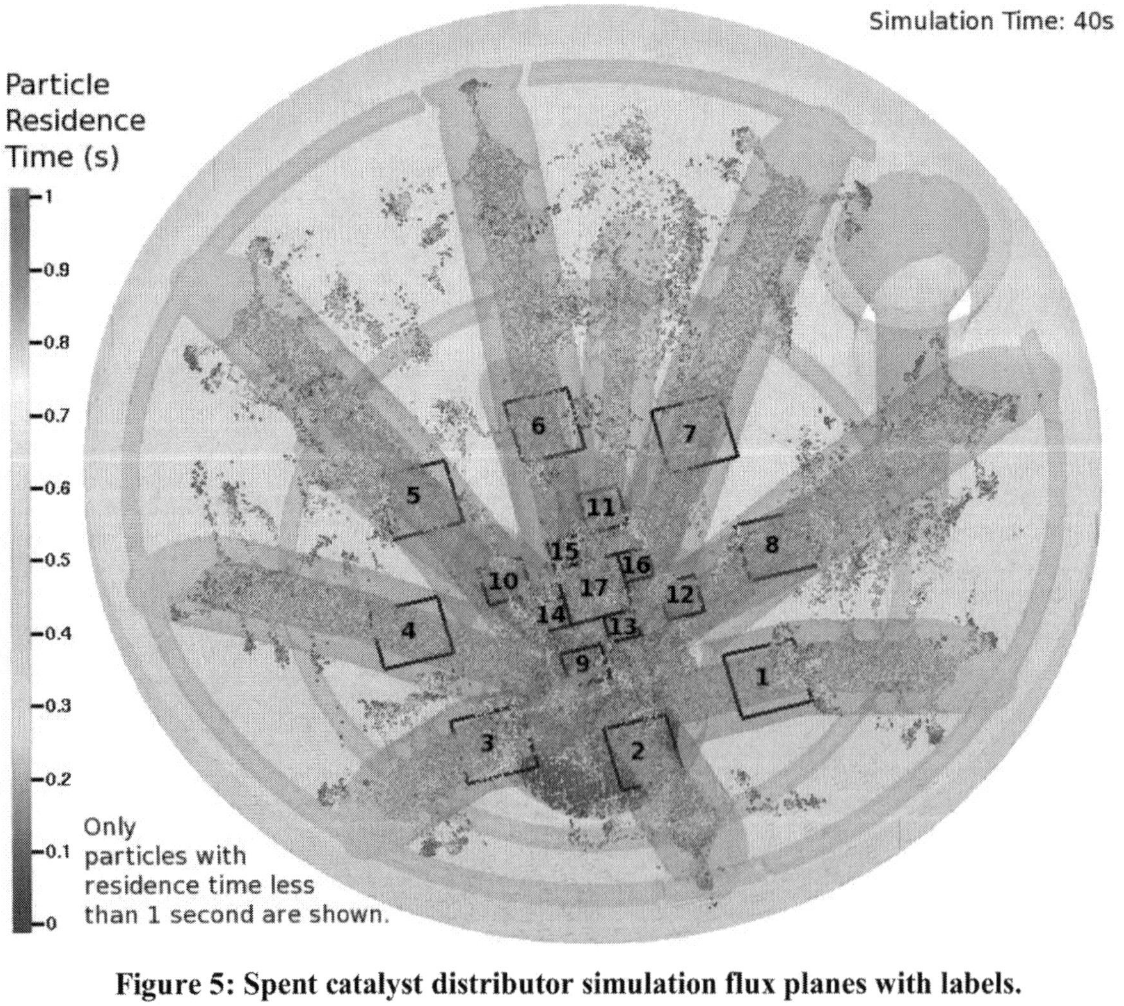

Figure 5: Spent catalyst distributor simulation flux planes with labels.

Regenerator Vessel Simulation

Computational Grid and Particles

The computational grid for the regenerator vessel simulation was coarser than that used for the spent catalyst distributor simulation. The flow of particles and gas through the individual holes in the distributor arm was not modeled, allowing for larger grid spacing and resulting in 237,999 real cells. This reduced number of cells resulted in a faster-running model, while still maintaining acceptable accuracy and resolution of the system geometry. A typical cell side-length was about 7 inches (0.18 m).

Multi-material particles, as described above for the spent catalyst distributor simulation, were used in the regenerator vessel simulation. An equilibrium particle species was defined for the initial catalyst bed, and a coked catalyst species (equilibrium catalyst containing carbon, hydrogen, sulfur & nitrogen) was defined for spent catalyst particles fed from the distributor. At steady operating conditions, from 200 s to 250 s, the simulation had 2.5 million computational particles, representing 507 tons (460,000 kg) of catalyst.

Boundary Conditions

Flow boundary conditions were defined at 196 distinct locations on the surfaces of the spent catalyst distributor arms, with particle and gas flow rates based on results of the spent catalyst distributor

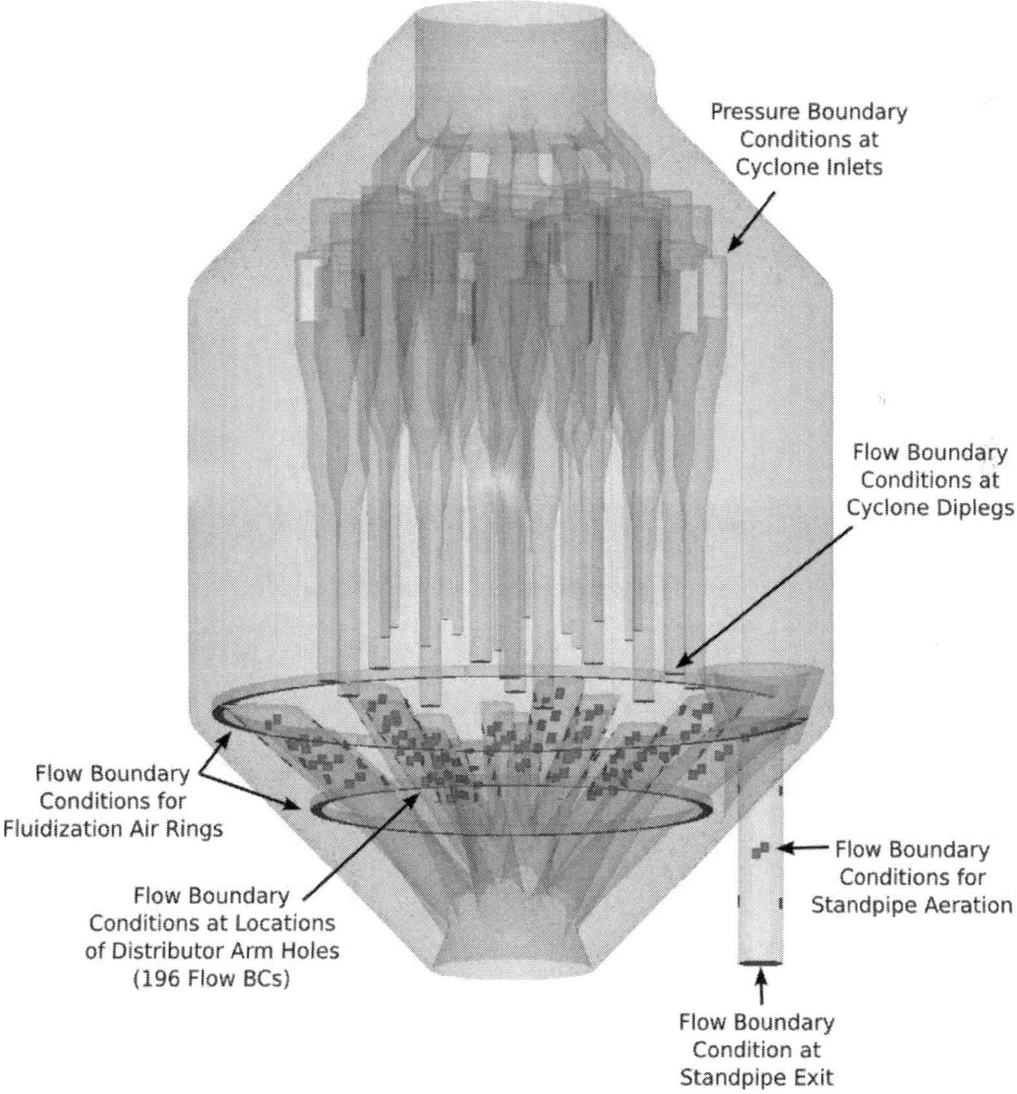

Figure 6: Regenerator vessel simulation boundary conditions

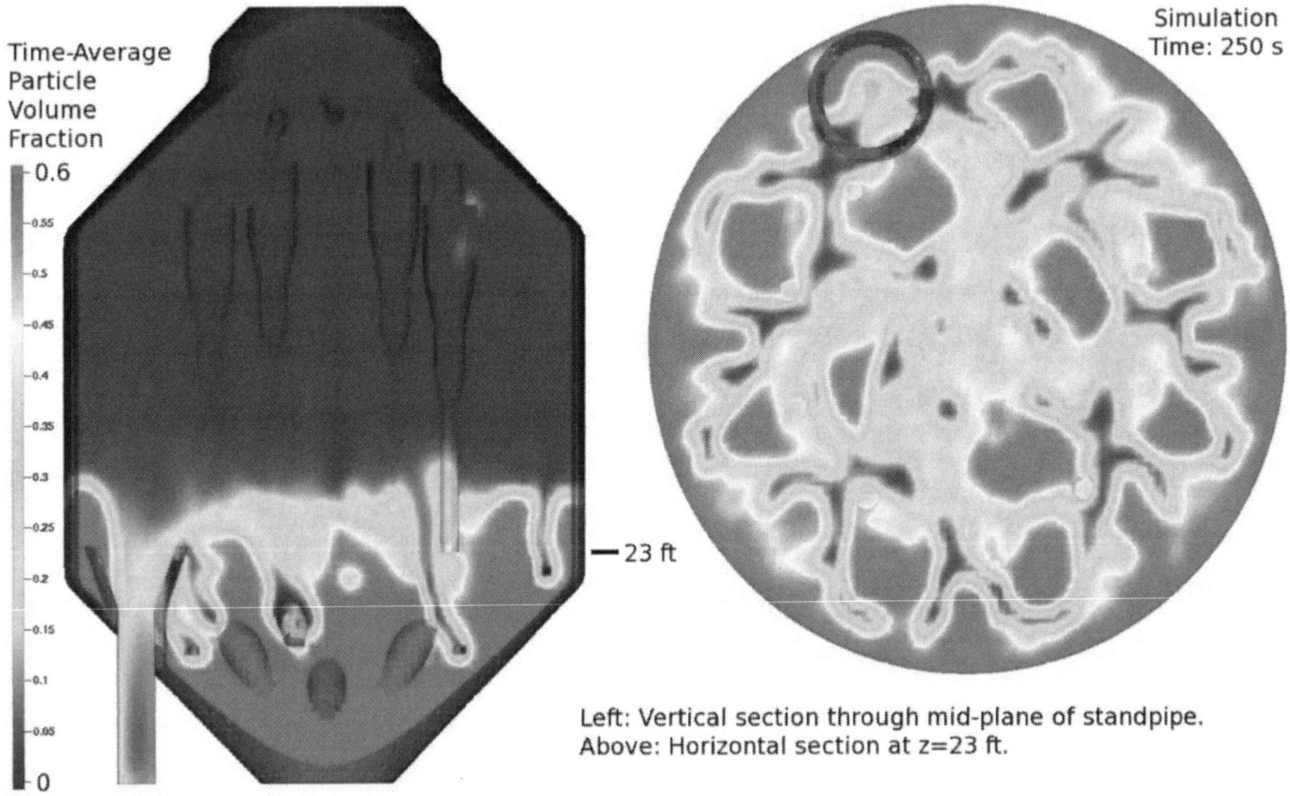

Figure 7: Regenerator vessel time-average particle volume fraction plots

simulation. Other boundary conditions were similar to those used in the spent catalyst distributor simulation, with parameters set to match actual operating conditions as closely as possible. Figure 6 shows the boundary conditions for the regenerator vessel simulation.

Simulation Results

The simulation of the regenerator vessel provided many interesting results, and was studied in greater detail than either of the preceding simulations. In this paper, we will highlight several specific results, namely: (1) general fluidization and particle flow behavior; (2) thermal behavior, particularly as it pertains to afterburn; and (3) gas species compositions throughout the system, including outlet gas compositions through the cyclones. Many other aspects of the system could be studied extensively and further insight into the operation of the regenerator would surely be gained.

The fluidization behavior of the system was studied with effectiveness of mixing in mind. Figure 7 shows the time-average particle volume fraction on vertical and horizontal cross-sections. Time-averaging was started at 220 s, once a pseudo-steady state was reached in the simulation, and was continously calculated up to 250 s. The bottom portion of the regenerator, i.e. below the level of the lower air ring, was found to be non-fluidized with stationary particles at close-pack. Gas from the air rings and distributor arms formed persistent paths of low-particle volume fraction. From the top view, numerous areas of high-volume fraction particles were present throughout the fluidized bed. Altogether, these results indicated less than optimal mixing behavior, and the gas-particle contact time in the simulation was likely impacted by the gas bypassing behavior observed.

Afterburn, which is a large temperature increase in the dilute phase (freeboard) above the fluidized bed, is a known issue for the regenerator simulated in this paper. The time-average fluid temperature predicted in the simulation is plotted in the upper-left hand corner of Figure 8. A 100 °F temperature difference was expected based on operational experience, and the simulation is showing a temperature difference of about this magnitude from the lower to the upper regions of the vessel.

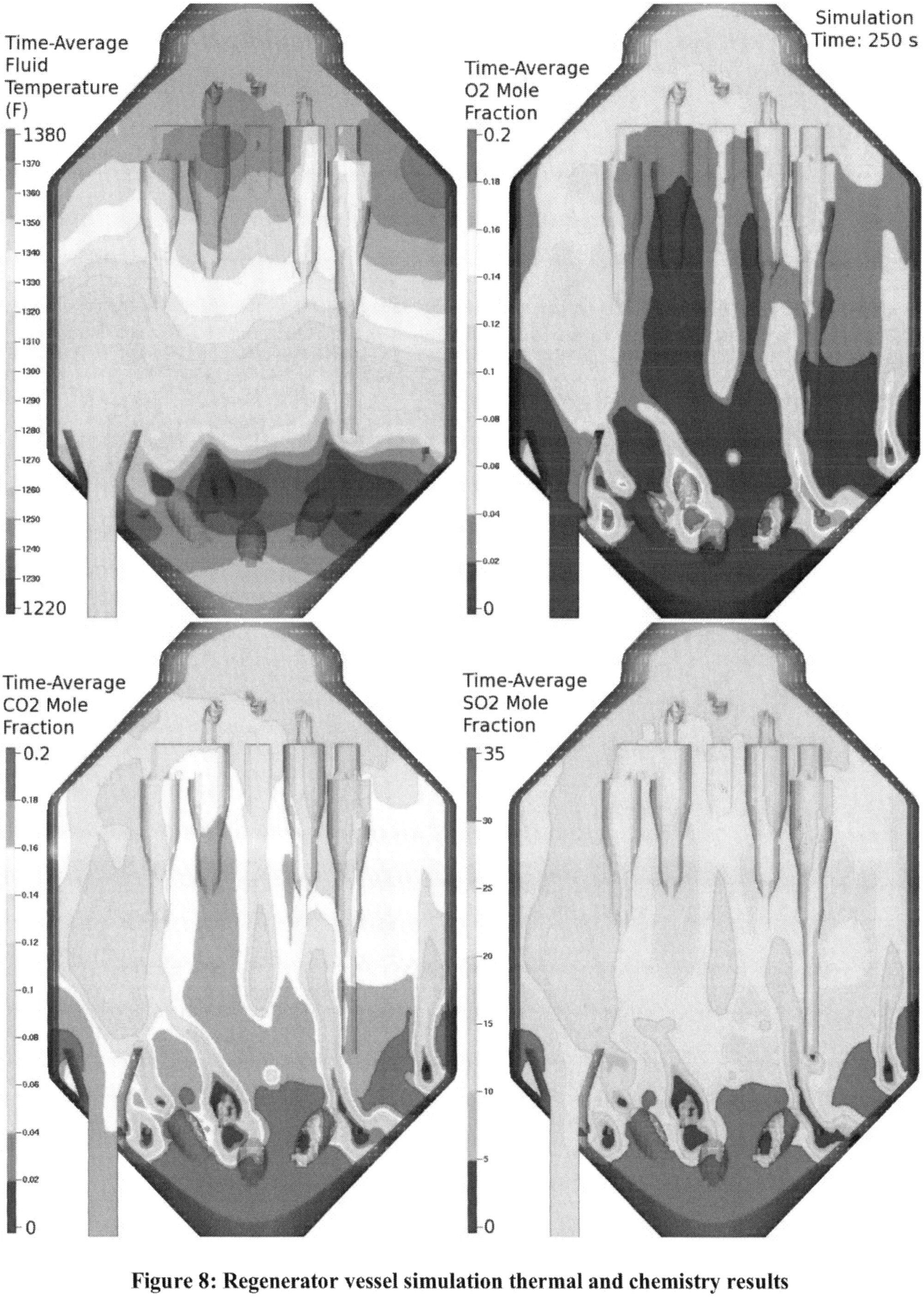

Figure 8: Regenerator vessel simulation thermal and chemistry results

Afterburn is undesirable because it negatively impacts the FCC catalyst deactivation rate and limits operating flexibility due to metallurgical constraints. Future simulations of this system could include analyses of proposed design changes to reduce afterburn. The current simulation gives a baseline against which future results can be compared.

The gas composition at the outlet of the regenerator was analyzed and compared with typical gas composition data from the regenerator. Vertical slices of time-averaged gas composition are shown in Figure 8 for O_2, CO_2, and SO_2. For O_2, a typical mole fraction value during unit operation is about 0.02. The simulation predicted higher concentrations than this in a significant portion of the upper half of the regenerator. It is suspected that the regions of high O_2 concentration were due to either: (1) bypassing of the gas exiting the distributor, or (2) the fact that combustion was neglected in the spent catalyst riser simulation. The CO_2 mole fraction near the cyclone inlet horns was near the expected value of 17. The SO_2 concentration, in parts per million, was also close to the expected outlet SO_2 concentration of 30 ppm.

Conclusions

The simulations of the commercial FCC regenerator provided many insights into the operation of the system. The thermal and chemistry results of the regenerator vessel simulation showed good agreement with real-world operating experience from the commercial regenerator. The multiple-simulation strategy used to analyze the overall regenerator system was successful in allowing an analysis of the entire system, while at the same time meeting project schedule goals. A single large simulation of the entire system would likely have taken much longer to yield the same amount of useful engineering information.

References

1. Andrews, M. J., and O'Rourke, P. J. (1996). "The Multiphase Particle-in-Cell Method (MP-PIC) Method for Dense Particle Flow", *Int. J. Multiphase Flow*, 22, 379-402.
2. O'Rourke, P. J., Zhao, P., Snider, D. (2009), "A model for collisional exchange in gas/liquid/solids fluidized beds", *Chemical Engineering Science*, 64, 1784-1799.
3. Parker, J. (2011), "Validation of CFD Model for Polysilicon Deposition and Production of Silicon Fines in a Silane Deposition FBR", *International Journal of Chemical Reactor Engineering*, 9, A40.
4. Snider, D. M., et al. (1998), "Sediment Flow in Inclined Vessels Calculated Using Multiphase Particle-in-Cell Model for Dense Particle Flows", *Int. J. Multiphase Flow*, 24, 1359-1382.
5. Snider, D. M. (2001). "An Incompressible Three-dimensional Multiphase Particle-in-Cell Model for Dense Particle Flows", *Journal of Computational Physics*, 170, 523- 549.
6. Snider, D. M. (2007). "Three fundamental granular flow experiments and CPFD predictions", *Powder Technology*. 176, 36-46.
7. Snider, D. and Banerjee, S. (2009) "Heterogeneous gas chemistry in the CPFD Eulerian–Lagrangian numerical scheme (ozone decomposition)", *Powder Technol.*, doi:10.1016/j.powtec.2009.04.023.
8. Snider, D., Guenther, C., Dalton, J., Williams, K. (2010), "CPFD Eulerian-Lagrangian Numerical Scheme Applied to the NETL Bench-top Chemical Looping Experiment", *1st International Conference on Chemical Looping*.
9. Snider, D. M., Clark, S. M., O'Rourke, P. J. (2011), "Eulerian-Lagrangian method for three-dimensional thermal reacting flow with application to coal gasifiers", *Chemical Engineering Science*, 66, 1285-1295.
10. Zhao, P., Snider, D., Williams, K. (2006), "Computational Particle-Fluid Dynamics Simulations of a Commercial-Scale Turbulent Fluidized Bed Reactor", *AIChE Annual Meeting*.

(427e) Computational Analysis of the Air Injection Effect On the Regeneration Process in FCC

Bumjoon Cha[1], Hyungtae Cho[1], Sung Real Son[2], Suk-Joon Kim[2] and Il Moon[1], (1)Chemical and Biomolecular Engineering, Yonsei University, Seoul, South Korea, (2)SK Innovation, Daejeon, South Korea

One of the problems associated with regeneration process in conventional fluid catalytic cracking (FCC) unit is the increase of flue gas temperature after leaving the dense catalyst bed, afterburning phenomena. Generation of excessive heat of combustion and presence of smaller number of catalyst particles to absorb this heat may degrade the quality of catalyst, and require additional fresh catalyst to maintain the proper working FCC unit. Hydrodynamic approach for gas-solid flows in fluidized bed is the best way to comprehend regeneration process including afterburning problem, and consequently a step-out improvement of the FCC regeneration. In this work, a regeneration process in a simple type regenerator was investigated by computational particle fluid dynamics (CPFD) numerical simulation which is Eulerian-Lagrangian solution method for fluid-solid multiphase flows. The combustion reaction was considered as a stoichiometric equation in the simulation scheme. The simulated temperature profile of catalyst and the particle-fluid coupled behaviour shows persuasive accuracy with experimental data. Figures 1 and 2 below represent simulated temperature profiles of gas and catalyst particle, respectively. The air distribution effect on regeneration tendency was also demonstrated for various types of air sparger and air inlet flow rate conditions. The results reveal that adjustment of air flow gives significantly better temperature distribution, and hence dramatically decrease ratio of the over-heated catalyst caused by afterburning effect.

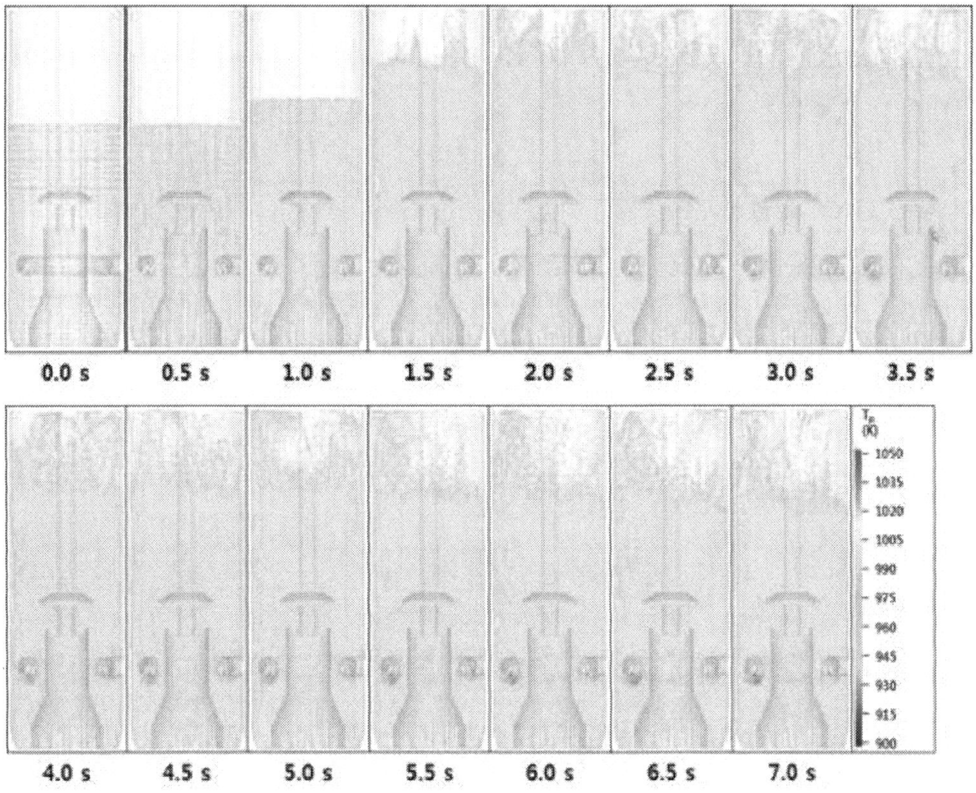

Figure SEQ Figure * ARABIC 1. Temperature profile of catalyst particles

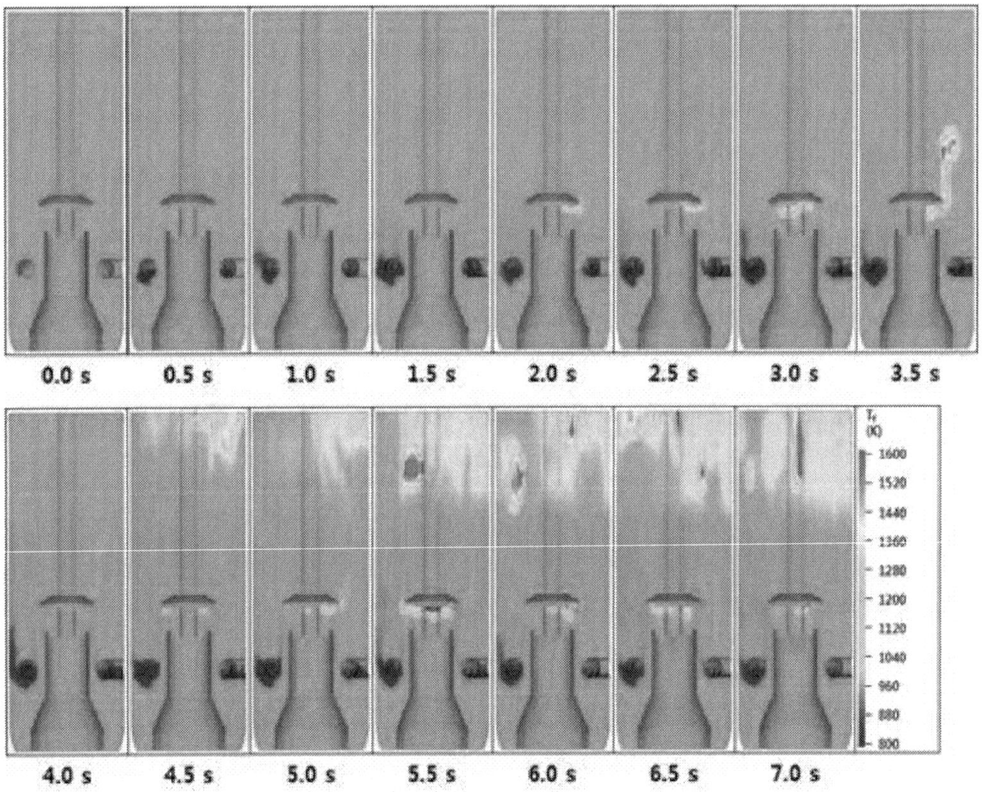

Figure SEQ Figure * ARABIC 2. Temperature profile of gas

Reference

[1] D. M. Snider, S. M. Clark and Peter J. O'Rourke, Eulerian–Lagrangian method for three-dimensional thermal reacting flow with application to coal gasifiers, *Chemical Engineering Science*, 66(6), 1285–1295 (2011).

[2] P. Zhao, P. J. O'Rourke and D. Snider, Three-dimensional simulation of liquid injection, film formation and transport, in fluidized beds, *Particuology*, 7(5), 337-346 (2009).

[3] P. J. O'Rourke, P. Zhao and D. Snider, A model for collisional exchange in gas/liquid/solid fluidized beds, *Chemical Engineering Science*, 64(8), 1784-1797 (2009).

[4] C. Y. Wen and Y. H. Yu, Mechanics of fluidization, *Chemical Engineering progress Symposium Series*, 62(62), 100-111 (1966).

[5] V. L. Zimont and Y. M. Trushin, Total combustion kinetics of hydrocarbon fuels, Combustion, Explosion, and Shock Waves, 5(4), 391-394 (1969).

[6] Yeong-Koo Yeo, Hee-Young Kim, In-Won Kim, Il Moon, Yonsoo Chung and V. V. Levdansky, Analysis of catalytic reaction systems under microwaves to save energy, *Korean Journal of Chemical Engineering*, 20(1), 1-7 (2003).

[7] V. A. Danilov, Jongkoo Lim, and Il Moon, Three-dimensional two-phase CFD model for DMFC design, *Journal of Power Source*, 162(2), 992-1002 (2006).

Modeling of temperature profile in ethylene polymerization fluidized bed reactor in condensed mode operation

YeFeng Zhou, LiJin Zhou※, BinBo Jiang, JingDai Wang, YongRong Yang

State Key Laboratory of Chemical Engineering, Department of Chemical and Biological Engineering, Zhejiang University, Hangzhou 310027, China

※SINOPEC YangZi PetroChemical Company Limited, Nanjing, 210048, China

Abstract

Condensed mode operation plays a critical role in raising polymer production in gas phase ethylene polymerization fluidized bed reactor (FBR), however, the study of the FBR in condensed mode operation attracts rare attention in academic circles. Based on classical emulsion-bubble two-phase model, this paper proposed a polyethylene FBR model in condensed mode operation by taking into consideration the adding condensed liquid and its evaporation. Two different flow regimes, the gas-liquid-solid (G-L-S) zone and the gas-solid (G-S) zone, are assumed to co-exist in the reactor model. Besides, the emulsion phase in the FBR is regarded as a plug flow in G-L-S zone and a well-mixed flow in G-S zone while the bubble phase is always treated as a plug flow in the whole fluidized bed. By modeling temperatures of the emulsion phase and the bubble phase, the temperature profile of the ethylene polymerization FBR is obtained. The simulation results show that bed temperature in G-L-S zone above gas distributor steps up sharply and then remain almost constant in G-S zone. What's more, G-L-S zone expands upward gradually with condensed liquid increased in recycling stream. The simulation results are later proved to fit industrial data satisfactorily. Based on the established model, effects of bubble size on heat transfer and on bed temperature profile have been also studied. In addition, the coexistence of G-L-S zone and G-S zone in one reactor provides new method for producing high-performance polymer product.

(457a) Effect of Particle Size On Hydrogen Production by Reaction of Liquid Water and Aluminum Powders

Hongqi Nie[1], Mirko Schoenitz[1] and Edward L. Dreizin[2], (1)NJIT, Newark, NJ, (2)Otto H. York Department of Chemical, Biological, and Pharmaceutical Engineering, New Jersey Institute of Technology, Newark, NJ

Abstract

Reaction of aluminum with water is of interest for hydrogen generation for a wide range of applications, from in-situ operated fuel cells to propulsion of underwater vehicles. This reaction is extensively studied and various aluminum-based composite materials and alloys are under development with the aims to accelerate the rate of hydrogen production and enable controlled operation of the hydrogen-generating systems. It would be particularly interesting to achieve the desired rate of hydrogen production at low operating temperatures, using liquid water as opposed to water vapor. However, the understanding of the rate-limiting processes is lacking for the reaction of pure aluminum and liquid water. Recent research by different authors showed, somewhat unexpectedly, that reaction dynamics are different for micron sized spherical aluminum powders with different size distributions. This result can be interpreted considering evolution of aluminum hydroxide layers and possible changes in their integrity and diffusion resistance when the layer thickness becomes comparable to its radius of curvature. In addition, porosity in such layers can develop when they are sufficiently thick; however, such thick layers may not necessarily grow on finer Al particles. This work attempts to understand the effects of particle size on its reaction dynamics in liquid water using experimental data from time-resolved micro-calorimetry. The experimental data tracking energy release caused by the Al/H_2O reaction are processed using the actual measured particle size distributions for three different Al powders. The reaction rate is assumed to be proportional to the particle surface; thus the entire heat flow measured by the calorimeter is partitioned among Al particles based on their size distribution. Evolution of particle sizes and thickness of the grown $Al(OH)_3$ layers are tracked during the micro-calorimetry experiment. The results are processed using a simplified reaction model and diffusion coefficients are found for particles of different sizes and at different reaction stages. Ideally, such coefficients should only depend on temperature, so that their observed dependencies on particle size and reaction progress serve to indicate where the modifications of the simplified reaction model are necessary. Furthermore, reaction rates are compared for individual particles of the same size which are present in aluminum powders with different size distributions. Once the correct reaction model is implemented, the reaction rates for such identical particles must coincide. Additional comparisons and measurements will be presented and discussed, aimed at establishing a corrected reaction model for aluminum and liquid water. Based on the developed reaction mechanism, a composition and morphology of an optimized, Al-based material for hydrogen generation will be proposed.

(457b) Aqueous Solubility, Octanol-Water Partition Coefficients and Carbon Adsorption of 1, 3, 3-Trinitroazetidine (TNAZ)

Veera Boddu, Environmental Processes Branch, US Army Engineer Research and Development Center, Champaign, IL, **Carmen Costales-Nieves**, Environmental Processes Branch, US ARMY ERDC-CERL, Champaign, IL, Rayan Zilner, Environmental Processes Branch, US Army ERDC-CERL, Champaign, IL and Reddy Damavarapu, Energetics and Warheads Division, U.S. Army - Armament Research, Development and Engineering Center, Picatinny Arsenal, NJ

A high energy explosive component, 1, 3, 3-Trinitroazetine (TNAZ), has the potential to replace TNT and other conventional explosives in various formulations. With a melting point of about 373 K, it is steam-castable (melt-castable) and is also a thermally stable material for up to 500 K. Its physical properties are important to estimate the environmental and toxicological consequences of residuals released from manufacturing, packaging, testing and training operations. Accurate values of these properties at ambient temperatures are very essential for estimating its environmental distribution, biotransformation, and the development of potential water treatment processes. In this study, the aqueous solubility (Sw) and octanol-water partition coefficient (Kow) were experimentally measured in the temperature range of 288 – 318 K. The experimental results were compared with literature data and predictions using SPARC model. Other physical properties including vapor pressure and Henry's law constant were also predicted using the SPARC. Preliminary isothermal adsorption of TNAZ on granular activated carbon (GAC) was also conducted in the above temperature range. The results show that the aqueous solubility increases with temperature and varies ~400-1000 mg/L in the temperature range.. Experimental solubility and octanol-water partition coefficients, SPARC predictions, and the results of the GAC adsorption studies will be presented.

(457c) Combustion Dynamics of Aluminum in Turbulent Flows

Amy Corcoran[1], Vern K. Hoffmann[2] and Edward L. Dreizin[1], (1)Otto H. York Department of Chemical, Biological, and Pharmaceutical Engineering, New Jersey Institute of Technology, Newark, NJ, (2)Mechanical Engineering, New Jersey Institute of Technology, Newark, NJ

Combustion dynamics of aluminum in turbulent flows

A.L. Corcoran, V.K. Hoffmann, E.L. Dreizin

New Jersey Institute of Technology

Predictive mechanisms for particle ignition and combustion rates are required in order to develop optimized propellant and energetic formulations using micron-sized metal powders, such as aluminum. Most current descriptions are based on laboratory experiments performed in stationary or laminar combustion configurations. However, turbulent environments exist in most applications and validity of the present descriptions for such environments has not been established. This experimental study is aimed to measure burn times versus size of fine aluminum powders burning in environments with different levels of turbulence. An initial laminar oxygen-acetylene flame is produced, and auxiliary tangential jets of air with adjustable flow rates are used to achieve different controlled levels of turbulence. Aluminum powder is forced into the center of the flame using a flow of nitrogen. A spherical powder with nominal particle size of 6.5 μm is used. Metal mass feed rate is adjusted using a custom-made powder feeder. The flame with burning aluminum is photographed using a Sony DSC-H50 camera. A spinning disk with multiple holes is used to interrupt the photo-exposure with a pre-set frequency. In the obtained images, the streaks of luminous burning particles appear dashed, and the particle burn times are determined by counting the dashes for each streak. Image processing produced statistical information on particle burn times. The particle burn times are correlated with the particle size distributions to obtain the burn times as a function of the particle size for flows with different turbulence levels. The data are compared with similar correlations available in the literature. Additional spectroscopic flame measurements will be performed to further characterize aluminum combustion at different flow conditions.

(457d) Evaluation of Predictive Capabilities for Pyrotechnics Using Different Thermodynamic Software Packages(INVITED)

Jan Puszynski[1], Paul Redner[2], Karl Oyler[3], Christopher Csernica[4], Jesse Sabatini[4], Jay Poret[4], Anthony Shaw[4], Neha Mehta[3], Gartung Cheng[3] and Deepak Kapoor[5], (1)Chemical and Biolgocial Engineering Department, South Dakota School of Mines and Technology, Rapid City, SD, (2)Advanced Materials Branch, US Army, RDECOM-ARDEC, Picatinny Arsenal, NJ, (3)Energetics and Warheads, US Army RDECOM-ARDEC, Picatinny Arsenal, NJ, (4)Pyrotechnics, US Army, RDECOM-ARDEC, Picatinny Arsenal, NJ, (5)Advanced Materials, US Army, RDECOM-ARDEC, Picatinny Arsenal, NJ

Thermodynamic codes play a valuable role in the downselection of the constituents of energetic systems. This is particularly true for thermite and other systems in which the ratio of fuel to oxidizer can be greatly varied and are not contained within the same molecule. If the correct code is used properly, it becomes a very important tool in identifying material systems with key thermophysical properties. Having knowledge of these properties will, in turn, reduce the time and number of experiments necessary to reach the desired endpoint because it will enable one to focus on higher-value materials.

This study explores the use of PEP2100, CHEETAH 6.0 and HSC Chemistry software packages to predict the adiabatic reaction temperature, enthalpy, and product species of various illuminant as well as nanothermite reacting systems. The predicted thermochemical data are compared with experimental observations and correlated to other characteristics that cannot be predicted using computational software, such as light output and color ratio.

(457e) Direct Calculation of 1-Octanol–Water Partition Coefficients From Adaptive Biasing Force Molecular Dynamics Simulations

Navendu Bhatnagar, Chemical Engineering, Wayne State University, Detroit, MI, Ganesh Kamath, Chemistry, University of Missouri-Columbia, Columbia, MO, Isaac Chelst, Chemical Engineering and Materials Science, Wayne State University, Detroit, MI and **Jeffrey J. Potoff**, Chemcial Engineering and Materials Science, Wayne State University, Detroit, MI

The 1–octanol–water partition coefficient log K_{ow} of a solute is a key parameter used in the prediction of a wide variety of complex phenomena such as drug availability and bioaccumulation potential of trace contaminants. In this work, the application of adaptive biasing force molecular dynamics simulations (ABF–MD)[1-4] to the calculation of 1-octanol-water partition coefficients is presented. Unlike traditional methods, such as thermodynamic integration and free energy perturbation, which produce relative free energy values, the ABF-MD method predicts absolute free energies of transfer and eliminates the need for reference solutes. Two approaches are evaluated; the direct transfer of the solute from 1–octanol to water phase, and separate transfers of the solute from the water or 1–octanol phase to vacuum, with both methods yielding statistically indistinguishable results. Simulations performed on a variety of system sizes are used to show that the calculated potential of mean force is remarkably insensitive to the microstructure of the solvent, a finding that suggests relatively small system sizes may be used to determine accurate free energies of transfer. The effect of various simulation parameters on the sampling efficiency is discussed.

Calculations are performed on a variety of systems, including n-alkanes, alcohols, chemical warfare agents, energetic materials and ionic liquids. The effect of wet vs. dry octanol and choice of water model is assessed for each system. The predictions of ABF-MD are found to be in excellent agreement with prior calculations performed with thermodynamic integration[5] or Gibbs ensemble Monte Carlo[6].

1. E. Darve and A. Pohorille, J Chem Phys **115** (20), 9169-9183 (2001).

2. E. Darve, M. A. Wilson and A. Pohorille, Mol Simulat **28** (1-2), 113-144 (2002).

3. J. Henin and C. Chipot, J Chem Phys **121** (7), 2904-2914 (2004).

4. D. Rodriguez-Gomez, E. Darve and A. Pohorille, J Chem Phys **120** (8), 3563-3578 (2004).

5. N. M. Garrido, A. J. Queimada, M. Jorge, E. A. Macedo and I. G. Economou, J Chem Theory Comput **5** (9), 2436-2446 (2009).

6. J. I. Siepmann and B. Chen, Journal of Physical Chemistry B **110** (8), 3555-3563 (2006).

(477a) Techno-Economic Evaluation of Integrated Renewable Acetic Acid, Glycolaldehyde, and Acetol Production From Biomass-Derived Pyrolysis Oil

André B. de Haan, Caecilia R. Vitasari and G. Wytze Meindersma, Chemical Engineering, Eindhoven University of Technology, Eindhoven, Netherlands

Acetic acid, glycolaldehyde, and acetol are bio-based chemicals which can be isolated from pyrolysis oil in an integrated biorefinery. This paper discusses the conceptual design of an integrated separation process for forest residue- and pine-derived pyrolysis oils. The process simulation was performed in Aspen Plus®, while the equipment cost was estimated in Aspen Process Economic Analyzer. The process was designed for a capacity of 200 kton pyrolysis oil per year and involves extraction, distillation, and evaporation. Water and 2-ethyl-1-hexanol are used as extraction solvents. All process step designs are based on data derived from experiments with forest residue derived pyrolysis oil. The designed process can isolate more than 99% glycolaldehyde and acetic acid and about two-thirds of acetol present in the oils. In comparison with the forest residue-based process (21 M€), the pine-based process requires a higher capital investment of 23 M€ and a slightly higher production cost 49 M€/a versus 48 M€/a, but is able to provide a higher revenue of 62 M€/a instead of 50 M€/a because pine-derived pyrolysis oil contains more acetic acid, glycolaldehyde, and acetol, which also makes it less sensitive to market price. The plant profitability increases with production capacity up to 300 kton pyrolysis oil per year. In this capacity, pine-derived pyrolysis oil yields approximately a three times higher return-on-investment (ROI) than forest residue-derived pyrolysis oil.

(477b) Clean Pyrolytic Sugars Solution

Marjorie Rover[1], Patrick Johnston[1], Laura Jarboe[2] and Robert C. Brown[3], (1)Center for Sustainable Environmental Technologies, Iowa State University, Ames, IA, (2)Department of Chemical and Biological Engineering, Iowa State University, Ames, IA, (3)Mechanical Engineering, Iowa State University, Ames, IA

This study focuses on the effective removal of contaminants from pyrolytic sugar to produce a suitable fermentation substrate. Pyrolytic sugars contain small percentages of other water soluble constituents such as low molecular weight acids, furfural, phenols, as well as other contaminants. A novel approach to "cleaning" the pyrolytic sugars prior to microbial fermentation has been developed and we will present our recent findings, innovative advances and strategies.

Bio-oil may be the most economical approach to advanced biofuel production. Produced from the fast pyrolysis of biomass, bio-oil contains hundreds of chemical compounds. Among the most commercially important components of bio-oil are sugars and phenols.

Significant progress has been achieved in the advancement of bio-oil collection, allowing for an effective bio-oil fractionation method for the separation of the pyrolytic sugars and the lignin derived constituents. A simple "water wash" removes the soluble sugars from the water insoluble components, providing a potential aqueous fermentable sugar stream.

This research focuses on various strategies and separation techniques to remove contaminants from the pyrolytic sugar stream.

(477c) 5-HMF Production in Piperylene Sulfone: A Bifunctional Solvent for Reaction and Separation

Christopher J Butch[1], Jakob Crowe[1], Gregory Marus[1], Pamela Pollet[2], Charles L. Liotta[2] and Charles A. Eckert[1], (1)School of Chemical and Biomolecular Engineering, Georgia Institute of Technology, Atlanta, GA, (2)School of Chemistry and Biochemistry, Georgia Institute of Technology, Atlanta, GA

5-Hydroxylmethylfurfural has garnered wide interest as a potential bio-based commodity chemical due to its multiple reactive groups and relatively easy formation from fructose and other sugars. For example HMF can be produced from fructose in DMSO in high yields; however the separation is difficult and wasteful. The literature contains numerous examples of HMF production in DMSO or DMSO and water, separated by extraction into methyl isobutyl ketone. When combined with heterogeneous catalysis from ion exchange resins, the yields become particularly attractive. However in each of these cases, despite the good yields, recovery of the product requires expensive liquid-liquid extraction processes, and often the DMSO is not recyclable, thus rendering the reaction impractical.

To address the difficulties associated with separating HMF from DMSO, we use piperylene sulfone, a solvent with very similar solvatochromic properties to DMSO. These similarities make piperylene sulfone a viable replacement as reaction medium. Piperylene sulfone is formed through a reverse cheletropic reaction between sulfur dioxide and 1,3-trans-piperylene. Superiorly to DMSO, piperylene sulfone can be decomposed to sulfur dioxide and piperylene at temperatures of ~110C which allows for facile separation of reaction products. The off-gassed sulfur dioxide and trans-piperylene can be re-reacted elsewhere to reform the piperylene sulfone, enabling recyclability.

We demonstrate a combined reaction separation system composed of fructose in piperylene sulfone with Amberlyst-15 as catalyst. We will present the results of a statistical design of experiments conducted to optimize conversion of HMF as well as the separation of HMF from piperylene sulfone. We shall also present opportunities for future research using this solvent system.

ENHANCED COMPOST-DRYING PROCESS OF SEWAGE SLUDGE WITH CARBONIZED SLUDGE

N. Kobyashi[1], *Y. Itaya*[2], *L. Li*[2], *A. Suami*[2], *M. Sawai*[3], *H. Hamabe*[3]

[1]*Department of Chemical Engineering, Nagoya University, Nagoya, Japan*

[2]*Department of Mechanical and Systems Engineering, Gifu University, Gifu, Japan*

[3]*Obu Unity Co. Ltd., Obu, Aichi, Japan*

Introduction

Sewage disposal is an essential environmental conservation technology in civil society but a municipal treatment process consuming a large amount of energy (Slim et al., 2008). In order to mitigate carbon dioxide emission and reduce energy consumption, excellent energy efficient process should be established including recycle of sewage sludge. Sludge as residue contains moisture as high as 80 percent level and is hard to dry. So heating value of sludge is too low to burn out it without using auxiliary fuel in incineration process.

In recent year, there are attempts drying, composting, carbonizing and/or gasifying sludge. A part of compost and carbonized char are utilized as soil conditioner or lagging of molten steel in steel industries while the most of dry sludge, compost, char and gasification gas are recycled to biofuel in coal firing or renewal power generation. In those processes, drying of sludge becomes an important key technology to improve total energy efficiency because it consumes the largest energy (Stasta, et al., 2006, Zhang et al., 2009).

Several fundamental studies relating to sludge drying have been done on moisture sorption equilibrium (Vaxelaire, 2001, Ruiz and Wisniewski, 2008), drying induced shrinkage and crack formation (Chen et al., 2004, Tao et al., 2005, Hsu et al., 2010), skin layer formation (Tao et al., 2006, Vaxelaire et al., 2002, Font et al., 2011), structure change (Tao et al., 2006). There are also researches on drying kinetics of sludge in different dryers like fluidized dryer (Shin et al., 2000), convective drying (Leonard et al., 2005) and solar and heat pump dryer (Slim et al., 2008). In recent year, R&D on drying of sludge by composting has been carried out (Zhao et al., 2010). This drying process utilizes exothermic heat generated during aerobic degradation of organic compounds for evaporation of high moisture as alternative energy to heating from the outside. It looks a promising self-energy process to remove water from sludge, but the drying rate is influenced by aerobic fermentation rate and air stream pattern in sludge layer. Reduction of drying and/or composting time and deodorization are important issues to treat promptly a large amount of sludge as well as to introduce it in urban area.

The present authors propose an innovative technology of sewage treatment system. The feature is summarized as: 1) drying and composting rate of sludge is enhanced by mixing with char, 2) The dried sludge is carbonized to produce char. 3) Yield char is used for deodorization and water disposal of BOD and COD reduction as well as mixing with sludge in composting/drying process. These processes can establish self-energy closed sewage disposal system to reduce energy consumption.

In the present work, drying and composting behaviors of sludge was studied to examine the effect of char addition. A fundamental drying experiment was carried out by hot air heating to confirm the enhancement effect of sludge drying due to char addition. The drying and composting rate of sludge and char mixture was determined by examination using a lab-scale of disposer (1 kg/day) and a pilot scale of down draft column whose volume was 50 m^3.

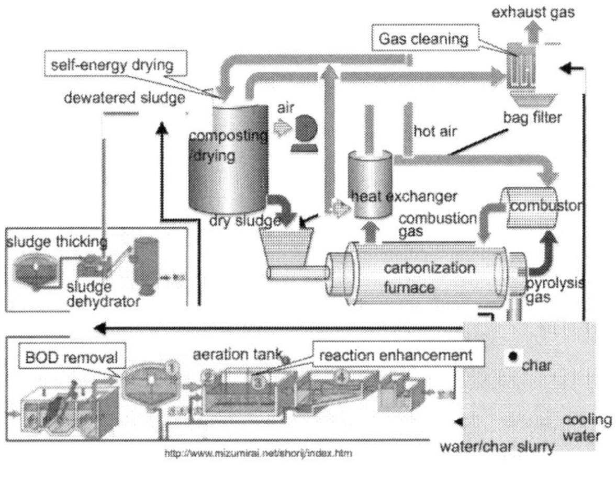

Fig. 1. Innovative self-closed sewage disposal and sludge carbonization system proposed

Enhanced Compost-Drying Process

Fig. 1 shows the outline of the enhanced sewage disposal and sludge carbonization system proposed in this study. Dewatered sludge separated from sewage is mixed with char powder and the mixture is subjected to composting in a down draft column. Then exothermic heat generated during composting induces simultaneously drying of sludge by air supplied to the column. The char particle may contribute to enhancement of composting and/or drying rate due to the effect of bacteria activation and water diffusion in an internal structure. Sequentially the dried sludge compost is decomposed to produce char in a kiln type of carbonization furnace by indirect heating using combustion of pyrolysis gas. The char is not only used in composting/drying processes of sludge, but also can be applied to agent for reduction of BOD and COD, enhancement of settlement of solid suspension in sewage treatment plant and gas cleaning by pre-coating on bag filter to remove ammonia and sulfur oxides causing offensive odor. Those techniques make it possible to establish self-energy closed treatment processes of sludge containing higher moisture content than 80% without heating from the outside and to reduce the total energy consumption in sewage disposal system.

Fundamental Drying Characteristic

Experiments

Fundamental drying behavior of sludge heated by hot air was examined preliminarily. A schematic drawing of a lab-scale of hot air drying examination setup is shown in Fig. 2. The setup consists of compressor, air reservoir, air heater and dryer. Stable hot air flow was maintained without fluctuation by using compressed air stored in reservoir from compressor. The flow rate of air was fixed at 0.35 m^3/min. The duct for hot air drying test was made of Styrofoam^{TM} board with 30 mm thick, and the size was 100x100 mm in inner cross section and 1200 mm in length. Wire mesh was placed to distribute hot air at the inlet of the duct. The sample holder was a shallow cylinder made of plastics whose diameter and depth were 30 mm and 5 mm. The side and bottom of the holder were insulated thermally by Styrofoam, and only the top surface was subjected by hot air heating and evaporation of moisture. Dewatered sewage sludge used as the sample was provided from a sewage disposal center. Char produced by carbonization of sludge was mixed at different contents with sludge to examine the effect of char on drying behavior. The holder packed with sludge or sludge/char mixture was jacked up at the test section 1000 mm far from the inlet of the duct, and then drying experiment was started. The weight of the sample and holder was measured by balance and the drying rate was determined by the time behavior of the weight.

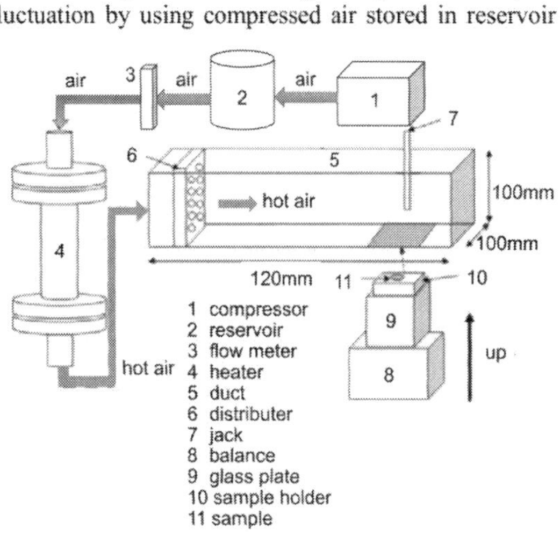

Fig. 2. A lab-scale of hot air drying setup

Drying characteristics

Time behavior of weight measured by balance was shown for different char contents in Fig. 3. The char content is defined by the ratio of char to initial wet sludge. The air temperature in this figure is 321 K. The curves look like that drying with higher char content completes in shorter time although the initial moisture content of each sample is different. In order to compare the drying behavior more in detail, drying characteristic curves were plotted on drying rate and moisture content. The moisture content is given by dry sludge basis here. The drying characteristic pattern showed no constant rate period in any char content because sludge forms immediately hard shell or skin on the surface as pointed out by the past references (Tao et al., 2006, Font et al., 2011) and it inhibits internal moisture diffusion. The curves reveal that higher char content results in greater drying rate through a whole drying period except lower moisture content of the sample with char content 0.05. This drying enhancement effect due to mixing char into sludge may advance moisture diffusion in the internal layer or shell layer formed by drying. The fast drying rate contributes to shortening of drying time. A similar drying characteristic was observed for drying experiment at another air temperature.

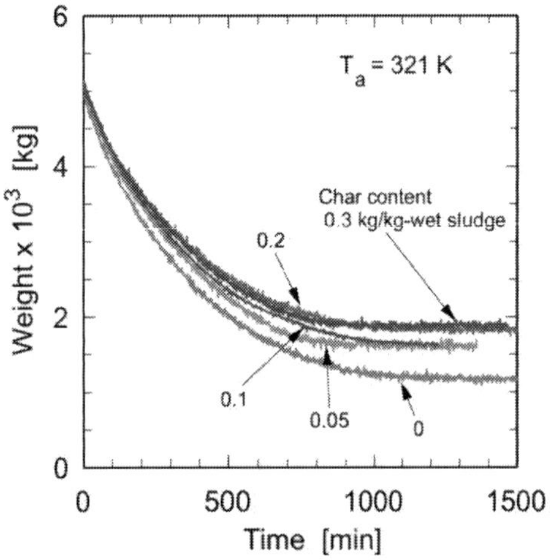

Fig. 3. Time behavior of sample with different char contents during hot air drying at 321 K

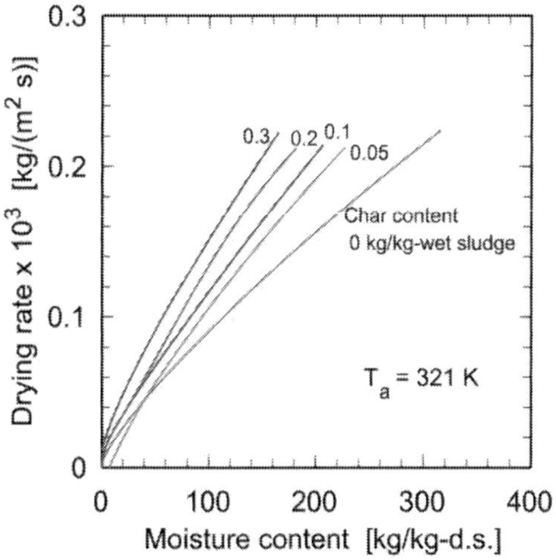

Fig. 4. Drying characteristic curves of samples for different char contents

Composting Characteristic

A lab-scale composting examination of sludge was carried out. The composting equipment used in this work is the one sold in market and the outline and picture are shown in Fig. 5. Bacteria seed for composting was preliminarily put into the vessel of the equipment. Composting rate depends on the kind of the bacteria seed, but the seed provided for composting of municipal wastes was used in this work as the most suitable bacteria was not known well and the main objective of this work is to study the possibility of drying only by exothermic heat in the composting step. Sludge or sludge/char mixture 1 kg was fed into the vessel in every day and the equivalent amount of compost to dry solid weight in the wet sample fed was taken out simultaneously. The moisture content of compost taken out was measured. These handling was repeated every day. The sample was stirred continuously by rotation blade during the examination.

Time behavior of moisture content during composting is compared between sludge and mixture of sludge and char in Fig. 6. The char content of the mixture was 0.11 kg-char/kg-wet sludge. The moisture content fell down gradually since beginning of experiment and achieved at steady state in 5 days for both the sample. It should be noted that the moisture content in the figure is given by the wet basis. The moisture content reduced to less than 30 % without any heating from outside. In this examination, little influence of char mixing was observed because the temperature and/or the composting rate in the vessel were too low to recognize the difference due to heat loss to the outside under cold atmospheric temperature in the winter season.

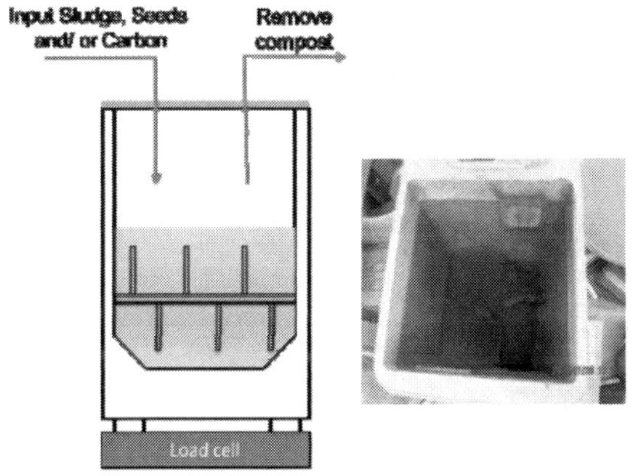

Fig. 5. Experimental setup of composting of sludge

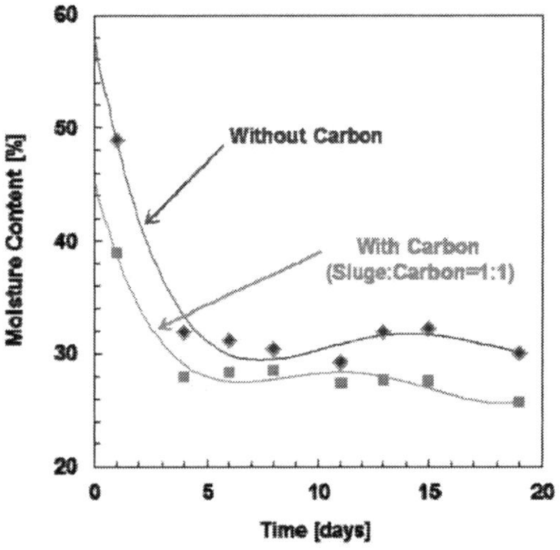

Fig. 6. Moisture content of sludge against time during composting examination

Conclusion

The innovative sewage disposal system was proposed to reduce total energy consumption by introducing self-energy closed processes for composting/drying and carbonization of sludge and char mixture. This study found the fundamental knowledge on an enhancement effect of char addition into sludge on drying and/or composting rate from hot air drying examination. In addition composting/drying could be performed with exothermic heat solely by aerobic decomposition and the rate was determined from the lab-scale and the pilot scale of examinations.

Acknowledgment

The present authors acknowledge New Energy and Industrial Technology Development Organization (NEDO) for financial supports to this study by "Research and Development Program for Innovative Energy Efficiency Technology (No. 10004805-0)".

Reference

1. Chen, J.B., X.F. Peng, T. Tao and D.J. Lee (2004), Thermal drying of wastewater sludge with crack formation, Water Sci. Technol., Vol. 50, pp. 177-182.
2. Hsu, J.P., T. Tao, A. Su, A.S. Mujumdar and D.J. Lee (2010), Model for sludge cake drying accounting for developing cracks, Drying Technology, Vol. 28, pp. 922–926.
3. Leonard, A., S. Blacher, P. Marchot, J.P. Pirard, M. Crine (2004), Measurements of shrinkage and cracks associated to convective drying of soft materials by X-ray microtomography, Drying Technology, Vol. 22, pp. 1695-1708.
4. Ruiz, T. and C. Wisniewski (2008), Correlation between dewatering and hydro-textural character-istics of sewage sludge during drying, Sep. Purif. Technol., Vol. 61, pp. 204-210.
5. Shin, Y.S., H.C. Kim and H.S. Chun (2000), Drying of water treatment process sludge in a fluidized dryer, Kor. J. Chem. Eng., Vol. 17, pp. 22-26
6. Slim, R., A. Zoughaib and D. Clodic (2008), Modeling of a solar and heat pump sludge drying system, Int. J. Refrig., Vol. 31, pp. 1156-1168.
7. Stasta, P., J. Boran, L. Bebar, P. Stehlik and J. Oral (2006), Thermal processing of sewage sludge, Appl. Therm. Eng., Vol. 26, pp. 1420-1426.
8. Tao, T., X.F. Peng and D.J. Lee (2005), Structure of crack in thermally dried sludge cake, Drying Technology, Vol. 23, pp. 1555-1568.
9. Tao, T., X.F. Peng and D.J. Lee (2006), Skin layer on thermally dried sludge cake, Drying Technology, Vol. 24, pp. 1047–1052.
10. Vaxelaire, J. (2001), Moisture sorption character-istics of waste activated sludge, J. Chem. Technol. Biotechnol., Vol. 76, pp. 377-382.
11. Vaxelaire, J. and J.R. Puiggali (2002), Analysis of the drying of residual sludge: from the experiment to the simulation of a belt dryer, Drying Technology, Vol. 20, pp. 989–1008.

12. Zhang, D.Q., P.J. He, L.Z. Yu and L. M. Shao (2009), Effect of Inoculation time on the bio-drying performance of combined hydrolytic-aerobic process, Bioresour. Technol., Vol.100, pp. 1087-1093.
13. Zhao, L., W.M. Gu, P.J. He and L.M. Shao (2010), Effect of air-flow rate and turning frequency on bio-drying of dewatered sludge, Waste Research, Vol. 44, pp. 6144-6152.

In -Situ Precipitation of $CaCO_3$ within Pulp Fibers Enhances Optical and Printing Properties of Paper

Pradeep Kumar, Yuvraj Singh Negi and Surendra Pal Singh

Department of Paper Technology, Indian Institute of Technology Roorkee, Saharanpur, India

Writing and printing papers are usually loaded with white pigment particles to improve many paper properties such as light scattering, smoothness, pore-size distribution, and ink-paper interactions. In the conventional method of loading, the filler particles are added directly into the papermaking stock before forming the paper sheet. In an alternative approach, the filler particles are placed within the lumen or the cell wall of the pulp fibers either by mechanical diffusion or by *in-situ* precipitation. Many researchers have found that the fiber-loading methods may offer several advantages over the conventional filler loading such as higher sheet strength, cleaner white water system, lower requirements of retention aids and improved formation of the sheet.

We have studied the effect of fiber-loading with $CaCO_3$ on paper sheets made from bagasse pulp. $CaCO_3$ particles were precipitated *in-situ* within the fiber cell wall and/or lumen by bubbling CO_2 gas through slurry of bagasse pulp fibers in a dilute aqueous suspension/solution of calcium hydroxide. Standard laboratory sheets were prepared from the filler-loaded pulp fibers. Strength and optical properties of the sheets were determined using standard test methods. The sheets were printed in the laboratory using an IGT printability tester. It was observed that the fiber loading by *in-situ* precipitation of $CaCO_3$ very significantly improved the light scattering coefficient of the bagasse pulp. As a consequence, the print through of the paper was decreased, which is particularly desirable in low grammage papers.

INTRODUCTION

Mineral fillers have been used in papermaking to reduce the cost and to improve brightness, opacity as well as the printability of the paper. In conventional method of filler loading, the filler particles are added directly into the stock. In finished paper, these filler particles occupy the inter-fiber voids in the sheet. Any attempt to increase filler content leads to reduced strength properties of the paper, cloudy paper machine back water, and dusting of paper during printing. Many of the drawbacks of conventional filler loading can overcome if the filler particles are placed within the fiber lumen or cell wall instead of the inter-fiber voids in paper sheet.

Methods for incorporating fillers inside fiber lumen or/and fiber cell wall have been studied extensively. Green et al. (1982) and Middleton and Scallan (1985) studied loading of fiber lumens with titanium dioxide particles by mechanical diffusion, in which the slurry consisting of softwood fibers and titanium dioxide was stirred over a period of time. Many other investigators followed similar approach using precipitated calcium carbonate (PCC) in place of titanium dioxide (Middleton et al. 2003; Miller and Paliwal 1985; Petlicki and van de Ven 1994).

Cellulosic fibers consist of a porous wall surrounding a central cavity (lumen). An estimated volume of pores in fibers is ≈ 1.5 cm^3/g fibers (Li et al 1995) and volume of the lumen is ≈ 0.2-0.4 cm^3/g fiber (Satyanarayana and Wypych 2007). Average size of the pores in the cell wall of unbleached softwood kraft pulp fiber is about 0.08 μm and that of the lumen is about 8 μm (Park et al. 2006). The various dimensions suggest that by mechanical dispersion the filler loading will be mainly into the fiber lumen and limited by the lumen volume. Therefore, the approach of lumen loading by mechanical diffusion is limited to rounded pigment particles having small enough particle size to pass through the pit apertures of the fibers to reach the lumen (Allan *et al.* 1992a). Alternatively, the filler particle could be incorporated within the cell wall voids (as high as 1.5 mL/g of pulp) by *in-situ* precipitation of insoluble inorganic materials. This concept of fiber loading or cell wall loading has been studied further by many researchers (Allan *et al.* 1992a, 1992b, 1998; Silenius 1996; Yoshida *et al.* 1987; Klungness et al. 1996, 2000; Subramanian et al. 2005; Chauhan et al. 2007; Kumar et al. 2009, 2011).

Although substantial disagreement exists in the findings of these researchers, the general view is that the paper made of pulp in which the filler has been incorporated within fibers exhibits greater tensile, burst, and tear strength than corresponding conventionally filled paper (Allan 1995; Miller and Paliwal 1985; Fujiwara 1993). Several other advantages of *in-situ* precipitation (Allan 1995) include easier white water management due to improved retention, reduction in the dosages of expensive polymeric retention aids, easier drying of paper due to reduced collapse of fiber voids, decreased abrasion damage to the forming fabric, less dusty paper, and smaller two-sidedness in paper because the filler is inside the fiber.

Klungness et al. (1996, 1999, 2000) found that the brightness and whiteness of pulp loaded by *in-situ* precipitation of calcium carbonate was less than that of the conventionally loaded pulp. Subramanian et al. (2005) from the study of blend of pulp and PCC composite fillers observed that adding PCC composite filler to the furnish enhanced light scattering properties of paper along with the tensile strength and Scott bond strength when compared with the conventional addition of PCC.

In the present work we have studied the effect of *in-situ* precipitation of $CaCO_3$ in bleached bagasse chemical pulp fiber on the optical and printing properties of the resulting paper sheets. It was observed that at equal ash levels, the bagasse pulp with calcium carbonate precipitated on fibers *in-situ* showed a significantly greater improvement in light scattering coefficient than precipitated calcium carbonate filler directly added to the pulp. The *in-situ* precipitation of filler also showed the desirable increase in print density and decrease in print through of printed sheets.

EXPERIMENTAL WORK

The pulps used in this study were never-dried bleached bagasse and hardwood pulps obtained from an integrated pulp and paper mill. The pulps were beaten in a PFI mill to 32 °SR. Calcium carbonate was precipitated *in-situ* within the pulp fibers by first saturating the pulp with calcium hydroxide solution and then carbonizing with pressurized carbon dioxide gas. 20 g pulp fibers with a calculated amount of a reagent grade calcium hydroxide powder were mixed with water in a beaker and stirred at 1200 rpm for 25 minute. The mass of calcium hydroxide was calculated to obtain a 15% $CaCO_3$ loading assuming its 100% conversion to $CaCO_3$, the resulted pulp. The amount of water added to the pulp was determined to obtain a desired consistency of 2% to 8% for the carbonizing reaction. After thorough mixing, the pulp slurry was transferred to a 2.5-L reaction vessel. The vessel was connected to a CO_2 gas cylinder, allowing flow of the gas into the vessel and obtaining a predetermined pressure (2-5 bar). After the reaction for a predetermined time, (3-12 minute), at the ambient temperature (≈ 30 °C), the vessel was depressurized. The pulp slurry was taken out from the vessel and washed thoroughly with running tap water over a 250-mesh screen until a clear filtrate was obtained to remove any $CaCO_3$ precipitated on the external surface of the fiber. The amount of precipitated $CaCO_3$ was determined by combustion of the pulp sample at 900 °C following the TAPPI T-413 standard method.

Standard hand sheets of 60 g/m^2 were prepared from the *in-situ* precipitated bagasse and hardwood pulps in a laboratory sheet former with a square cross section, 165mm x 165mm, according to the standard method SCAN C: 26. The hand sheets were air dried in contact with glaze plates. The hand sheets were internally sized by AKD to give $Cobb_{60}$ sizing test values of about 22 g/m^2. For comparison, standard handsheets were also prepared with direct addition of a commercial filler grade precipitated calcium carbonate (PCC-60) pigment and a retention aid (Percol47) obtained from a local chemicals supplier.

The brightness, opacity, specific scattering coefficient, and specific absorption coefficient of handsheets were determined from the measurements on Elerpho spectrophotometer (L&W Model 071). The other properties such as ash content and grammage of the handsheets were determined using appropriate SCAN methods. The physical properties of the pulps used in the study are given in Table 1.

Table 1. Physical Properties of the Pulps Used

Property*	Bagasse	Hardwood
Ash, %	0.61	0.66
Tensile index, kNm/kg	58.9	65.6
Burst index, kPa m^2/g	3.73	4.73
Tear index, mNm^2/g	3.96	5.34
Folding endurance	1.39	1.47
Brightness, %	73.5	73.0
Scattering coefficient, m^2/kg	21.2	36.7
Absorption coefficient, m^2/kg	0.52	0.73

The morphological characteristics of the fibers were determined using the L&W Fiber Analyser. The average fiber properties for bagasse and hardwood fibers are given in Table 2.

Table 2. Fiber Morphology

S. No.	Property	Bagasse	Hardwood
1.	Pulp freeness, °SR	32	32
2.	Mean fiber length (length weighted), mm	0.769	0.552
3.	Mean width (length weighted), μm	22.2	16.0
4.	Fiber Coarseness, μg/m	122.3	82.5
5.	Mean shape, %	88.3	92.6
6.	Number of vessels per 100,000 fibers	5245	264
7.	Number of vessels per gram, g^{-1}	452,705	35,628
8	Vessel area / fiber volume, μm^{-1}	0.000935	0.000182
	Mean vessel area, μm^2	155936	162724

Printing of handsheets

Handsheets of bagasse and hardwood pulps both loaded conventionally and by in-situ-precipitation were printed in an IGT printability tester (Model AIC_25). Test strips of the required length of 315 mm were prepared by butt-joining two handsheets with a thin adhesive tape at the backside. The paper strips were printed at varying amount of ink present on the printing disk while maintaining the other printing conditions constant as shown in Table 3.

A sufficient amount of ink was initially applied to the IGT inking unit and distributed uniformly on different rolls of the inking unit. During a test print, the printing disc was weighed, using an analytical balance with a least count of 0.1 mg, before and after the inking, and again after the printing. From these weights, it was possible to determine the amount of ink present on the printing disc before the printing and the amount of ink transferred to the paper during the printing.

The printing disc was cleaned after printing each strip using a solvent (cyclohexane). No fresh ink was applied to the inking unit after the initial application. This ensured that the amount of ink applied to the printing disk gradually decreased with printing of each strip. For each type of paper sheets, a series of strips was printed with varying amounts of ink on the printing disc. The initial amount of ink applied to the inking unit was adjusted such that the printing disk received anoint of ink equivalent to about 30 g/m^2 of the disc surface in the first inking. This amount decreased gradually with continued printing of paper strips. The prints were allowed to air-dry overnight. For calculation of print density and print through, various reflectance values on the test strips were measured using L&W Elerpho-071/070 brightness tester.

Table 3. Printing conditions in IGT printability tester

Atmospheric conditions during printing	(Temp=27±1°c; RH=65% ±5%)
Printing ink	Standard black offset ink supplied by the IGT. The ink was of the non-drying, oil-based type containing carbon black as pigment.
Printing speed	1 m/s
printing force	300 N
Printing disk	IGT Rubber covered disk, 50 mm
Backing material	None

RESULTS AND DISCUSSION

Effect of Filler on Optical Properties of Handsheets

Fig. 1 shows the effect of increasing amount of filler on the specific scattering coefficient, brightness, and opacity of the bagasse and hardwood handsheets. At zero filler level, the scattering coefficient of bagasse handsheets was much lower than that of hardwood handsheets. Direct filler loading resulted in marginal increase in the scattering coefficient for both bagasse and hardwood handsheets, while the fiber loading by *in-situ* precipitation resulted in a much greater increase, particularly for the bagasse handsheets. Both the bagasse and the hardwood pulps had nearly the same values of the scattering coefficient for ash content greater than about 12%.

The brightness and opacity of the handsheets made of bagasse fibers loaded by *in-situ* precipitation showed a very significant improvement because of the large increase in its scattering coefficient. To eliminate the effect of variation in the grammage of the handsheets, the values of opacity used in Fig. 1 were calculated for a grammage of 60 g/m^2 using the values of s and k, and the simplified Kubelka-Munk equations (Biermann 1996).

Offset printing behaviour of bagasse paper sheets loaded by in-situ-precipitation

Print density and print through were determined for the paper strips printed using different values of ink on the printing disk. Print density is the optical contrast between the printed and unprinted areas on the paper surface. Print through is an undesirable phenomenon in printing, in which the print is visible from the reverse side of the printed paper. The print density and the print through are usually determined from the measurement of reflectance values of printed and unprinted areas from the front and reverse sides of a print (SCAN P-36):

$$PD = \log (R_{\infty P}/R_P)$$

$$PT = 100 \times \log\left(\frac{R_{\infty B}}{R_{pB}}\right)$$

Where,

PD = print density

PT = print through

$R_{\infty P}$, $R_{\infty B}$ = reflectance of unprinted paper with a backing of opaque pad of the same paper on the printed side and on the reverse side respectively

R_P, R_{PB} = reflectance of the printed paper backed with an opaque pad of the unprinted paper on the printed side and on the reverse side respectively

The reflectance measurements were made on an L&W Elrepho-074 instrument corresponding to ISO Y-C/2°. An average of five measurements was recorded for each printed strip.

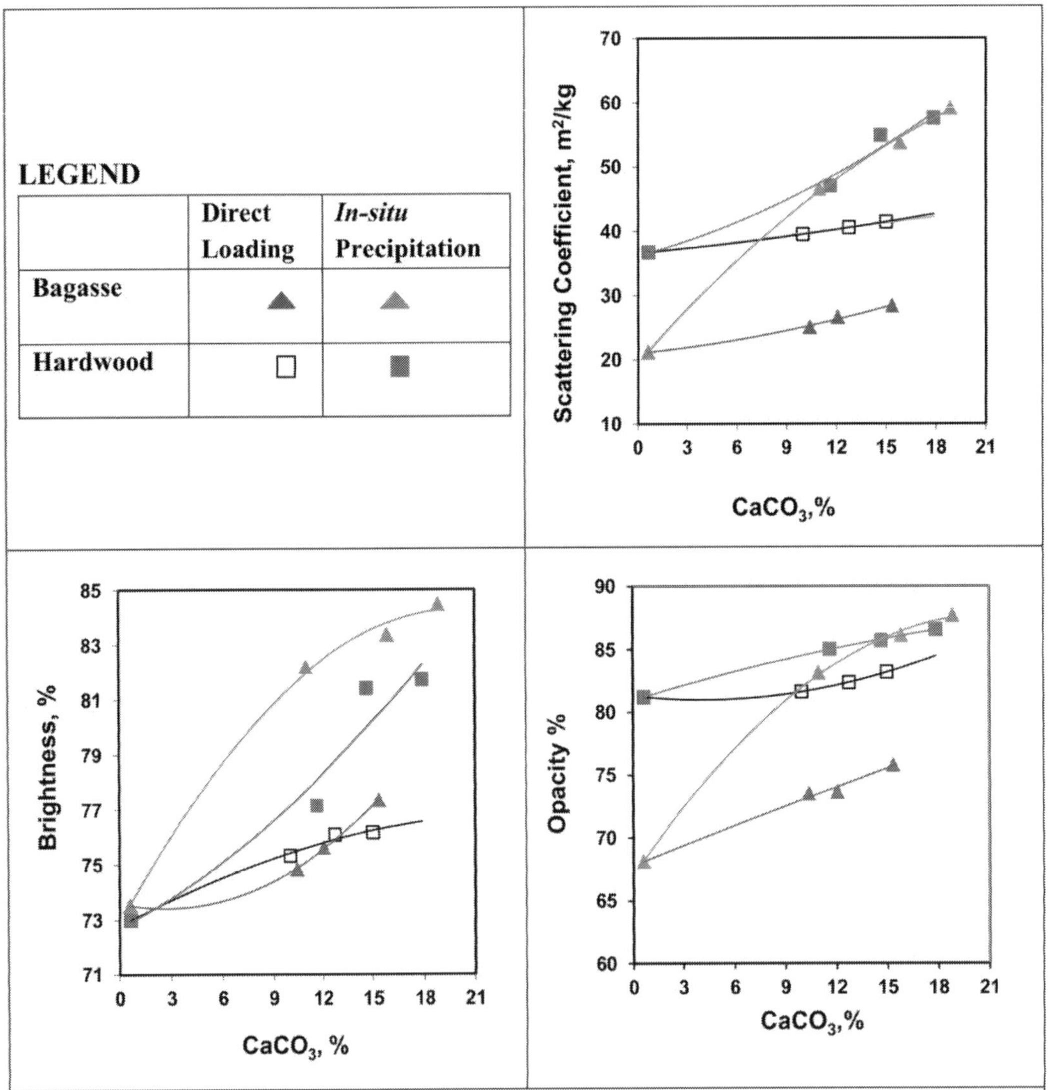

Fig. 1. Effect of filler on optical properties of the handsheets at different $CaCO_3$ levels. Opacity values were calculated for a constant grammage of 60 g/m^2 using simplified Kubelka-Munk equation.

Fig. 2 shows the variation of print density with varying values of amount of ink on the printing disk (x) and amount of ink on the paper (y) for both the bagasse and the hardwood handsheets. Here, x and y are expressed as g of ink per m^2 of disk or print area. An observation of the figures revealed the following points:

1. For any given ink transferred to the paper, the print density was greater for bagasse sheets than for the hardwood sheets.
2. For hardwood pulp, the print density at given amount of ink transferred to the paper was nearly the same for unfilled sheets as well as for loaded sheets, either by direct loading or by *in-situ* precipitation.
3. For bagasse pulp, the print density at given amount of ink transfer was different for unloaded and loaded sheets. The print density was lower of the loaded sheets than for the unloaded sheets. Further, the print density was more for the sheets loaded by *in-situ* precipitation than the sheets loaded directly.

Fig. 3 shows the variation of print through values for bagasse and hardwood sheets as function of x and y. For hardwood pulp, the print through decreased with loading of filler, but with little difference in the manner of filler loading, whether direct loading or loading by *in-situ* precipitation. On the other hand, for bagasse sheets, the effect of filler addition in reducing the print through was very significant, and more so for the loading by *in-situ* precipitation.

By increasing the amount of ink on the disk or the paper, the desirable improvement in print density is obtained along with the undesirable increase in the print density. Fig. 4 shows the print through as function of print density for bagasse and hardwood sheets. It was observed that for a given print density, the bagasse sheets loaded by *in-situ* precipitation had nearly the same values of print through as the hardwood sheets at equal filler level.

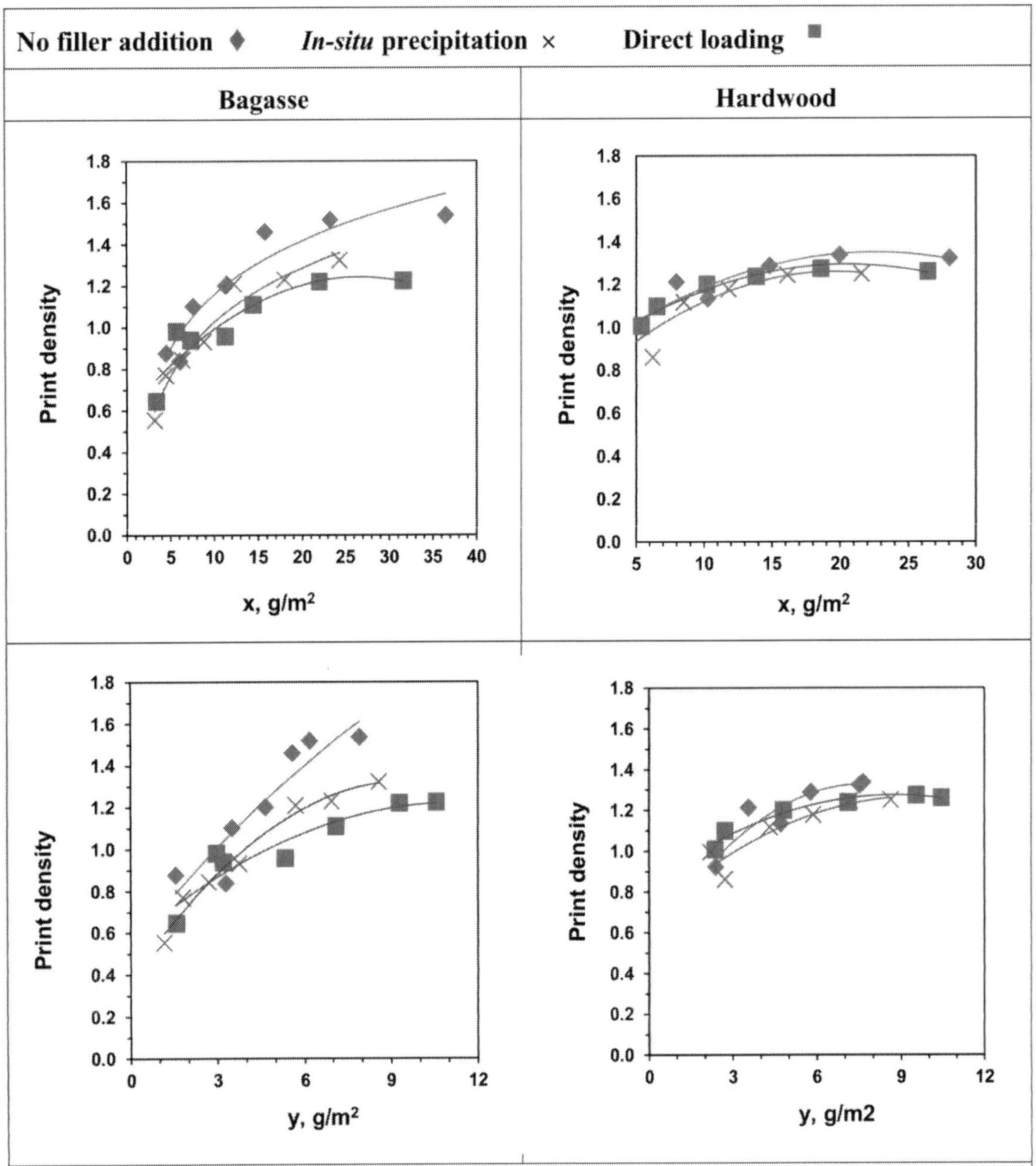

Fig. 2. Print density as a function of amount of ink on the printing disc and the amount of ink transferred to the paper.

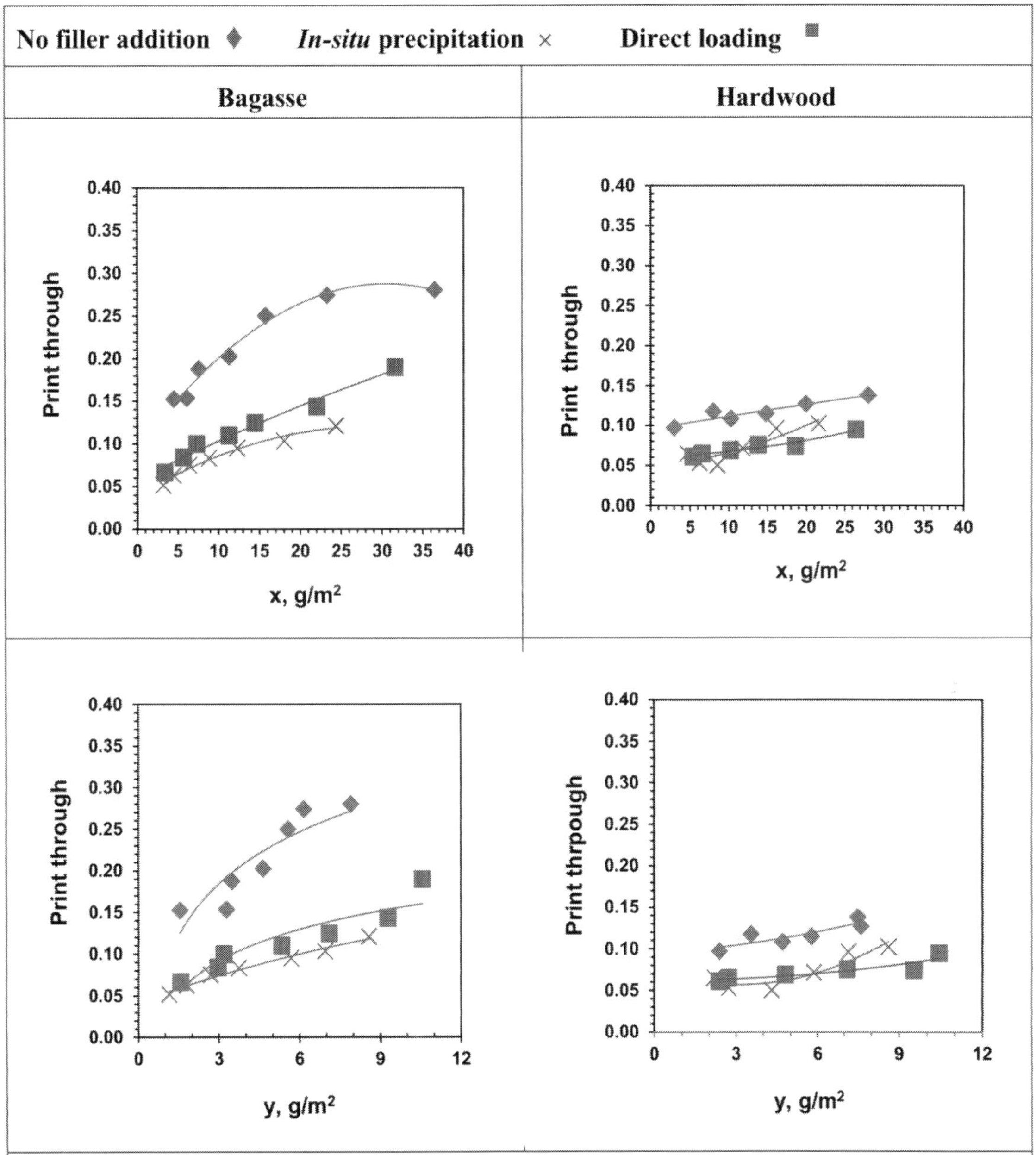

Fig. 3. Print through as a function of amount of ink on the printing disc and the amount of ink transferred to the paper.

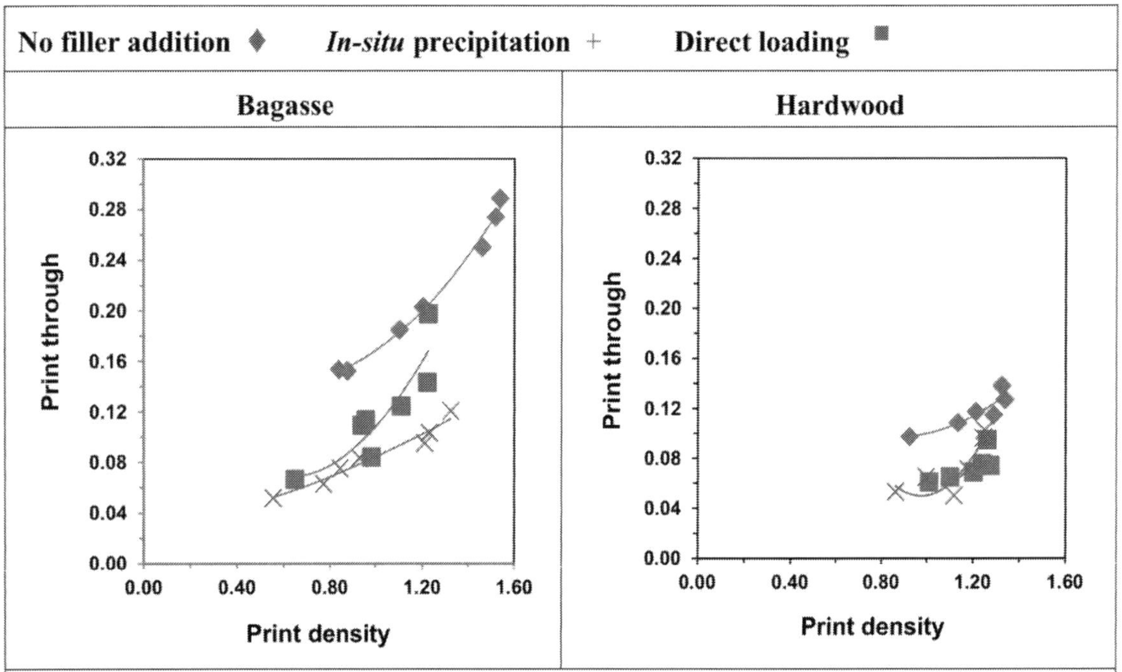

Fig. 4. Print through as a function of print density.

CONCLUSIONS

1. *In-situ* precipitation of calcium carbonate on bagasse fibers resulted in a very significant increase in specific scattering coefficient and consequently large improvements in opacity and brightness of the handsheets made of such pulp.
2. At the same level of filler loading, the scattering coefficient of *in-situ* precipitated pulp was much greater than that of the directly loaded pulp.
3. The improvement in optical properties of pulp by *in-situ* precipitation of calcium carbonate on pulp fibers was significantly higher for bagasse pulp than for the hardwood pulp.
4. *In-situ* precipitation of calcium carbonate caused a drop in strength properties of bagasse pulp. At equal ash level, the strength loss was more for loading by *in-situ* precipitation than by the direct filler loading. The strength loss of *in-situ* precipitated bagasse pulp could be recovered to a large extent by refining the in-situ precipitated bagasse pulp once again.
5. During offset printing, the ink transferred to paper sheet increased on loading of the paper with filler. However, the ink transfer was less for papers loaded by in-situ precipitation than for direct loaded papers. This difference in ink transfer behaviour could be due to difference in the positioning of the filler particles. In direct loaded sheets, the filler particles, present mainly in the inter-fiber voids, increase surface smoothness and fineness of the porous structure to facilitate easy ink transfer. On the other hand, in sheets loaded by *in-situ* precipitation, the filler particles present inside the fiber have little influence on these properties.

6. For *in-situ* precipitation of filler in bagasse fiber showed favourable effects on both the print density and the print through. For a given amount of ink on the printing disk or on the paper, the print density was more and the print through was less for *in-situ* precipitation of filler when compared with the direct filler loading

REFERENCES CITED

1. Allan, G. (1995). "Paper for the 21st Century," *Paper Age* (1), 10-15.
2. Allan, G., Carrol, J. P, Jimeney, G., and Negri, A. R. (1998). "Enhancement of the optical properties of a bagasse newsprint furish by fiber-wall-filler," *Cellulose Chem. Technol.*, 32, 339-347.
3. Allan, G., Carroll, J. P., Negri, A. R., Raghuraman, M., Ritzenthaler, P., and Yahiaoui, A. (1992a). "The microporosity of pulp: The precipitation of inorganic fillers within the micropores of the cell wall," *Tappi J.* 75(1), 175-178.
4. Allan, G., Negri, A. R., and Ritzenthder, P. (1992b). "The microporosity of pulp: The properties of paper made from pulp fiber internally filled with calcium carbonate," *Tappi J.* 75(3), 239-244.
5. Biermann, C. J. (1996). *Handbook of Pulping and Papermaking*, 2nd Ed., Academic Press.
6. Chauhan, V. S, Singh, S. P., and Bajpai, P. K. (2007) "Fiber loading of hardwood by *in-situ* precipitation of aluminosilicate," *BioResources*, 2(4), 560-571.
7. Fujiwara, K. and Morikawa, M. (1992). "Preparation of Magnetically Lumen-Loaded Pulp by In Situ Chemical Reaction", *Proc. Pan-Pacific Pulp Pap. Technol. Conf. (Tokyo), Sept. 8-10*,Part B, 131-134.
8. Green, H. V., Fox, T. J., and Scallan, A. M. (1982). "Lumen-loaded paper pulp," *Pulp & Paper Canada* 83(7), T203-T207.
9. Klungness, J. H., Ahmed, A., Ross-Sutherland, N., and AbuBakr, S. (2000). "Lightweight, high-opacity paper by fiber loading: filler comparison," *Nordic Pulp and Paper Research Journal,* 15(2), 345-350.
10. Klungness, J. H., Caulfield, D. F., Sachs, I. B., Sykes, M. S. and Tan, F. (1993). "Method for fiber loading a chemical compound," *U.S. Patent* 5, 223, 090.
11. Klungness, J. H., Pianta, F., Stroika, M. L., Sykes, M., Tan, F., and AbuBakr, S. (1999). "Lightweight high opacity paper: process cost and energy use reduction," AIChE *Symposium Series*, 95(322), 99-102.
12. Klungness, J. H., Syhes M.S., Tan,F.,Abubakr,S.,and EisenNassen,J.D.,(1996) "Effect of fiber loading on paper properties," *Tappi J.*79(3)297-301.
13. Kumar P, Gautam, S. K., Kumar, V., and Singh, S. P. (2009). "Enhancement of optical properties of bagasse pulp by *in-situ* filler precipitation," *BioResources* 4(4), 1635-1646.
14. Kumar, P., Negi, Y. S., and Singh, S. P. (2011). "Offset printing behavior of bagasse and hardwood paper sheets loaded by *in-situ*," *BioResources* 6(1), 207-218.
15. Middleton, S. R. and Scallan, A. M. (1985) "Lumen-loaded paper pulp: Mechanism of filler-to-fiber bonding," *Colloids and Surfaces* 16, 309-322.

16. Middleton, S. R., Desmeules, J., and Scallan, A. M. (2003). "Lumen loading with calcium carbonate fillers," *JPPS*, 29(7), 241-246.
17. Miller, M., and Paliwal, D. (1985). "The effects of lumen loading on strength and optical properties of paper," *J. Pulp Paper Sci.* 11(3), 84-88.
18. Petlicki, J., and van de Ven, T. G. M. (1994). "Kinetics of lumen loading," *J. Pulp Paper Sci.* 20(12), 375-382.
19. Silenius, P. (1996). "Preparation of filler containing papermaking materials by precipitating calcium carbonate *in-situ* in the presence of cellulosic materials," *Licentiate Thesis:* Lappeenranta University of Technology, Department of Chemical Technology, Lappeenranta, Finland.
20. Subramanian, R., Maloney, T., and Paulapuro, H. (2005). "Calcium carbonate composite fillers," *Tappi J.* 4(7), 23-27.
21. van de Ven, T. G. M. (1989). "Physicochemical and hydrodynamic aspects on fines and filler retention," In: Baker, C. F. and Punton, V. W. (eds.), *Fundamentals of Papermaking,* Transactions of the ninth fundamental research symposium held at Cambridge, Mechanical Engineering Publications Limited, London, Vol. 1, 471-494.
22. Yoshida, T., Kaneko, K., and Kazumori, K. (1987). "Manufacture of paper in neutral condition," *Japanese Patent* 62-162098.

(493a) A Preliminary Study On Uncertainties in CFD Simulations of a Pilot-Scale Circulating Fluidized Bed Riser

Tingwen Li, URS Corp. National Energy Technology Laboratory, Morgantown, WV, Aytekin Gel, Alpemi Consulting LLC, Phoenix, AZ, Mehrdad Shahnam, Computational Science Division, National Energy technology Laboratory, Morgantown, WV and Madhava Syamlal, Office of Research and Development, National Energy Technology Laboratory, U.S. Department of Energy, Morgantown, WV

In this work, we present a preliminary study on numerical approximation uncertainties in the CFD simulation of a pilot scale 12-inch circulating fluidized bed riser available at NETL. As the first step of a systematic Uncertainty Quantification (UQ) study, we compare the numerical results of CFD simulations with different numerical settings and configurations including grid and temporal resolution, averaging time for mean flow field, boundary conditions (inflow and wall boundary conditions) and operating conditions (gas flow rate and solids circulation rate) carried out by the open-source CFD code-MFIX. Differences between numerical results are analyzed with respect to axial pressure gradient, solids concentration and solids velocity to estimate the numerical approximation uncertainty in the predicted flow hydrodynamics causing by the aforementioned factors. In addition, the experimental data measured through various experimental techniques are used to compare against the CFD results for validation. Through this exercise the primary contributors to one of the most significant source of total uncertainty in numerical simulations, i.e., numerical approximation uncertainty will be identified and the efforts in the future will be directed to reduce these uncertainties.

(493b) Filtered Model for Energy Transport in Gas-Solid Flows

Kapil Agrawal, William Holloway and Sankaran Sundaresan, Chemical Engineering, Princeton University, Princeton, NJ

We employ a kinetic-theory[1] based two-fluid model to develop a filtered two-fluid model[2] for energy transport in gas-solid flows. The filtering procedure gives rise to terms describing the filtered interphase heat transfer and filtered thermal diffusion, which need to be constituted in order to close the filtered transport equations. In this work, the closure for these terms is accomplished by performing fine-grid simulations of the two-fluid model in a two-dimensional periodic domain.[3] Filtered thermal diffusion is investigated by imposing a lateral mean gradient in the temperature of each phase. Interphase energy transport is investigated by prescribing a heat source(sink) in the solids(gas) phase, such that the energy content of the mixture is preserved and a local temperature difference is established between the two phases. We find the filtered interphase heat transfer coefficient to be as much as two orders of magnitude smaller than the microscopic interphase heat transfer coefficient. The variation of the filtered transport coefficients with respect to filtered particle volume fraction, and scaling with respect to filter size, is discussed.

[1] Gidaspow, D. *Multiphase Flow and Fluidization*; Academic Press: San Diego, CA 1994.

[2]Agrawal, K., Loezos, P.N., Syamlal, M., Sundaresan, S. The role of meso-scale structures in rapid gas-solid flows. *J. Fluid Mech.* 2001, 445, 151-185.

[3] Igci, Y., Andrews, A.T. IV, Sundaresan, S., Pannala, S., O'Brien, T. Filtered Two-Fluid Models for Fluidized Gas-Particle Suspensions. *AIChE J.* 2008, 54, 1431-1448.

(493c) A Numerical Method for Large-Scale Dense Particulate Flows Based On an Improved MP-PIC Method and Extended Lattice-Boltzmann Scheme

Radompon Sungkorn, Chemical and Materials Engineering, University of Alberta, Edmonton, AB, Canada and Jos Derksen, Department of Chemical & Materials Engineering, University of Alberta, Edmonton, AB, Canada

Detailed simulation of large-scale, dense particulate flows is extremely demanding. Prohibitively high computational expenses arise from large numbers of particles and their interactions. In order to be able to gain insight into such flows, numerical methods have been proposed in the literature. A multiphase particle-in-cell (MP-PIC) method offers a satisfactory level of accuracy with reasonable computational expenses [1]. In the MP-PIC method, particles with identical properties (e.g., particle diameter, velocity) are represented by a parcel and tracked in a Lagrangian manner. At the particle volume fraction near close-packing, conditions are imposed to prevent the particle volume fraction beyond close-packing. Effects of particle collisions are modeled based on the local particle distribution and their properties. Recently, O'Rourke *et al.* [2] proposed a collision procedure based on a Bhatnagar, Gross, and Krook (BGK) approximation to the collision terms in a particle distribution function transport equation. A collision damping time which represents the rate toward the local average particle velocity was introduced.

We present a numerical method for large-scale dense particulate flows with applications to sedimenting suspensions. The locally averaged conservation equations for the fluid phase are discretized using an extended lattice-Boltzmann (LB) scheme. A MP-PIC method with an improved collision procedure is used to track parcels of particles. A new collision damping time is derived from the discrete kinetic equation for the particle distribution function. The LB scheme and the MP-PIC method are coupled resulting in the so-called four-way coupling. The computer code for the present method is implemented and parallelized using the message passing interface (MPI) library.

The present method is used to carry out simulations of dense sedimenting suspensions in a cubic periodic domain. In order to evaluate the accuracy of the present method, the predicted average and variance of particle settling velocity are compared to the experimental data from the literature [3]. In addition, the excellent speedup on a parallel computing platform of the present method demonstrates its potential for simulations of industrial-scale particulate flows.

References

[1] Snider D. M. An incompressible three-dimensional multiphase particle-in-cell model for dense particle flows. J. Comp. Phys. 2001;170:523-549.

[2] O'Rourke P. J., Snider D. M. An improved collision damping time for MP-PIC calculations of dense particle flows with applications to polydisperse sedimenting beds and colliding particle jets. Chem. Eng. Sci. 2010;65:6014-6028.

[3] Guazzelli E., Hinch J. Fluctuations and instability in sedimentation. Annu. Rev. Fluid. Mech. 2011;43:97-116.

(493d) Parametric Analysis of the Effects of Particle Properties On the Granular Flows of Cylindrical Particles

Yu Guo, Chemical Engineering, University of Florida, GAINESVILLE, FL, Carl R. Wassgren, Mechanical Engineering, Purdue University, West Lafayette, IN, William R. Ketterhagen, Pfizer Global Research and Development, Pfizer Inc., Groton, CT, Bruno C. Hancock, Pharmaceutical Development, Pfizer Worldwide Research and Development, Groton, CT, Benjamin D. James, Chemical Engineering, University of Florida, Gainesville, FL and Jennifer Sinclair Curtis, University of Florida, Gainesville, FL

In this work, the simple shear flows of cylindrical particles are simulated using the Discrete Element Method (DEM) to study the effects of particle properties (e.g., particle aspect ratio, coefficient of restitution, friction coefficient and Young's modulus) on the granular shear flow behavior and the constitutive relations of stresses. It is found that a decrease in the coefficient of restitution can cause a decrease in the stresses while an increase in the apparent friction coefficient, which is defined as the ratio of shear stress to normal stress. This is because the loss of kinetic energy increases after interparticle collisions with smaller coefficients of restitution.

The friction coefficient has a significant impact on the granular flow. At low solid volume fractions, smaller stresses are obtained for the particles with friction due to the energy dissipation through the sliding friction. However, the system-spanning force chains are suddenly formed as a critical solid volume fraction is reached for the frictional particles, leading to a sharp increase in the stresses. This critical solid volume fraction decreases as the particle aspect ratio increases and/or the friction coefficient increases, illustrating the system-spanning force chains and sharp stress increases occur at a lower solid volume fraction for the larger aspect ratio particles with larger friction coefficient. When the force chains are formed, the relative movement of particles becomes more difficult, so that some particles move and rotate together like a solid body. As a result, the vortex-like circulation of particles is obtained for the dense flows with the frictional particles. Due to this circulating motion of particles, the apparent friction coefficient is significantly reduced as the solid volume fraction increases. In contrast, the system-spanning force chains, sharp stress increase and vortex structure are not observed for the frictionless particle systems.

The Young's modulus of particles determines the contact stiffness when collisions occur. In the dilute granular flows for which the force chains are not formed yet, the stresses are independent of the Young's modulus. However, as the solid volume fraction increases and the force chains are formed, the contacts become frequent and long lasting, and thus the stresses strongly depend on the contact forces, which increase as the Young's modulus increases. In addition, the apparent friction coefficient for both dilute and dense systems is not dependent on the Young's modulus in the range of Young's moduli considered.

(493e) Coupled CFD and DEM Simulation of Fluid-Granular Systems

Dalibor Jajcevic[1], Charles Radeke[1], Georg Scharrer[1] and Johannes G. Khinast[2], (1)Research Center Pharmaceutical Engineering GmbH, Graz, Austria, (2)Institute for Process and Particle Engineering, Graz University of Technology, Graz, Austria

Coupled CFD and DEM Simulation of Fluid-Granular Systems

Dalibor Jajcevic[1], Charles Radeke[1], Georg Scharrer[1], Johannes G. Khinast[2]

1: Research Center Pharmaceutical Engineering, Graz, Austria

2: Graz University of Technology, Institute for Process and Particles Technology, Austria

Granular flows are extremely important for the pharmaceutical and chemical industry. The understanding of the impact of particle size and related effects on the mean, as well as on the fluctuating flow field, in granular flows is critical for design and optimization of powder processing operations. Computational Fluid Dynamics (CFD) has become essential to the design and development of many engineering devices. Although initially limited to specialized engineering fields, in the last decades it became a commonly-used tool for analyzing many complex technical problems that involves fluid flows, heat transfer, chemical reactions etc. In the last few years Discrete Element Model (DEM) simulations have been increasingly used to study and analyze flows of granular systems with the main emphasis on granular flows where the interaction between particles and a fluid phase may be neglected. Instead only particle-particle interactions are considered. In pharmaceutical industry, fine granular materials (usually smaller than 3 mm) are generally used, but large enough that the gas drag overcomes the gravity. The particles are moved and mixed leading to a non-uniform and unsteady flow situation, which is beneficial for processes such as coating, granulation, draying etc. Thus, a deep understanding of the fluid-particle interaction is important to overcome severe difficulties in the design and scale-up of the industrial devices.

Applying above mentioned simulation techniques a simulation of the fluid-granular systems in a multiphase setup is possible. In the modeling, the drag force in only accelerating force acting on a particle and thus plays an important role in the coupling between gas and solid phases, see Van der Hoef et al. [1] and Wei Du et al. [2]. In literature, several widely used models are available, such as Gidaspow et al. [3], Syamlal and O'Brien [4], Di Felice [5] etc. Authors, such as Wei Du et al. [2], reviewed the models and showed that the Gidaspow model gives the best agreement with the experimental observation both qualitatively and quantitatively. Gidaspow et al. [3] derivate the model combining Ergun [7] equation for dense regimes and a correlation proposed by Wen and Yu [8] for the more dilute regimes. In the last few years common used model is a model proposed by Beetstra et al. [6]. Using a similar approach as Hill et al. [9], Beetstra et al. [6] derived the model based on a wide range of data for Reynolds numbers up to 1000 showing proper limiting behaviour and is therefore believed to be more practical, see Deen et al. [10].

In the first steep of the presented work the model proposed by Gidaspow et al. [3] and the relative new model derivate by Beetstra et al. [6] are investigated. The models are implemented via user function in the commercial CFD code AVL-Fire in an Eulerian two-phase setup (a gas and a solid phase). The user function allows defining the momentum interfacial exchange between the phases, whereas the solid phase is treated as non-moved, therefore a coupling between CFD and DEM code in this steep was not required. Taking into account Ergun [7] equation the pressure drop is calculated and compared with results of the CFD simulation for the Reynolds numbers up to 2000, four particles diameters and three different void fractions. The results show that both models can successfully predict the pressure drop through a porous media composed of monodisperse particles. Nevertheless, the model introduced by Beetstra et al. [6] shows increasing pressure drop behaviour with a decreasing particle diameter, but still in an acceptable range and is further used in this work.

In the second step, a coupled simulation between in-house developed DEM code and the commercial CFD code AVL-Fire was realized. In order to be able to simulate real-life granular systems, a new technology was implemented and a high-performance DEM code was developed based on the newly available "Compute United Device Architecture" (CUDA) technology, see Radeke et al [11]. The DEM code can simulate millions of particles on readily available hardware. The simulations can be directly visualized by a GUI using OpenGL display features. Based on the computational efficiency of our method the development of large-scale simulations is highly convenient. The data exchange between the codes was realized applying AVL Code Coupling Interface (ACCI). ACCI is a software component to perform a co-simulation with an arbitrary number of instances of different simulation programs. The codes concurrently simulate the same time interval and pass information to each other continuously. The results obtained in the simulation were verified using experimental data of Link et al. [12]. The measurement is obtained in an experimental study of the flow in a pseudo-2D spout-fluid bed taking into account different operating conditions. A good agreement with the experimental data confirms the accurateness of used coupling methodology.

1. References

1. Van der Hoef M. A., Ye M., van Sint Annaland M., Andrews A. T., Sundaresan S., Kuipers J. A. M., "Multi-Scale Modeling of Gas-Fluidized Beds", In: Advances in Chemical Engineering, vol. 31, pp. 65

2. Wei Du, Xiaojun Bao, Jian Xu, Weisheng Wei „Computational fluid Dynamics (CFD) modeling of spouted bed: Assessment of drag coefficient correlations", Chemical Engineering Science 61, pp. 1401-1420, Elsevier 2005

3. Gidaspow D, "Multiphase Flow and Fluidization", Continuum and Kinetic Theory Description, 1994

4. Syamlal M. and O'Brien, T.J., "Simulation of granular layer inversion in liquid fluidized beds", International Journal of Multiphase Flow 14, pp. 473-481, 1988

5. Di Felice, R., "The voidage functions for fluid-particle interaction system", International Journal of Multiphase Flow 20, pp. 153-159, 1994

6. Beetstra R., Van der Hoef M. A., Kuipers J. A. M., "Numerical study of segregation using a new drag force correlation for polydisperse systems derived from Lattice-Boltzmann simulations", Chem. Eng. Sci. 62, pp. 246-255, 2007

7. Ergun S., "Fluid flow through packed columns", Chemical Engineering

Progress 48, pp. 89–94, 1952

8. Wen Y.C. and Yu Y.H., "Mechanics of Fluidization", Chemical Engineering

Progress Symposium Series 62, pp. 100–111, 1966

9. Hill R.J., Koch D.L., Ladd J.C., " Moderate-Reynolds-numbers flows in ordered and random arrays of spheres", Journal of Fluid Mechanics 448, pp. 243–278, 2001

10. Deen N.G., Van Sint Annaland M., Van der Hoef M. A., Kuipers J.A.M., "Reviewof discrete particle modeling of fluidized beds", Chemical Engineering Science 62, pp. 28-44, 2007

11. Radeke C, Glasser B., Khinast J., „Large-scale Mixer Simulations Using Massively Parallel GPU Architectures", Chemical Engineering Science, Volume 65, Issue 24, pp. 6435–6442, 2010

12. Link J., Zeilstra C., Deen N., Kuipers H., „Validation of a Discrete Particle Model in a 2D Spout-Fluid Bed Using Non-Intrusive Optical Measuring Techniques", Canadian Journal of Chemical Engineering 82, pp. 30-36, 2004

Numerical study on large scale discrete element modeling for fine particles in a fluidized bed

Mikio SAKAI, Naoto SEKIMURA, Tatsuya ITOI
Department of Nuclear Engineering and Management
School of Engineering
The University of Tokyo

Introduction

Fluidized beds are widely employed in industrial operations, ranging from the pharmaceutical and food industry, to processes such as catalytic cracking of petroleum, combustion and biomass gasification. The fluidized beds are typical gas-solid flow systems. The flow in the fluidized beds is too complicated to fully characterize experimentally. Hence, numerical simulation might contribute to understanding the complex phenomena. The gas-solid flows have often been simulated by combining the discrete element method (DEM) (Cundall and Strack [1]) with computational fluid dynamics (CFD) (Tsuji et al. [2]). The DEM is a Lagrangian approach where each individual particle is calculated based on Newton's second law of motion, and is used to compute the solid particle behavior, enabling the granular flow characteristics to be investigated at the particle level. Since the development of the DEM-CFD method, various gas-solid flows have been computed so far.

Although the numerical approach is promising for understanding granular flows, the DEM has a critical problem, namely, the number of calculated is substantially restricted because of the excessive calculation cost. The number of calculated particles that could be handled in recent studies was at most a few hundreds of thousands. This is because the DEM simulations using an excessive number of particles cannot be completed within a practical time period. On the other side, over billion particles were required in industries. Consequently, the existing DEM is difficult to be applied to industrial systems.

A new large-scale DEM simulation model, referred to as the coarse grain model (Sakai and Koshizuka [3]; Sakai et al. [4,5]) was proposed to solve this issue. A coarse grain particle represents a group of original particles. Therefore, a large scale simulation can be performed by using a smaller number of calculated particles than is physically present. In the present study, we address the coarse grain model to simulate fine particles involving the van der Waals force. Verification of the coarse grain model was made by comparing the simulation with the results obtained from the original particle system.

The coarse grain model

Solid phase

At first, we briefly address the coarse grain model for non-cohesive particles which was previously developed to model contact, drag and gravitational forces. The details of the modeling are well addressed in the literature [4]. There are l^3 original particles in the coarse grain particle whose size is l times larger than the original particle. The translational motion of the coarse grain particle is assumed to be the same as that of the group of original particles. Therefore, the velocity and displacement of the coarse grain particle is assumed to be the average of those of the original particles. As far as the rotational motion is concerned, the original particles existing in the coarse grain particle are assumed to rotate around each center of mass with equal angular velocity. The contact force acting on the coarse grain particle was estimated under the assumption that the kinetic energy of the coarse grain particle agrees with that of the original particles. In addition, when a binary collision of the coarse grain particles occurs, the binary collisions due to all the original particles (i.e. l^3 binary collisions) are assumed to occur simultaneously. The contact force acting on the coarse grain particle is evaluated using a spring, a dash-pot and a friction slider, as for the existing DEM. The drag force and external force (for example gravitational force) are modeled by balancing the coarse grain particle with the group of the original particles. Spherical particles are used by the same way as many existing DEM-CFD simulations. Consequently, the following relationship is obtained between the coarse grain particles and the original particles:

$$\begin{aligned} m_{CGM}\dot{\boldsymbol{v}}_{CGM} &= \boldsymbol{F}_{f_{CGM}} - V_{CGM}\nabla p + \sum \boldsymbol{F}_{C_{CGM}} + \boldsymbol{F}_{g_{CGM}} \\ &= l^3\overline{\boldsymbol{F}}_{f_O} - l^3 V_O \nabla p + l^3 \sum \overline{\boldsymbol{F}}_{C_O} + l^3 \overline{\boldsymbol{F}}_{g_O} \end{aligned} \quad (1)$$

where m, $\boldsymbol{v}$, l, $\boldsymbol{F}_f$, V, p, $\boldsymbol{F}_C$ and $\boldsymbol{F}_g$ indicate solid mass, solid velocity, coarse graining ratio, drag force, particle volume, pressure, contact force and gravitational force, respectively. The subscripts of CGM and O refer to the coarse grain model and original particle.

The normal and tangential components of the contact force were expressed as:

$$\boldsymbol{F}_{C_{nCGM}} = l^3\left(-k\overline{\boldsymbol{\delta}}_{n_O} - \eta\overline{\boldsymbol{v}}_{n_O}\right) \quad (2)$$

$$\boldsymbol{F}_{C_{tCGM}} = \begin{cases} l^3\left(-k\overline{\boldsymbol{\delta}}_{t_O} - \eta\overline{\boldsymbol{v}}_{t_O}\right) & \left(\left|\boldsymbol{F}_{C_{tCGM}}\right| < \mu\left|\boldsymbol{F}_{C_{nCGM}}\right|\right) \\ -\mu\left|l^3\overline{\boldsymbol{F}}_{C_{nO}}\right|\overline{\boldsymbol{v}}_{t_O}/\left|\overline{\boldsymbol{v}}_{t_O}\right| & \left(\left|\boldsymbol{F}_{C_{tCGM}}\right| \geq \mu\left|\boldsymbol{F}_{C_{nCGM}}\right|\right) \end{cases} \quad (3)$$

where k, δ and η are the stiffness, the overlap and the damping coefficient, respectively. The translational motion of the coarse grain particle is assumed to agree with that of the original particles. Hence, the CGM subscript can be replaced by O in this equation.

The drag force $\boldsymbol{F}_f$ acting on the coarse grain particle is written as:

$$\boldsymbol{F}_{f_{CGM}} = \frac{\beta}{1-\varepsilon}\left(\boldsymbol{u}_f - \boldsymbol{v}_{CGM}\right)V_{CGM} = l^3\frac{\beta}{1-\varepsilon}\left(\boldsymbol{u}_f - \overline{\boldsymbol{v}}_O\right)V_O \quad (4)$$

where ε, $\boldsymbol{u}_f$ and β are the volume of the particle, the void fraction, the fluid velocity, and an interphase momentum transfer coefficient.

The following relationship is obtained for the rotational motion between the original and coarse grain particles:

$$\dot{\omega}_{CGM} = \frac{1}{l}\overline{\dot{\omega}}_{O} \tag{5}$$

Eq. (5) ensures that the rotational energy is in agreement between the coarse grain particle and the group of original particles.

The van der Waals force is modeled based on the assumption that the potential energy of the coarse grain particle is the same as that of the original particles [5]. When coarse grain particles i and j interact, binary interaction of all the original particles in the coarse grain particles is assumed to have occurred. Hence, the relationship between the potential energy of the coarse grain particle and the original ones is given by:

$$\int \boldsymbol{F}_{ipCGM}\left(h_{CGM}\right) dh_{CGM} = l^3 \int \overline{\boldsymbol{F}}_{ipO}\left(h_O\right) dh_O \tag{6}$$

where h indicates the intersurface distance. The force is integrated between the interface distance of the solids. The original particles are assumed to be located at even intervals according to the location of the coarse grain particles. When the coarse grain particles become closer in the horizontal direction, h_O is assumed to be correspondingly shorter. The relationship between the intersurface distance of the coarse grain particles and the original particles is expressed as:

$$h_O = \frac{h_{CGM}}{l} \tag{7}$$

From Eqs. (6) and (7), in general, a distant force acting on the coarse grain particle can be expressed as:

$$\boldsymbol{F}_{ipCGM}\left(h_{CGM}\right) = l^2 \overline{\boldsymbol{F}}_{ipO}\left(h_O\right) = l^2 \overline{\boldsymbol{F}}_{ipO}\left(\frac{h_{CGM}}{l}\right) \tag{8}$$

Consequently, the van der Waals force Fvdw acting on the coarse grain particle is given as:

$$\boldsymbol{F}_{vdwCGM} = l^2 \overline{\boldsymbol{F}}_{vdwO} = l^3 \frac{H_A d_{sCGM}}{6h_{CGM}^2}\boldsymbol{n} = l^2 \frac{H_A d_{sO}}{6h_O^2}\boldsymbol{n} \tag{9}$$

where $\boldsymbol{n}$ is the normal vector.

Eventually, the equation of the coarse grain model, which takes into account the drag, contact, gravitational and van der Waals forces is given as:

$$\begin{aligned} m_{CGM}\dot{\boldsymbol{v}}_{CGM} &= \boldsymbol{F}_{fCGM} - V_{CGM}\nabla p + \sum \boldsymbol{F}_{CCGM} + \boldsymbol{F}_{gCGM} + \boldsymbol{F}_{vwCGM} \\ &= l^3 \overline{\boldsymbol{F}}_{fO} - l^3 V_O \nabla p + l^3 \sum \overline{\boldsymbol{F}}_{CO} + l^3 \overline{\boldsymbol{F}}_{gO} + l^2 \overline{\boldsymbol{F}}_{vwO} \end{aligned} \tag{10}$$

This model can be applied in both 2D and 3D simulations. When this model is used in a 2D simulation, we assume that movement of the solid particles is restricted to the x and y directions. Even in a 2D simulation, the particle shape is modeled to be spherical, that is, a three dimensional object. Accordingly, Eq. (10) is consistent even in 2D simulations.

Gas phase

According to the volume average technique, the governing equations for the gas phase are the fluid continuity and Navier-Stokes equations for an incompressible fluid, given by:

$$\frac{\partial \varepsilon}{\partial t} + \nabla \cdot (\varepsilon \boldsymbol{u}_f) = 0 \tag{11}$$

$$\frac{\partial(\varepsilon \rho_f \boldsymbol{u}_f)}{\partial t} + \nabla \cdot (\varepsilon \rho_f \boldsymbol{u}_f \boldsymbol{u}_f) = -\varepsilon \nabla p + \boldsymbol{f} + \nabla \cdot (\varepsilon \boldsymbol{\tau}_f) + \varepsilon \rho_f \boldsymbol{g} \tag{12}$$

where ρ_f, $\boldsymbol{f}$, $\boldsymbol{\tau}_f$ and $\boldsymbol{g}$ are the fluid density, drag force, the molecular viscous stress tensor expressed by Newton's law of viscosity and the acceleration due to gravity, respectively. $\boldsymbol{f}$ is defined according to Newton's third law of motion:

$$\boldsymbol{f} = \frac{\sum_{i=1}^{N_{grid}} \boldsymbol{F}_{f_{CGM}}}{V_{grid}} \tag{13}$$

where V_{grid} and N_{grid} are the volume of CFD grid and the number of solid particles located in that region. $\boldsymbol{F}_f$ is given by Eq. (4).

A proper drag model for the description of β is vital in DEM-CFD simulations. A combination of the Ergun [6] and Wen-Yu [7] correlations is often used in the literature for evaluating drag forces in fluidized beds. Therefore, this combination is employed in this study. The void fraction of 0.8 is adopted as the boundary between these two regimes. β is given by:

$$\beta = \begin{cases} 150 \dfrac{(1-\varepsilon)^2}{\varepsilon} \dfrac{\mu_f}{d_s^2} + 1.75(1-\varepsilon) \dfrac{\rho_f}{d_s} |\boldsymbol{u}_f - \boldsymbol{v}| & (\varepsilon \le 0.8) \\ \dfrac{3}{4} C_d \dfrac{\varepsilon(1-\varepsilon)}{d_s} \rho_f |\boldsymbol{u}_f - \boldsymbol{v}| \varepsilon^{-2.65} & (\varepsilon > 0.8) \end{cases} \tag{14}$$

where μ_f, ds and C_d are the viscosity, particle diameter and drag coefficient for an isolated particle, respectively. C_d depends on the particle Reynolds number (Re_s). C_d is given by:

$$C_d = \begin{cases} \dfrac{24}{Re_s}\left(1 + 0.15 Re_s^{0.687}\right) & (Re_s \le 1000) \\ 0.44 & (Re_s > 1000) \end{cases} \tag{15}$$

where Re_s is written as:

$$Re_s = \frac{|\boldsymbol{u}_f - \boldsymbol{v}| \varepsilon \rho_f d_s}{\mu_f} \tag{16}$$

Algorithm

The solid phase was first solved to yield the particle positions and velocities. Next, the volume fraction of the gas phase in each computational grid was estimated based on the particle positions. Then, the gas-phase hydrodynamic equations were solved to give the gas velocity and pressure fields by using the known values of particle velocity and the void fraction at the present time. This procedure was iterated until the calculation was finished.

Table 1 Calculation Conditions

		Case 1	Case 2	Case 3	Case 4
Solid phase					
Coarse graining ratio	-	1.0	2.0	3.0	1.0
Original particle size	μm	200	200	200	600
Calculated particle size	μm	200	400	600	600
Number of particles	-	90000	22500	10000	10000
Hamaker number	J	1.0×10^{-20}			
Time step	sec	2.0×10^{-6}			
Gas phase					
Number of grids	-	10×120			
Grid size	mm	3.0×3.0			
Time step	sec	1.0×10^{-5}			

Numerical simulation

In the current study, 2D DEM-CFD simulations are performed to show that cohesive particles can be simulated by the coarse grain model [5]. The coarse grain model is shown to reproduce the results obtained by simulations of the original particles.

Calculation conditions

The domain was rectangular with size 30 mm×360 mm. The spherical particles were initially packed randomly. The cell size was chosen to be large relative to the particle diameter. The number of grids was 10×120. The gas was injected from the bottom side by changing the superficial velocity.

The particle density was 800 kg/m^3. The value of the spring constant was set to be softer than that estimated for the actual material. Coefficients of restitution and friction were 0.9 and 0.3, respectively. The gas density and viscosity were 1.0 kg/m^3 and 1.8×10^{-5} Pa sec. The same values of the physical properties were used in all the simulations.

Four kinds of simulations were performed in each case. Table 1 shows the calculation conditions. We simulate cohesive particle behavior by setting the Hamaker number 1.0×10^{-20} J. In each case, the coarse graining ratio was set to be 2.0 or 3.0. The simulation results obtained with/without the coarse grain model using the same calculated particle were also compared, *i.e.*, Case 3 and Case 4 in Table 1. In this study, the coarse grain model is shown to simulate the original particles accurately, despite using a smaller number of large-size modeled particles. It is also illustrated that the original particle behavior cannot be simulated by the use of large-sized particles without the coarse grain model.

Results and Discussion

Figure 1 shows typical snapshots of the simulation results, taken at 5, 10, 15 and 20 sec. The bed height became lower as the superficial velocity decreased. The bubbles became larger as they moved up the bed, and were also larger when the superficial velocity was higher. A few particles stuck to the walls. The particle behavior obtained using the coarse grain model was in quantitatively agreement with the results obtained from the original particle system. On the other hand, the solid particles barely moved in Case 4, implying that simply using large particles cannot reproduce the behavior of small particles. Figure 2 illustrates the relationship between pressure drop and superficial velocity. Very similar results were obtained for Cases 1 to 3, and the minimum fluidization velocity was estimated to be 0.022 m/s. The original particle behavior could be simulated by the coarse grain model because the drag, contact, and van der Waals forces were modeled suitably. The coarse grain model was shown to effectively model the cohesive particle system. Hence, adequacy of the coarse grain model was proven by the comparison in the 2D systems.

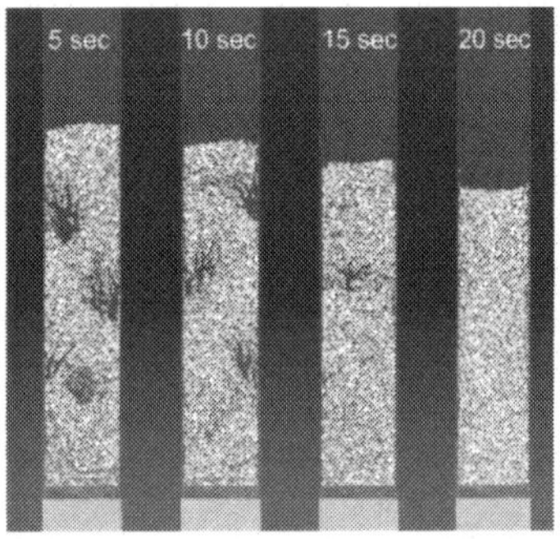

(a) Case 1: Original system

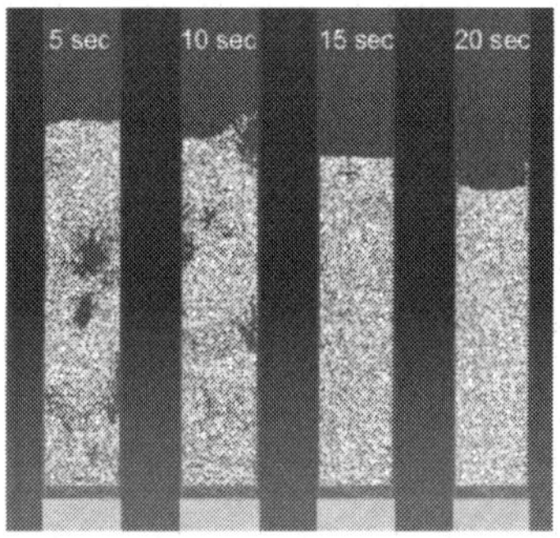

(b) Case 2: $l = 2.0$

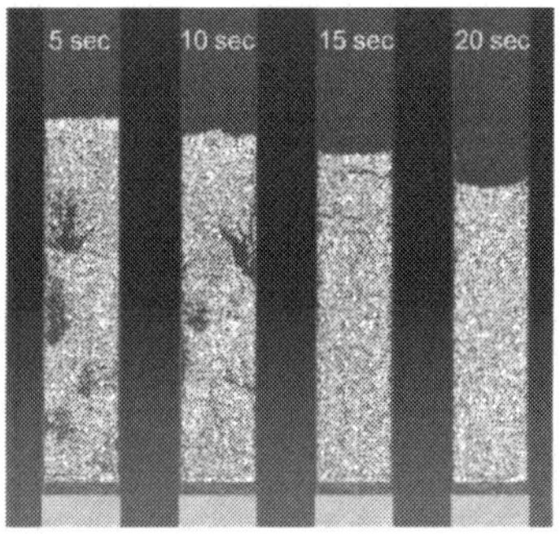

(c) Case 3: $l = 3.0$

Fig. 1 Typical Snapshots of the Simulation Results

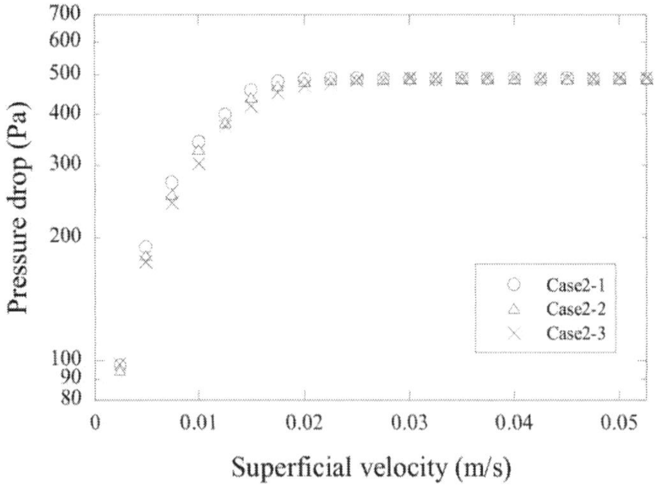

Fig. 2 Bed Pressure Drop versus Superficial Velocity

Conclusion

We show the coarse grain model to simulate fine particles involving the van der Waals force. Verification of the modified coarse grain model was made by comparing the simulation with simulation results obtained from the original particle system.

This model can contribute to the simulation of Geldart A or C particles in a large-scale powder system. This approach can be extended to simulate particles which are cohesive due to other forces, *e.g.,* liquid bridge, electrostatic forces, etc.

Acknowledgement

This study was financially supported by a Grant (22760579) from the Ministry of Education, Culture, Sports, Science and Technology (MEXT), Japan.

References

1. Cundall, P.A., Strack, O.D.L., Geotechnique 29 (1979) 47-65.
2. Tsuji, Y., Kawaguchi, T., Tanaka, T., Powder Technol. 77 (1993) 79-87.
3. Sakai, M., Koshizuka, S., Chem. Eng. Sci. 64 (2009) 533-539.
4. Sakai, M., Yamada, Y., Shigeto, Y., Shibata, K., Kawasaki, V.M., Koshizuka, S., Int. J. Num. Meth. Fluids 64 (2010) 1319-1335.
5. Sakai, M., Takahashi, H., Pain, C.C., Latham, J-P., Xiang, J., Adv. Powder Technol. 23 (2012) 673–681.
6. S. Ergun, Chem. Eng. Progr. 48 (1952) 89-94.
7. C.Y. Wen, Y.H. Yu, Chem. Eng. Progr. Symposium Series 62 (1966) 100-111.

(493g) A Physics-Based Cellular Automata Approach for Modeling Industrial Granular Flow Problems

Martin C. Marinack Jr. and C. Fred Higgs III, Mechanical Engineering, Carnegie Mellon University, Pittsburgh, PA

Granular flows are prevalent in nature and industrial sectors where they present complex problems due to their inherently nonlinear and multiphase behavior. In solids processing industries such as pharmaceuticals, food processing, and fertilizer production, granular flow phenomena (e.g. segregation, jamming, etc.) can adversely affect product effectiveness, yield, and cost. Thus, the accurate prediction of particle flow inside of solids processing equipment, such as hoppers, silos, mixers, and rotating drums, becomes important in being able to design against these effects. The discrete element method (DEM) provides for a rigorous physical treatment of particle interactions, and as such has been shown to be the "gold standard" for simulating many granular systems. One possible supplement to a DEM-based design or engineering analysis framework is cellular automata (CA). CA provides a platform for obtaining ultra-fast first-order predictions of the behavior of granular flow systems. Several prior physics-based CA models have been developed by the authors to predict granular flow inside of annular Couette shear cells. However, in this work, the CA modeling framework is extended to more industrially relevant solids processing problems and geometries. Results from CA models are compared against data from the literature and comments are offered about the merits of leveraging the capabilities of CA for solving problems in industrially-relevant solid processing applications.

(517a) Dissipative Particle Dynamics Simulations of Nanopowder Dispersion for Applications in Solid Propellants

Naida Lacevic[1], David Reese[2], Lori Groven[2], Darren Naud[3] and Steven F. Son[2], (1)NextGen Aeronautics, Inc, Torrance, CA, (2)School of Mechanical Engineering, School of Aeronautics & Astronautics, Purdue University, West Lafayette, IN, (3)DMD Systems, Los Alamos, NM

Metal nanoparticle addition into solid propellants has potential to significantly improve performance, but poor control over the final microstructure of the propellant has so far greatly limited practical applications. Here we present a computational and experimental study of nanoscale mixing processes to understand the processing-dispersion relationship of nanoscale metal fuel in composite propellants. We use multiscale simulations informed from experimental mixing conditions and compare final microstructure with microstructures obtained from experiments via SEM. We find that increasing nanoparticle volume fraction increases tendency for agglomeration. We also describe the effect on this tendency produced by altering the interaction between nanoparticles at constant agitation amplitude. The dissipative particle dynamics model capture the basic mechanisms retarding dispersion at higher mass fraction of nanoparticle filler, and indicate the degree of surface treatment of nanoparticles needed in order to achieve dispersion. These studies provide guidance to further improve incorporation of nanoparticle into propellants formulation and have direct applicability to more complex propellant systems. The methodology presented here could be applied to any mixing of nanopowder into highly viscous matrices.

Authors acknowledge support from Armament Research, Development and Engineering Center, Picatinny Arsenal. Work is performed under DOD SBIR program.

Numerical Modeling and Simulation Analysis of Casting of Energetic Material in a Projectile Controlled by the ACH Solidification and Mechanical Vibration Technologies

Ruslan Mudryy[1] and Laurentiu Nastac[2]

1. U.S. ARMY, RDECOM-ARDEC, Armaments Engineering and Technology Center, Picatinny, NJ 07806, USA, Phone: 973-724-5473
email: ruslan.s.mudryy@us.army.mil

2. THE UNIVERSITY OF ALABAMA, Department of Metallurgical and Materials Engineering, P. O. Box 870202, Tuscaloosa, Al, 35487, USA, Phone: 205-348-4844

Abstract

This paper investigates the solidification shrinkage problem in casting of energetic material in a projectile. Special challenges that arise in casting highly viscous energetic materials in projectiles of complex geometries are addressed. An active cooling and heating (ACH) control solidification technology and mechanical vibration technology are applied to achieve unidirectional solidification during casting and to reduce shrinkage porosities. Comprehensive numerical model is developed to simulate the heat and mass transfer processes during melt casting of energetic materials, as well as the resulting thermal stresses induced.

The design parameters of the proposed technologies are developed via numerical modeling. In particular for the first technology, the media heating and cooling temperatures as well as the number of cooling/heating sections and their time sequences are optimized based on the mold diameter, thermo-physical properties of the energetic material, and mold thickness and type. For the second technology, a resonant frequency of mechanical vibration was determined that helps to remove air entrapped during pouring process. It is demonstrated that the proposed technologies can successfully be applied to significantly minimize the solidification related defects (including macro-shrinkage and shrinkage porosities) in cast energetic materials.

Introduction

Casting processes are widely employed in the manufacture of products with intricate shapes and, in particular, in applications where the material of the final product is sensitive to machining. It is especially useful in processing of energetic materials. The cooling conditions applied in the casting process can affect the quality of the final cast in terms of void formation, residual stress distributions, and mold separation. Substantial shrinkage is also observed due to the density change upon solidification [1-3]. Residual stresses are known to be closely related to the formation of cold cracks and hot tears during casting. The formation of a gap between the mold and the cast material is of critical importance due to its deleterious effect on heat removal and on crack formation. In the casting of energetic materials all these defects can significantly impair the detonation velocity, Gurney energy, and insensitive munitions characteristics of the formulation, and lead to catastrophic accidents in explosives handling [4-5]. Imposition of carefully controlled cooling condition is thus critical in optimizing the cast quality that could help avoid such destructive effects.

In the current work, a comprehensive numerical model was developed using ANSYS'S FLUENT and ANSYS MECHANICAL [6] software to accurately simulate the transport phenomena as well as induced thermal stresses encountered in the casting process of recently developed explosive, consisting RDX-binder mixtures. An enthalpy method, successfully exploited by many authors [7-8], was used to simulate the solid/liquid phase change process. Instead of working entirely in terms of the temperature of explosive material, an enthalpy function is defined which represents the total heat content per unit mass of the material. The advantage of such a reformation of the problem is that the necessity to track the position of the solid/liquid interface is eliminated. Shrinkage effects and the resulting velocities induced in the melt have been largely neglected in the literature due to the difficulties involved in multiphase pressure-velocity coupling, and the interaction between free surface dynamics and solidification volume change [9]. In order to track the shrinkage shape that is critical in simulation of explosive casting process, NOVAFLOW&SOLID software [10] was used. Effective shrinkage was calculated at each time step and the volume of the solidified material was then subtracted from the liquid phase in the control volumes that contain the interface.

Accurate materials characterization is also important for modeling solidification process of explosive material. A successful simulation tool relies upon a complete material property database, which is not available in the literature for novel energetic materials. Experimental measurements of physical properties such as thermal conductivity, specific heat, thermal expansion coefficient, and liquid viscosity were conducted using Diamond TMA, Hot Disk Instrument, and Brookfield viscometer. The stress-strain relationship was measured using simple compression technique developed in-house [11].

We propose new technologies (ACH and MV) that can be used for melt cast process of energetic materials in order to improve product quality. Optimal cooling parameters predicted by numerical simulations can be easier controlled using the ACH technology. Maintaining higher temperature along a riser and gradually decreasing temperature at the bottom part will keep solid front flat propagating upward. This helps to reduce the excessive thermal stress formed due to large temperature gradient and provides control of shrinkage shape. The developed numerical model was also used to predict the frequency of vibrations for MV technology, which will help to remove air bubbles entrapped in the explosive material and to decrease the separation gap between the mold and the casting. New design of riser was proposed after calculation of thermal modulus distribution inside the projectile.

Numerical Model

The studied energetic material IMX-101 is assumed to be isotropic, incompressible and Newtonian fluid with a very high viscosity due to the high volume fraction of solid particles. The constituents of IMX-101 are DNAN (2, 4-dinitroanisole), NQ (Nitroguanidine) and NTO (3-Nitro-1, 2, 4-triazol-5-one). Of these three constituents only DNAN undergoes solid/liquid phase change while NQ and NTO remain in solid, crystalline form during the entire process. The solidifying melt was modeled as a single material with temperature-dependent density, thermal expansion coefficient and Young's modulus, with all other properties remaining the same in both solid and liquid phases. The variation of density with temperature was described by Boussineq approximation. The numerical model was based on solving the system of governing equations for conservation of mass, momentum, and energy:

$$\frac{\partial \rho}{\partial t} + \nabla \cdot (\rho \vec{u}) = 0 \quad (1)$$

$$\frac{\partial \rho \vec{u}}{\partial t} + \nabla \cdot (\rho \vec{u} \vec{u}) = -\nabla p + \nabla \cdot \left[\mu \left(\nabla \vec{u} + \nabla \vec{u}^T\right)\right] + \rho_\infty \vec{g} \beta_l (T - T_\infty) + \frac{(1 - f_l)}{(f_l^3 + \varepsilon)} A_{mush} \vec{u} \quad (2)$$

$$\frac{\partial \rho h}{\partial t} + \nabla \cdot (\rho \vec{u} h) = \nabla \cdot (k \nabla T) \quad (3)$$

where A_{mush} is the mushy zone constant, f_l the liquid fraction, ε =0.001 is used to prevent division by zero, and $h = c_p T + f_l \Delta H$, the specific enthalpy of the melt. The enthalpy method was employed to model the solidifying melt, while the SIMPLEC algorithm was used for pressure-velocity coupling. Second-order discretization schemes were used for momentum and energy equations and the pressure staggered option (PRESTO!) was adopted for discretizing the pressure term. In the absence of detailed visco-elastic-plastic material properties, the material was treated as an isotropic thermo-elastic material:

$$\frac{\partial}{\partial t}\left(\rho \frac{\partial \vec{w}}{\partial t}\right) = \nabla \cdot \left[\mu \left(\nabla \vec{w} + \nabla \vec{w}^T\right)\right] + \nabla \left(\lambda \nabla \cdot \vec{w} + (3\lambda + 2\mu)\beta_s T\right) - \left\{\nabla \cdot \left[\frac{\sigma_{ij}^d \sigma_{kl}^d}{\bar{\sigma}^2}\left(\frac{9G^2}{H' + 3G}\right)\nabla \vec{w}\right]\right\} + \vec{b} \quad (4)$$

where $\lambda = \frac{\upsilon E}{(1+\upsilon)(1-2\upsilon)}$, μ are Lame's coefficients, H' is the plastic modulus, G is the shear modulus and b is the body force.

In order to analyze the shrinkage and void formation caused by the density change during solidification, the algorithm employed must be capable of tracking a moving free surface. The volume of fluid (VOF) method is employed in this work since it can handle free surface movement and has been previously applied to study solidification shrinkage. For the casting problem considered, only two phases, i.e., IMX-101 and air is present in the system. When the density change upon solidification is taken into consideration, the governing equations for conservation of mass, momentum, and energy will have the additional source terms due to the density difference between the solid and liquid phases of IMX-101.The resulting system of algebraic equations is solved sequentially in each time of iteration, using NOVAFLOW&SOLID commercial software.

Experimental Results

Figure 1 presents the mold filling and solidification defects in a 120 mm mortar. As it can be seen from Figure 1, the air bubble (entrapped during the mold filling) and the separation and shrinkage formed during solidification are severe. These defects are detrimental to the quality of the 120 mm mortar. Optimization of the process design as well as the design of improved rigging systems is required to minimize/eliminate these defects. Some of the key design aspects are presented in the "Numerical Simulation Results" section below.

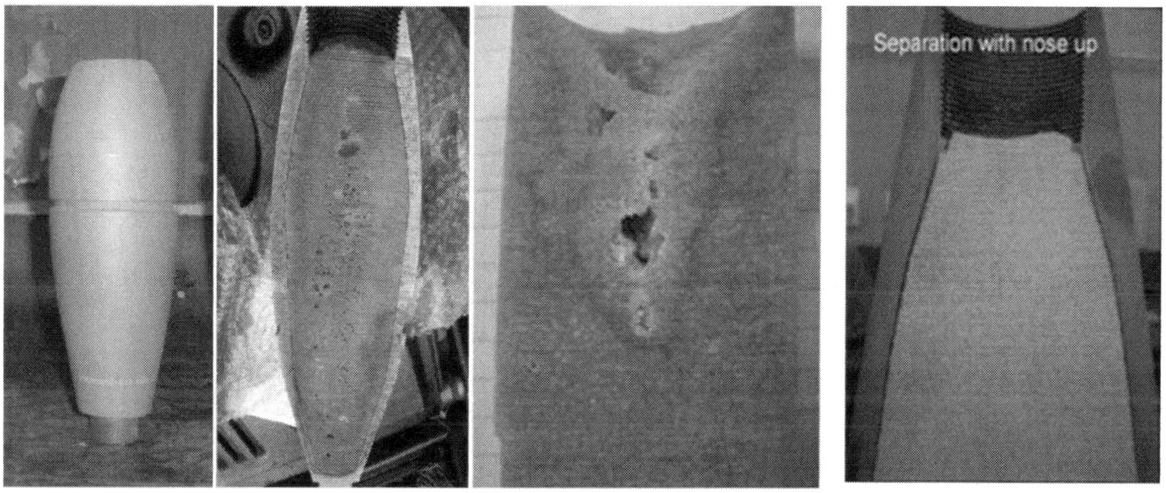

(a) (b) (c) (d)

Figure 1. Casting position (a) and pouring and solidification defects (air bubble and shrinkage cavities ((b) and (c)) and separation (gap) (d)) in the 120 mm mortar.

Numerical Simulation Results

Mold filling and Solidification Simulation Results for the current Process Design. NOVAFLOW&SOLID [11] and ANSYS'S FLUENT [6] were used in this study to perform all casting simulations.

The mold geometry of the mortar is shown in Figure 2a. Figure 2b illustrates the distribution of air bubbles entrapped in the mortar during filling as well as the shrinkage porosities (e.g., micro-shrinkage) and macro-shrinkage distribution at the end of solidification in the original design (process and rigging system).

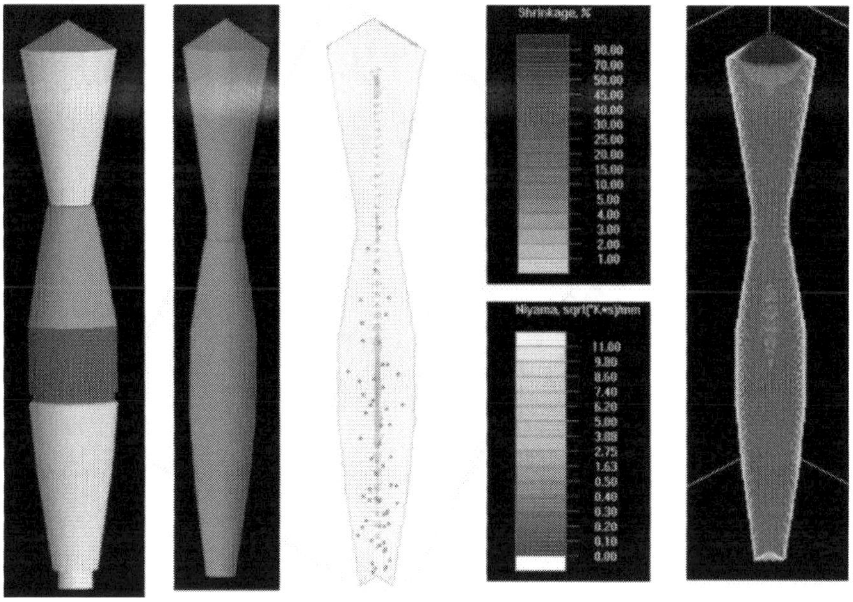

(a) (b)

Figure 2. Geometry (a) and predicted shrinkage porosities and macro-shrinkage (b) in an 120 mm mortar.

Figure 3 show the predicted liquid fraction, velocity and air bubble distributions during casting. The predictions of air bubbles shown in Figure 2b and 3c are in reasonable agreement with the experimental results in Figure 1. Predicted temperature profile, solidification time and thermal modulus profiles at the end of solidification for a 120 mm mortar are presented in Figure 4. Figures 2b and 4 show that two liquid pools would form during solidification and thus the solidification shrinkage cannot be fed efficiently by the top riser (funnel), which is then conducive to the formation of the shrinkage porosities and macro-shrinkage defects in the main body of the mortar. These predictions demonstrate the mold filling solidification defects presented in the 120 mm mortar (see figure 1) that are due to the (1) mortar geometry (e.g., necking is occurring between the top riser (funnel) and the main body of the mortar) and (2) process related deficiencies (e.g., top pouring, un-optimized mold heating and water cooling profile). Both of these problems cannot be address entirely by the current process design. Similar changes in the process are provided in the following sub-section.

Figure 5 show the liquid fraction and temperature profile in the original setup using a preheat riser and water cooling. The simulation was performed using ANSYS'S FLUENT. Special boundary conditions as a function of time were developed using user defined functions (UDFs). The simulation results shown in Figure 5 confirms the solidification defects predicted by NOVAFLOW&SOLID casting software.

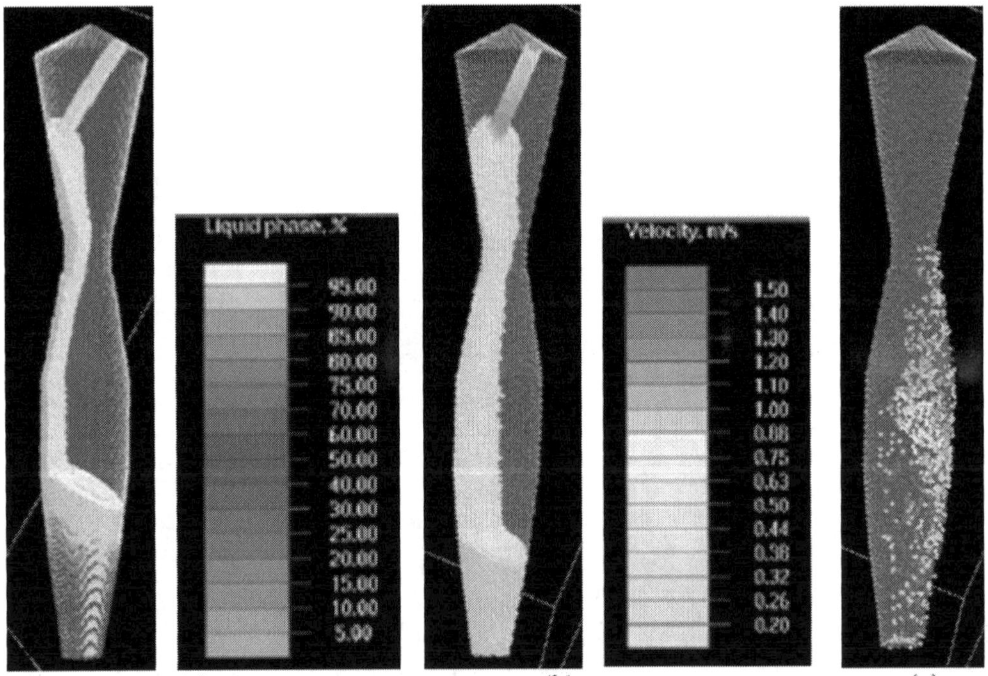

(a) (b) (c)

Figure 3. Predicted liquid fraction profile, velocity magnitude and air bubble distribution (color shown different air bubble sizes) profiles during mold filling of a 120 mm mortar.

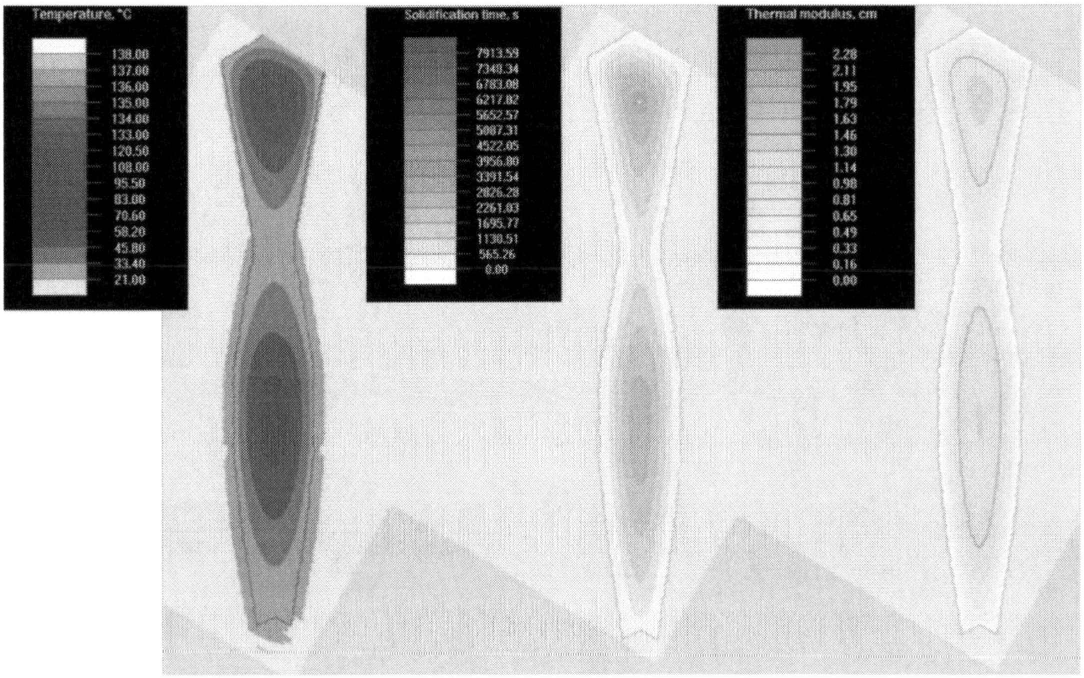

(a) (b) (c)

Figure 4. Predicted Temperature profile, solidification time and thermal modulus profiles for a 120 mm mortar.

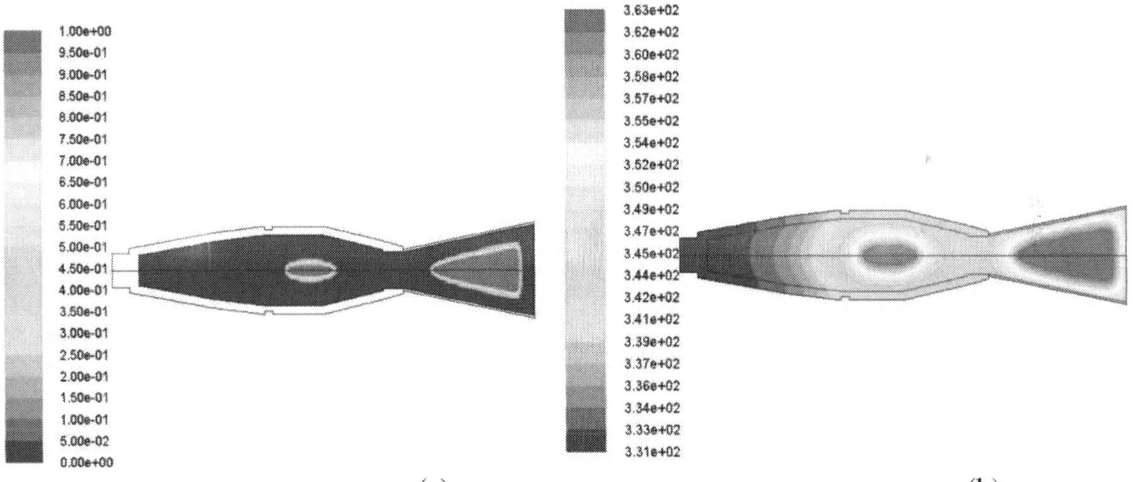

(a) (b)

Figure 5. Liquid fraction and temperature profile at time t = 13796s: original setup (preheat riser and water cooling).

Proposed Design Improvements. The new design consists of several improvements including the geometry modification of the top riser, mold preheating and application of mechanical vibration as well as electrical tape heating in the neck area between the top riser and the mortar (see Figures 6-9). As shown in Figures 7-9, the active heating and the modified riser can be applied to improved feeding and therefore eliminate the macro-shrinkage in the neck area.

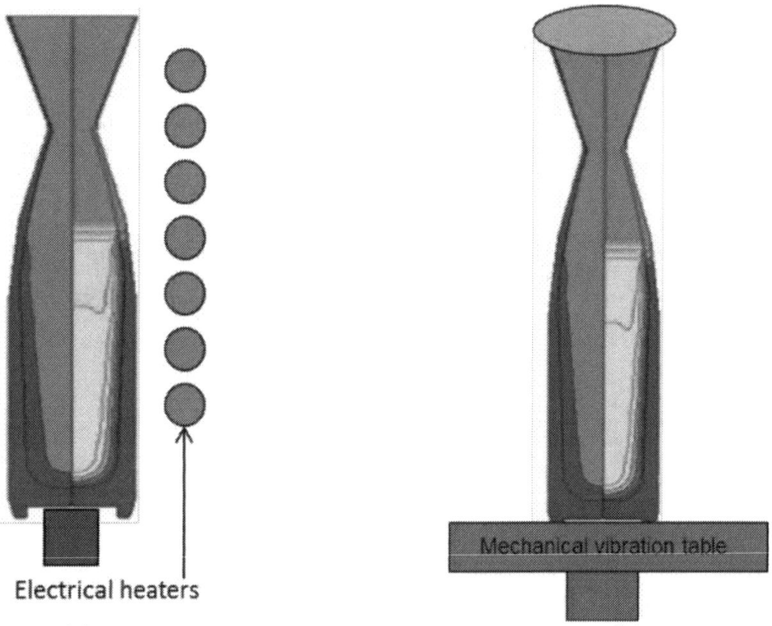

(a) (b)

Figure 6. Active cooling and heating (a) and mechanical vibration (b).

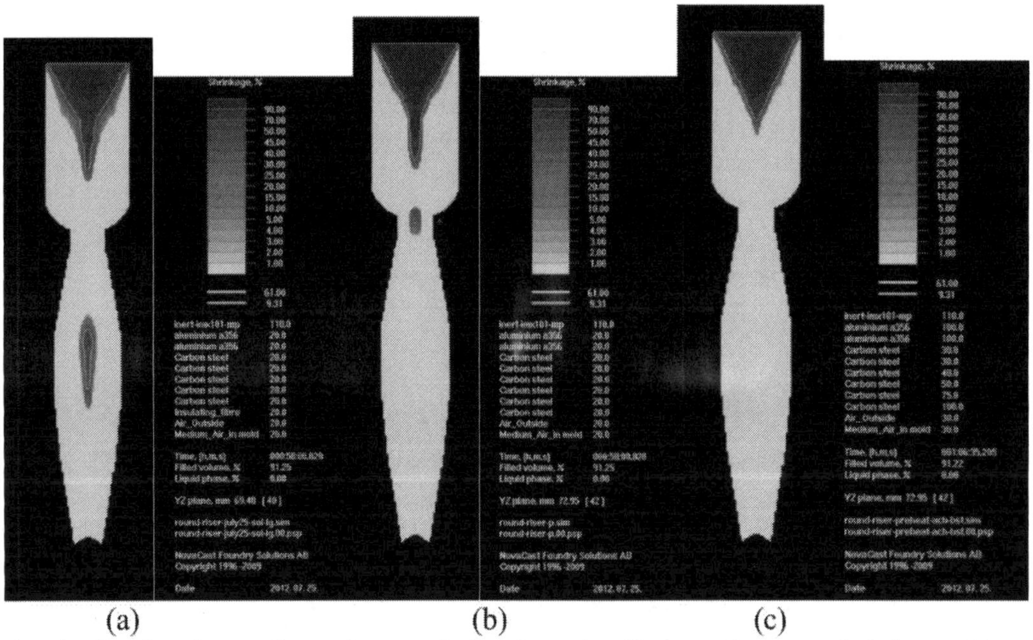

(a) (b) (c)

Figure 7. Active heating and top riser optimization: Prediction of macro-shrinkage: (a) no mold preheating and no-tape heating; (b) 0.25KW tape heating and no-mold preheating and (c) 0.12 KW tape heating and mold preheating

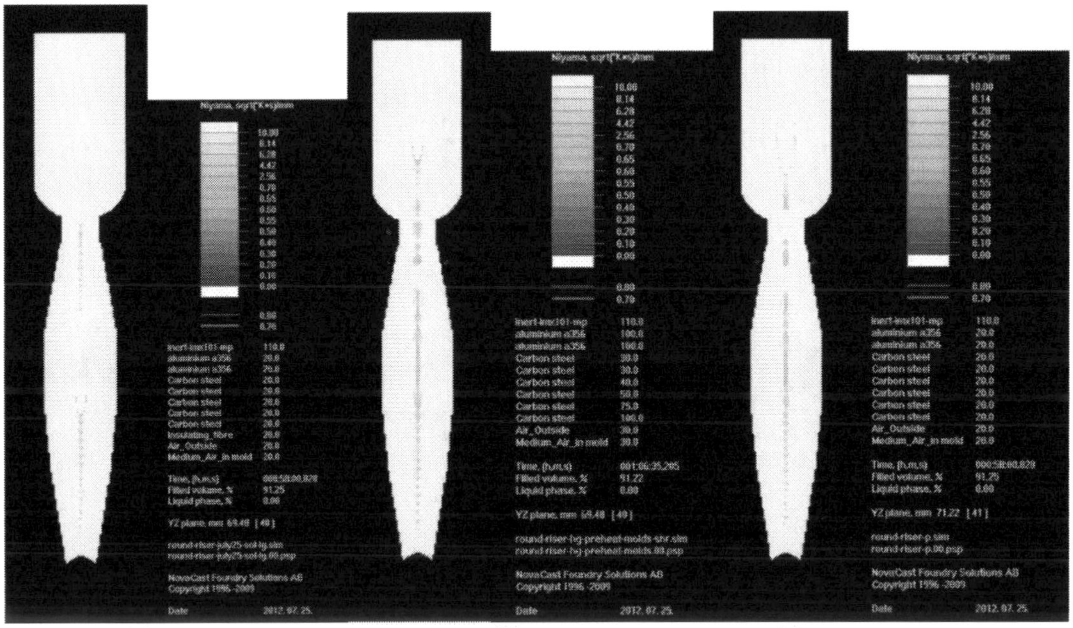

(a) (b) (c)

Figure 8. Active heating and top (round) riser optimization: Prediction of micro-shrinkage: (a) no mold preheating and no-tape heating; (b) 0.25KW tape heating and no-mold preheating and (c) 0.12 KW tape heating and mold preheating.

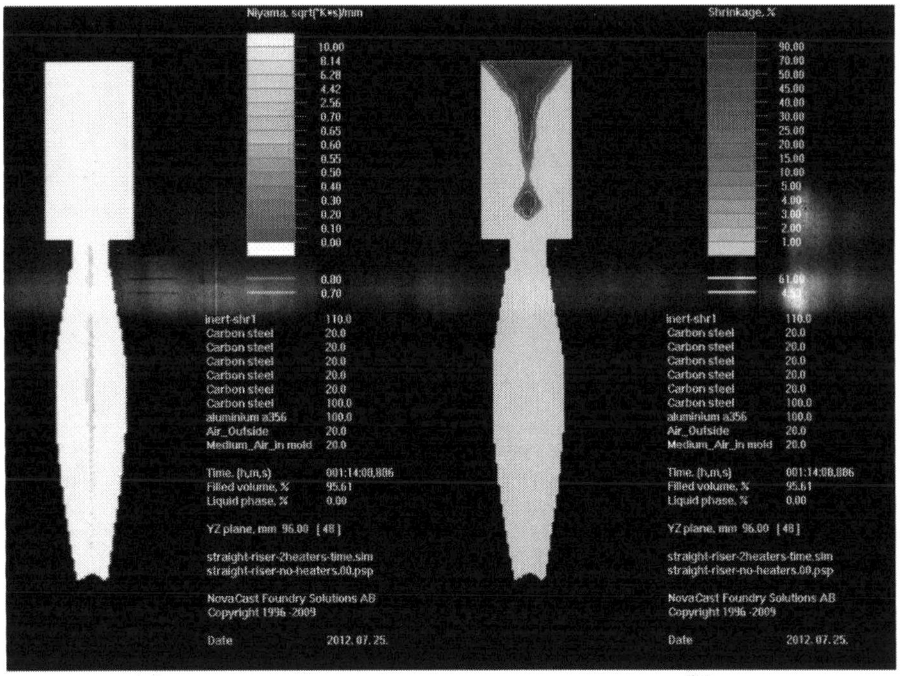

(a) (b)

Figure 9. Active heating and top (straight) riser optimization (0.12KW tape heating and mold preheating): Prediction of micro-shrinkage (a) and macro-shrinkage (b).

Figures 10-11 present the liquid fraction and temperature profile for two different process conditions. Similar conditions for bottom mold cooling and electrical tape heating of the side mold were used in both cases. The preheat mold temperature was 25 deg. C for the case in Figure 10 and 50 deg. C for the case in Figure 11.

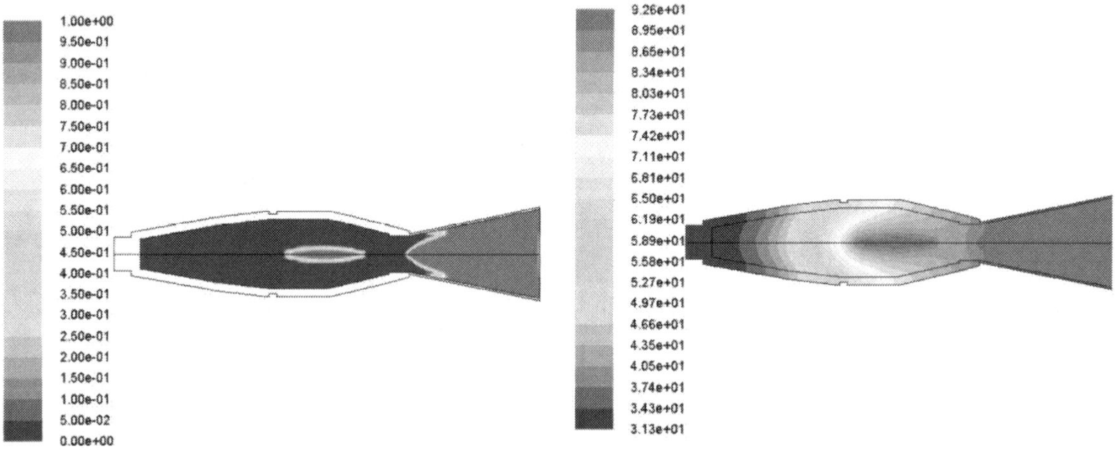

Figure 10. Liquid fraction and temperature profile at time t = 4538s: mold temperature = 25 deg. C, bottom cooling (heat flux coefficient is 50 W/m2/K) and side heating (electrical tape heating (flux components = 75 W/m2, 25 W/m2, 1 W/m2)).

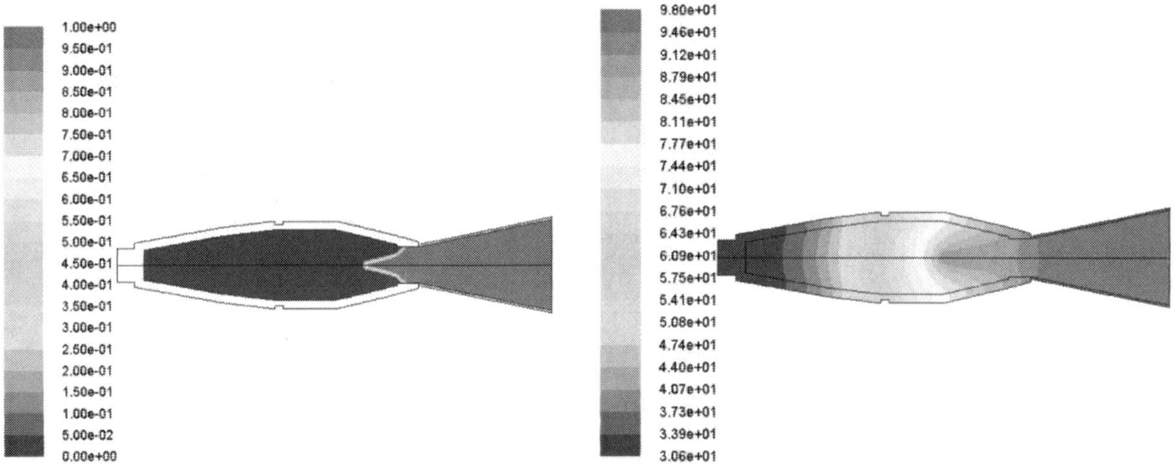

Figure 11. Liquid fraction and temperature profile at time t = 11624s: mold temperature = 50 deg. C, bottom cooling (heat flux coefficient is 50 W/m2/K) and side heating (electrical tape heating (flux components = 75 W/m2, 25 W/m2, 1 W/m2)).

Mechanical vibration – "Modal Analysis" using ANSYS MECHANICAL [6]: To determine the resonant frequency of the entire casting system, which will need to be mechanically vibrated using a mechanical vibration (MV) table, a modal analysis was performed in ANSYS MECHANICAL software. It was determined that the resonant frequency of the entire casting system is 342 Hz. The MV table is thus safe to mechanically vibrate the casting system for the energetic materials components since it has a frequency range between 30 and 47 HZ that is regulated by air pressure between 30 and 90 PSI. Also, the acceleration of the system is between 0.7 G and 1.9 G, again regulated by power by changing the air pressure between 30 and 90 PSI.

Concluding Remarks and Future Work

Numerical simulations have been performed in NOVAFLOW&SOLID and ANSYS to improve the design and process conditions of energetic materials cast into a projectile. The proposed improvements in the casting and solidification processing of the energetic materials led to significant decrease in solidification shrinkage defects in the cast projectiles.

The improved process design consists of using both mechanical vibration (MV) and active cooling and heating (ACH) technologies with proper modifications in the top riser geometry, pouring rates, mold preheat temperature, and water cooling conditions. Also, modal analysis in ANSYS MECHANICAL demonstrated that the MV table is indeed safe to mechanically vibrate the entire casting system.

Future work will include the optimization of the electrical tap heating and water cooling conditions using ANSYS'S FLUENT and its experimental validation as well a study regarding the MV effects on the solidification of energetic materials cast into projectiles.

References

1. D. Sun, S. Annapraganda, S. Garimella, and S. Singh, "Analysis of Gap Formation in the Casting of Energetic Materials", *Numerical Heat Transfer A*, 51:415-444, 2007.
2. D. Sun and S. Garimella, "Numerical and Experimental Investigation of Solidification Shrinkage", *Numerical Heat Transfer A*, 52:145-162, 2007.
3. R. Mudryy, "Thermo-Mechanical Modeling of Melt Casting of Explosives", *Proc. of AIChe Annual Meeting*, Salt Lake City, Utah 2010.
4. U. Teipel, Energetic Materials, Wiley-VCH, Berlin, Germany, 2005.
5. R. Mudryy, "Solidification Modeling of DNAN Based Explosive Compositions", *Proc. of 42nd International ICT Conference*, Karlsruhe, Germany, June 27- July 01, 2011.
6. ANSYS'S FLUENT and ANSYS MECHANICAL: www.ansys.com
7. A. Wood, S. Ritchie, and G. Bell, "An Efficient Implementation of the Enthalpy Method", *International Journal for Numerical Methods in Engineering*, vol.17, pp. 301-305, 1981.
8. V. Voller and C. Swaminathan, "Generalized source-based method for solidification phase change", *Numerical Heat Transfer B*, vol. 19, pp. 175-189, 1991.
9. D. Sun, S. Garimella, S. Singh, and N. Naik, "Numerical and Experimental Investigation of the Melt Casting of Explosives", *Propellants, Explosives, Pyrotechnics*, vol. 30, pp. 369-380, 2005.
10. NOVAFLOW&SOLID and NOVASTRESS software, NOVACAST AB, Sweden: www.novacast.se.
11. D. Wiegand, "Mechanical failure of composite plastic bonded explosives and other energetic materials", *Proc. 12th Int. Detonation Symposium*, San Diego, CA, 11-16 Aug 2002, Office of Naval Research, Department of Navy, pp. 744-750.

SQUEEZE FLOW RHEOMETRY

Dilhan M. Kalyon[1], *Halil Gevgilili*[1] *and Hansong Tang*[2]

Stevens Institute of Technology
Currently with: City University of New York, New York, NY[2]

Introduction

The accurate determination of the parameters of wall slip and the shear viscosity material function of energetic suspensions is a must for realistic simulation of their processing, especially when continuous processing methodologies are utilized (Fig. 1). However, due to the difficulty

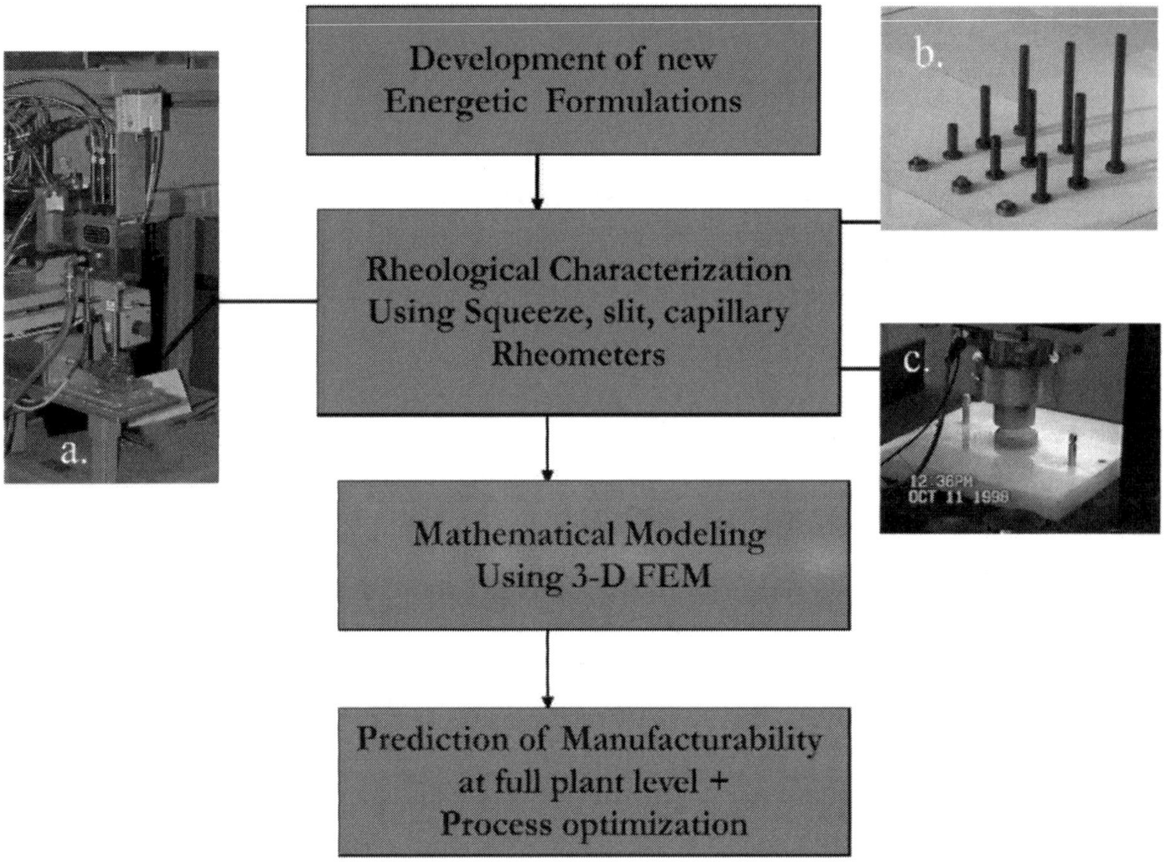

Fig. 1 Role of rheological characterization in energetic

of handling energetic materials at various relevant shear rate ranges novel methodologies are necessary for the characterization of parameters of slip and shear viscosity. The available methods include capillary rheometry using multiple capillaries with differing diameters and length over diameter ratios to allow Bagley and wall slip corrections to be carried out [1, 2], rectangular slit flow using an adjustable gap slit rheometer to allow the determination of wall slip velocity values from dependence of the flow curves on the surface/volume ratio of the slip rheometer [3], and squeeze flow rheometry [4-18] as shown in Fig. 1 a-c. The squeeze flow rheometry offers significant advantages in the characterization of rheological behavior of highly filled suspensions including energetic materials [17]. The rheological characterization allows the determination of the parameters of viscoplastic constitutive equations and wall slip, which can then be used in the mathematical modeling of the flow and deformation occurring in continuous processing equipment, including single and twin screw extruders and shaping through dies [19-28]. Here a review of the inverse problem solution methodologies [29, 30] involving squeeze flow is made. The advantages of the utilization of squeeze flow in combination with other viscometric flows including rectangular slit, steady torsional and capillary flows to enable the determination of parameters of wall slip and shear viscosity for energetic suspensions are outlined. Various types of squeeze flow rheometers that can be field implemented are also described.

Squeeze flow

In squeeze flow the fluid to be characterized is contained in the gap, h (t), formed by two horizontal parallel circular disks that are of same radius, R, and are both initially at rest (Fig. 2).

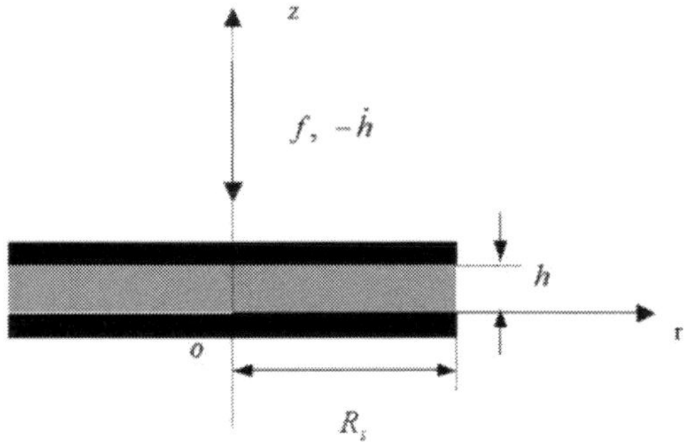

Fig. 2 Schematics of the squeeze flow

The fluid subsequently undergoes squeezing action by either applying a constant force F to one of the two disks and keeping the other stationary or by moving one of the two disks at a given cross-head speed and keeping the other stationary. The first mechanism generates a plate velocity, $\dot{h}(t)$, that is time dependent and which can be monitored, while the second mechanism assumes that the cross-head speed of the disk, $\dot{h}$, can be controlled.

Two types of squeeze flow rheometers are shown in Fig. 3. The drive mechanisms of the

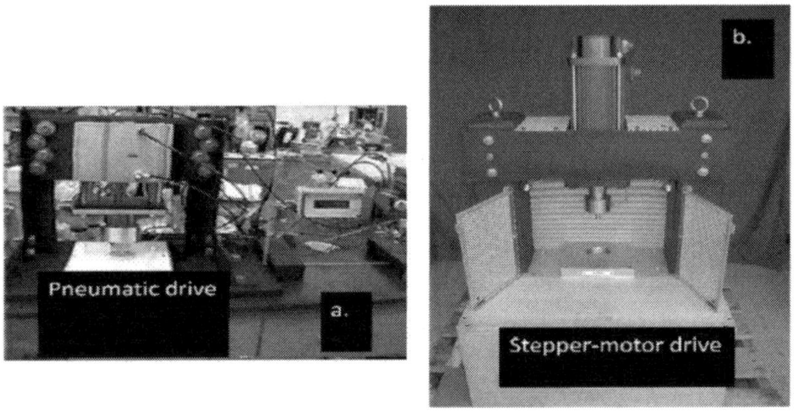

Fig. 3 Two different types of squeeze flow rheometers

two rheometers are different. The first rheometer (Fig. 3a) involves a pneumatic drive, i.e., the top disk is driven via the application of a gas pressure (typically N_2). Although the pressure of the gas can be controlled the cross-head speed cannot be controlled since the resistance of the fluid to deformation (rheological behavior of the fluid being squeezed) and the conditions that are applied determine the change in cross-head speed as a function of time, $\dot{h}(t)$. Although the cross-head speed of this first rheometer (Fig. 3a) cannot be controlled it can be monitored via a strain sensor. A pressure transducer monitors the normal force being exerted as a function of time, $f(t)$, and an embedded computer collects all the data and provides the analysis for the determination of the parameters of the viscoplastic constitutive equation and wall slip. In the second squeeze flow rheometer (Fig. 3b) a stepper motor is available and thus the targeted cross-head speed of the top disk, $\dot{h}$, can be imposed and accurately maintained.

Analysis of the squeeze flow

Under the typical creeping flow conditions (Reynolds number<<1) the inertial terms in the equations of motion can be neglected. The governing equations for axisymmetric flow of incompressible fluids under quasi-steady state and isothermal conditions become [18]:

$$\frac{1}{r}\frac{\partial r v_r}{\partial r} + \frac{\partial v_z}{\partial z} = 0, \tag{1}$$

$$\frac{1}{r}\frac{\partial r \tau_{rr}}{\partial r} + \frac{\partial \tau_{rz}}{\partial z} - \frac{\tau_{\theta\theta}}{r} + \frac{\partial p}{\partial r} = 0, \tag{2}$$

$$\frac{1}{r}\frac{\partial r \tau_{rz}}{\partial r} + \frac{\partial \tau_{zz}}{\partial z} + \frac{\partial p}{\partial z} = 0. \tag{3}$$

Here p is the pressure, the velocity components in the radial, r, and axial, z, directions are v_r and v_z, respectively, and τ_{rr}, τ_{zz}, $\tau_{\theta\theta}$ and τ_{rz} are the normal and shear components of the deviatoric stress tensor. If the rate of movement of the plate, $\dot{h}(t)$, is sufficiently slow to allow viscoelastic effects to be considered to be negligible, generalized Newtonian fluid constitutive behavior is applicable, i.e., $\underline{\underline{\tau}} = -\eta(II)\underline{\underline{\Delta}}$, where $\underline{\underline{\tau}}$ and $\underline{\underline{\Delta}}$ are the stress and rate of deformation tensors, respectively. $\eta(II)$ is the shear viscosity material function varying as a function of the second scalar invariant, II, of the rate of deformation tensor. The squeeze flow is subject to the following non-linear wall slip condition, i.e., slip velocity, U_s, versus the shear stress, τ_{rz} [2, 14, 18, 24, 26-28, 31]:

$$U_s = \beta \, |\tau_{rz}|^{sb}, \tag{4}$$

where the Navier slip coefficient, β, and the slip exponent, sb, are parameters of the wall slip behavior. The slip coefficients for the top and bottom disks, i.e., β_t and β_b respectively, can be different due to differences in roughness or the materials of construction of the two plates. For 1-D flow the shear stress for the Herschel-Bulkley type viscoplastic constitutive equation becomes $\tau_{rz} = \pm\tau_y - m\,|dv_r/dz|^{n-1}(dv_r/dz)$ for $|\tau_{rz}| \geq \tau_y$ (- sign is used for negative shear stress, τ_{rz}) and the shear rate $(dv_r/dz) = 0$ for $|\tau_{rz}| < \tau_y$. Thus, for the Herschel-Bulkley fluid subject to wall slip there are five parameters (m, n, τ_y, β, sb) that need to be determined for representing the shear viscosity along with the wall slip condition. For concentrated suspensions, sb can be estimated from the shear viscosity material function of the binder and the properties of the particles [24] and thus typically only the other four parameters need to be characterized. The solution of the squeeze flow problem for the Herschel-Bulkley fluid also provides the solutions for Newtonian fluid with n=1 and $\tau_y = 0$; Bingham fluid with n=1; power-law behavior with $\tau_y = 0$; all subject to either no slip or slip at the wall.

The normal force acting on the top plate, f, is:

$$\begin{aligned} f &= \int_0^R 2\pi(p+\tau_{zz})\big|_{z=h}\, r dr \\ &= \int_0^R \pi\left(\left(-\frac{dp}{dr}\right)r^2 + 2\tau_{zz} r\right)\Bigg|_{z=h} dr. \end{aligned} \tag{5}$$

The solution of Equations (1)—(5) can be achieved analytically (upon simplification) to determine the normal force, f, acting on the moving plate (i.e., [4- 18]). Numerical analysis of the squeeze flow of viscoplastic fluids [11, 14, 18] is also available. The parameter estimation can be formulated as an inverse problem solution [18]:

$$\textit{Find} \quad \{\alpha\} = \{\alpha \mid J(F, F^e) = \min_\alpha J(F, F^e)\}, \tag{6}$$

$$\textit{Subject to} \quad f_i = \int_0^R 2\pi(p+\tau_{zz})_i\big|_{z=h}\, r dr, \quad i = 1, 2, \cdots, M. \tag{7}$$

Associated with FEM based solution procedure, the minimization of the Eq. (6) can be carried out. The derivatives of objective functions can be determined as [18]:

$$\frac{\partial J}{\partial m} \approx \frac{J_{m+\Delta m, n, \tau_y, \beta} - J_{m-\Delta m, n, \tau_y, \beta}}{2\Delta m}, \tag{8}$$

$$\frac{\partial J}{\partial n} \approx \frac{J_{m,n+\Delta n,\tau_y,\beta} - J_{m,n-\Delta n,\tau_y,\beta}}{2\Delta n}, \tag{9}$$

$$\frac{\partial J}{\partial \tau_y} \approx \frac{J_{m,n,\tau_y+\Delta\tau_y,\beta} - J_{m,n,\tau_y-\Delta\tau_y,\beta}}{2\Delta \tau_y}, \tag{10}$$

$$\frac{\partial J}{\partial \beta} \approx \frac{J_{m,n,\tau_y,\beta+\Delta\beta} - J_{m,n,\tau_y,\beta-\Delta\beta}}{2\Delta \beta}. \tag{11}$$

A typical determination of the four parameters for a 20-40% by volume suspension of spherical glass particles incorporated into a poly(dimethyl siloxane) based binder, which has led to the following comparison of the experimental data with the results of the solution of the inverse problem for the four parameters, is shown in Fig. 4 [18] below:

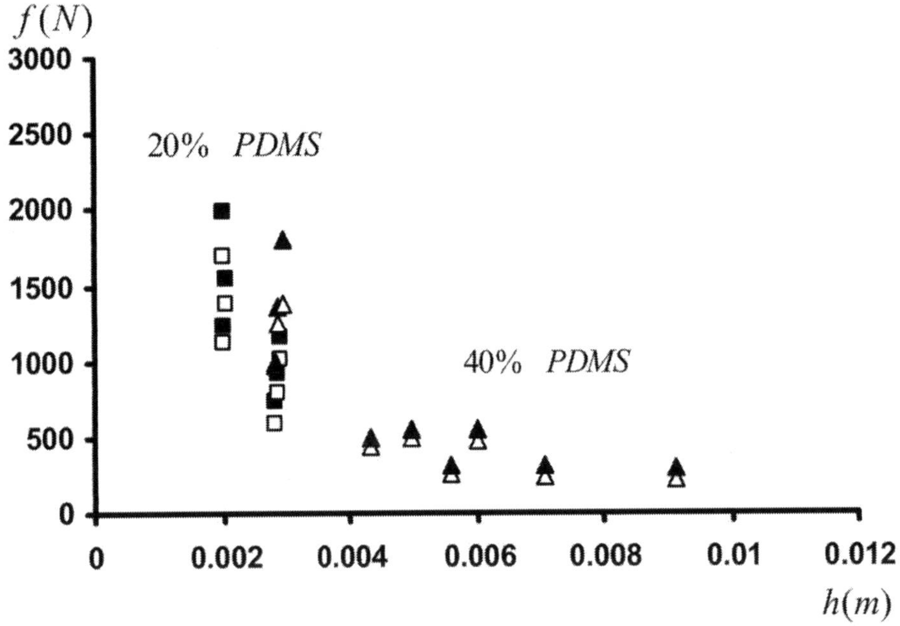

Fig. 4 Typical comparison of the experimentally observed normal force versus gap for two suspensions of poly(dimethyl siloxane), PDMS, with 20-40% by vol. of spherical glass particles.

The acceptable agreement is indicative of the utility of the squeeze flow rheometry in rheological characterization of concentrated suspensions in general and energetic suspensions in particular.

References

1. Mooney M., Explicit formulas for slip and fluidity, J. Rheol. 2 (1931) 210-222.
2. Yilmazer U. and Kalyon D. M., Slip effects in capillary and parallel disk torsional flows of highly filled suspensions, J. Rheol. 33 (1989) 1197-1212.
3. Kalyon D. M. and Gokturk H. S., "Adjustable Gap Rheometer," U.S. Patent and Trademark Office # 5,277,058, Jan. 11, 1994.
4. Scott J. R., Theory and application of parallel-plate plastometer, Trans. Int. Rubber Ind. 7 (1931) 169-175.
5. Covey G., B. Stanmore, Use of the parallel-plate plastometer for the characterization of viscous fluids with a yield stress, J. Non-Newt. Fluid Mech. 8 (1981) 249- 260.
6. Zhang W., N. Silvi, J. Vlachopoulos, Modeling and experiments of squeezing flow of polymer melts, Intern. Polym. Proc. 10 (1995) 155-164.
7. Adams M. J., I. Aydin, B. J. Briscoe, S. K. Sinha, A finite element analysis of the squeeze flow of an elasto-viscoplastic paste material, J. Non-Newt. Fluid Mech. 71 (1997) 41-57.
8. Lawal, A., Kalyon D. M., Squeeze flow of viscoplastic fluids subject to wall slip, Polymer Eng. Sci. 38 (1998) 1793-1804.
9. Laun H. M., M. Rady, O. Hassager, Analytical solutions for squeeze flow with partial wall slip, J. Non-Newt. Fluid Mech., 81 (1999) 1-15.
10. Denn M.M., Marrucci G., 87 (1999) "Squeeze flow between finite plates", J. Non-Newtonian Fluid Mech., 175–178.
11. Lawal, A., Kalyon D. M., Compressive squeeze flow of generalized Newtonian fluids with apparent wall slip, Intern. Polym. Proc. 15 (2000) 63-71.
12. Smyrnaios D. N., J. A. Tsamopoulos, Squeeze flow of Bingham plastics, J. Non-Newt. Fluid Mech, 100 (2001) 165-189.
13. Roussel N., C. Lanos, Plastic fluid flow parameters identification using a simple squeezing test, Appl. Rheol. 13 (2003) 132-141.
14. Tang H. and Kalyon D. M., "Estimation of the Parameters of Herschel-Bulkley Fluid under Wall Slip Using a Combination of Capillary and Squeeze Flow Viscometers", Rheologica Acta, 43 (2004) 1 80-88.
15. Engmann J., C. Servais, A. Burbidge, Squeeze flow theory and applications to rheometry: A review, J. Non-Newt. Fluid Mech. 132 (2005) 1- 27.
16. Karapetsas G., J. Tsamopoulos, Transient squeeze flow of viscoplastic materials, J. Non-Newtonian Fluid Mech. (2006) 133, 35-56.
17. Kalyon D. M., Tang, H. and Karuv B., "Squeeze flow rheometry for rheological characterization of energetic formulations", Journal of Energetic Materials, 24 (2006) 195-202.
18. Kalyon D. M. and Tang H., "Inverse problem solution of squeeze flow for parameters of generalized Newtonian fluid and wall slip", Journal of Non-Newtonian Fluid Mechanics, 143 (2007) 133-140.
19. D. Gotsis, A. Ji and D. M. Kalyon, "3-D Analysis of the Flow in Co-rotating Twin Screw Extruders," Society of Plastics Engineers, ANTEC Technical Papers, 36, 139-142(1990).
20. Lawal A. and D. M. Kalyon, "Non-isothermal Model of Single Screw Extrusion of Generalized Newtonian Fluids," Numerical Heat Transfer, 26 (1), 103-121 (1994).
21. Lawal, A. and Kalyon D. M., "Non-Isothermal Extrusion Flow of Viscoplastic Fluids

with Wall Slip", International Journal of Heat and Mass Transfer, 40 (1997) 3883-3897.
22. Lawal, A. and Kalyon D. M., "Viscous Heating in Nonisothermal Die Flows of Viscoplastic Fluids with Wall Slip," Chem. Eng. Sci., 52 (1997) 8, 1323-1337.
23. Malik M. and Kalyon, D.M., "Three-dimensional Finite Element Simulation of Processing of Generalized Newtonian Fluids in Counter-rotating and Tangential Twin Screw Extruder and Die Combination", Int. Polym. Processing, 20 (2005) 398- 409.
24. Kalyon D. M., Apparent slip and viscoplasticity of concentrated suspensions, J. Rheol. 49 (2005) 621-640.
25. Kalyon D. M. and Malik M., "An integrated approach for numerical analysis of coupled flow and heat transfer in co-rotating twin screw extruders", International Polymer Processing, 22 (2007) 293-302.
26. Tang H. and Kalyon D. M., "Unsteady circular tube flow of compressible polymeric liquids subject to pressure-dependent wall slip", J. Rheology, 52 (2008) 507-526.
27. Tang H. S. and Kalyon D. M., "Time-dependent tube flow of compressible suspensions subject to pressure dependent wall slip: Ramifications on development of flow instabilities", Journal of Rheology 52 (2008) 5, 1069-1090.
28. Kalyon D.M. and Malik M., "Axial laminar flow of viscoplastic fluids in a concentric annulus subject to wall slip", Rheologica Acta (2012) 51:805–820.
29. Yeow, Y. L., Taylor J. W., Obtaining the shear rate profile of steady laminar tube flow of Newtonian and non-Newtonian fluids from nuclear magnetic resonance imaging and laser Doppler velocimetry data, J. Rheol. 46 (2002) 351-365.
30. Yeow Y. L., Leong Y. K., Shear rate and wall slip velocity functions of polyvinyl chloride melts based on slit die viscometry data, Polym. Eng. and Sci. 44 (2004) 153-162.
31. Aral B. and Kalyon D. M., Effects of temperature and surface roughness on time-dependent development of wall slip in steady torsional flow of concentrated suspensions, J. Rheol. 38 (1994) 957-972.

(517e) Propellant Dough Screening using Squeeze Flow Characterization

Michael Fair, Robin Crownover and Joseph Rokosz, US Army ARDEC, Picatinny Arsenal, NJ

The benefit of the squeeze flow durometer is to remove variability in propellant processing by providing a quantitative means of assessing when a propellant mix is at the proper consistency for further processing (e.g. extrusion, cutting, etc). Currently the mixer operator determines when a propellant mix is ready by handling the material and making a qualitative decision as to whether the propellant is properly mixed and contains the correct level of residual solvent. While a properly trained operator may be able to consistently determine when a mix is ready for further processing, results vary between operators, introducing batch to batch propellant variation. The durometer decreases operator variability through more accurate and reproducible means of measurement. This presentation covers the method of data collection and interpretation for the squeeze flow durometer, as well as results showing how consistent solvent level aids in further propellant processing.

(521a) Scale-up of Capsule Filling Operations Using a Nozzle-Dosator Machine

Marcos Llusá[1], Eva Faulhammer[2], Stefano Biserni[3], Vittorio Calzolari[3], Simon Lawrence[4], Georg Scharrer[5] and Johannes G. Khinast[6], (1)RCPE GmbH, Graz, Austria, (2)Technical University Graz, Graz, Austria, (3)MG2, Bologna, Italy, (4)GSK, Harlow, United Kingdom, (5)Advanced Simulation, Research Center for Pharmaceutical Engineering, Graz, Austria, (6)Institute for Process and Particle Engineering, Graz University of Technology, Graz, Austria

Scale-up of capsule filling operations using a nozzle-dosator machine

Introduction

This paper discusses the methodology for the development and scale up of capsule filling operations using a research capsule filling unit (Labby, MG2) and an industrial capsule filling machine (Planeta, MG2).

The research capsule filling machine (Labby) has only one nozzle dosator and can be operated using relatively small quantities of powder. At the moment of submitting this abstract, nozzle dosators that correspond to capsule size 3 and 5 have been investigated and the powder used is microcrystalline cellulose (Avicel 102). The scope of the study will be enlarged to include cohesive as well as free flowing powders and also larger nozzles (size 0 and 1). The main finding using the research capsule filling machine is that powder net weight in capsules increases as we increase the speed of capsule filling. It was evident to the experimenter that, although subtle, a different level of mechanical vibration caused by the driving engine is associated to the different speeds of operation of the capsule filling machine. The latter causes different degrees of powder densification or porosity (Woodhead, 1980) in the bowl (as we could also assess), which easily explains the different net powder weight in capsules. Nozzle dosators operate on a volumetric basis: the nozzles collect a specified volume of powder from a rotating bowl and subsequently transfer the powder into an open capsule. The following correlation is observed: sampling denser powders leads to larger net weight in capsules.

When the operations are carried out in an industrial capsule filling machine (Planeta, MG2), the effect of speed on powder net weight was negligible. In fact, all MG2 production machines have a weight control system available. This system automatically changes the dimension of the dosing chamber in the nozzle to keep the target weight whenever it is necessary (i.e. when powder density inside the rotary bowl changes). The weight control system makes it possible to automatically adjust the average net weight in the capsules, maintaining it always as close as possible to the target weight ("theoretical net weight"). This process is called "automatic adjustment".

In addition to the development of capsule filling operations, the work presented here is relevant to current good manufacturing practices (cGMP) in the pharmaceutical industry. Mechanical vibration is an environmental variable that, just as relative humidity, affects powder properties and the quality of powder based products. Therefore, just as relative humidity, it should be carefully assessed and controlled.

Methods

1. Capsule filling: Performed using MG2 equipment (Labby and Planeta).

2. Capsule weight: groups of 100 capsules manufactured with different nozzles: 22 mg (nozzle size 5), 50 mg (nozzle 3), 120 mg (nozzle 3) are weighed.

3. Powder density in the bowl: A graduated glass cylinder is fixed to the driver of the rotating powder bowl (therefore bowl and glass cylinder are exposed to equivalent vibration). The powder densification can be visually observed and measured in the glass cylinder.

4. Vibration of the research capsule filling machine: Assessed using Laser Doppler Vibrometer (LDP).

Results

1. Effect of capsule filling speed (and vibration) on powder density in the bowl

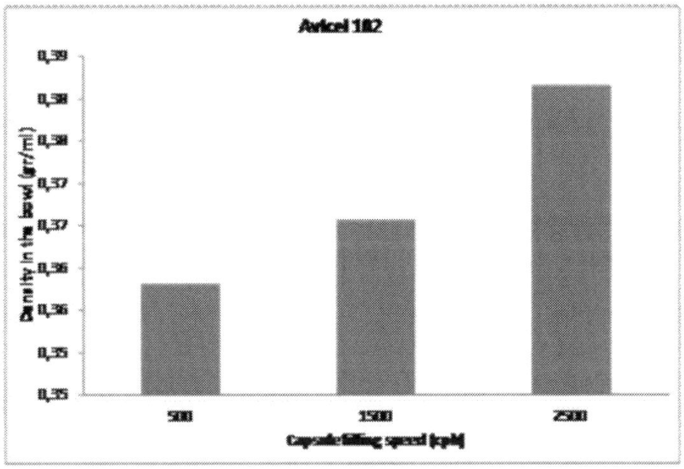

Figure 1: This figure shows that the powder density (measured in a glass cylinder that is fixed on the driving wheel of the rotating bowl) increases as the speed of the capsule filling machine increases.

2. Effect of capsule filling speed (and vibration) on capsule net weight

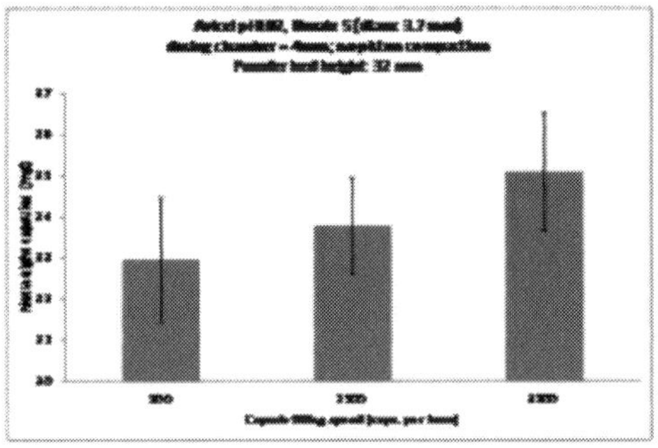

Figure 2: This figure shows an increasing trend of powder net weight in capsules as the speed of the capsule filling machine increases. Data obtained using a nozzle size number 5.

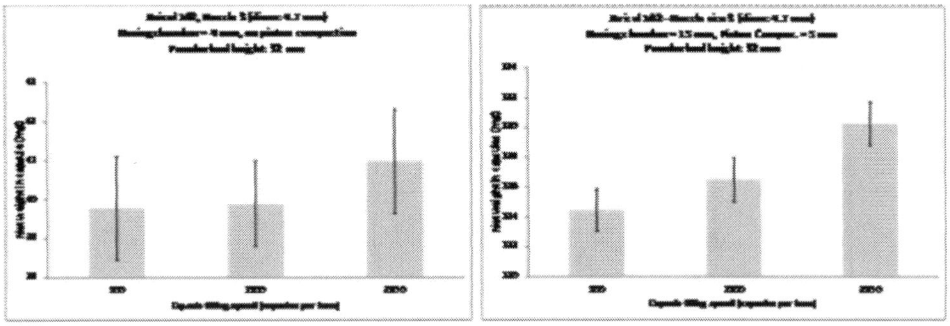

Figures 3 (left and right): These figures show an increasing trend of powder net weight in capsules as the speed of the capsule filling machine increases. Data obtained using a nozzle size number 3 and two different dosing chamber spaces.

Conclusion

Higher capsule filling speed of the research capsule filling machine (Labby) is coupled to more intense vibration of the engine, which leads to a further densification of the powder in the bowl. As a result, the powder mass collected in the volume of the dosing chamber of the nozzle dosator (and hence powder net weight in capsules) is larger. Although the effect is less evident (and not statistically significant), there is a clear trend for low doses (~20 mg). As we increase the volume of powder collected by the nozzle (~120 mg), the effect of powder bed densification on capsule net weight becomes more obvious. However, the effect of capsule filling speed on powder net weight in capsules is not observed in the industrial scale (Planeta) due to the corrective actions of the weight control system.

Acknowledgments: To the financial and scientific support from MG2 and GSK.

References

Philip John Woodhead. Doctoral Thesis: "The influence of powder bed porosity variations on the filling of hard gelatin capsules by a dosator system". University of Nottingham, 1980.

(521b) Determination of Residence Time Distribution in Rotary Calciner

Yijie Gao[1], Alberto Cuitino[2], Benjamin J. Glasser[1], Fernando J. Muzzio[1], Marianthi G. Ierapetritou[1], Jean Beeckman[3], Natalie A. Fassbender[3] and William G. Borghard[3], (1)Department of Chemical and Biochemical Engineering, Rutgers, The State University of New Jersey, Piscataway, NJ, (2)Department of Mechanical and Aerospace Engineering, Rutgers, The State University of New Jersey, Piscataway, NJ, (3)ExxonMobil, Annandale, NJ

Determination of Residence Time Distribution in Rotary Calciner

Yijie Gao, Alberto M. Cuitino, Benjamin J. Glasser, Fernando J. Muzzio, Marianthi G. Ierapetritou

Dept. of Chemical and Biochemical Engineering, Rutgers University, Piscataway, NJ 08854

Jean W. Beeckman, Natalie A. Fassbender, William G. Borghard

Process Research Laboratories, ExxonMobil Research & Engineering Co., Annandale, NJ 08801

Continuous calcination is a widely used process in large scale catalyst manufacturing. Many works have been done recently that target on the estimation of material flow in the rotary drum, which is the first critical step on the quality control of the rotary calcination process. However, reliable prediction of material flow is not available yet[1]. In this study, we investigated the influence of operation and material properties on the RTD in a rotary calciner. In order to accurately predict the residence time distribution (RTD) of different types of catalyst particles, the Saeman's model[2] was used for the prediction of the bed depth distribution and the mean residence time (MRT) whereas the Taylor's dispersion model[3, 4]was applied to estimate the MRT and the axial dispersion coefficient through RTD fitting process.

The two important indices for continuous calcination are: (1) the residence time of the particles inside the calciner, and (2) the characteristic time of calcination. Since this unit operation requires that particles stay for a minimum residence time inside the system to guarantee complete calcination, long residence time and limited axial dispersion is preferred. This work results in three very important findings. Firstly, for the investigated catalyst particles, we found that large particle repose angle, small incline angle, and descending rotary speed led to long MRT and limited axial dispersion of particles. As a result, we suggest using slender particles with higher angle of repose in continuous calcination process. Secondly, based on the experimental observations, a reliable method was developed to provide reasonable RTD estimations for different types of catalyst particles, which can be conveniently used in the quality control and performance improvement of practical rotary calcination process. Finally, since the calciner performance is determined by the competition between the residence time and the characteristic time of calcination in the cross-sectional directions, a detailed study on the latter aspect is necessary before a thorough guidance of operation and material selection can be proposed. Therefore, the cross-sectional heat transfer and calcination process in a rotary cylinder is suggested for further investigation[5].

Reference

1. Liu XY, Zhang J, Specht E, Shi YC, Herz F. Analytical solution for the axial solid transport in rotary kilns. *Chemical Engineering Science*. 2009; 64(2): 428-431.

2. Saeman WC. Passage of solids through rotary kilns: factors affecting time of passage. *Chemical Engineering Progress*. 1951; 47: 508-514.

3. Sudah OS, Chester AW, Kowalski JA, Beeckman JW, Muzzio FJ. Quantitative characterization of mixing processes in rotary calciners. *Powder Technology*. 2002; 126(2): 166-173.

4. Gao Y, Vanarase A, Muzzio F, Ierapetritou M. Characterizing continuous powder mixing using residence time distribution. *Chemical Engineering Science*. 2011; 66(3): 417-425.

5. Chaudhuri B, Muzzio FJ, Tomassone MS. Modeling of heat transfer in granular flow in rotating vessels. *Chemical Engineering Science*. 2006; 61(19): 6348-6360.

(521c) Studying the Effects of Fluidized Bed Scale On the Solid Dynamics Using Sophisticated Optical Probe

Faraj M. Zaid, Department of Chemical and Biological Engineering, Missouri University of Science and Technology, Rolla, MO and Muthanna H. Al Dahhan, Chemical and Biological Engineering, Missouri University of Science and Technology, Rolla, MO

Studying the Effects of Fluidized Bed Scale on the Solid Dynamics Using Sophisticated Optical Probe

Faraj Zaid, and Muthanna H. Al Dahhan, Department of Chemical and Biological Engineering, Missouri University of Science and Technology, Rolla, MO

Gas-solids fluidized-bed reactors have the found widespread applications in industry due to a number of advantages. The radial and axial profiles of solids holdup and particle velocities in these reactors are important fluid dynamic parameter. In this work, sophisticated optical probes that can measure simultaneously both solids concentration and velocity and their time series fluctuation have been used. The effects of the bed size of these parameters have been investigated using two different fluidized beds (0.14 m and 0.44 m in ID). The radial and axial solids holdup and particle velocity profiles have been studied at the same three dimensionless axial and radial positions in both units. Air was used as the fluidizing gas and different particle sizes glass beads (ρ_s=2500 kg/m^3, d_p=210 and 70 μm) and copper (ρ_s=5300 kg/m^3, d_p=200 μm), as were used as solids material. The bed aspect ratio before fluidization was set at two. Both fluidized beds were equipped with a porous-plate (pore size: 40 μm) distributor. The effect of the bed scale on solids dynamics has been studied at a range of different superficial gas velocities *U*.

Using simple and reliable calibration method, the measured solids concentrations by the probe can convert to solids holdup. In both fluidized beds the solids holdups increased towards the wall. Same trends were reported by Mabrouk et al. (2005). Additionally, it was found that the radial particle velocity profile was nearly parabolic in both fluidized beds. The dimensionless group proposal by Glicksman (1988) and Glicksman, et al (1993) for hydrodynamics similarity have been also examined in this work using local measurements. The detailed result and analysis will be discusses and presented in this talk.

References

[1] Mabrouk et al., "Scale effects on fluidized bed hydrodynamics" Inter. J. of Chemical Reactor Eng 3, Article A18; 2005.

[2] Glicksman (1988), "Scaling relationships for fluidized beds" Chem. Eng. Sci. 43

[3] Glicksman et al (1993) Simplified scaling relationships for fluidized beds. *Powder Technol.* 1993; 77:177-199.

Keywords: Fluidized beds, optical probe, solids holdup, spatial distribution, particle velocity

(521d) Residence Time Distribution As Predicted by a Kinematic Model for Mass Flow Powder Discharge From a Conical Hopper

Luke Fullard, Institute of Fundamental Science, Massey University, Palmerston North, New Zealand, Clive E. Davies, School of Engineering and Advanced Technology, Massey University, Palmerston North 4442, New Zealand and Graeme Wake, Massey University, Auckland, New Zealand

The mixing and residence time distribution of the elements comprising a powder inflow to a silo with a conical hopper is relevant and important in some industrial powder handling and storage systems, and could also provide a platform for testing the validity of different approaches to the determination of the velocity distribution in the hopper.

In this paper, we have followed a kinematic model for the velocity profile, [see Nedderman, R M. And Tüzün, U., A Kinematic Model for the flow of granular materials, Powder Technology, 1979 22 243; Choi, J. Kudrolli, A. Bazant, M. Z., Velocity profile of Granular flows inside silos and Hoppers, J. Phys.: Condens. Matter 17 (2005) S2533-S2548], and used this to estimate mixing parameters and the residence time distribution for the hopper. We have specifically considered a case where, initially, powder is deposited in the hopper and cylindrical bunker above it, in horizontal layers. When flow is initiated, any given layer is unable to keep its radius constant as it travels into the hopper, and hence, the volume of powder would decrease unless the horizontal structure is broken. The kinematic model allows us to specify how this displaced powder is repositioned in the layer. Residence time information is extracted from the reconfigured layers as they emerge from the hopper exit. We note that the parameters in the kinematic model must be fitted for a particular system.

Plans for further work in this area are briefly discussed, including a possible tracer experiment, and a model in which the velocity profile of the hopper is established using stress field theory (as in Nedderman, R M., Statics And Kinematics of Granular Materials, Cambridge University Press).

(521e) Effect of Water and Magnesium Stearate On the Strength of Ceramic Compacts

Sarang Oka, Department of Chemical and Biochemical Engineering, Rutgers, The State University of New Jersey, Piscataway, NJ, Alisar Tuncer, Mechanical and Aerospace Engineering, Rutgers, The State University of New Jersey, Piscataway, NJ, Alberto Cuitino, Mechanical Engineering, Rutgers University, Piscataway, NJ and Fernando Muzzio, Chemical & Biochemical Engineering, Rutgers University, Piscataway, NJ

Effect of Water and MgSt on the Strength of Ceramic Compacts

Sarang Oka, Alisar Tuncer, Alberto Cuitinio, Fernando Muzzio

Rutgers University

The strength of the catalyst support is critical in any process that employs the material since it more often than not is subject to intense process conditions as well physical stress. Alumina-Magnesia is a common catalyst support material compacted to various geometries. The penultimate step in its manufacturing process is the compaction of the milled and mixed granules. To maximize the strength of the compacts it is thus necessary to understand and quantify the effect of each material component as well as unit operation that leads up to the final formulation. The work focuses on the effect of quantity and timing of addition of water and MgSt to milled extrudates on the strength of the compacts. In addition, the PSD of the milled material also seems to play a dominant role on the final compact properties. The work explores the effect of the aforementioned parameters as well as their interplay on the strength of the compacts.

The addition of water to the milled ceramic material is an exothermic process and an immediate subsequent addition of the lubricant may compromise the strength of the compacts due to the fusing of the lubricant to the ceramic material. Thus, the effect of introducing a time delay in the addition of the lubricant is investigated.

Finally, past observations suggested that the strength of the compacts may be affected if the milled material is allowed to age after the addition of the water and hence the effect of material ageing is examined.

Water levels ranging from 0 to 25% at 5% intervals are explored. MgSt quantities ranging from 0% to 5% are studied while the effect of PSD ranging from a d90 of 700um to 1400um is quantified.

It is observed that the addition of water increases the bulk density of the material and this translates into a gain in strength of the compacts. An excess however may promote lumping and agglomeration and determining an optimum amount is thus the key. An increase in the amount of MgSt seems to have a detrimental effect on the strength of the compacts, with the compacts yielding at much lower stresses with higher amounts of lubricant. It is thus recommended to employ minimum amount of MgSt while ensuring homogeneity in mixing. The minimum amount of lubricant required for a formulation increases with an increase in water and decrease in particle size.

A decrease in particle size increases compact strength and decreases sensitivity to MgSt.

Neither ageing nor the introduction of a time delay seemed to have any effect on the compact strength. The tabletability profiles stayed intact in both the above cases.

(521f) Measurement of Particle Impact Velocities in Vibrationally Fluidized Granular Flows

Kamyar Hashemnia[1], Amirhossein Mohajerani[2] and Jan. K. Spelt[2], (1)Mechanical and Industrial Engineering, University of Toronto , Toronto, ON, Canada, (2)Mechanical and Industrial Engineering, University of Toronto, Toronto, ON, Canada

A probe consisting of a high-speed laser displacement sensor was developed to measure directly the surface-normal impact velocities of spherical steel and porcelain particles in the vibrationally-fluidized granular flow produced by a tub vibratory finisher. Measurements were made in various locations and directions within the flow (Fig. 1).

Vibratory finishing is widely used to deburr, polish, burnish, harden and clean metal, ceramic and plastic parts. In a tub vibratory finisher, a container filled with granular media is oscillated by an eccentric rotating shaft so that it develops a vibrationally-fluidized circulatory bulk flow of the media that is largely two-dimensional. The media have both a large-scale bulk flow velocity and a much smaller-scale local impact velocity. Therefore, work-pieces that are entrained in the flowing media are subjected to the repetitive, high-frequency impacts of the surrounding particles.

Within a vibratory finisher, surface interactions such erosive wear and plastic deformation are affected strongly by the velocity, frequency, and direction of particle impacts. High impact velocities can cause fracture and fragmentation while low impact velocities can make the process less efficient. The impact velocity is also closely related to the breakage of granular materials in many processes such as drug tablet coating within rotating drums, bulk materials handling, food processing, and particle attrition in vibratory finishing. Large particle impact velocities may also result in excessive erosion of machine components in processes such as vibratory sieving and mixing.

Measurement of local quantities such as the impact velocity of vibrating particles in a vibrationally-fluidized granular media is challenging, because of the relatively small scale of the motion and difficulties in designing probes capable of measuring local quantities without disturbing the media. Therefore, most existing research in the field of flowing granular media has focused on the bulk flow of the media. In some of these studies, discrete element modeling (DEM) simulations have been used to predict the bulk flow behavior of granular media and the numerical results have been verified experimentally. Generally, these investigations have shown that discrete element simulations give reasonable predictions of bulk flow velocity and volume fraction in both fluidized and non-fluidized flows.

Only a few studies have focused on the local behavior of the media in granular flows. In most of these studies, DEM predictions of collision scale quantities such as impact velocity, collision frequency or impact energy were not validated experimentally. Therefore, it is not known if the commonly used approaches and parameters in DEM simulations yield correct predictions of the impact velocity. This remains a significant limitation since the impact forces that govern erosion, wear and fracture in granular flows are fundamentally linked to the particle impact velocity and kinetic energy.

There are several different methods to measure particle velocity in fluidized beds: laser-Doppler velocimetry, photographic and video techniques, optical fiber probes and laser displacement sensors. Photographic methods have been used to measure bulk flow properties, but the limited spatial resolution of these approaches makes them unsuited to the measurement of local impact velocities. Laser-Doppler velocimetry is restricted to the low solid concentrations (loose media). Optical fiber probes have been used to measure bulk flow velocities and the void fraction of media passing transversely across the end of the sensor. Some designs can measure displacements along the optical probe axis using a correlation with the amount of reflected light, but such devices become inaccurate when the light reflected by the particle varies because of a significant transverse velocity across the probe tip that causes the particle to move out of the incident beam.

Therefore, laser displacement sensors were chosen to measure the impact velocity of 6 mm diameter steel and porcelain balls inside a relatively packed granular flow produced in a tub vibratory finisher. Laser displacement sensors measure the distance to an object using triangulation. Laser light reflected from the object is concentrated on a linearized charge-coupled device such that its position depends on the distance to the object. The accuracy of the impact velocity measurement using the laser displacement sensor was assessed in a drop test. The disruption to the particle bulk flow was minimized by enclosing the sensor in a streamlined elliptical tube.

The displacement signals from the laser sensors were analyzed to obtain the probability distribution functions of the impact velocity of the particles. The output was also interpreted to give the frequency with which media passed the sensor and hence a measure of the particle packing. Both the impact velocity and the packing density were found to vary significantly with the orientation of the laser probe (direction in which the laser was pointing) and its location in the tub vibratory finisher. The impact velocity was found to vary by up to 38% depending on the orientation of the laser at a single location in the tub, and by up to 24% at various locations within the tub. The packing density varied by up to 55% with orientation at a single location and by 46% at different locations. The average impact velocity of the lighter and stiffer porcelain balls was 15% greater than that of the steel balls for the same tub vibration. These laser sensor impact velocity measurements compared reasonably well with those obtained in a previous study using an impact force sensor.

The measurements can help to elucidate the distribution of impact energy within a vibratory finisher and improve the understanding of processing speed and uniformity. The data can also serve to guide the development of DEM simulations that attempt to predict local impact velocities in vibrationally fluidized beds of granular materials.

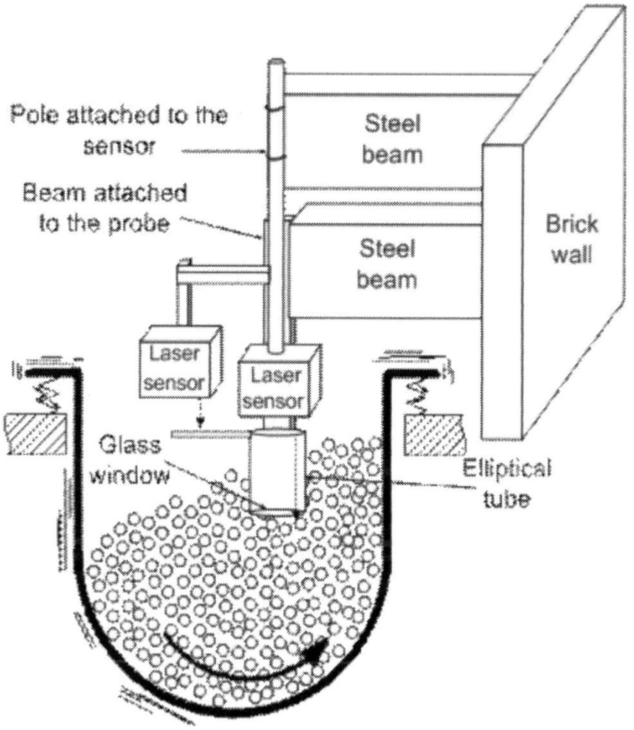

Fig. 1: Schematic of the test apparatus used to measure the impact velocity of steel and porcelain balls in the vertical direction in the tub vibratory finisher.

(521g) Rate-Based Separation by Exploiting Cohesive Interactions

Diana M. Liévano, Department of Chemical Engineering, University of Pittsburgh, Pittsburgh, PA and Joseph McCarthy, Chemical and Petroleum Engineering, University of Pittsburgh, Pittsburgh, PA

Achieving an efficient separation of individual components in a granular mixture is as much of a challenge as achieving a complete mixture. This is particularly true, if one want to utilize methods that are either low in energy intensity or environmentally benign, or both. This study focuses on a novel separation technique, analogous to chromatography for a fluid system, which we call 'granular chromatography'. In general, chromatography takes advantage of the preferential adsorption of one component, within a mixture, to the selected adsorbent. Here, we exploit a granular equivalent whereby particles preferentially adhere to a vibrated chute. Specifically, we select a fluid that through the combined effect of capillary and viscous forces causes preferential adhesion of one of the particle types in a granular mixture. Hence, separation is predicated on differences between the relative importance of the "work of wet adhesion" and the vibration energy for the particle sin question. Preliminary DEM simulations determine parameters that yield separation for both size and density variable granular mixtures. The effectiveness of this separation method can be verified experimentally and using simulations. The 'granular chromatography' theory developed verifies these results and allows for prediction of separation in different systems.

(583a) Design of Better Particle Coatings with Respect to Attrition Resistance

Gabrie Meesters, Chem Eng, Delft Univ Technology, Delft, Netherlands; DSM Biotechnology Center, DSM, Delft, Netherlands, Giacomo Perfetti, DSM Composite Resins, Zwolle, Netherlands and P. Van Hee, Food Specialties, DSM, Delft, Netherlands

Many coated particles show wear due to transport after production. In this research we have focused on assessing this wear, in particular attrition and have related this attrition resistance to the particle coating properties. By measuring the coating properties on the particle, as casted films but also as spray coated films a clear correlation is found between resistance to attrition and the visco-elastic properties of the coating. This finding will make the optimization easier when attrition wear needs to be reduced, since polymer will not have to be tested randomly, which is often the case, but it can be done much more direct by looking into the visco-elastic properties of the coating. Several HPMC and PVA coatings were tested and the most of the results will be presented. The research opens a new field of assessing coating properties without applying them on to coated particles, which can reduce the time of product development a lot.

(583b) Improved Recovery and Dissolution of Poorly Water-Soluble Drug Nanoparticles From Dried Nanocomposite Microparticles

Anagha Bhakay, Chemical engineering, New Jersey Institute of Technology, Newark, NJ, Mohammad Azad, Chemical, Biological, and Pharmaceutical Engineering, New Jersey Institute of Technology, Newark, NJ, Rajesh N. Dave, Chemical Engineering, New Jersey Institute of Technology, Newark, NJ and Ecevit Bilgili, Chemical, Biological and Pharmaceutical Engineering, New Jersey Institute of Technology, Newark, NJ

Incorporation of poorly water-soluble drug nanoparticles into solid dosage forms entails drying of nanosuspensions as an intermediate step, which can lead to loss of their large surface area and may lead to poor bioavailability. In this study, nanoparticle recovery of poorly water soluble drugs from nanocomposite microparticles (NCMPs) redispersed in water was investigated as a function of various excipients. Nanosuspensions of griseofulvin (GF), considered as a model poorly water soluble drug, were prepared by wet media milling and subsequently dried through coating on Pharmatose® carrier particles in a fluidized bed processor. A polymer (hydroxypropyl cellulose, HPC), a surfactant (sodium dodecyl sulfate, SDS), their combinations at various concentrations and a superdisintegrant (croscarmellose sodium, CCS) were used as stabilizers during milling. The NCMPs obtained from the fluidized bed coating were redispersed in water using various methods of agitation to recover the GF nanoparticles. The redispersion methods studied had a slight impact on the recovery of the nanoparticles when SDS was present in the formulation. The HPC–SDS combination resulted in a synergistic effect in the formation of finer nanoparticles as well as in fast redispersion of the nanoparticles of five different poorly water soluble drugs from the NCMPs. Poor redispersion of the nanoparticles and slower dissolution were observed when SDS was removed from the formulations and addition of dispersants like mannitol also could not recover the GF nanoparticles. However, incorporation of a novel additive such as milled CCS and a mixture of milled/unmilled CCS led to enhanced GF nanoparticle recovery and dissolution rate from surfactant-free NCMPs.

(583c) One-Step Fabrication of Agent-Loaded Biodegradable Microspheroids for Drug Delivery and Imaging Applications

Michael J. Heslinga, Gabriella Willis and **Omolola Eniola-Adefeso**, Chemical Engineering, University of Michigan, Ann Arbor, MI

Non-spherical particles may offer advantages over conventional spherical systems for vascular wall-targeted imaging and drug delivery applications for the treatment of several diseases. This work describes the novel fabrication of agent-loaded poly(lactic-co-glycolic acid) (PLGA) spheroid carriers via oil-in-water (O/W) and water-in-oil-in-water (W/O/W) emulsion solvent evaporation (ESE) methods, respectively. The versatility of this technique is demonstrated via the loading of a variety of therapeutics including paclitaxel, bovine serum albumin, and cadmium sulfide fluorescent nanospheres into PLGA prolate spheroids. The encapsulation efficiency for spheroids made via O/W emulsions is found to be highest at low aqueous phase surfactant concentrations and high oil phase volume fractions while encapsulation efficiency for spheroids made from W/O/W is highest at high aqueous phase surfactant concentrations and basic aqueous phase pH values. We show that rods fabricated via the W/O/W method display less polydispersity than ones via the O/W. Overall, particle volume polydispersity in the ESE method can be minimized with high stir rate and high PLGA concentration while aspect ratio polydispersity can be minimized via use of high stir rate, as well as high aqueous phase Tris concentration and pH. The ESE technique is an attractive alternative over recently described methods for fabrication of non-spherical particles due to its simplicity in setup, high particle yield and adaptability to a variety of biodegradable polymers and therapeutics.

(583d) Particle Engineering Via Dry Coating of Micronized API Powders for Improved Dissolution of Directly Compacted Tablets with High Drug Loading

Xi Han, Otto H. York Department of Chemical, Biological and Pharmaceutical Engineering, New Jersey Institute of Technology, Newark, NJ, Chinmay Ghoroi, Department of Chemical, Biological and Pharmaceutical Engineering, New Jersey Center for Engineered Particulates, Newark, NJ and **Rajesh N. Dave**, Chemical, Biological and Pharmaceutical Engineering, New Jersey Institute of Technology, Newark, NJ

Our recent study showed that simultaneous micronization and surface modification through fluid energy milling (FEM), a continuous process, can improve both the flow and dissolution rate of the active pharmaceutical ingredient (API) powders. In this paper, the improved performance of direct compacted tablets with high drug loading of micronized dry coated API powders (20 µm) is examined. As a benchmark comparison, larger, 50 µm API powders dry coated without micronization using a continuous comil process are also considered. Blends with 30, 60 and 70% drug loading, containing API powders that are micronized or non-micronized and dry coated or uncoated are prepared and evaluated. The results show that the blends containing dry coated API powders, even micronized ones, have excellent flowability and high bulk density, which are required for direct compaction. In contrast, blends containing uncoated APIs have poor flow and lower bulk densities. As the drug loading increases, the difference between dry coated and uncoated blends is more pronounced, which is clearly illustrated through a bulk density-FFC phase map. Powder compressibility tests showed improved performance of dry coated powder blends, indicating a good potential for direct compaction. Consequently, tablets are formed through direct compaction for micronized API blends; tablets prepared from dry coated APIs exhibit superior compactibility and dissolution profiles, particularly for higher drug loadings. Dissolution time, t_{80}, for 30 or 60 % drug loaded tablets from 20 µm dry coated ibuprofen, is well under 5 min, illustrating the advantages of the dry coating during API micronization.

(583e) Controlled Release From Self-Assembled Polymer Microparticles That Are Responsive to pH and Temperature

James M. Myrick, Department of Chemical & Biomolecular Engineering, Clarkson University, Potsdam, NY, Sitaraman Krishnan, Department of Chemical and Biomolecular Engineering, Clarkson University, Potsdam, NY and Frederick A. Sexton, New World Pharmaceuticals, Rumson, NJ

Controlled release of biologically active molecules is of significant interest in the pharmaceutical and biomaterials fields. Benefits of controlled release mechanisms include adjustable release rate, improved pharmacokinetics, and reduction of undesirable side-effects. Among controlled delivery systems, oral delivery using biocompatible polymer particle excipients has received significant attention [1]. In this study, particles were formed from a selectively water-soluble cellulose ether with a lower critical solution temperature (LCST) close to body temperature [2]. Above the LCST, hydrophobic polymer interactions dominate resulting in self-aggregation of polymer chains. The polymer chains were cross-linked above the LCST, and formed hydrated microgel particles when cooled below the phase transition temperature. Particles were encapsulated with a carboxylated polymer, resulting in a pH-responsive core-shell configuration. Controlled release from the particles was investigated using hydrophilic small-molecules as entrapped solutes. The efficacy of particle formulations was measured through release kinetics performance in comparison to aqueous solutions of the hydrophilic molecules. Sustained release from the particles and control was analyzed *in vitro* at temperatures above and below the LCST using a horizontal static diffusion cell. The empirical data were found be in good agreement with a mathematical model developed for *in vitro* release kinetics experiments.

REFERENCES

1. Kost, J.; Langer, R. Responsive polymeric delivery systems. *Adv. Drug Deliv. Rev.* **2001**, *46*, 125-148.
2. Sexton, F. A.; Krishnan, S.; Vendra, V. K. Sustained release of nutrients *in vivo*. *U. S. Pat. Appl. Publ.*, US 20090155409 A1 20090618 (2009).

(583f) Wet Coating of Geldart-C Type Particles in a Rotating Fluidized Bed in a Static Geometry

Philippe Eliaers and Juray De Wilde, Materials and Process Engineering, UCL, Louvain-la-Neuve, Belgium

Fine particle coating has important applications in the pharmaceutical and food/feed industry. Two main routes can be used: wet- and dry coating. Fluidized bed technology is widely used for wet particle coating. The performance of conventional (i.e. gravitational) fluidized beds is, however, limited when coating cohesive Geldart-C type particles. In this work, a rotating fluidized bed in a static geometry (RFB-SG) [1] was used for coating fine particles with an aqueous polymer solution. Proof of concept was given and it was shown that high-G operation allows removing important limitations of conventional fluidized beds.

A 24 cm diameter, 5 cm length chamber, equipped with 72, 0.2 mm gas inlet slots was used. The particles to be coated were fed via the front side of the chamber, whereas the liquid solution was injected using a 15° spray mounted centrally in the chamber and directing toward the outer wall of the reactor (Figure 1). The particles with a mean diameter of 70 micron and a density of 260 kg/m^3 were fed at 2 g/s. The liquid solution was heated up to 90°C prior to injection and the liquid droplets leaving the spray nozzle were on average 65 micron. The liquid flow rate and liquid-solid contact time were varied. The liquid solution was injected after establishing a stable rotating particle bed. Experiments were carried out at different air flow rates, 250 and 400 Nm3/h. The air fed was heated by means of an electric resistance to average feed temperatures between 55°C and 70°C. Batch-wise and continuous particle coating was studied.

The operating conditions were found to be critical for both the particle bed stability and the quality of the particle coating process. Within a range of operating conditions, coated particles could be produced in a reproducible way. Figure 2 shows an example of coated powder that could be produced and its particle size distribution. A comparison with the original powder is made. Some agglomeration was observed, but the agglomeration could be controlled by means of the gas, liquid and solids flow rates, as illustrated in Figure 3. The role of the operating conditions and the chamber geometry are discussed in detail.

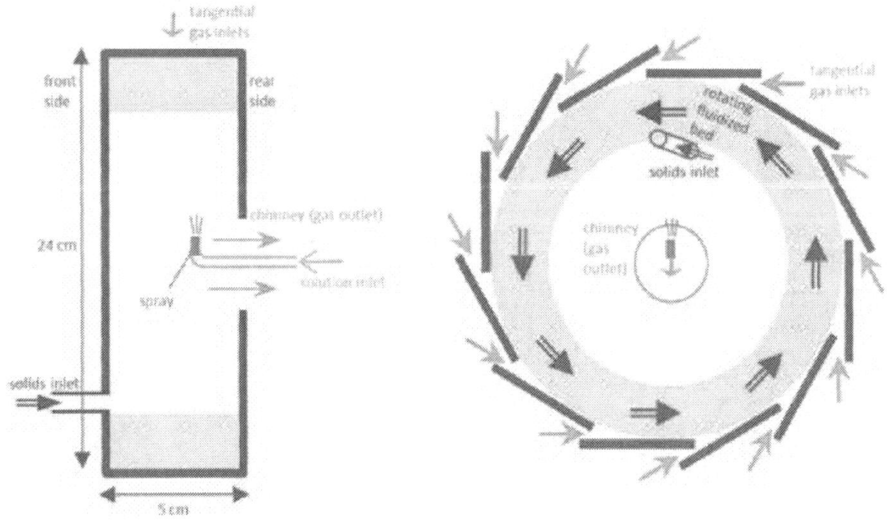

Figure 1. Schematic representation of the RFB-SG particle coater, top and front view.

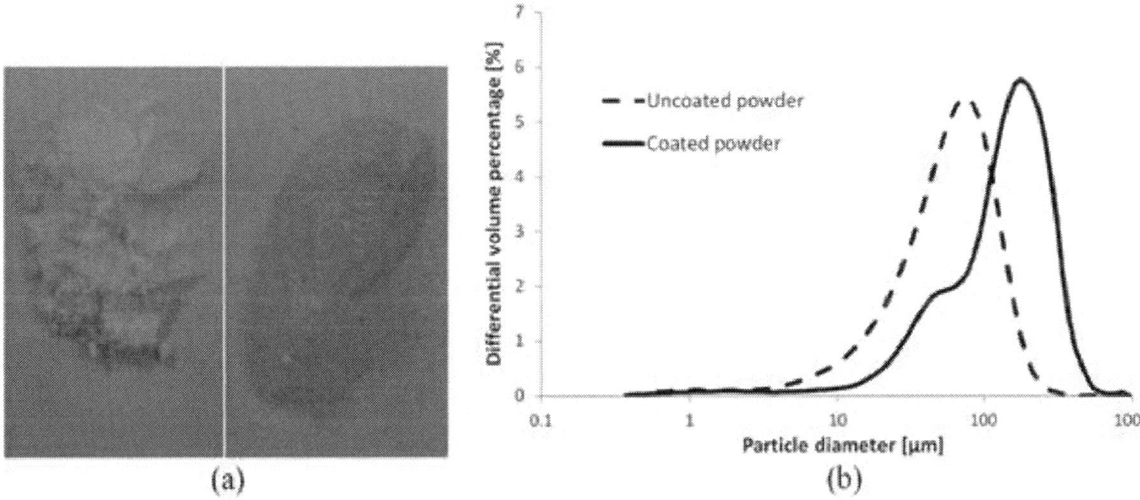

Figure 2. (a) Visual comparison of the uncoated (left) and coated (right) powder; (b) Particle size distribution of the uncoated and coated powder. Solids feeding rate: 2 g/s, liquid solution feeding rate: 1 g/s, air flow rate: 250 Nm3/h at 55°C.

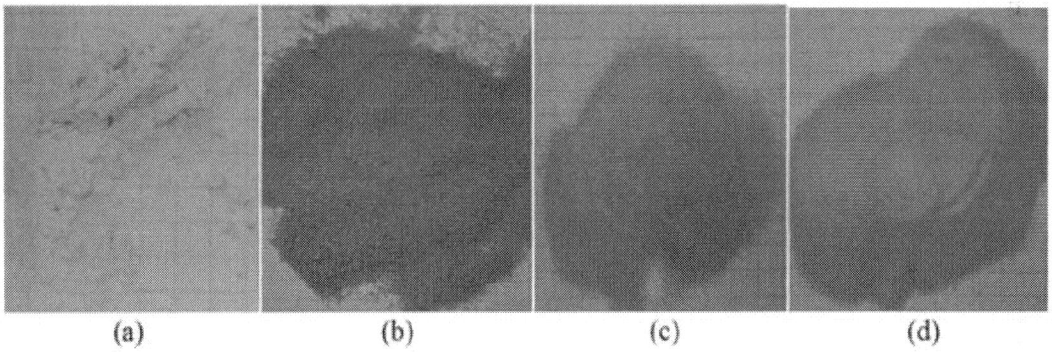

Figure 3. (a) Uncoated powder; (b)-(d) Coated particles of different granulometry produced with decreasing liquid-solid contact time.

[1] J. De Wilde and A.de Broqueville: Rotating fluidized beds in a static geometry: Experimental proof of concept. AIChE Journal, 53, p. 793–810, 2007.

(583g) Solvent-Free Beta-Carotene Nanoparticle Manufature

Phong Huynh, Rutgers University, New Brunswick, NJ and Paul Takhistov, Food Science, Cook College, Rutgers, The State University of New Jersey, New Brunswick, NJ

Most nutraceutical compounds are poorly-water soluble. Their low solubility decreases the adsorption rate in living organisms leading to their low bioavailability. Utilization of nanoparticles is a promising way to improve the solubility of hydrophobic compounds. Some traditional methods for decreasing particle size include pearl or jet milling, where particles are broken down through grinding or collisions under high pressure. These mechanical processes not only require high energy input but also raise a concern of milling media residues. The high pressure homogenizer approach applies implosion forces and collision of particles to generate nanosuspensions. This method requires microsuspensions as starting material and consumes high energy. Among several emulsion-based techniques for preparing nanoparticles, solvent diffusion practice is a novel approach in which a poorly-water soluble compound is transferred into nanoemulsion droplets of a partially water-soluble organic solvent. The compound then crystallizes because the solvent diffuses out of the emulsion droplets. The key point of proposed emulsion-diffusion technology is that the phase transition occurs within an isolated nanoemulsion droplet. We propose a "green" and scalable method for preparing nanosuspensions of highly hydrophobic compounds using FDA GRAS ingredients to create nanoparticles of poorly-water soluble material. b-carotene is selected as a model hydrophobic nutraceutical. Triacetin, a partially-water soluble triacetate compound, is used as the dispersed phase of nanoemulsions. The influence of surfactant, water concentration, and homogenization time on particle size and stability is investigated. The impact of surfactant on diffusion flux of triacetin is studied. Kolmogorov theory is applied to reveal the breakup mechanism of emulsion droplets under shear and predict their size. A mathematical model is built to discover the size of emulsion droplet during the formation of nanosuspensions.

(631a) Relationship Between Solution Metastable Limit and Solubility

Michael Tomasini, Biomedical Engineering, Rutgers University, Piscataway, NJ and M. Silvina Tomassone, Chemical and Biochemical Engineering, Rutgers University, Piscataway, NJ

In pharmaceutical crystallization, and especially in the creation of nanoparticles, the degree of supersaturation and the mechanism of producing supersaturation can have a significant impact on the resultant solid particle properties. For the creation of small particles it is beneficial to suppress the growth phase of crystallization in favor of the nucleation phase. One way to do this is to operate very close to the metastable limit. A difficulty that arises is that the metastable limit of a drug in a particular solvent is not an easy quantity to obtain. Therefore, we seek a method to determine the metastable limit such that we can operate in the metastable zone at a high level of supersaturation but not exceed the metastable limit. It has been empirically observed that the metastable limit curve (the metastable limit as a function of temperature) often mirrors the functionality of the solubility curve. If this were generally true, evaluation of the metastable limit curve would be a straightforward shift of the solubility curve. With the aim of exploring the functional relationship between the metastable limit curve and the solubility curve, we used gauge-cell Monte Carlo (MC) to simulate nucleation and determine the metastable limit and solubility of model drug compounds.

We performed simulations on four different model drug compounds with increasing solubility to determine the solubility curve and metastable limit curve for each compound. We determined that the mestastable limit curve and solubility curves show similar traits, both exhibiting a logarithmic functionality. We also computed the metastable zone width (MSZW) for each compound, the difference between the metastable limit curve and the solubility curve. When normalized to the drug solubility, the MSZW shows a decrease with increasing solubility indicating that as a more soluble drug will have more molecules in solution, it will therefore undergo more molecular collisions and nucleate at lower supersaturations. We also relate the computed collision frequency to the MSZW to allow for prediction of the metastable limit curve from the solubility curve. The results show similar trends to experimental results of drug nucleation.

(631b) Measuring the Energetic Heterogeneity of Pharmaceutical Powders and the Effects On Powder Processing and Formulation

Daniel J. Burnett, Surface Measurement Systems, Allentown, PA and Jiyi Khoo, Surface Measurement Systems, Alperton, United Kingdom

INTRODUCTION

Most pharmaceutical solids are energetically heterogeneous. This heterogeneity could be unintentional due to different crystal facets with different functional groups, co-crystal formation, impurities, defect sites, amorphous regions, or co-processed mixtures of materials. Additionally, surface energy heterogeneity or homogeneity could be engineered to produce desired powder properties. Therefore, it is often difficult or even misleading to characterize the surface energy of these materials with a single value. Recent advances in Inverse Gas Chromatography (IGC) theory and applications have developed new methods for determining surface energy distributions for powders. This relatively new approach now allows the direct measurement of dispersive surface energy at a specific surface coverage or determination of surface energy profiles over a range of surface coverages.

EXPERIMENTAL

The current study investigates the surface energetics of various pharmaceutically relevant materials using IGC technology. Finite concentration IGC experiments allow for the determination of surface energy distributions which more accurately describe the anisotropic surface energy for real materials. Various materials, including Indomethacin (with different processing routes), mannitol (with different surface treatments), lactose (with different amounts of fines), and aspirin (different milling conditions) were studied. The surface energy distributions were then correlated to powder performance properties including wet granulation, powder flow and dissolution behavior.

RESULTS

The affect of processing route on the surface properties of partially amorphous Indomethacin has been investigated. Quench-cooled, milled, and crystalline Indomethacin exhibited different surface energy profiles. In particular, amorphous Indomethacin was energetically most heterogeneous. Additionally, mannitol was studied to relate surface energy distributions to powder flow properties due to changes in surface chemistry. After surface treatment with a silanizing agent, mannitol showed a significantly more homogeneous and hydrophobic surface. These differences were reflected in their powder flow characteristics and wettability during granulation. Finally, the affect of milling was studied on aspirin powders. Milling significantly increased both the average surface energy and surface energy distribution, thus resulting in faster dissolution behavior.

CONCLUSIONS

Surface energy heterogeneity profiles were measured on a wide range of pharmaceutical materials including excipients and APIs. Processing route, surface treatment, and milling conditions showed a dramatic affect on the surface energy values for these materials. These changes in surface energy in turn showed a correlation with bulk power properties like powder flow, dissolution and wet granulation behavior. Therefore, measuring and controlling surface energy values can be an important characteristic for engineering pharmaceutical powders with desired performance properties.

(631c) Formation of Itraconazole-Succinic Acid Cocrystals by CO_2 Antisolvent Cocrystallization

Courtney A. Ober and Ram B. Gupta, Chemical Engineering, Auburn University, Auburn, AL

Cocrystals of itraconazole, an antifungal drug with poor bioavailability, and succinic acid, a water-soluble dicarboxylic acid, were formed by gas antisolvent (GAS) cocrystallization using pressurized CO_2 to improve itraconazole dissolution. In this study, itraconazole and succinic acid were simultaneously dissolved in a liquid solvent, tetrahydrofuran, at ambient conditions. The solution was then pressurized with CO_2, which decreased the solvating power of tetrahydrofuran and caused precipitation of itraconazole-succinic acid cocrystals. A fundamental understanding of cocrystallization in a CO_2-expanded liquid was obtained by modeling the equilibrium concentrations of drug and former in solution as a function of pressure using the Peng-Robinson equation of state and AspenTech software. The cocrystals prepared by GAS cocrystallization were compared to those produced using a traditional liquid antisolvent, n-heptane, for size and surface morphology, crystallinity, thermal behavior, chemical structure, clinical relevance, and stability. Scanning electron microscopy, powder X-ray diffraction, differential scanning calorimetry, and Fourier transform infrared spectroscopy analyses showed that itraconazole-succinic acid cocrystals with physical and chemical properties similar to cocrystals produced using a traditional liquid antisolvent technique can be prepared by CO_2 antisolvent cocrystallization. The dissolution profile of itraconazole was significantly enhanced through GAS cocrystallization with succinic acid, achieving over 90% dissolution in less than 2 hours. The cocrystals appeared stable against thermal stress for up to 4 weeks under accelerated stability conditions, showing only moderate decreases in their degree of crystallinity but no change in their crystalline structure. This study shows the utility of an itraconazole-succinic acid cocrystal for improving itraconazole bioavailability while also demonstrating the potential for CO_2 to replace traditional liquid antisolvents in cocrystal preparation, thus making cocrystal production more environmentally benign and scale-up more feasible.

(631d) Preparation of Co-Ground Drug-Superdisintegrant Nanosuspensions As a Precursor to the Production of Fast Dissolving Surfactant-Free Nanocomposite Microparticles

Mohammad Azad[1], Afolawemi Afolabi[1], Rajesh Davé[2] and Ecevit Bilgili[2], (1)Chemical, Biological, and Pharmaceutical Engineering, New Jersey Institute of Technology, Newark, NJ, (2)Chemical, Biological and Pharmaceutical Engineering, New Jersey Institute of Technology, Newark, NJ

About 40% of the newly discovered drug compounds in the pharmaceutical industry exhibit poor water solubility, which usually leads to poor drug bioavailability. The dissolution rate and bioavailability can be improved by producing nano-sized drug crystals as nanoparticle suspensions (nanosuspensions). In the manufacture of various solid dosage forms, nanosuspensions are dried to prepare nanocomposite microparticles, which are then incorporated into the dosage forms via standard pharmaceutical unit operations. During drying of the nanosuspensions, formation of drug nanoparticle aggregates, solid bridges and agglomerates, and fused (hard) aggregates leads to incomplete recovery of the nanoparticles and ensuing slower dissolution. As a class of dispersants, surfactants play a critical role in fast recovery and dissolution of drug particles from the nanocomposite microparticles (Bhakay et al., 2012). However, surfactants can cause several issues such as physical instability of the nanosuspensions via Ostwald ripening and toxicity especially in inhalation applications, etc. Therefore, surfactants should be eliminated or used sparingly to minimize all potential negative impact. Hence, production of surfactant-free nanosuspensions as a precursor material for nanocomposite microparticle production is of utmost interest to formulators.

Nano-sized superdisintegrants (SDIs) can be used in nanocomposite microparticles as a novel class of dispersants for fast recovery and dissolution of drug nanoparticles (Bilgili et al., 2012). SDIs are water-insoluble, swellable, cross-linked polymers that are traditionally used for fast disintegration of tablets. Unfortunately, the smallest size of SDIs available is in the several microns range. To the best knowledge of the authors, there is no information available in the literature about the nano-sized SDIs. In this study, two commonly used SDIs, sodium starch glycolate (SSG) and crospovidone (CP), were milled in a wet stirred media mill. Aqueous colloidal suspensions of these SDIs were produced in the presence or absence of stabilizers. To incorporate SDIs in the nanosuspensions of drug, SDIs were also co-ground with the drug. Fenofibrate (FNB) was used as a model poorly water-soluble drug. FNB was milled in the presence of an anionic surfactant (sodium dodecyl sulfate, SDS), and was also co-ground with SSG or CP in the absence of SDS. In both cases, a soluble biopolymer (hydroxypropyl cellulose, HPC) was used as a steric stabilizer. We have successfully demonstrated the feasibility of preparing stable SDI and FNB-SDI suspensions using HPC as the stabilizer. The particle size distribution of the FNB-SSG suspension was similar to that of the FNB-SDS suspension. This finding suggests that colloidal SDIs may replace the surfactants or minimize their use in nanocomposite microparticles. The potential implications of the findings are discussed in terms of the redispersion and dissolution of the drug nanoparticles.

References

1. Bhakay, A., Dave, R., Bilgili, E. 2012. Recovery of BCS Class II Drugs during Aqueous Redispersion of Core-Shell Type Nanocomposite Particles Produced via Fluidized Bed Coating. Powder Technol., doi: 10.1016/j.powtec.2011.12.066, in press.

2. Bilgili, E. A., Bhakay, A., Azad, M., Dave, R. 2012. System and Method for Superdisintegrant-Based Composite Particles for Dispersion and Dissolution of Active Pharmaceutical Agents. Provisional Patent, Application No. 61625082, Filed on April 16, 2012.

(631e) Wet-Media Milling of Colloidal Drug Suspensions Stabilized by Means of Charged Nanoparticles

Michael Juhnke, Technical R&D, Novartis Pharma AG, Basel, Switzerland and Edgar John, Technical R&D, Novartis Pharma AG, Basle, Switzerland

INTRODUCTION

Tailoring poorly water-soluble drugs into products with colloidal drug particle size is a proven formulation approach to improve general issues in drug delivery [1]. Colloidal drugs are often produced by wet-media milling technology, due to the availability from pre-clinical to production scale, reliable up-scaling and regulatory acceptance [2]. Colloidal drug suspensions obtained by wet-media milling can be converted by suitable techniques into regular drug products, e.g., tablets and injectables. Stabilization of colloidal drug suspensions is provided by steric and/or electrostatic stabilization using polymers and/or surfactants [2, 3]. Beside these conventional stabilization mechanisms, colloidal particles can be also stabilized by charged nanoparticles providing high electrical charge to the otherwise negligibly charged colloidal particles [4, 5].

EXPERIMENTAL

We investigated the stabilization of colloidal drug particles by charged nanoparticles in order to explore the mechanism of nanoparticle-stabilized colloids as a complementary method compared to conventional stabilization mechanisms. The stabilization mechanism was investigated with the simultaneous size reduction of drug particles by wet-media milling. Aqueous colloidal drug suspensions were produced for a poorly water-soluble drug compound with constant process parameters, pH and content of a non-ionic polymer to facilitate wetting of the hydrophobic drug compound. Stabilization of colloidal drug particles was investigated for positively and negatively surface functionalized charged nanoparticles from silica and polystyrene. Particle size and shape was characterized by dynamic light scattering (DLS) and scanning electron microscopy (SEM). Stabilization of colloidal drug suspensions was characterized without further preparation by rotational and oscillational rheometry. The structural arrangement of charged nanoparticles in relation to colloidal drug particles was determined by cryo transmission electron microscopy (cryo-TEM).

RESULTS

Colloidal drug particle size and shape were identified for all compositions investigated with comparable characteristics. Hence, stabilization of colloidal drug suspensions is obviously determined by the charged nanoparticles for the compositions investigated. Through variations in nanoparticulate matter, size, sign of charge and concentration, colloidal drug suspensions with pronounced stability were produced, in contrast to a strongly agglomerated colloidal drug suspension in gel state without addition of charged nanoparticles. Upon addition of negatively charged polystyrene nanoparticles up to a lower critical concentration, stabilization of negligible charged drug colloids is facilitated into liquid state and almost Newtonian shear behavior. A further increase in nanoparticle concentration resulted in pronounced stability of the colloidal drug suspension without any detectable thixotropy. Interestingly, the lower critical nanoparticle concentration to facilitate drug colloid stabilization is in good agreement with the results obtained by Chan and Lewis [5] for binary mixtures of negligibly charged colloidal silica stabilized by charged polystyrene nanoparticles. In contrast to Chan and Lewis [5] and Zhang et al. [6], we have identified no preferred enrichment of charged nanoparticles near colloidal drug particles for a suspension that undergoes pronounced stabilization.

CONCLUSIONS

We have demonstrated the successful application of charged nanoparticles to facilitate stabilization of a colloidal drug compound produced by an industrial relevant production technology. The application of charged nanoparticles can potentially replace conventional molecularly dissolved ionic stabilizers and opens up new opportunities in drug colloid stabilization without need for drug specific anchor segments.

REFERENCES

[1] Singh A, Worku ZA, van den Mooter G, *Expert Opin Drug Delivery* 8 (2011) 1361-1378.

[2] Merisko-Liversidge E, Liversidge GG, *Adv Drug Delivery Rev* 63 (2011) 427-440.

[3] Bhakay A, Merwade M, Bilgili E, Dave RN, *Drug Dev Ind Pharm* 37 (2011) 963-976.

[4] Tohver V, Smay JE, Braem A, Braun PV, Lewis JA, *Proc Nat Acad Sci* 98 (2001) 8950-8954.

[5] Chan AT, Lewis JA, *Langmuir* 24 (2008) 11399-11405.

[6] Zhang F, Long GG, Jemian PR, Ilavsky J, Milam VT, Lewis JA, *Langmuir* 24 (2008) 6504-6508.

(631f) Interactive Mixtures for Dissolution Rate Enhancement of Cefuroxime Axetil

Namita Dalal, Graduate School of Pharmaceutical Sciences, Duquesne University, Pittsburgh, PA and Ira Shea Buckner, Pharmaceutics, Duquesne University, Pittsburgh, PA

INTRODUCTION

The poor aqueous solubility of newly developed drug molecules is often associated with slower dissolution rate in biological fluids, insufficient and inconsistent biological exposure and sub-optimal efficacy in patients when delivered via the peroral route of administration. One of the methods commonly employed to enhance the dissolution rate is to increase the surface area available for dissolution. In principle, for a given dose of drug, increasing the surface area by decreasing the particle size will lead to a proportional increase in dissolution rate. However, formulations containing micronized drug particles do not always exhibit an improvement in dissolution rate; even reduced rates have been reported. This has been attributed to the agglomeration of fine, primary particles (due to electrostatic and van der Waals forces), which decreases the surface area available for dissolution.

One approach used to take better advantage of this particle size reduction procedure is the preparation of interactive mixtures. Interactive mixtures result from the adherence of fine particles of one constituent to the considerable larger particles of the other (carrier particle). On contact with the dissolution medium, the carrier particles (water-soluble or hydrophilic) dissolve rapidly and also facilitate wetting of the drug particles, releasing the dispersed drug particles into the dissolution medium. This method is believed to maximize the fraction of the particles surface that is in contact with the dissolution medium, thus improving the dissolution rate. This approach should increase the dissolution rate of poorly-soluble drugs such as cefuroxime axetil (CFA).

EXPERIMENTAL

The current study investigates the possibility of dissolution rate enhancement of CFA (commercially available, amorphous API, < 53 µm size range) by forming an interactive mixture with coarse carrier particles of sucrose (500-1000 µm size range). Interactive mixtures with 10% w/w drug loading were prepared using a high shear centrifugal planetary mixer (Thinky USA, Inc.) which was able to deagglomerate the drug particles and cause them to adhere to the sucrose carrier particles. Physical mixtures containing the same amount of drug were also made using a laboratory-scale end-over rotation V-blender for comparison. This mixing method did not effectively break up the CFA agglomerates and the absence of interaction was assessed using optical microscopy. Mechanical stability of the mixtures was determined by sieving methods. Dissolution studies were performed as per the USP 32/NF 27 method for cefuroxime axetil tablets using 0.07N hydrochloric acid (900 ml) as the dissolution medium with the basket apparatus. Samples were analyzed using a validated UV spectroscopic assay. Dissolution studies were performed for the CFA agglomerates as well. The intrinsic dissolution rate of the API itself was measured for reference using the stationary disk method.

RESULTS

The intrinsic dissolution rate (IDR) of amorphous CFA was found to be 0.0765±0.009 mg/cm^2/min. Interactive mixtures of the model drug, CFA and sucrose carrier particles showed a significant improvement in dissolution rate as compared to physical mixtures of the same and CFA agglomerates. The amount dissolved with respect to time was modeled using the Wagner model (assuming exponential surface area changes). The rate constant determined by this method for the CFA agglomerates, physical mixture and interactive mixture with sucrose ($n=3$ each) were 0.0010±0.0000065 min^{-1}, 0.0014±0.000013 min^{-1} and 0.0254±0.0012 min^{-1}, respectively. In addition the effective specific surface area (SSA) was calculated using the rate constant and IDR values. The SSA was found to be 331.61±15.39 cm^2/g, 17.64±1.26cm^2/g and 12.96±0.84 cm^2/g for the interactive mixture, physical mixture and CFA agglomerates respectively. These results indicate that interactive mixing resulted in a significant increase in the fraction of the particles surface area that was available for dissolution leading to an increase in the dissolution rate. The considerably slower dissolution of CFA agglomerates and the physical mixture can be attributed to the fact that a large fraction of the particles surface was inaccessible to the dissolution medium. This problem was overcome by distributing primary particles of the API over the surface of the water-soluble carrier particles in the interactive mixture.

CONCLUSION

Agglomeration of fine particles is a particular problem for hydrophobic drugs which can limit the amount of drug that can dissolve and as a result contribute to the poor oral bioavailability of such drugs. In addition agglomeration can also to negate the particle size reduction procedure that is commonly employed to improve dissolution rate. This study demonstrated that deagglomeration of the CFA agglomerates and distribution of the primary particles of the drug over the surface of the carrier particles significantly enhanced the dissolution rate by increasing the effective surface area of the fine drug particles exposed to the dissolution medium and thus available for dissolution. It is expected that interactive mixtures can help to overcome limitations associated with the use of poorly-soluble, finely powdered drugs. This approach can be further extended to other poorly-soluble drugs in cases where it is critical to maximize the amount dissolved in the gastrointestinal tract.

(631g) Formation of Stable Nanocarriers by Tuning Pharmaceutically Active Ingredient Properties Via an *in Situ* Salt Precipitation

Nathalie M. Pinkerton[1], Arnaud Grandeury[2], Andreas Fisch[2], Jörg Brozio[3], Bernd Riebesehl[2] and Robert K. Prud'homme[1], (1)Chemical and Biological Engineering, Princeton University, Princeton, NJ, (2)Novartis Pharma AG, Basel, Switzerland, (3)Pharmaceutical & Analytical Development, Novartis Pharma AG, Basel, Switzerland

NanoCarriers (NCs) are a promising drug delivery route, which offer increased active pharmaceutical ingredient (API) circulation times and enhanced bioavailability. When formed via rapid precipitation techniques such as flashnanoprecipitation, NCs have higher API loading than their post-loaded delivery counterparts. The aqueous solubility and crystallinity of the API affects NC formation and stability. Generally, hydrophobic materials (logP > 3.5) can be successfully formulated as NCs because sufficient supersaturation can be achieved for nucleation-dominated particle growth. With weakly hydrophobic API, however, the required supersaturation cannot be achieved resulting in low encapsulation efficiency or loss of particle size control. Additionally, with weakly hydrophobic and highly crystalline API, instabilities can occur due to redissolution and recrystallization of the drug compound. To modify API properties for successful encapsulation into NCs, two common routes exist. The first is to modify the drug molecular structure by creating a prodrug. The prodrug is considered a new compound and must undergo FDA approval. The other is to create a salt of the drug, which does not create a new compound and thus circumvents the need for new FDA approval. The salt technique has been used for the formation of bulk salts to improve dissolution behavior, but never to aid in the formation and stability of NCs. We demonstrate the successful ion pairing of APIs to hydrophobic counterions to enable NC formation by rapid precipitation. No additional preparation steps prior to the precipitation process are needed, as the hydrophobic salt is formed *in situ* during the NC self-assembly. The rules for successful nanocarrier formation are presented using three model APIs, cinnarizine, clozapine and α-lipoic acid. This technique expands the types of APIs that can be successfully encapsulated in NC form for drug delivery.

(634a) Application of a Kinetics Model to Simulate Biomass Gasification Using Oxygen and Steam

Sudhanya Banerjee[1], Jordan Tiarks[1], Song-Charng Kong[1], Terry Meyer[1], Robert C. Brown[1] and Brian Anderson[2], (1)Mechanical Engineering, Iowa State University, Ames, IA, (2)Chemical Engineering, West Virginia University, Morgantown, WV

Application of a Kinetics Model to Simulate Biomass Gasification Using Oxygen and Steam

Sudhanya Banerjee[1], Jordan Tiarks[1], Song-Charng Kong[1], Terry Meyer[1] , Robert C. Brown[1] and Brian Anderson[2]

[1]Mechanical Engineering, Iowa State University, Ames, IA

[2]Chemical Engineering, West Virginia University, Morgantown, WV

Abstract: The use of biomass as an energy source has been studied extensively in recent years and special attention is paid to biomass gasification. Chemical kinetics in the gasifier plays a pivotal role in determining the product gas composition. In this study, a model based on ASPEN Plus is set up to simulate the biomass gasification process, and a set of kinetics reactions is chosen to predict the syngas composition. Simulations are carried out for conditions of 21%, 30%, 40% and 50% oxygen levels in the gasifying agent on wet basis. Different types of biomass feedstock are used for the study, including various woody biomass and corn stover. Due to the differences in feedstock composition and operating conditions, the syngas composition varies and the model is able to predict these variations. The simulations results agree well with experimental data which show increases in hydrogen and carbon monoxide as the oxygen level increases in the gasifying agent.

(634b) Gasifier Performance Modeling Including Mass Transfer Effects

Ronald W. Breault, Computational Sciences Division, National Energy Technology Laboratory, Morgantown, WV and Phil Nicoletti, URS COrporation, Morgantown, WV

In the development of multiphase reacting CFD codes, a number of simplifications were incorporated into the codes and models. One of these simplifications was the use of a simplistic mass transfer correlation for the faster reactions and omission of mass transfer effects completely on the moderate speed and slow speed reactions such as those in a fluidized bed gasifier. Another problem that has propagated is that the mass transfer correlation used in the codes is not universal and is being used far from its developed bubbling fluidized bed regime when applied to riser reactors. These problems are true for the major codes, which at this time does not consider mass transfer effects.

To alleviate this problem, a mechanistic based mass transfer coefficient algorithm has been developed based upon earlier work by Breault et al. [1, 2, 3 and 4]. This fundamental approach uses the local hydrodynamics to predict a local, time varying mass transfer coefficient. The algorithm previously developed has been generalized to work on cells of any volume. As was the case for the non-generalized cell volume, the predicted mass transfer coefficients and the corresponding Sherwood numbers agree well with literature data and are typically about an order of magnitude lower than the correlation noted above. The incorporation of the new mass transfer model into the C3M code gives the expected behavior for all the gasification reactions evaluated in the paper. The new algorithm should be used such that the conversions are not over predicted. Additionally, its behavior with changes in solids flow rate are consistent with the changes in the hydrodynamics.

1. 1. Breault, R.W, "A review of gas–solid dispersion and mass transfer coefficient correlations in circulating fluidized beds", Powder Technology 163 (2006) 9–17
2. 2. Breault, R. W. and Guenther, C. P., "Mass transfer in the core-annular and fast fluidization flow regimes of a CFB", Powder Technology 190 (2009) 385–389
3. 3. Breault, R. W. and Guenther, C. P., "Mass transfer coefficient prediction method for CFD modeling of riser reactors", Powder Technology 203 (2010) 33–39
4. 4. Breault, R. W., Li, Tingwen, and Nicoletti, P., "Mass Transfer Effects in a Gasification Riser", submitted to Powder Technology

(634c) Effects of Steam and Oxygen On Syngas Composition Derived From Biomass Gasification

Cuong Van Huynh and **Song-Charng Kong**, Mechanical Engineering, Iowa State University, Ames, IA

Experiments are conducted on a pilot-scale, pressurized bubbling fluidized bed biomass gasifier to investigate effects of oxygen and steam on syngas composition. The oxygen content in the enriched air varies from 21, 45, to 80 vol% on dry basis, corresponding to 21, 30, and 40 vol% on wet basis. The bed temperature is maintained at 800 °C for all tests. Three different biomass feedstock with nitrogen contents varying from 0.5 to 1.4 wt% are used for study. Syngas composition is measured using a combination of micro gas chromatograph, a modified IEA Tar Protocol, and Karl Fischer Titration. Results indicate that oxygen-enriched air and steam gasification favors the production of combustible gas components including hydrogen, carbon monoxide, methane, and lighter hydrocarbons. When 40% oxygen is used, hydrogen and carbon monoxide can increase by up to 70% and 34%, respectively, for particular feedstock. Overall, it is found that oxygen and steam gasification is most effective for feedstock with low nitrogen and moisture contents. Results also show that ammonia and NOx concentrations in syngas increase as oxygen enrichment increases. The lower heating value of syngas can increase by as much as 43% for the feedstock studied. When the oxygen level increases from 21% to 40%, the H_2/CO ratio also increases to a moderate level.

Economical Study of Bioparticles used for Denitrification in Fluidized Bed Bioreactors (FBBR)

Ahmed Eldyasti[1]*; George Nakhla[1,2]*; and Jesse Zhu[2]

[1]Department of Civil and Environmental Engineering, [2]Department of Chemical and Biochemical Engineering, The University of Western Ontario, London, Ontario, Canada N6A 5B9

*Corresponding author
Telephone: 519-661-2111 ext. 85470
Fax: 519-850-2921
Email: gnakhla@eng.uwo.ca

ABSTRACT:

Economical study of different bioparticles for four Denitrifying Fluidized Bed Bioreactors (DFBBRs) using granulated maxi-blast plastic, granulated multi-blast plastic, natural zeolite, and lava rock were investigated to evaluate the feasibility of employing these media as carrier media or "bioparticles" for denitrification in a fluidized bed bioreactors. The work showed that the multi-blast plastic and natural zeolite achieved nitrogen and organic removal efficiencies of 99% and 87%, respectively whereas, the maxi-blast plastic and lava rock achieved removals efficiencies of 94% and 77%, respectively. Low biomass yields were observed for multi-blast plastic and natural zeolite of 0.12g VSS/g COD, while the yield for both the maxi-blast plastic and lava rock was 0.19 g VSS/g COD. Comprehensive economic evaluation of the fluidization energy consumption of each media considering the initial capital cost, operating and running cost, and replacement cost showed that the multi-blast plastic and maxi-blast plastic media were cheaper than natural zeolite and lava rock. Considering the average sphericity of each media, the particles with average sphericity of 0.9 i.e. multi-blast plastic and natural zeolite achieved nitrogen and organic removal efficiencies of 99% and 87%, respectively with a fluffy protruding biofilm whereas the particles with average sphericity of 0.5 i.e. maxi-blast plastic and lava rock achieved removals efficiencies of 94% and 77%, respectively with a patchy biofilm. Low biomass yields were observed for multi-blast plastic and natural zeolite with a sphericity of 0.9 of 0.12g VSS/g COD, while the yield for both the maxi-blast plastic and lava rock with sphericities of 0.5 and 0.6 respectively was 0.19 g VSS/g COD. Annualized unit media costs are mostly governed by fluidization energy costs which constitute 76%-97% of the total costs i.e. capital and media replacement costs are insignificant over the long-term. Considering denitrification performance and annualized overall media costs including fluidization energy, media loss, and capital cost, Multi-Blast (MB) plastic particle is the most suitable and economic media for the denitrification process in DFBBR, with a 20 years annualized capital and running cost of \$1.56-\$1.72 per kg, 30% cheaper than natural zeolite and 50% cheaper than lava rock.

KEYWORDS: Circulating fluidized bed bioreactor, Denitrification, Zeolite, energy properties, Suspended solids removal, Biomass yield.

Introduction

Supporting carrier particles characteristics i.e. size, shape, density, porosity, roughness, and surface area play a significant role in the attachment/detachment rate, hence impacting biological nutrient removal performance in order to meet the effluent quality requirements and reduce the recirculation flow rate which governed by the upflow velocity of bioparticles. In unison, bioparticles affect the investment of capital construction and operation cost significantly (Tang and Fan, 1989).

As the increasing application of FBBR system, deeper understanding on operational discipline, operation optimization, bioparticles structural characteristics, and energy consumption are required to improve the treatment efficiency and operation stability coupled with an optimum fluidized energy running cost.

Thus, a comprehensive evaluation of different bioparticles i.e. granulated maxi-blast plastic (MX), granulated multi-blast plastic (MB), natural zeolite (NZ), and lava rock (LR) as a biofilm carrier for DFBBR was studied with a synthetic municipal wastewater for 180 days including (a) physical properties, (b) biofilm and microbial morphology, (c) denitrification performance, and (d) energy consumption. This study also aims at evaluating the construction and operation cost of different bioparticles and select the optimum bioparticle for DFBBR system over a 20 years operation period.

Materials and Methods

Experimental setup

The DFBBRs, depicted in Figure 1a, comprise four identical plexi-glass reactors with heights of 100 cm and diameters of 2.54 cm ID corresponding to a volume of 507 ml. A layer of stainsteel balls were added to the bottom of each reactor to insure a uniform liquid distribution and approximately 40 g (dry weight) of each particle were used. Each reactor was working in a fluidized bed bioreactor regime using a liquid recirculation. The synthetic wastewater was pumped into each FBBR through peristaltic pumps (Masterflex I/P, Masterflex AG, Germany) and flowed upward through the stainsteel balls. Along with the feed, the liquid were also recycling using a calibrated peristaltic pump (Masterflex I/P, Masterflex AG, Germany) to fluidize the media at a same shear force governed by the bed voidage and monitor the energy consumption in order to evaluate the running cost of each bioparticles. The detailed operating conditions of the DFBBR, as given in Table 1, were controlled by maintaining liquid recirculation. Superficial liquid velocities varied from of 0.8±0.1 to 2.7±0.1 cm/s were maintained in MX, MB, NZ, and LR bioreactors respectively which were between the MX, MB, NZ, and LR minimum fluidization velocities (u_{mf}) of 0.20, 0.21, 0.36, and 0.4 cm/s and the terminal settling velocities (u_t) of 2.6, 3.2, 5.3, and 5.7 cm/s, respectively.

Anoxic condition was monitored in the DFBBRs by oxidation reduction potential (ORP) measurement. The reactor ORP, pH, and temperature were monitored using ORP meters (DPD1P5, pHD Sc Digital Differential pH/ORP Sensor, Loveland, USA), installed along the columns. The process temperature was maintained at 20±3°C throughout the study.

Four types of particles i.e. granulated maxi-blast plastic (Maxi-Blast Inc., Canada), granulated multi-blast plastic (Maxi-Blast Inc., Canada), natural zeolite (Zeolite ore, Bear River Zeolite Inc., USA), and lava rock (Zeox Mineral Materials Corp., USA) were used in this study (Figure 1b) in order to imitate the artificial and natural particles with different dry and true density varied from 719 kg/m^3 to 1142 kg/m^3 and 1363 kg/m^3 to

2685 kg/m^3, respectively. A same average diameter (d_m) of 600-850 μm for all media and a specific surface area determined by BET (Micrometrics ASAP 2010, Micrometrics Co., USA) varied from 0.72 m^2/g to 15.5 m^2/g were used as summaries in Table 2.

Moreover, an abrasion test was performed experimentally in order to measure the losses of each media. A sample of 50 g of the washed medium was inserted to a 1000 ml flask with a 350 ml of distilled water and then mixture was stirred for 30 min at 500 rpm by an agitator with two blades in a diameter of 70 mm.

Results and Discussion

Physical Properties of the support media

Physical properties of the four media used in this study are summarized in Table 2. The bulk density (ρ_b) of all particles were in a close range of 720-1100 kg/m3, while the true (wet) density (ρ_t) of the artificial particles had a significantly lower true density of 1360-1550 kg/m^3 compare to the natural particles of 2500-2680 kg/m^3. Moreover, considering the densities and the slight difference in the particles shapes as showing in Figure 1b and c, the upflow velocity at 200% bed expansion of bare particles for MX and MB was 0.8 and 1.55 cm/sec corresponding to 16 Q_{feed} and 33 Q_{feed}, respectively while the upflow velocity of NZ and LR was 2.4 and 2.6 cm/sec corresponding to 52 Q_{feed} and 56 Q_{feed}, respectively. After the biofilm attached to all particles and the DFBBRs runs at steady state condition, the upflow velocity on the day of 120th at 200% bed expansion of bioparticles for MX, MB, NZ, and LR was dropped to 0.8, 0.8, 1.3, 1.7 cm/sec corresponding to 16 Q_{feed}, 16 Q_{feed}, 27 Q_{feed}, and 36 Q_{feed}, respectively as shown in Figure 2. It is noteworthy to mention that the recirculation ratios of the artificial particles; which were used to fluidized the medium and affects the energy consumption and running cost;

were 50% less than the recirculation ratios used for natural particles. Hence, the energy consumption of MX and MB was 50% less than NZ and LR during the steady state denitrification process.

Additionally, the abrasion loss of MX, MB, NZ, and LR; calculated from the weight loss during agitation under high shear forces; was 2.8%, 6.4%, 18%, and 5.2%, respectively. Therefore, the maintenance and replacement cost of natural zeolite will be significantly higher than other particles.

Nutrient Removal

The DFBBRs columns were tested at an average flow rate of 20 L/d with synthetic municipal wastewater for 180 days to evaluate the difference between MX, MB, NZ, and LR media. In order to ensure attainment of the steady-state conditions in all DFBBRs and monitor the organic and nitrogen removal, the attached and suspended biomass in each column are measured and depicted in Figure 3 and Figure 4, respectively. As noticed from the data in the figure, the coefficient of variation (COV) for attached biomass of MX, MB, NZ, and LR columns were varied from 4% to 10%. Although it is arguable that suspended VSS concentrations varied more widely, as reflected by COV of 14% to 19%, this process is indeed a fixed-film system and 99.99% of the biomass inventory in each DFBBR is in the form of attached biomass. Moreover the denitrification activity per gram media of MX, MB, NZ, and LR depicted in Figure 3b demonstrates that the SDNR coefficients of variation were 1.1%, 4.2%, 3.6%, and 2.9%, respectively. Therefore, the attached biomass and biomass activity reached steady-state.

Figure 5 shows the performance of each DFBBR with respect to chemical oxygen demand (COD) and nitrogen removal efficiencies. As illustrated in Table 3 and Figure 5c, the COD removal efficiencies of MX and LR were in a range of 78% while the MB and NZ media achieved a COD removal of 88% at an organic loading rate (OLR) of 5.9±0.5 kg COD/ m^3.d. In case of MX and LR medium, the COD removal efficiency decreased sharply to 78% due to the higher amount of effluent VSS of 23 mg/l as a result of thinner biofilm thickness and sticky biofilm morphology (Figure 1c) and high biomass detachment rate coefficient. As despite in Figure 1c, the biofilm enlargement of MX bioparticles was limited due to a very sharp particle edges with a discrete biofilm thickness. Moreover, the biofilm of LR bioparticles was also limited due to a very high roughness of the lava rock particles whereas the biofilm morphology and enlargement of MB and NZ were very high with an extended biofilm thickness. As concluded in Figure 1c, the biofilm thickness of MX, MB, NZ, and LR were 104, 555, 684, and 134 μm, respectively. The first-order detachment rate coefficients (d^{-1}), reported in Table 1, were calculated using Equation 1, where the total daily amount of biomass (as VSS) leaving the reactor effluent (X_l) was divided by the total amount of attached biomass (X_m) available in the reactor estimated as the product of particles in the reactor and attached biomass concentrations (Nakhla and Suidan, 2002; Patel et al., 2005).

$$b' = \frac{QX_1}{MX_m} \quad \text{.................................. Equation (1)}$$

As apparent from Table 1, the detachment rates of MB and NZ were 0.05 d^{-1} and 0.01 d^{-1} which were significantly lower than MX and LR rates ranged from 0.15 d^{-1} to 0.17 d^{-1}.

Energy consumption and running cost

Comprehensive economic analysis for each media was conducted considering three scenarios for particles abrasion including the initial capital cost, fluidization energy cost, and replacement cost every 1 year, 5 years, and 10 years intervals for scenario a, b, and c, respectively as shown in Figure 5. It must be asserted that the abrasion media losses of MX, MB, NZ, and LR for 2.8%, 6.4%, 18%, and 5.2% were assumed to occur over periods of 1 year, 5 years, and 10 years, respectively for scenarios a, b, and c. The economic study for the DFBBR system was conducted over a 20 years operational period with an interest rate of 6%. Cost for MX, MB, NZ, and LR media of $1.25, $1.25, $0.25, and $1 per pound (lb), respectively were used. Energy consumption for fluidization was based on the measured recirculation flow rate measured (assuming the same head loss) of 16 Q_{feed}, 16 Q_{feed}, 27 Q_{feed}, and 36 Q_{feed} for MX, MB, NZ, and LR, respectively with a pumping efficiency of 60%, and a unit cost of $0.1/KWH.

As summarized in Table 4, the total annualized unit costs for MX, MB, NZ, and LR were essentially insensitive to media replacement costs, with the artificial media 30# cheaper than natural zeolite despite a much lower initial cost, and 50% cheaper than lava rock (Table 4). The fluidization energy costs is the predominate component accounting for 76%-97% of the total annualized cost. For NZ and LR, fluidization energy costs constituted 97% and 93% of the annualized cost, respectively, while for MX and MB, it was only 84% of the total annualized cost. Hence, considering the abrasion loss, fluidization energy, and the total annualized unit cost per kg for the MB, the artificial multi blast particles (MB) were at least 30% lower than the other natural media used in DFBBR. Although the performance of the two artificial media (MX and MB) was comparable in term of COD and nutrient removal, considering a higher sludge yield of the MX highlights the advantages of the MB media with a lowest unit annualized cost ranging from \$1.56-\$1.72 per kg.

Conclusions

Biofilm structure, reactor performance, and energy consumption were studied for four Denitrifying Fluidized Bed Bioreactors (DFBBRs) employing natural and artificial particles. The DFBBRs were operated at an OLR of 5.9±0.5 kg COD/ m^3.d and NLR of 1.2 kg N/m^3.d. The principal finding of this study are:

- Considering the average sphericity of each media, the particles with average sphericity of 0.9 i.e. multi-blast plastic and natural zeolite achieved nitrogen and organic removal efficiencies of 99% and 87%, respectively with a fluffy protruding biofilm whereas the particles with average sphericity of 0.5 i.e. maxi-blast plastic and lava rock achieved removals efficiencies of 94% and 77%, respectively with a patchy biofilm.
- Not only will the shear forces and detachment rate affect the biofilm morphology, but also the particles sphericity is an important factor for bioparticles technology. Generally particles with low sphericity exhibited higher biofilm detachment rates, as well as thinner and denser biofilms.
- Low biomass yields were observed for multi-blast plastic and natural zeolite with a sphericity of 0.9 of 0.12g VSS/g COD, while the yield for both the maxi-blast plastic and lava rock with sphericities of 0.5 and 0.6 respectively was 0.19 g VSS/g COD.

- Annualized unit media costs are mostly governed by fluidization energy costs which constitute 76%-97% of the total costs i.e. capital and media replacement costs are insignificant over the long-term.
- Considering denitrification performance and annualized overall media costs including fluidization energy, media loss, and capital cost, Multi-Blast (MB) plastic particle is the most suitable and economic media for the denitrification process in DFBBR, with a 20 years annualized capital and running cost of \$1.56-\$1.72 per kg, 30% cheaper than natural zeolite and 50% cheaper than lava rock.

ACKNOWLEDGEMENTS

The authors gratefully acknowledge Trojan Technologies, Canada, Natural Science and Engineering Research Council of Canada (NSERC), Ontario Center of Excellence (OCE), Canada, and City of London, ON, Canada for their endless support and interest at every stage of this research project.

References

Al-Degs Y, Khraisheh M, Allen S, Ahmad M. 2000. Effect of carbon surface chemistry on the removal of reactive dyes from textile effluent, Water research 34(3):927-935.

Alves CF, Melo LF, Vieira MJ. 2002. Influence of medium composition on the characteristics of a denitrifying biofilm formed by *Alcaligenes denitrificans* in a fluidised bed reactor, Process Biochemistry 37:837–845.

APHA, AWWA, WEF. 1998. Standard Methods for the examination of water and wastewater. 20th Edition, American Public Health Association, Washington D.C., US.

Arnaiz C, Gutierrez JC, Lebrato J. 2006. Support material selection for anaerobic fluidized bed reactors by phospholipid analysis, Biochem. Eng. J. 27:240–245.

Arnaiz C, Buffiere P, Elmaleh S, Lebrato J, Moletta R. 2003. Anaerobic digestion of dairy wastewater by inverse fluidization: the inverse fluidized bed and the inverse turbulent bed reactors, Environ. Technol. 24:1431–1443.

Balaguer MD, Vicent MT, Paris JM. 1997. A comparison of different support materials in anaerobic fluidized bed reactors for the treatment of vinasse, Environ. Technol. 18:539–544.

Buffiere P, Bergeon JP, Moletta R. 2000. The inverse turbulent bed: a novel bioreactor for anaerobic digestion, Water Res. 34:673–677.

Chowdhury N, Nakhla G, Zhu J, Islam M. 2009. Pilot-scale experience with biological nutrient and biomass yield reduction in a liquid-solid circulating fluidized bed bioreactor, Water Environ. Res. 82(9):772-781.

Cui Y, Nakhla G, Zhu J, Patel A. 2004. Simultaneous carbon and nitrogen removal in anoxic-aerobic circulating fluidized bed biological reactor (CFBBR). Environ. Technol., 25:699-712.

Durmaz, B, Sanin FD. 2001. Effect of carbon to nitrogen ratio on the composition of microbial extracellular polymers in activated sludges. Water Sci. Technol. 44:221-229.

Gjaltema A, Vinke JL, van Loosdrecht MCM, Heijnen JJ. 1997. Abrasion of suspension biofilm pellets in airlift reactors: Importance of shape, structure and particle concentration. Biotechnol. Bioeng. 53:88–89.

Grady J, Daigger GT, Lim HC.1999. Biological Wastewater Treatment. Marcel. 1099 p

Hakulinen R, Salkinoja-Salonen M. 1982. Treatment of pulp and paper industry wastewater in an anaerobic fluidized bed reactor .Proc. Biochem. 18-22.

Hao-Ran J, Xun Z, Xiou-He L, Shu-Ping Z, Xiao-Wei C, Wung-Wai T. 1983. Porous kaolinite bead: A novel but economical solid support material for biomass in fluidized bed technology, Biotechnology and Bioengineering, 5(2):85-88.

Henze M, Gujer W, Mino T, Matsuo T, Wentzel MC, Marais GR. 1995. Activated Sludge Model No. 2. IAWQ Scientific and Technical Report No. 3, IAWQ, London, England.

Islam M, Nakhla G, Zhu J, Chowdhury N. 2008. Impact of Carbon to Nitrogen Ratio on Nutrient Removal in a Liquid Solid Circulating Fluidized Bed Bioreactor (LSCFB). Process Biochemistry, 44(5):578-583.

Jeannette S, Thomas S, Dennis P, Santoro M, Keenan A. 1997. Biological upflow fluidized bed denitrification reactor demonstration project - Stanford, CT, USA., Water Sci. Technol. 36:139-146.

Jeris JS. 1983. Industrial wastewater treatment using anaerobic fluidized bed reactors. Wat. Sci. Technol., 15:169-176.

Kida K, Morimura S, Sonoda Y, Obe M, Kondo T. 1990. Support media for microbial adhesion in an anaerobic fluidized-bed reactor, J. Ferm. Bioengng, 69:354–359.

Lazarova V, Manem J. 1995. Biofilm characterization and activity analysis in water and wastewater treatment, Wat. Res. 29(10):2227-2245.

Leenen E, Dos Santos VA, Grolle K, Tramper J, Wijffels R. 1996. Characteristics of and selection criteria for support materials for cell immobilization in wastewater treatment, Water Res. 30:2985–2996.

Li T, Bai R, Liu J. 2008. Distribution and composition of extracellular substances in membrane-aerated biofilm. J. Biotechnol. 135:52-57.

Maqueda C, Perez-Rodriguez JL, Lebrato J. 1995. Anaerobic digestion of wastewater and solid waste using raw clays as supports, Fresenius Environ. Bull. 4:129–134.

Metcalf and Eddy. 2003. Wastewater Engineering, Treatment and Reuse. 4th Ed. McGraw-Hill, New York, 622-623 p.

Miqueleto AP, Dolosic CC, Pozzi E, Foresti E, Zaiat M. 2010. Influence of carbon sources and C/N ratio on EPS production in anaerobic sequencing batch biofilm reactors for wastewater treatment, Bioresource Technology, 101:1324–1330.

Mulcahy LT, Shieh WK. 1987. Fluidization and reactor biomass characteristics of denitrification fluidized bed biofilm reactor (FBBR). Water Res. 21:451-458.

Nakhla G, Suidan MT. 2002. Determination of biomass detachment rate coefficients in anaerobic fluidized-bed-GAC reactors. Biotechnol. Bioeng., 80:660-669.

Patel A, Nakhla G, Zhu J. 2005. Detachment of multi species biofilm in circulating fluidized bed bioreactor, Biotechnol. Bioeng. 92:427–437.

Ras M, Lefebvre D, Derlon N, Paul E, Girbal-Neuhauser E. 2011. Extracellular polymeric substances diversity of biofilms grown under contrasted environmental conditions, Water Research, 45(4):1529–1538.

Rittmann BE. 1982. The effect of shear stress on biofilm loss rate. Biotechnol. Bioeng. 24:501-506.

Rockey JS, Forster CF. 1983. The use of an anaerobic expanded bed reactor for the treatment of domestic sewage. Environ. Technol. Lett. 3:487-496.

Sanz I, Fernández-Polanco F. 1989. Anaerobic treatment of municipal sewage in UASB and AFBR reactors. Environ. Technol. Lett. 10:453-462.

Sehraa G, Jewell WJ. 1984. High rate conversions of soluble organics with a thermophilic anaerobic attached film expanded bed. J. WPCF. 56:226-232.

Shieh WK, Keenan DK. 1986. Fluidized bed biofilm reactor for wastewater treatment. In: Fiechter, A. (Ed.), Advances in Biochemical Engineering/Biotechnology. Springer-Verlag, Berlin, 132-168.

Tang W, Fan LS. 1989. Hydrodynamics of a three-phase fluidized bed containing low-density particles, AIChE Journal, 35(3):355–364.

Tijhuis L, van Loosdrecht MCM, Heijnen JJ. 1994. Formation and growth of heterotrophic aerobic biofilms on small suspended particles in airlift reactors. Biotechnol. Bioeng. 44:595–608.

Van Loosdrecht MCM, Tijhuis L, Wihdieks AMS, Heijnen JJ. 1995. Population distribution in aerobic biofilms on small suspended particles. Wat. Sci. Technol. 31:163–171.

Xing XH, Jun BH, Yanagida M, Tanji Y, Unno H. 2000. Effect of C/N values on microbial simultaneous removal of carbonaceous and nitrogenous substances in wastewater by single continuous-flow Fluidized-bed bioreactor containing porous carrier particles, Biochemical Engineering Journal, 5:29-37.

Ye F, Peng G, Li Y. 2011. Influences of influent carbon source on extracellular polymeric substances (EPS) and physicochemical properties of activated sludge, Chemosphere, 84:1250–1255.

Zhang TC, Bioshop PL. 1994. Density, porosity, and pore structure of biofilms, Wat. Res. 28(11):2267-2277.

Table 1: Operational conditions of DFBBRs

Parameter	Media			
	MX	MB	NZ	LR
Influent flow, Qin (L/d)	20±2			
Average organic loading (kg COD/m^3·d)	5.9±0.5			
Average nitrogen loading (kg N/m^3·d)	1.2±0.1			
Average phosphorus loading (kg P/m^3·d)	0.10±0.02			
HRT (hr)	0.6			
Recirculation ratio (Q_r /Q_{in})	16	16	27	36
Upflow velocity (cm/sec)	0.8	0.8	1.3	1.7
Avg. attached biomass (mg VSS/g media)	21	43	46	30
Biomass (g VSS)	8.4	17.2	17.6	12.4
Food/microorganisms ratio (g COD/g VSS·d)	0.35	0.19	0.17	0.24
Detachment rates (d^{-1})	0.167	0.047	0.011	0.145
Estimated SRT (d)	18	42	50	24

Table 2: Physical properties for different media were used in DFBBRs.

Parameter	Media			
	MX	MB	NZ	LR
Bulk Density (kg/m^3)	719	810	944	1142
True Density (kg/m^3)	**1363**	**1554**	**2496**	**2685**
Internal Porosity (%)	7	10	16	13
External Porosity (%)	40	40	46	44
Total Porosity (%)	47	50	62	57
BET Surface Area (m^2/g)	0.72	0.86	15.5	0.85

Table 3: Influent and effluent characteristics for different media

Parameter	Influent*	Effluent*			
		MX	MB	NZ	LR
DO (mg/L)	-----	0.30	0.30	0.30	0.30
ORP (mv)	-----	-79	-80	-85	-80
pH	7.70	8.21	8.44	8.84	8.28
Alkalinity**	255	355	364	367	357
TCOD (mg/L)	149±3	50±4	34±6	33±4	49±4
SCOD (mg/L)	123±9	36±5	20±4	21±5	33±8
TBOD (mg/L)	88±11	31±5	20±6	22±4	28±9
SBOD (mg/L)	76±14	20±8	12±4	13±3	18±10
NO_3-N (mg/L)	30±4	1.4±0.6	0.1±0.6	0.1±0.4	1.5±0.4
NO_2-N (mg/L)	0.013±0.03	0.09±0.06	0.06±0.06	0.15±0.02	0.08±0.07
TN (mg/L)	30.4±1.9	3.4±1.0	2.3±0.5	2.2±0.8	3.6±1.2
PO_4-P (mg/L)	1.0±0.2	0.7±0.1	0.6±0.2	0.6±0.1	0.7±0.4
TP (mg/L)	1.2±0.4	0.9±0.4	0.8±0.3	0.8±0.1	0.8±0.7
TSS (mg/L)	17±2	30±3	20±3	19±3	29±4
VSS (mg/L)	14±2	22±3	14±3	16±3	23±3
mgVSS/g media	-----	21±2	43±4	46±2	30±3
C:N:P			5:1:0.1		

*Average ± SD (a number of samples 40 with a frequency of a sample every 4 days); **(mg $CaCO_3$/L)

Table 4: Cost analysis of each media used in DFBBR

	MX	MB	NZ	LR
% Capital cost [a]	15%	15%	2%	6%
% Replacing Cost [b]	1%	2%	1%	1%
% Operating Cost [c]	84%	83%	97%	93%
Total annual cost ($/kg per year)[*]	**1.56**	**1.58**	**2.27**	**3.15**

[a] Capital cost percentage from the total annual cost considering an interest of 6%

[b] Material replacing cost percentage from the total annual cost considering a replacement interval of 5 years.

[c] Operating cost percentage from the total annual cost considering an efficiency of 60%

[*] Total annual cost of each media considering a total performance at 40 g media

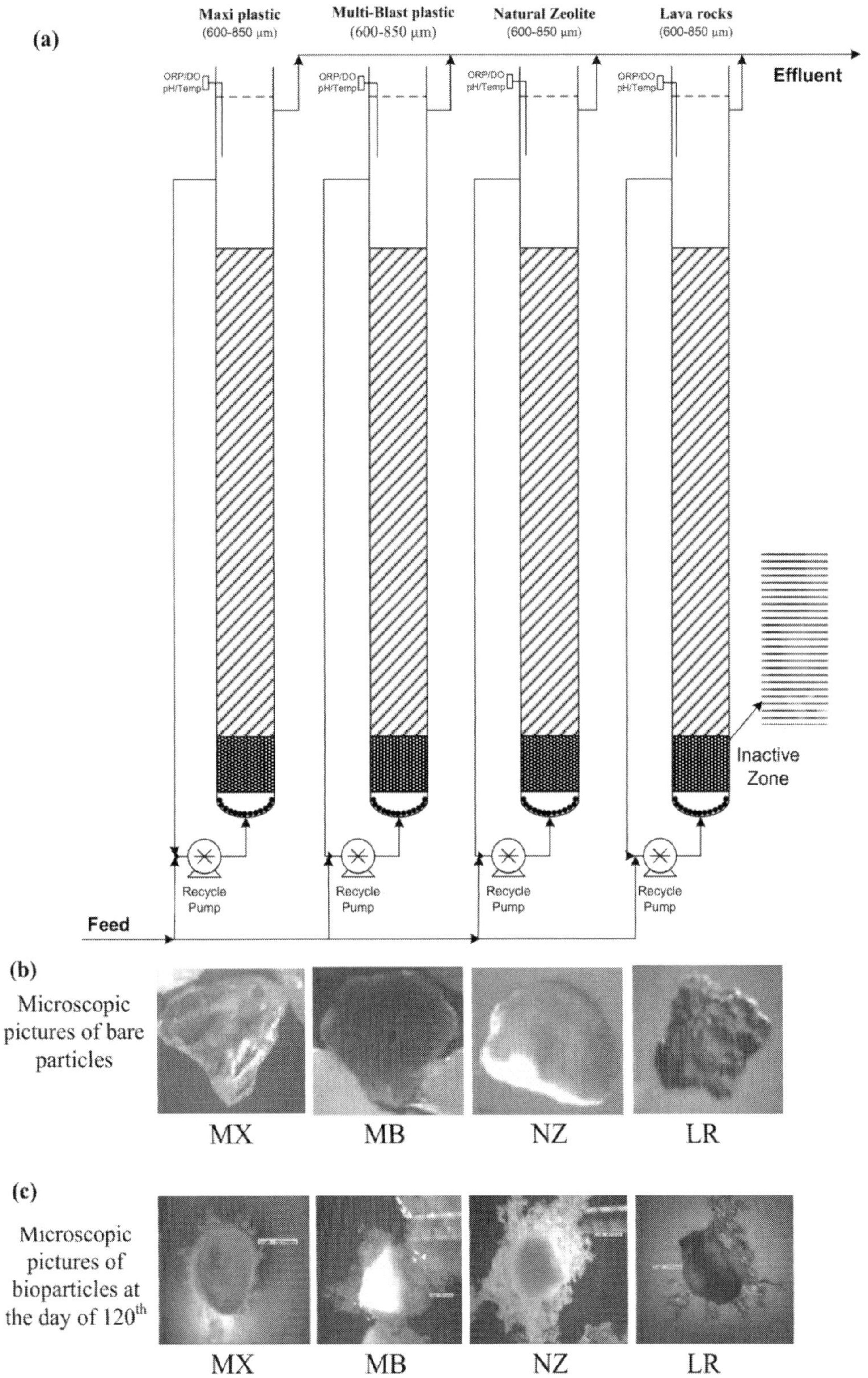

Figure 1: (a) Schematic diagram of DFBBRs, (b) Microscopic picture of bare particles, (c) Microscopic picture of bioparticles at the day of 120th.

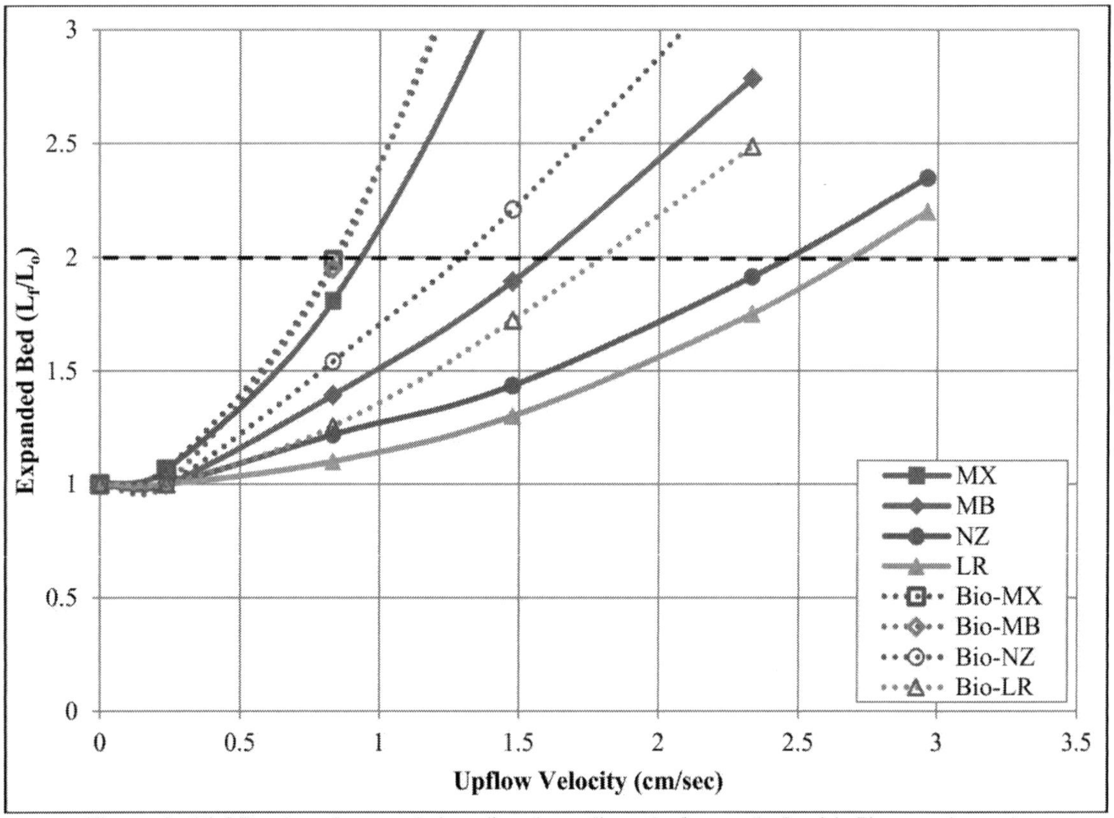

Figure 2: Fluidization characteristics of each medium (before and after biofilm attachment)

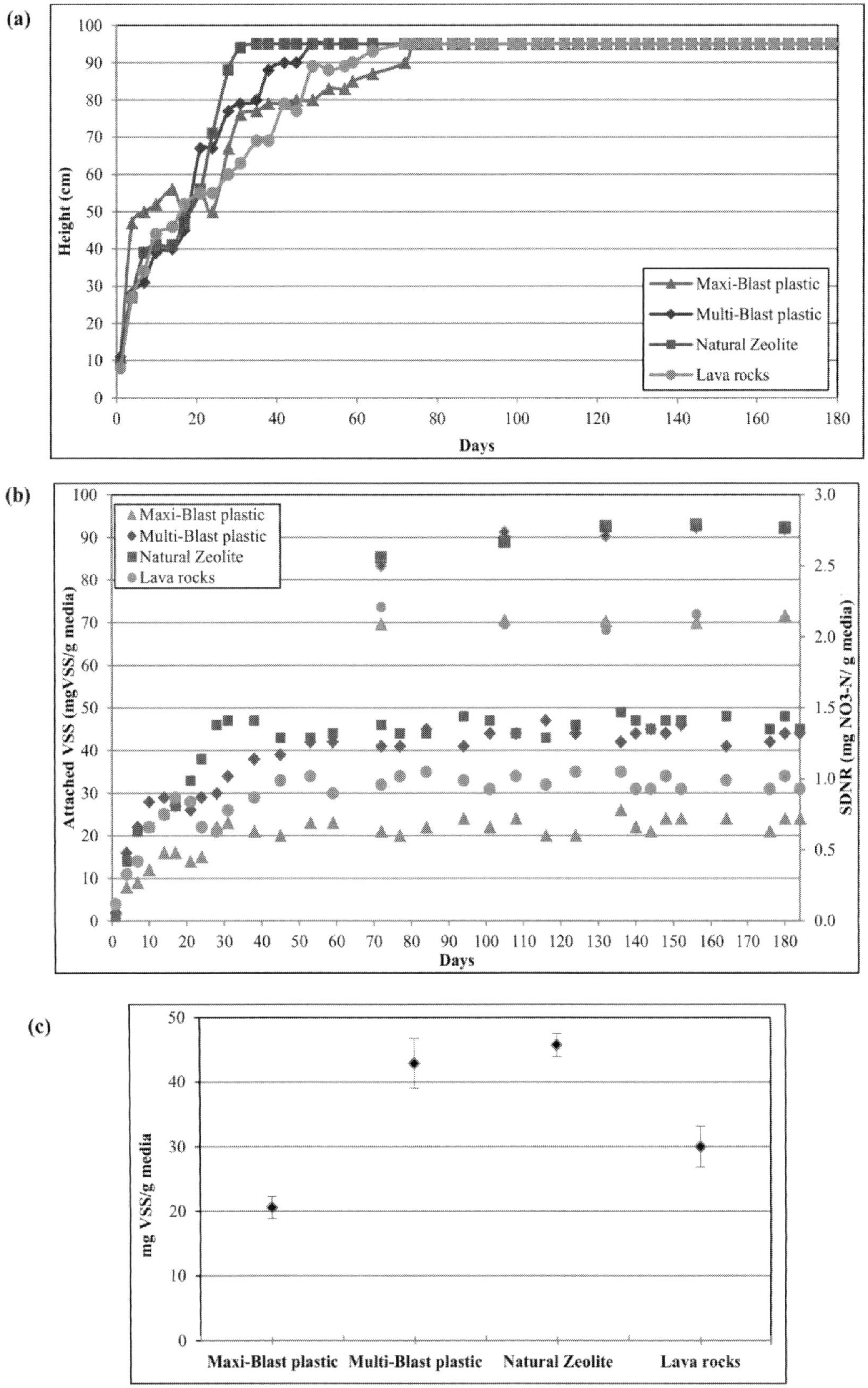

Figure 3: (a) change in heights of expanded bed, (b) attached VSS inventory, and (c) Average and SD of the attached VSS of each media

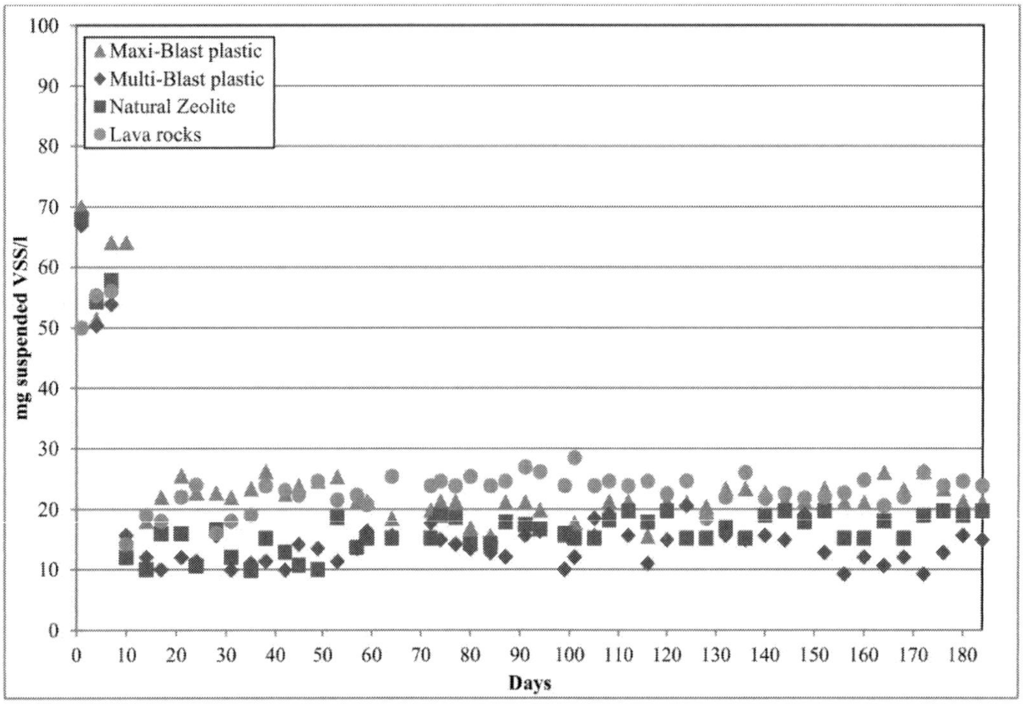

Figure 4: Suspended VSS for each media

(a)

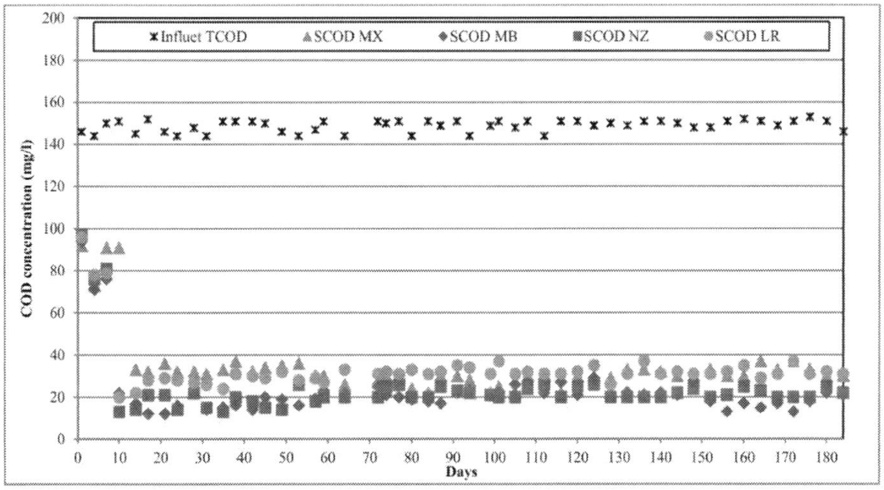

(b)

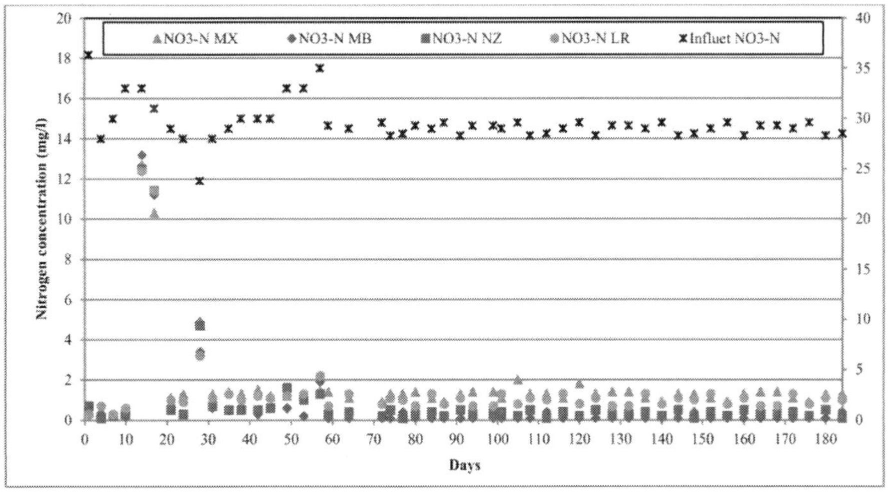

(c)

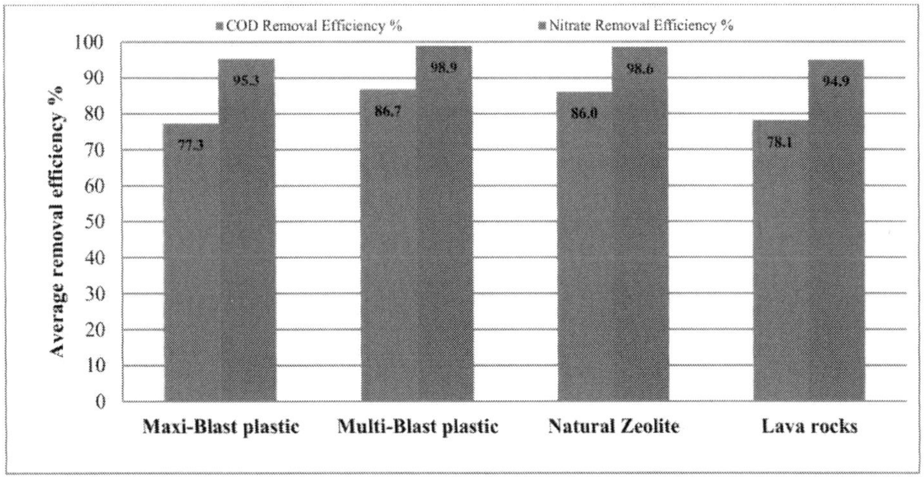

Figure 5: (a) TCOD for influent and SCOD for effluent of each media, (b) Nitrate concentration of influent and effluent for different DFBBRs, and (c) Organic and Nitrogen removal efficiency for each media

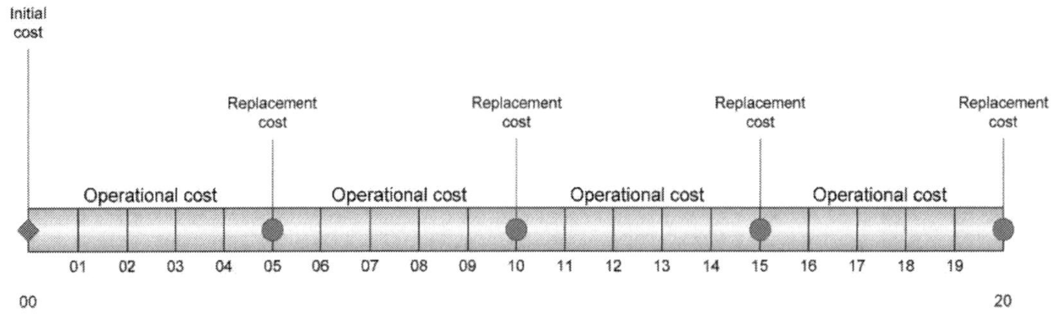

Figure 6: Cash-flow diagram of each media used in DFBBR

(634f) Modeling Effects of Drag Force On Biomass Particles in a Fast Pyrolysis Reactor

Qingang Xiong, Mechanical Engineering, Iowa State University, Ames, IA

Modeling Effects of Drag Force on Biomass Particles in a Fast Pyrolysis Reactor

Qingang Xiong and Song-Charng Kong

Department of Mechanical Engineering, Iowa State University, Ames 50011-2161, IA, USA

Email addresses: xiong@iastate.edu (Q. Xiong), kong@iastate.edu(S. Kong)

Abstract Various drag models are implemented in this study to investigate their effects on biomass particle in a fast pyrolysis reactor. Thermochemical conversion of biomass to produce fuels via fast pyrolysis has become increasingly important. In this study, an open-source CFD code, OpenFOAM, is used to model the evolution of biomass particles in a fast pyrolysis reactor. In the simulation, both gas and biomass particles are treated as interpenetrating continua and the kinetic theory of the granular flow is employed to close the solid stress. Biomass particles are fed from the side and the product gas exits from top. In order to focus on the issue of drag relations, a simplified lumped reaction model is used to represent the pyrolysis process. Drag relations derived from homogeneous fluidization and closures from heterogeneous analysis (e.g., EMMS/Matrix model) are compared in terms of product yields and distributions. It is found that homogeneity-based drag relations, due to their lack of considering sub-grid structures in each computational cell, give relative poor predictions while for those heterogeneity-based models, the simulation appears to be better. Detailed results about the effects of model formulation will be presented. Such results indicate that for gas-solid flows, regardless of chemical reactions, sub-grid structures need to be considered in inter-phase drag force within the framework of the two-fluid model.

(634g) Effects of Acid-Washing Treatment On Pyrolysis Products of Biomass in a Downer Reactor

Tongli Ding[1], **Songgeng Li**[2], Wenli Song[2] and Weigang Lin[2], (1)Institute of process engineering, Chinese Academy of Sciences, Beijing, China, (2)Institute of Process Engineering, Chinese Academy of Sciences, Beijing, China

With an ever-increasing demand for energy and a growing concern on global climate change, biomass has gained much attention as an alternative to fossil fuels in recent decades due to its renewability and CO_2 neurality. Pyrolysis is attractive in the utilization of biomass for energy purpose. In this work, expertimental study has been conducted in a bench-scale CFB downer reactor with an emphasis on the effects of acid-washing treatment on pyrolysis product distribution and product properties. It is found that the removal of inorganic species increases the yield of liquid product and reduces the char formation and gas yield. CO and CO_2 are the dominant components in the gaseous product, which account for over 90%. The concentration of CO in the gaseous product increases after the acid treatment while the CO_2 concentration decreases. Oxygen and water contents in the liquid product is decreased on acid treatment, which leads to relatively high heating value and viscosity. It is found that volatiles in the char derived from the acid treated wheat staw is higher than that from the raw wheat straw. This may suggest a longer residence time is needed for the pyrolysis of the acid treated wheat straw in order to obtain the maximal yield of volatile matter.

Historical Review of Pneumatic Conveying and Solids Processing World Wide

George E. Klinzing
University of Pittsburgh

Trying to determine just when the first pneumatic conveying system was designed and built is a bit of a challenge. In the process of forming the organization C Handling and Processing Systems (CHoPS) Haim Kalman made reference to solids handling in the Bible. The late Ming Leung often would say nothing could beat the age and historical significance of the ancient Chinese culture. We can accept these claims but with the printed page references can be made about 150 years of history. One of the earliest to be cited is a case of grain unloading from ships in Russia. The German history can point back to the early 1920's following the intense study of fluid dynamics that occurred some 10 to 20 years earlier. Looking back further in 1785 an account appears on the analysis of a dust explosion in the handling of flour at a flour mill in Turin, Italy. Reading this account one is struck by how ideal the conditions were noted for the perfect storm for the occurrence of a dust explosion.

The beginning of rather modern pneumatic conveying and solids processing began about the same time in Japan and Germany in the late 1950's through the 1960's followed by activity in the United States.

In Japan Professor Inoya at Nagoya University assembled a team of students to begin to address the challenging area of solids processing from all levels. His lab produced many of the researchers in this field throughout Japan for the next 20 years or more. In fact it was challenging for other Japanese researcher who did not have the Nagoya stamp. In pneumatic conveying Professor Morikawa of Osaka University began to research pneumatic conveying and added a well known student to his research tear, Yukata (Tak) Tsuji. Morikawa and Tsuji did many projects together.

Professor Tsuji followed Professor Morikawa as the Principal Professor at Osaka and since has retired being following by Professor Tanaka. They have produced a large volume of students who are world renown in the modeling and simulations area of solid processing and pneumatic conveying. After retirement from Osaka, Professor Tsuji became the technical director of the Hosokawa Research Institute. Recently he stepped down from this position.

In Japan there were a few other hot spots of pneumatic conveying research. Professor Tomita of Kyushu Institute of Technology contributed extensively to the literature. Professor Tashiro was a student of his and his now at Kurume Technical University. I have had the pleasure of Professor visiting in my laboratory for his sabbatical.

At the large Nihon University Professor Ochi has done pneumatic conveying research. He also spent a summer research sabbatical in my laboratory. He returned to Japan and along with his research became the Dean of the School of Engineering at Nihon University.

At Kyoto University Professor Masuda concentration on trying to understand the electrostatic phenomena that occurs in pneumatic conveying. He did some very basic work and also has explored measurement devices flow measurement in pneumatic conveying.

In Germany the solids processing concept and its study began with Professor Rumpf of Karlsruhe University along with other senior colleagues especially Professor Barth who concentrated his studies in pneumatic conveying provide a rich source of students in solid processing. Almost all of the researchers in this solids processing through 1960's to 1990's came out of Karlsruhe. Many of the existing researchers can also trace their academic lineage back to Karlsruhe.

Professor Barth's untimely death did not stop the research machine from going and researchers such Lothar Reh, Farid Rizk, Karl Sommer, Edward Muschelknautz, Wolfgang Krambrock, Manfred Weber, Matias Bohnet came from this lineage.

Other notable from Karlsruhe who populated Europe's elite of solids processing were Kurt Leschonski, Sunil DiSilva, Klaus Schonert, Hans Loeffler, Otto Molerus, and Jorg Schwedes.

The Karlsruhe group spread throughout Europe taking professorships or senior research positions in industry. Many started in industry and the moved on to universities.

Leschonski – Clausthal Technical University
DeSilva – Postgrun Institute in Norway
Molerus – University of Erlangen
Rizk – BASF
Somers – Munich Technical University
Muschelknautz – University of
Weber – Karlsruhe University
Bohnet – Braunschweig Technical University
Sommer – Technical University of Munich
Krambrock – Waschle, Zeppelin

In the U.S. the first developments in pneumatic conveying came out of Fred Zenz who did some very creative experiment work with a simple vacuum cleaner. The Zenz plots of pressure drop vesus gas flowrates have the named after him. His book with Othmer on Fluidization and Fluid Particle Systems from 1960 is a classic.

Much of the U.S. work did not come out of one font as was in Germany with Karlsruhe. With the importance of energy and coal, fluidizations studies were plentiful and many industrial companies and research organization developed extensive expertise in this field which is ancillary to pneumatic conveying. The work of Wen-Ching Yang of Westinghouse and Ted Knowlton at GTI and later PSRI are most noteworthy. Novel and creative experiments and devices were developed being placed almost immediately on the industrial scale. Expertise developed in the solids processing and fluidization area at Exxon, Texaco, Shell and others. Amrit Agrawal from Union Carbide was their resident expert in pneumatic conveying design with a specialty in dilute phase transport. Dow Chemical developed similar expertise under the leadership of Karl Jacob and Shrikant Dhodapkar.

The area of consulting engineering cannot be forgotten when it comes to pneumatic conveying. Paul Solt after many years with Fuller Company carried out extensive testing on a wide variety of materials and designs. Presently he is teamed with Jack Hilbert in having webinars through the magazine Bulk Solids Engineering. Paul Solt's "Point to Ponder" in the same magazine is classic with series now being sold separately. Solt did extensive short course offerings with the Center for Professional Development, AICHE and the University of Wisconsin. Jenike Johanson is now also offering consulting services in pneumatic conveying.

As I mentioned before the field of pneumatic conveying did not spring from a single source but had a variety of different researchers at universities approach the topic. E. Michalides did some interesting work on the solids friction factors. Charlie (Shao Lin) Soo developed unique mathematical analyses for pneumatic conveying with a heavy emphasis of electrostatic generation. Jennifer Sinclair Curtis now at the University of Florida probed both simulation and modeling proficiencies as well as experimental verification in her work.

Australia had a significant effect on the verification and usage of large scale test facilities and design. Peter Arnold at the University of Wollongong began this work and was always in close communication with Alan Roberts of Newcastle University. Alan specialized in bin and hopper design as well as mechanical transport with belt conveyors. Peter Wypych who studied with Peter Arnold moved the progress forward in large scale pneumatic conveying. He always has real data and real experiments when it comes to pneumatic conveying. Peter Wypych guided a number of very talented students in his laboratory, Bo Mi and Renfu Pan come quickly to mind. Dr. Pan has returned to China and leading a larger effort in China at this time. About 15 to 20 years ago Newcastle University made an attractive offer to Mark Jones who was then with the Stan Mason group at Caledonia University in Glasgow, Scotland. Others than worked closely with the Mason group were Mark joined the Alan Roberts team and carried on what had initially been done by John Chambers to become a real force not only in Australia but around the world. He has had a number of gifted students in Ken Williams and Toby Krull. It is unusual not to see this group at the international conferences on pneumatic conveying around the world.

In Great Britain in the early 1970s Stan Mason developed a strong pneumatic conveying facility and expertise at Thames Polytechnic in London. He attracted David Mills as a student who later developed into a trusted colleague in pneumatic conveying. David has published widely and lectures often. He has had a close collaboration with Dr. Agarwal in India in the transport of fly ash. Alan Reed also was part of the Thames group and took a real leadership role in research and administration later moving to Greenwich University where he works closely with Mike Bradley. They have a large consultancy that goes under the name of the Wolfson Group. While in Scotland, Stan Mason attracted two researchers who developed some interesting basic concepts in measurements in pneumatic conveying: Don McGlinchey and John Pugh. They continue to develop novel flow measurements devices.

At Cambridge University in Great Britain Professor Ronald Nedderman supervised a very talented student, Ken Konrad, in theoretical and experimental analysis of dense phase conveying using the basic principles of soil mechanics. There work is classic and the real beginning to the basic understanding of dense phase flow. Ken Konrad moved to Virginia Tech to teach and then moved to Exxon changing fields somewhat.

A little later in the time line researcher in Israel began to explore the intricacies of pneumatic conveying lead by Professor Haim Kalman. There were other researchers in two phase flow system in Israel who approached the topic in a fundamental basic analysis such as Hetroni. Haim spurred on his research to attrition and wear and produced a number of students who advances these topics but blossomed into other avenues such as Avi Levy who took the modeling and simulation torch further in a very aggressive manner. Haim questioned the basic premises on which most of pneumatic conveying is built and examined previous data assiduously. Through this kind of approach he was able to develop some generalized correlations based on fundamental parametric analysis.

(639b) Electrostatic Charging of Particles in a Straight Pipe

Poom Bunchatheeravate, Chemical Engineering, University of Florida, Gainesville, FL, Yusuke Fuji, Kyoto University, Kyoto, Japan, Jennifer Sinclair Curtis, University of Florida, Gainesville, FL and Shuji Matsusaka, Department of Chemical Engineering, Kyoto University, Kyoto, Japan

Title: Electrostatic charging of particles in a straight pipe

Poom Bunchatheeravate[1], Yusuke Fuji[2], Jennifer Curtis[1], and Shuji Matsusaka[2]

[1]Department of Chemical Engineering, University of Florida, Gainesville, FL 32611. USA

[2]Department of Chemical Engineering, Kyoto University, Kyoto 615-8510. Japan

As two surfaces come into contact, electrons are transferred between the two surfaces. This phenomenon leads to one surface gaining additional electrons and become negatively charged, while the other surface becomes positively charged. Contact electrification is a common occurrence in dilute pneumatic conveying. When particles are transferred via pipelines, particles collide with each other and the bounding surfaces causing charge to build up on the particles. Excessive charge build up can lead to several issues ranging from particle agglomeration to deadly dust explosions. Under the correct piping conditions, it is possible to control the particle charge as the particles travel through the conveying line. This study propose an experimental technique for measuring the particle charging properties on a small experimental scale, as well as a model which can predict the particle charging behavior when conveyed through a straight pipe of any given length and diameter. The experiments involved measuring the particle charge at the inlet and the outlet of a 1m pipe. The charge of the inlet particles was modified using a particle charger. The particle charge exiting the 1m pipe is directly related to the particle charge when entering the 1m pipe. Knowing this relationship, a particle charging profile can be generated which shows the evolution of particle charge as particles travel through a pipe with the same material and diameter. The particle charging profile yields the equilibrium charge and the characteristic length. Both of these parameters depend upon the electrical interaction between the particle material and the pipe wall material.

Power Considerations of Pneumatic Convey Systems

by

Jonathan Thorn, M.S.
Mac Process
7901 NW 107th Terrace
Kansas City, MO 64157
thornj@macprocessinc.com

Submission for 2012 AICHE Annual Meeting – Particle Technology Forum (03C)

Introduction

Significant research has been conducted in regards to understanding the gas flows and pressures required to pneumatically convey materials, however the true power requirements to operate a pneumatic conveying system are less well defined. The work performed by a pneumatic conveying system (actual work) comes from the compression of the conveying gas (volume and pressure) and originates at the compression device. The operation of the compression device encompasses several inefficiencies which serve to define the needed energy input (apparent work) to operate a pneumatic conveying system. Rotary airlock leakage is a common source of lost supply gas that is generated by the compression device but is not used in the convey line to transport material. Delivery of the convey gas from the compression device to the feedpoint will create pressure drop (especially when gas flow control devices are employed) and in turn elevates the supply pressure required. Finally, the compression device itself will carry an absolute compression efficiency. These factors combine to generate an apparent work that is drastically different than the actual work required to convey the material. Therefore, it is the application of the apparent work analysis that will determine if a process if being operated efficiently.

Observing the apparent work required to operate a pneumatic conveying system, whether it be dilute phase or dense phase in nature, exposes a power minimization curve. By gaining control over key operation parameters such as the input gas volume (and consequently the convey velocity), it becomes clear that reducing the convey velocity to a point will improve the operating efficiency of the system (for a fixed rate and distance). However, the minimization curve shows that eventually conveying will not be supported (plugging) or the conveying will actually become less efficient. The minimization curve differs significantly depending on the type of convey system, the type of air controls employed and the compression device. Using these concepts to understand the operational factors that directly impact the apparent work and power minimization curve will allow the peak efficiency of the system to be realized.

Apparent Work Concept

The apparent work to pneumatically convey materials accounts for gas volume and pressure generated but not used in directly in the conveying of materials through the pipeline (actual work). The identification and quantification of excess flow, pressure and subsequently power will vary by system type and equipment used as described in later sections. However, the basic concepts as described here are universal.

The actual work to operate a pneumatic convey system stems directly from the basic design parameters of the system. Material characteristics (entrainment velocity, friction factor) and system characteristics (rate, distance, line size, airflow) will define the pressure required to overcome the resistance in the pipe and transport material. The rate of actual work (power) for compression can then be identified by where k=1.4 for air in an adiabatic process.

$$W = nC_pT\left[\left(\frac{P_2}{P_1}\right)^{\frac{k-1}{k}} - 1\right] \quad (1)$$

Assuming ideal gas laws, and substituting $k=C_p/C_v$ and $R=C_p-C_v$ the expression can be rewritten as

$$\dot{W} = \frac{QP_1k}{k-1}\left[\left(\frac{P_2}{P_1}\right)^{\frac{k-1}{k}} - 1\right] \quad (2)$$

constituting the rate of work (power) required to compress the convey volume up to the convey pressure.

Apparent work is then the actual work influenced by any factor involved in generating or delivering the required airflow and pressure that does not directly effect the movement of material. Rotary airlock leakage is a common source of excess volume generated that allow feeding of material into the convey line but not the actual movement in the pipe. Rotary airlock valves will vary in size and construction but once manufactured will have a fixed pressure versus leakage relationship similar to an orifice. Therefore, the convey pressure of the pneumatic convey system will directly impact the leakage volume and the total supply volume required.

$$Q_S = Q_C + Q_L \quad (3)$$

Delivery of the air volume from the compression device to the feedpoint is a common source of pressure drop the compression device must overcome although no material is moved during the pressure reduction. Air alone pressure loss can occur simply from the movement of air through supply lines or because the air volume must be tightly controlled and therefore metered through a control valve. In the case of a communal compressed air supply, the convey air is normally generated at pressures from 6-7 bar and then regulated to a pressure significantly less for use. Therefore, we must differentiate between communal air supplies (compressed air) and dedicated air supplies (PD blower, compressor) that are generating only the volume and pressure needed for conveying. Clearly the apparent work will be minimized when the air alone loss is minimized and a dedicated compression device is used to only generate the pressure needed to drive the system.

$$P_2 = P_S = P_C + dP_{air} + dP_{control} \quad (4)$$

The selection of the compression device and its inherent compression efficiency (η) will strongly influence the apparent power. Typically the type of compression device will have the largest impact but the model selection within a type and even the performance curve of a particular model can contribute. Taking a dilute phase example, Positive Displacement (PD) blowers are common dilute phase compression devices due to their low cost and simple construction. The compression efficiency of a PD blower will vary from approximately 50-70% depending on the discharge pressure up to 1 bar and will be approximately 55% at 0.8 bar. A dynamic centrifugal compressor (turbo blower) is approximately 70% efficient at 0.8 bar. Through selection of the type of compression device a reduction of >20% apparent work can be realized. Within a PD blower range, to perform a certain duty, one model may perform at an elevated efficiency compared to another. In the example above, a range of models may produce efficiencies from 52-58% resulting in a 10% variance in of apparent work depending on model selection. And in the case of the turbo blower, it may generate the same flow (constant pressure) at different speeds using its flow modification features because of the operating location on the performance curve. The resulting efficiency at different speeds may vary from 60-70% efficient resulting in a 15% variance simply by tuning. The efficiency of a compression device at a particular operating point may not be easily identified but this value has a large impact on the apparent work and the resulting efficiency of the system.

Combining the influences of leakage, additional pressure inputs and compressor efficiency the actual power can be modified to represent the apparent power required for the system.

$$Power_{app} = \dot{W} = \frac{\frac{(Q_C + Q_L)P_1 k}{k-1}\left[\left(\frac{P_C + dP_{air} + dP_{control}}{P_1}\right)^{\frac{k-1}{k}} - 1\right]}{\eta} \quad (5)$$

Experiment

Experiments were conducted on a testing apparatus that allows several types of conveying through a common pipe. The piping has lengths available from 75 to 300 m in horizontal length, 7 m vertical length and quantity (7-13) 90 deg turns depending on the route chosen. Metalicine based polyethylene pellets weighing 550 kg/m^3 and approximately 4500 micron were conveyed through a 4" Sch10 stainless steel pipe. A high pressure rotary airlock capable of pressures up to 4 bar was used to feed the material into the convey line for both the dilute phase and dense phase experiments. Pressure measurements were recorded by transducer at the discharge of the compression device and at the feedpoint. Total airflow values were measured by a hot wire anemometer style mass flow meter while leakage flows (and subsequently convey airflows and velocities) were calculated based on the convey pressure. Power values were collected from the digital output of a variable frequency drive. Equilibrium was to found to be achieved when conveying conditions were maintained for greater than 15 minutes and followed by 15 additional minutes used to average the data.

System Configurations

Several pneumatic convey system configurations were studied to deduce what factors most strongly influence the apparent power to operate the system.

- Pressure dilute phase fed by rotary airlock and powered by a PD blower
- Pressure dilute phase fed by rotary airlock and powered by a Turbo blower
- Pressure dense phase fed by rotary airlock and powered by communal compressed air
- Pressure dense phase fed by rotary airlock and powered by a dedicated compressor

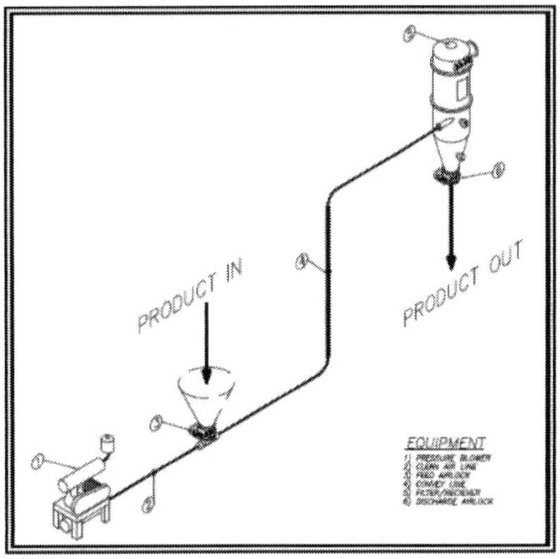

Figure 1. Pressure dilute phase

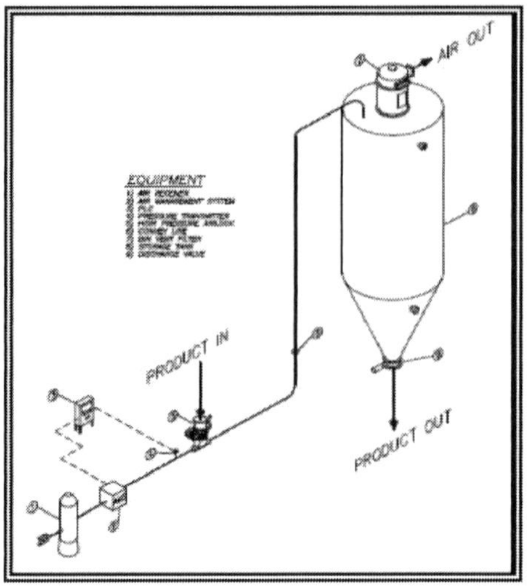

Figure 2. Pressure Continuous Dense Phase

Pressure Dilute Phase

Pressure dilute phase uses gas flow (Q_S) created by the compression device and supplies it to the rotary airlock where material is dropped into the stream (see Figure 1). A portion of the gas exits the rotary airlock as leakage (Q_L) and the remainder flows through the line as convey gas (Q_C). As long the resulting superficial gas velocities are greater than the minimum entrainment velocity for that material then material will be transported. The compression device then operates at a pressure sufficient to overcome the resulting resistances associated with the gas and material flow. The apparent power in this arrangement is influenced by: 1) rotary airlock leakage, 2) pressure drop in gas supply line, 3) efficiency of the compression device and 4) excess velocity above the minimum.

Rotary airlock leakage is considered a necessary sacrifice to allow the introduction of the material into the system. For a specific duty the leakage rate can be considered constant (at a specific pressure) when using equipment in good repair. The clean air line pressure drop will vary depending on the line size but is constant for a fixed pipe size and generally constitutes a small percentage of the required pressure to operate the system. Efficiency of the compression device plays a large role in the apparent power formula. In the efficiency comparison given above for a PD blower and turbo blower, the required power to operate a system can vary greatly. In Figure 3 we see the measured power input for the two compression devices performing under the same duties. The Turbo blower demonstrates significantly reduced power input as compared to the PD blower. Excess velocity in the convey line is also a significant contributor to apparent power. In addition to the compression of the excess gas flow, higher transport velocities result in elevated pressure drops for both gas and material flows.

Rate	Convey Distance	Convey Pressure	Total Airflow	PD Blower Power	Turbo Blower Power
mTPH	m	bar	m^3/hr	KW	KW
4.2	200	0.38	332	20.8	16.3
6.1	200	0.52	358	29.4	22.4
7.7	200	0.66	391	43.1	29.0

Figure 3. Chart power requirements for PD and Turbo blowers

Figure 4 shows the measured and calculated apparent power for a range of pick-up velocities in relation to the actual power. These results will be further discussed in relation to a power minimization curve.

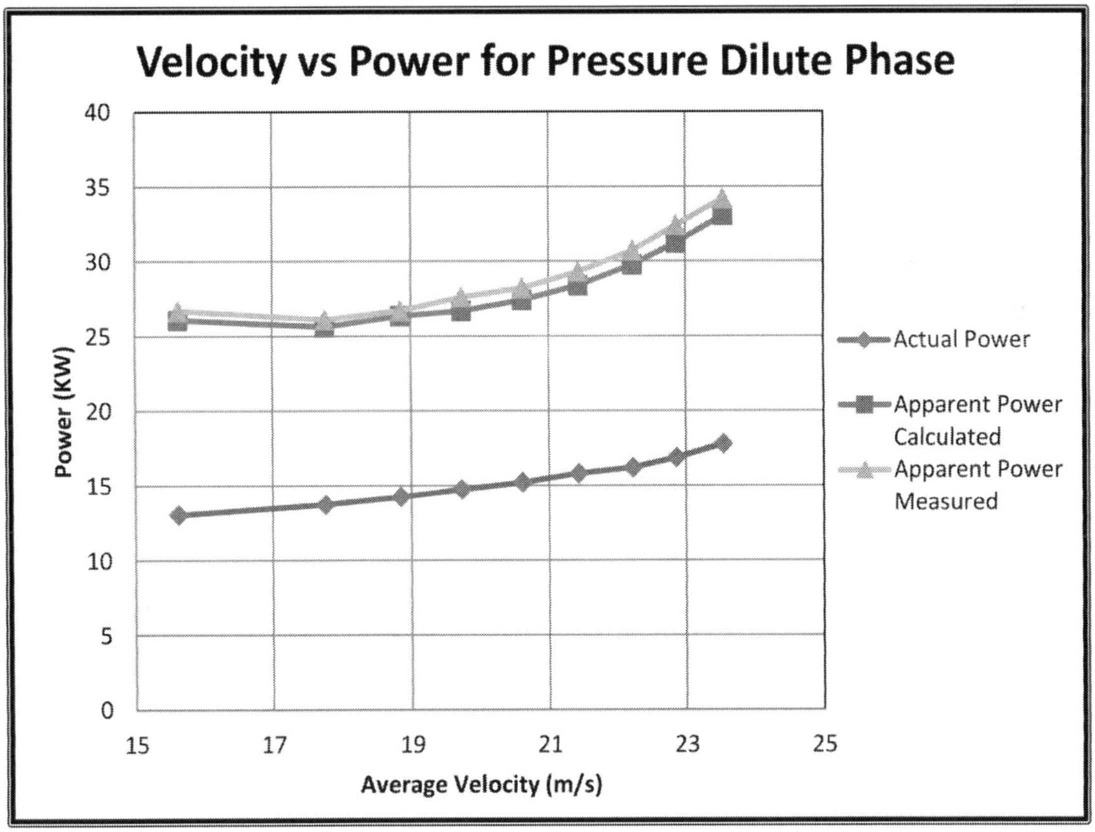

Figure 4. Actual and apparent power curves for pressure dilute phase

Pressure Dense Phase

Pressure dense phase flow can be created using feed from a rotary airlock if the gas flow amount is tightly controlled (often called continuous dense phase – see Figure 2). Dense phase flow generates significantly higher pressures than dilute phase and requires a compression device to suit. Communal compressed air (6-7 bar) from a compressed air header is a common to way to drive this type of a system but a dedicated compressor can also be used with significant differences in the resulting apparent power. The desired convey gas volume is combined with the expected leakage volume and delivered through a gas metering valve. By passing the gas through a metering valve, a resulting pressure drop occurs of which the compression device must overcome. Losses in the clean air line are present but are generally negligible. The apparent power in this arrangement is influenced by: 1) supply pressure (communal or dedicated), 2) metering valve loss (dedicated), 3) rotary airlock leakage and 4) convey velocity.

The supply pressure of the compressed gas has a strong influence on the apparent power. In the case of communal gas provided at an elevated pressure, loss through the metering valve is not relevant. The compressor has already provided an excessively high gas pressure which is regulated down at the use point. With a dedicated compressor the loss through the metering valve goes directly to apparent power yet this methodology is significantly more efficient having not compressed the gas to an unnecessarily high header pressure. The supply pressure for a dedicated compressor is simply the convey losses (P_C) and metering valve losses ($dP_{control}$) combined. Similar to dilute phase the rotary valve in dense phase is generally fixed for a certain duty however the leakage volume becomes a much larger portion of the total air (due to lower convey velocities and higher pressures). The conveying velocity will have a direct influence the convey pressure and thus the total supply gas volume (Q_S) is reduced by the convey volume (Q_c) and increased via the leakage (Q_L). Figure 5 shows the relationship between power and velocity as the velocity is reduced.

Power Minimization Curve

Power for a specific conveying duty (constant rate, distance & line size) can be compared to the velocity of the conveying and will ideally generate a curve with an inflection point yielding a minimum. The premise is that a velocity reduction will inherently reduce the convey gas volume and, until the convey pressure increases drastically, there will be a net reduction in actual power. However, the inflection point for apparent power is more visible due to the mitigating factors of airlock leakage, compression efficiency and the like.

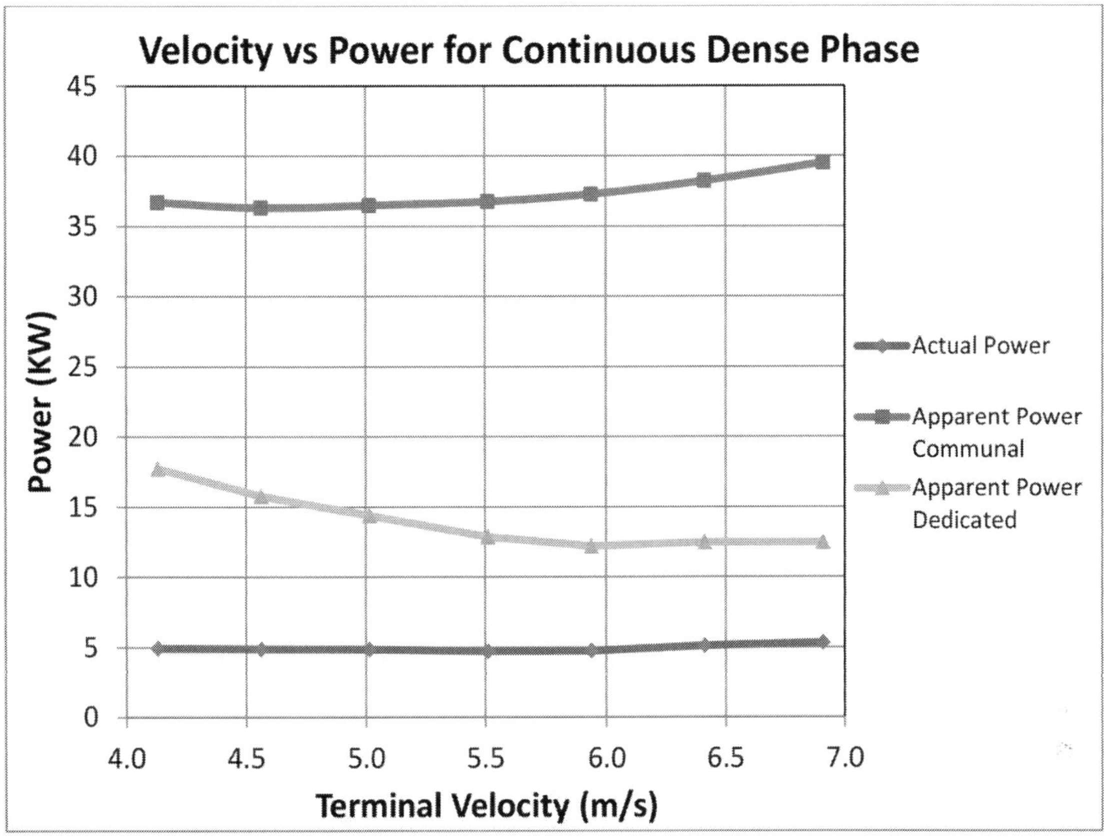

Figure 5. Actual and apparent power curves for pressure continuous dense phase

The apparent power curve will be distinctly different for the dense phase and dilute phase modes of conveying. Velocity reduction in dilute phase will result in a pressure reduction and thus a combined power reduction (until the minimum entrainment velocity is reached). A slight pressure increase can be observed around the velocity minimum (see Figure 4), offsetting the gas flow reduction in the apparent power. The actual power does not reach a minimum before because further reductions in velocity will fail to sustain conveying and thus actual power minimization occurs when the velocity is minimized. However, we see a minimization occur in the apparent power around 18 m/s average velocity when leakage and compression efficiency become a controlling factor. We also see the apparent power is approximately 2x the actual power for conveying under these conditions.

Velocity reduction in dense phase causes material to accumulate in the convey line and results in a convey pressure increase. In Figure 5 we see the actual power stays approximately constant with a slight inflection around 6 m/s terminal velocity. This suggests that the work required to convey material in dense phase flow is approximately constant although further work would be required to make this claim. The apparent power of the communal gas arrangement shows a minimum around 4.5 m/s. Since supply pressure is considered constant in this case, the total supply gas (Q_s) is the controlling variable and the inflection point occurs only when the leakage rate increases faster than the convey volume decreases. The apparent power for a dedicated compressor reveals a minimum around 6 m/s; similar to the actual power. Although rotary valve leakage significantly increases the apparent power in the dedicated compressor scenario, the supply pressure appears to be a controlling factor in the minimization. In these experiments the apparent power was 2-3x higher when a dedicated compressor scenario was employed and 7-8x higher with a communal compressor.

Conclusions

The experiments performed establish that the apparent power to operate a pneumatic conveying system is significantly greater than the actual work being performed; i.e. the transport of materials through pipelines. Dilute phase systems appear to suffer most from inefficient modes of compression and from convey velocities in excess of the minimum required. Dense phase systems appear to be subject to the tendency to use communal air supplies and the high rates of rotary airlock leakage (continuous dense phase). In this case, using a dedicated compressor in place of a communal supply reduces the apparent power.

For a specific conveying arrangement, it is possible to establish a convey velocity that minimizes the apparent power required to operate the system. For pressure dilute phase systems the minimization occurs at or around the minimum convey velocity. For pressure dense phase using a rotary airlock, minimization occurs at a higher velocity when using a dedicated compressor as compared to drawing from a communal gas supply.

Nomenclature

C_p – heat capacity at constant pressure	Q – gas flow
C_v – heat capacity at constant volume	Q_C – conveying gas flow
k – adiabatic expansion coefficient	Q_L – leakage gas flow
P_C – convey pressure	Q_S – total supply gas flow
P_S – supply Pressure	T - temperature
P_1 – compressor inlet pressure	W - work
P_2 – compressor discharge pressure	$\dot{W}$ - rate of work
dP_{air} – pressure drop air alone	η - overall compression efficiency
$dP_{control}$ – pressure drop of air controls	

References

"Compressor Efficiency Definitions", K. Ueno, PhD, and R. E. Bye, VAIREX Corporation, K. S. Hunter, PhD, University of Colorado, May 12th, 2003.

www.engineeringtoolbox.com, "Horsepower required to Compress Air".

(639d) Effect of Outlet Gas Velocity and Solids Loading On Erosion in 2nd Stage Cyclones

S. B. Reddy Karri, Particulate Solid Research, Inc., Chicago, IL

Effect of Outlet Gas Velocity and Solids Loading on Erosion in 2nd Stage Cyclones

S. B. Reddy Karri*, Ray Cocco and Ted M. Knowlton

Particulate Solid Research, Inc.

4201 W. 36th Street, Chicago, IL 60632, USA

ABSTRACT

Severe erosion in the lower cone and in the upper dipleg of second stage cyclones have been observed in commercial cyclones. The main objective of this study is to shed light on the mechanism by which this erosion takes place, and how different design and operating parameters affect the erosion. Experimental data on how parameters such as the cyclone inlet solids loading and gas outlet velocity affect second stage cyclone erosion are presented. The outlet gas velocity was varied by changing the size of the vortex tube diameter. The effect of a vortex stabilizer on cyclone cone erosion is also discussed.

INTRODUCTION

Petroleum refineries have increased their focus on improving unit reliability, and reducing operational and maintenance costs. Refiners operating fluid catalytic cracking (FCC) units are under increasing pressure to improve the profitability by increasing run lengths between major turnarounds. Because their cyclones have high efficiencies, FCC process operators are now concerned with longer campaign durations, and would like to improve cyclone reliability. The 2008 NPRA survey and other surveys (1,2) revealed that FCCU cyclone reliability (3,4) was a major concern for refineries.

Particulate Solid Research, Inc. (PSRI) conducts research and testing programs that address common concerns of member companies. These programs are guided by elected technical advisors from member companies. One (1) recent PSRI program studied and benchmarked different FCC cyclone technologies, since cyclone erosion and reliability are highlighted as the major concerns of FCC operation for member companies in recent surveys. The most pervasive problem is erosion in the secondary cyclone, particularly in the lower cone and in the transition to the dipleg, which is the focus of the study.

1. Solomon Associates LLC , 2006 Solomon FCC Events Determining TAR Timing Survey, 2006
2. Grace Davison Survey, FCC Conference in Dublin, California, USA, 2002
3. Zenz, F.A., Cyclone Separators, in Manual on Disposal of Refinery Wastes Volume on Atmospheric Emissions, Chapter 11, American Petroleum Institute, Pub. No. 931, Washington, D.C. (1975).
4. Tenney, Ed., FCC Cyclone Problems And How They Can Be Overcome With Current Designs, Presented at the Grace-Davison FCC Technology Conference, Toledo, Spain (1992).

(639e) An Experimental Study of Uniflow Cyclone Design Parameters

U. Muschelknautz, Environmental, Process and Energy Engineering, MCI - The Entrepreneurial School, A-6020 Innsbruck, Austria

An Experimental Study of Uniflow Cyclone Design Parameters

Ulrich Muschelknautz, Paul Pattis, Michael Reinalter, Michael Kraxner, Tobias Kofler, Matthias Aigner, Gregor Eichwalder

MCI Department of Environmental, Process and Energy Engineering

Maximilianstraße 2, A-6020 Innsbruck, AUSTRIA

Uniflow cyclones have gas and particles passing through them in only one direction. In principle they can be used in all industrial applications where conventional reverse flow cyclones are applied. Advantageous compared to reverse flow cyclones is their compact design and their easy implementation into piping systems. These characteristics minimize manufacturing and installation costs.

Even though their principle is known since a long time, not as much literature can be found on how to design uniflow cyclones. Approved calculation models for uniflow cyclones, which are valid for a wide range of applications, are missing up to now.

MCI intends to develop design fundamentals for this cyclone type on the basis of systematic experimental studies. Results from these investigations provide useful insights into the operation of such devices and means to design them. The influences of the main geometry and operation data of uniflow cyclones on their performance are shown.

INFLUENCE OF THE INLET VANE GEOMETRY ON THE UNIFLOW CYCLONES PERFORMANCE

M. Kraxner[1], U. Muschelknautz[1], S. Wechner[1], S. Ackermann[2], V. Greif[2], J. Bolda[2]

[1]MCI – The Entrepreneurial School, Department for Environmental, Process & Energy Engineering, Maximilianstraße 2, A-6020 Innsbruck, AUSTRIA
[2]MANN+HUMMEL GmbH, Brunckstraße 15, D-67346 Speyer, GERMANY

Introduction

Uniflow cyclones used for the separation of particles from gases are characterized, compared to standard reverse flow cyclones, by its compactness and its low energy consumption. The compact construction gives plant operators the possibility of easy implementation in piping systems. Therefore emissions of fine dust can be reduced by a low fabrication effort. Uniflow cyclones can be used as an ownstanding or as an upstream dedusting device. The usage as an upstream deduster can boost the lifetime of downstream fabric filters by a multiple. Multicyclones consisting of many parallel arranged uniflow cyclones achieve high separation efficiencies due to the small design size of one single cyclone cell. They are used for highly efficient dedusting procedures on constricted space.

The inlet vanes guide the gas-particle flow into a rotating vortex. The strength of this vortex is determined by the mean vane angle α_m of the inlet vanes. The vane angle is designed to achieve sufficient centrifugal forces on the particles and therefore sufficient separation efficiency. The standard shape of the inlet vanes is a spiral along the core. Experiments show, that for a given vane angle α_m the separation efficiency as well as the pressure drop can be improved significantly by modifying the vane shape and the vane length. Experiments show that the application of this improvement in upstream uniflow cyclones leads to a strong increase of the downstream filter life time.

(651a) Toward Design of Semiconductor Ternary Quantum Dots with Optimal Optoelectronic Function

Sumeet C. Pandey, Department of Chemical Engineering, University of Massachusetts, Amherst, MA, Xu Han, Chemical Engineering, University of Massachusetts Amherst, Amherst, MA and Dimitrios Maroudas, Chemical Engineering, University of Massachusetts Amherst, Amhest, MA

Heterogeneous semiconductor quantum dots (QDs), such as core/shell QDs, are semiconductor nanocrystals with non-uniform compositional distribution that have attracted significant technological interest in optoelectronic and photovoltaic device fabrication. Compositionally, it is very common for such QDs to be ternary semiconductor nanocrystals; random alloys of such ternary QDs exhibit size-dependent excitonic properties and allow for exceptional tunability of their band gaps by modulation of the nanocrystal composition. Band structure engineering based on such controllable parameters enables the development of nanomaterials with optimal function for photovoltaic and light-emitting devices. Synthesis of such semiconductor nanocrystals has followed various colloidal chemistry routes.

In this presentation, we investigate the possibility of rigorous single-step synthesis routes to minimally strained core/shell-like QDs with optimal optoelectronic function by studying systematically the distribution of atomic species in ternary QD systems, emphasizing on surface segregation as a potential means for self-assembly of core/shell-like semiconductor QDs. Toward this end, we have combined atomic-scale computations of full relaxation of semiconductor nanocrystals, first-principles electronic band structure calculations, and continuum mass transport theory and numerical modeling. The computational results determine the thermodynamic stability of core/shell QDs synthesized by two-step synthesis routes (growth of a shell on a core QD), elucidate the effects of compositional distribution on the electronic band structure of ternary QDs, and provide interpretations to X-ray photoelectron spectra (XPS) and photoluminescence (PL) emission spectra of ternary QDs.

We have computed the equilibrium concentration profiles in QDs of $ZnSe_{1-x}Te_x$, $In_xGa_{1-x}As$, and $ZnSe_{1-x}S_x$ according to a computational analysis of full nanocrystal relaxation and report results over a broad range of x and QD diameter (from ~1 to ~6 nm). The computational procedure consists of coupled compositional, structural, and strain relaxation of the semiconductor nanocrystals. The analysis is based on Monte Carlo (MC) simulations according to extended classical valence-force-field descriptions of interatomic interactions, which have been parameterized according to first-principles DFT calculations of segregation energies using slab supercells. The equilibrium profiles reflect the different tendencies of the QD constituents to segregate on the QD surface: $ZnSe_{1-x}Te_x$ and $In_xGa_{1-x}As$ form core/shell-like structures with Te- and In-deficient cores, respectively, and Te- and In-rich shells, respectively, characterized by compositionally graded core/shell interfacial regions, while the equilibrium compositional distribution in $ZnSe_{1-x}S_x$ QDs resembles a random alloy with only a weak tendency for Se surface segregation. We have determined the electronic structure of $ZnSe_{1-x}Te_x$ and $ZnSe_{1-x}S_x$QDs for various concentration distributions (core/shell, reverse core/shell, and random alloy) over the entire compositional range ($0 \leq x \leq 1$). We have predicted the dependence of the band gap as a function of x in all cases and demonstrated that, for the materials examined, the thermodynamically stable concentration distributions lead to QDs with tunable band gaps.

In the dilute limit of low x, the computed equilibrium concentration profiles are explained fully by a phenomenological species transport theory that we have developed, which accounts for Fickian diffusion of species in ternary QDs and drift due to the thermodynamic driving force for species surface segregation. The predictions of the continuum theory provide a powerful tool for the design of novel ternary QDs for applications in optoelectronics and photovoltaics. We have derived analytical and numerical solutions to the boundary-value problems for species transport in ternary QDs at equilibrium, steady state, and during transient periods. We have used the predictions for the time dependence of the near-surface-region concentration to fit XPS data for ternary $ZnSe_{1-x}S_x$ QD near-surface compositional changes during annealing at temperatures ~100 °C from ZnSe/ZnS core/shell initial configurations. The theoretical predictions provide excellent fits to the XPS data and can be used to evaluate diffusion coefficients of species interdiffusion in the QDs and the strength of the thermodynamic driving force for species surface segregation, which are the two fitting parameters used in this analysis.

(651b) Surface Mineralization and Characterization of Palladium Nanoparticles On Genetically Engineered Tobacco Mosaic Virus (TMV) Templates

Alexander Freer, Lucas Guarnaccio, Kristin Wafford, Johanna Smith, Jayne Steilberg, James Culver and Michael Harris, Chemical Engineering, Purdue University, Lafayette, IN

Certain innate properties of the tobacco mosaic virus (TMV) make it ideal as a biotemplate for nanoparticle formation. Genetic modifications of the TMV coat proteins allow for an increase in the selective deposition and controlled growth of different metals onto the surface of the virus. In the current process, TMV is coated with uniform layers of palladium metal particles in aqueous solution under very mild conditions. Palladium nanowires of 300nm in length and 30-40nm in diameter have been created with this process. Tunneling electron microscopy (TEM) and small angle x-ray scattering (SAXS) are used to charaterize the thickness and uniformity of the metal surface. Procedures to control the higher order assembly of the TMV help to limit the uncontrolled aggregation of individual nanorods and arrange the TMV into nanostructures to be used in multiple applications ranging from catalysts to sensors to battery electrodes and other devices in nanoelectronics. This presentation will focus on the aqueous coating procedure and the characterization of the resulting nanowires with TEM and SAXS.

(651c) Infrared Active Photocatalysis Through Morphological Control of Semiconductor Nanoparticles/Metal Tip Heterostructures

Doh C. Lee and Chaewon Pak, Department of Chemical and Biomolecular Engineering, Korea Advanced Institute of Science and Technology (KAIST), Daejeon, South Korea

Semiconductor nanocrystals in strong quantum confinement regime exhibit interesting size-dependent bandgap. The bandgap tunability offers a new opportunity for bandgap engineering; for example, exciton dynamics can be tailored by designing dimensions of heterostructure semiconductor nanocrystals. We synthesized PbSe/CdSe/CdS core/shell/shell nanocrystals with varying morphology, e.g., pyramid and tetrapod. These nanocrystals show varying electron-hole exchange interactions and radiative recombination lifetimes depending on their dimensions. For instance, compared to PbSe nanocrystals, PbSe/CdSe/CdS heterostructure nanocrystals show nearly 2 orders of magnitude longer radiative recombination lifetime. Using this semiconductor light absorber may help increase quantum efficiency in case an exciton has to be dissociated into an electron and a hole. By growing metal tips at the corners of the structures, we prepared semiconductor/metal nanojunction and used the nanomaterials as photocatalysts for reduction of various dyes. Interestingly, the morphology change leads to the difference in the tip growth pattern, and we demonstrate infrared active photocatalysis using the colloidal heteronanostructures.

(651d) Tunable Localized Surface Plasmon Resonances in Tungsten Oxide Nanocrystals

Karthish Manthiram[1,2] and A. Paul Alivisatos[2,3], (1)Department of Chemical and Biomolecular Engineering, University of California, Berkeley, CA, (2)Materials Sciences Division, Lawrence Berkeley National Laboratory, Berkeley, CA, (3)Department of Chemistry, University of California, Berkeley, CA

Free charge carriers, which are dielectrically-confined to a nanoparticle, participate in resonant, collective oscillations known as localized surface plasmon resonances (LSPRs). Although the vast majority of investigations of LSPRs have been conducted on noble metals, LSPRs have also been observed in heavily-doped semiconductors, in which the LSPR energy may be tuned by adjusting the stoichiometry. Transition metal oxides, such as tungsten oxide, are interesting candidates for LSPR hosts as they exhibit fascinating properties arising from the unique character of their outer-d valence electrons. We demonstrate that nanoscale oxygen-deficient tungsten oxide supports a strong LSPR mode, based on theoretical prediction of the LSPR energy from Mie-Gans theory and experimental measurement of the LSPR energy as a function of the refractive index of the medium. We also demonstrate that the LSPR energy may be tuned by heating the tungsten oxide nanoparticles in air. The identification of a tunable LSPR mode in tungsten oxide opens up the possibility of rationally designing plasmonic tungsten oxide nanoparticles for photocatalysis, bioimaging, and gas sensing.

Laser- Drawn Features on Nanoparticle Films

Sanjeev Kumar Kandpal, Aimee Co, Kody Allcroft, Michael D. Mason,

Douglas W. Bousfield, David J. Neivandt

Department of Chemical and Biological Engineering

The University of Maine, Orono, Maine -04473

Introduction

The present work explores the silver (Ag) nanoparticles film-Laser interaction based applications in the field of printed electronics, sensors, anti-counterfeiting, optical grating. Ag has the highest conductivity among metals even in its oxide form, making it the most sought after material for providing conductivity in electronics. Ag in nanoparticle size range show interesting optical, physical and electronic properties like melting point depression, large absorption and scattering cross section when compared to bulk silver, have a color that depends upon the size and the shape of the particle, high surface to volume ratio, tunable surface Plasmon resonance (SPR) peak wavelength, metal enhanced fluorescence (MEF) (Geddes *et al.*, 2003), surface enhanced Raman scattering (SERM) (Li & Peng, 2010), antibacterial application, diagnostic application, enhanced thermal and conductivity application. In the present work, we fabricated three-dimensional features on the Ag nanoparticles film using a laser and the resulting structures showed interesting fluorescent properties. We are also looking into fabricating structure on polymer-Ag nanoparticles films and characterize them for the conductivity, optical, fluorescence properties. These structures might serve as conductive paths on polymer-nanoparticle film, paving the way for an alternative method of fabricating printed circuits.

We are looking into the mechanisms behind the structure formation and our initial investigation into the process suggests that it is a combination of multiple processes like liquid melt ejection and sintering. The structure formation starts when the film is exposed to a laser beam (using the objective to focus the laser beam) and the Ag particles in the film absorb the light which is converted to heat due to the efficient photo-thermal conversion property of metal nanoparticles. The heat generated in nanoparticles starts dissipating away from the particles into the surrounding matrix. If this heat is sufficiently high, it can melt the surrounding matrix and surface of metal nanoparticles. To give a feel of melting point depression; a 20nm Ag nanoparticle has a melting point of 150°C where as bulk Ag has 960°C (Buffat & Borel, 1976). If we subject the nanoparticle film with a large flux then we can create a small volume where the temperature is much higher than the melting point of the film material, leading to its vaporization in that focal volume. These hot vapors try to expand in all directions but are unable to do so at the film surface, which leads to pressure buildup on the focal volume in the film. The buildup of pressure is released by pushing the liquid melt around the focal volume on the film surface, where it solidify. This buildup around the focal volume gives raise to structure formation if we move the film under the laser; the structure pattern formed will be the pattern in which the film was moved under the beam. If the flux used is not very high and particles are very close to each other then the heat dumped into the film will melt the surface of the particle and these melted surface run into each other and solidify, the process is known as sintering.

Materials and Methods

Synthesis

Silver nanospheres were made using single-pot redox chemical techniques according to published material (Turkevich et al., 1951; Pillai and Kamat, 2004). In this we make use of a soluble metal salt (silver nitrate), a reducing agent (sodium citrate) and a stabilizing agent (excess sodium citrate). Synthesis starts with the nucleation step, followed by nanoparticle growth and the stabilizing agent which caps the particle leaving a negatively charged surface helping to avoid aggregation (Rivas *et al.*, 2001). We start with cleaning all the glassware with concentrated nitric acid, 600mg of sodium citrate dissolved in 160 ml of 18MΩ pure water in a 500 ml round bottom flask. This was then brought to a temperature of 95°C in an oil immersion bath. Simultaneously, 40mg of silver nitrate was dissolved in 40 ml of 18MΩ ultrapure water in a glass beaker. This was then added to the round bottom flask containing sodium citrate solution and heated to the same temperature (95°C) under constant stirring. The temperature was maintained until the reaction was complete, usually less than 1 hr. The heating, stirring is stopped and solution is allowed to cool to room temperature. To impart electrostatic stability to nanoparticles, excess of stabilizing agent was added. The size of the nanoparticles are measured using a Dynamic light scattering technique (Malvern, Zetasizer Nano-ZS), a Transmission electron microscope, and UV-Vis Spectroscopy. Fig. 2a shows a transmission electron micrograph (TEM) image of the freshly prepared silver nanoparticles. From the image data the average diameter was estimated to be approximately 100 nm.

Nanoparticles Film Making

Nanoparticles solution synthesized using the method described above, is very dilute in nature. To increase the number density of particles, the solution was subjected to rotation vaporization using Rotary Evaporator (Rotovap) instrument (BUCHI Rotavapour R-205). The solution was concentrated from 0.013wt% to 0.102wt%; the solution was then characterized using DLS and UV-Vis spectroscopy. After this was done, 100μl of this solution was drop casted on 22x22mm cover slip and allowed to air dry.

Exposing film to laser and characterization them

The drop casted film was then subjected to 532nm continuous wave laser, the film was scanned under the laser in a raster pattern using an x-y scanning stage and the scan stage was controlled using LabVIEW software. Laser exposed films were characterized using an optical microscope, CLSM (Confocal Laser Scanning Microscope), SEM (Scanning Electron Microscope), AFM (Atomic Force Microscope), XRD (X-ray Diffraction), and EDAX (Energy-dispersive X-ray spectroscopy).

Result & Discussion

Characterization of particles

The nanoparticle solution was characterized using DLS, UV-Vis spectroscopy and TEM. The DLS graph shows that the freshly made Ag particles are around 50nm in radius and Rotovap concentrated are around a micron. These numbers are complemented by TEM images and they also tell us about the shape of the particles. In UV-Vis graph the flat red curve shows the aggregation effect of particles and black curve is the typical UV-Vis curve of 100nm Ag particles. These data suggests a tremendous aggregation effect in Ag particles when they are concentrated using the Rotovap.

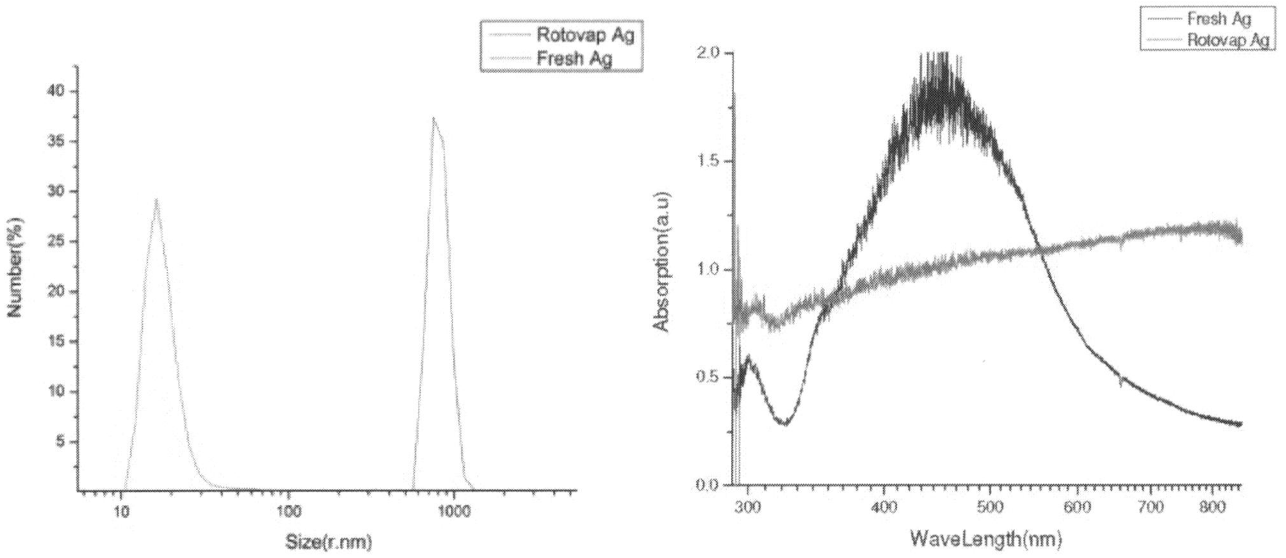

(a) **(b)**

Figure1: Comparison of DLS (a), UV-Vis spectrum (b) of freshly made and Rotovap concentrated silver.

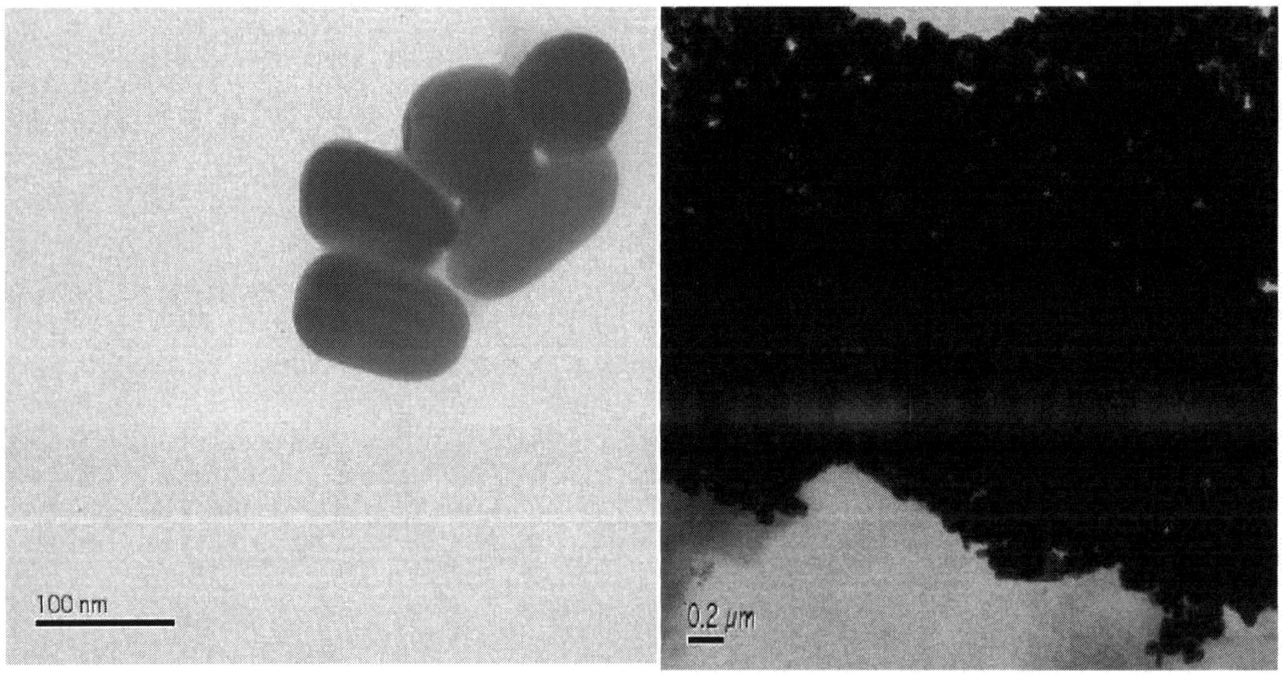

(a) **(b)**

Figure2: TEM image of freshly made silver (a), TEM image of Rotovap concentrated silver (b).

Drop Casting and Exposing

The Figure: 3a shows the digital image of 100μl drop cast of Rotovap concentrated Ag solution. The interesting feature of this drop cast film is formation of an inverse coffee ring. The film was latter exposed to 532nm CW DPSS laser using Figure: 3b optical step-up, which has provision to expose the film to laser and record the video simultaneously.

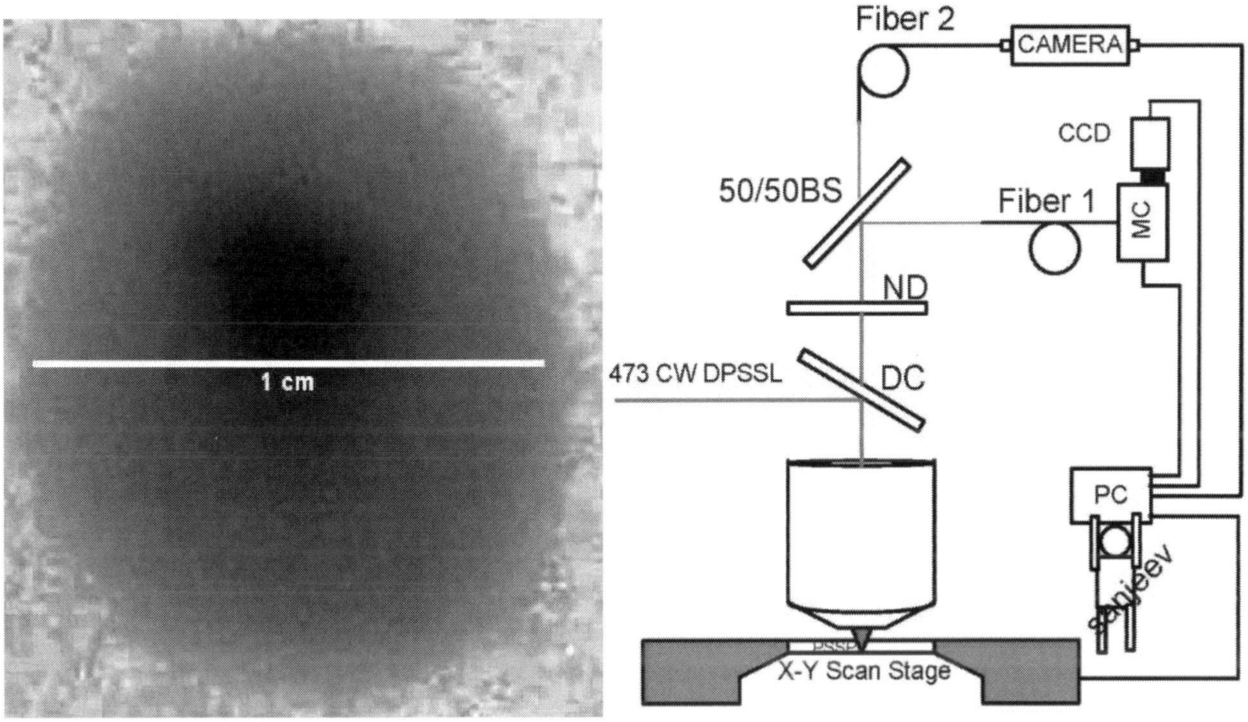

(a) (b)

Figure3: Drop cast image of silver (a) , Optical setup (b) , Transmission mode optical image of laser exposed silver drop cast image (right).

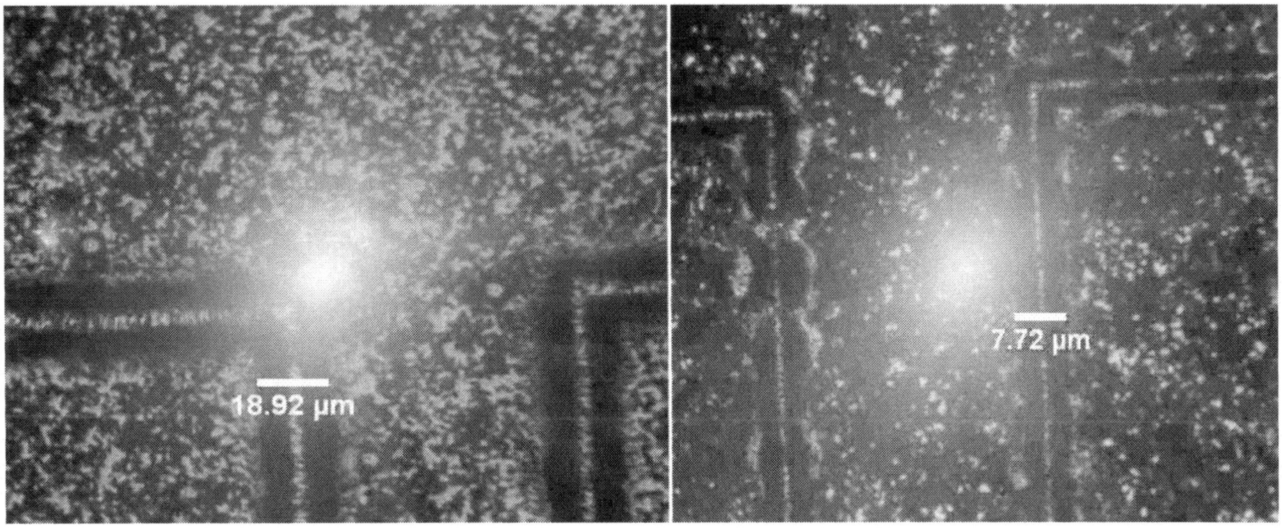

(a) (b)

Figure4: Transmission mode optical image of laser exposed silver drop cast film, using 0.4NA Objective (a), 0.65NA Objective (b).

In Figure: 4 we see a black and green speckle pattern, the black regions are the Ag particles and the green regions are the empty space between the particles. The reasons they look green are because of transmission mode imaging and camera setting. The raster patterns in the images are the regions which were exposed to the laser. It is observed that there is vacant region (bright green region) at the center of the pattern and ridges to the sides (dark regions). We observed a decrease in the width of the Laser Induced Structures (LIS) when we used a higher numerical aperture (NA) objective, which is due to the higher NA objective having the ability to focus the light to become very small volume. The resolution we were able to achieve was 10 times better than most traditional printing techniques.

Drop Cast Film on Paper Substrate

To explore the compatibility of LIS with a paper based substrate, we tried two different papers (copy paper, magazine paper). 100μl of concentrated Ag solution was drop cast onto the copy paper, allowed to air dry, and then subjected to 532nm CW laser.

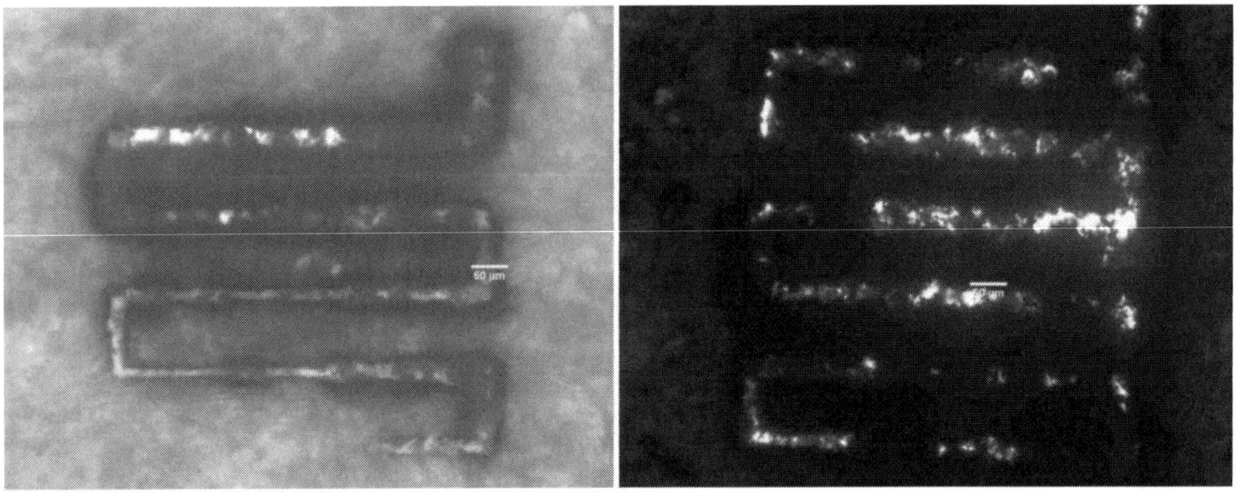

(a) (b)

Figure5: Laser exposed Copy paper, Plain (a), with Ag particles (b)

We observed the damage of the copy paper in the areas where laser was exposed (Figure5a) and the major damage when the paper was coated with Ag particles. In the Figure5, the much brighter regions are the laser damaged regions and they look bright because of light passing through that area effectively (transmission mode imaging). The reason for the enhanced damage of the Ag particle coated paper is the low optical yield of Ag particles. The light absorbed by the particles is effectively converted into heat and this heat is transferred to the surrounding material, which causes the damage to the paper substrate. We then tried coated paper (magazine paper) to overcome the damage and then we compared the coated paper without Ag particle to the coated paper with Ag particle. We observed that there was no damage to coated paper without Ag particles (Figure6a) but there was damage to the coated paper with Ag particles, however the damage was less than the uncoated copy paper. Hence, coated paper is resistant to laser damage.

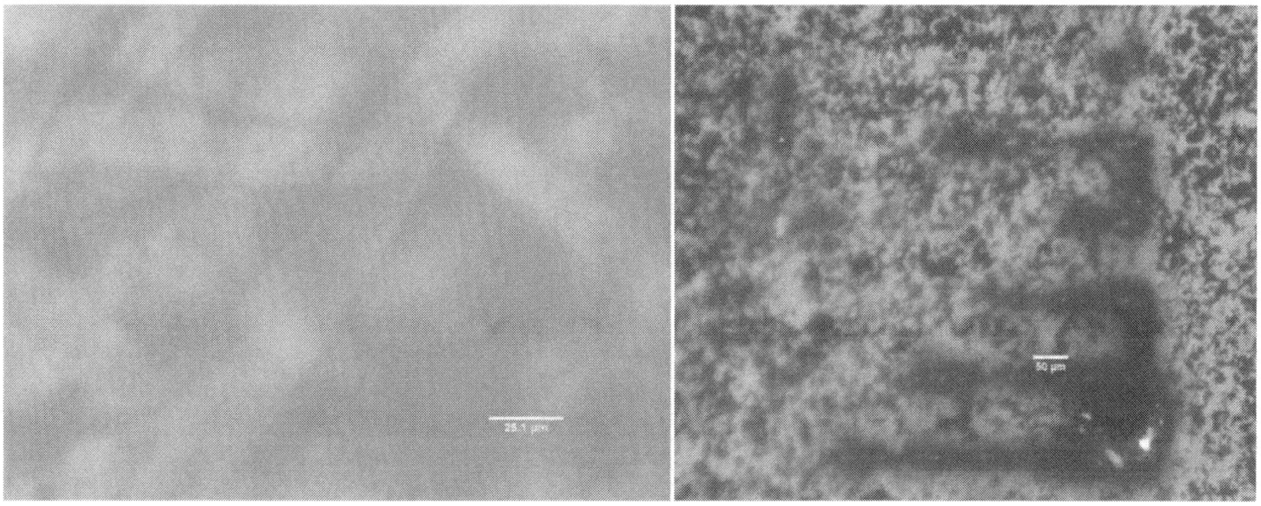

(a) (b)

Figure6: Coated Paper without Ag particles (a), Coated Paper with Ag particles (b)

Characterization of LIS

AFM Characterization

To know the exact dimensions of these structures in three different directions, we took an AFM image of the LIS. We made 'PSSP' LIS on concentrated Ag drop cast film on glass as our sample to do characterization.

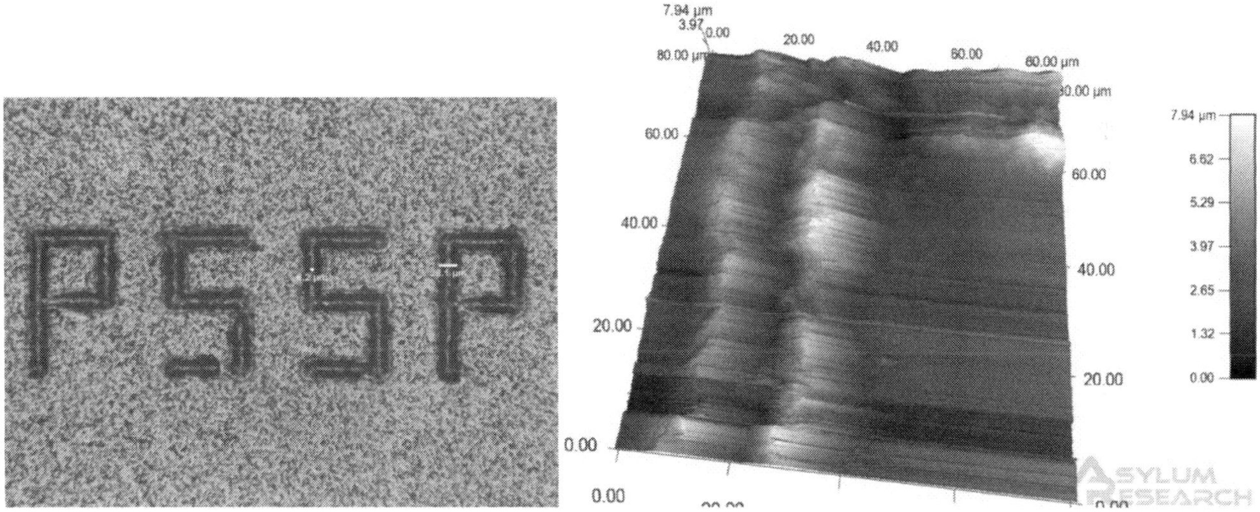

(a) (b)

Figure7: 'PSSP' LIS (a), AFM image of tail of letter 'P'

The LIS was characterized using AFM non-contact mode and we found that the structures are 8µm high, 25µm in width and there is a cavity at the center of the structure and ridges to the sides. The AFM data confirms that there is a physical change happening under the laser.

Confocal Characterization

In this class of characterization, 'PSSP' LIS was excited with certain wavelengths of laser and the signal from the structure was collected using photomultiplier detector.

(a) Exi=488nm, Coll: 500-530nm (b) Exi=543nm, Coll: 560-620nm (c) Exi=633nm, Coll: 650-750nm

Figure8: Confocal images of LIS.

We observed a strong fluorescent signal at the edges of the LIS, and the fluorescent was over a broad range of spectrum. To quantify the fluorescent signal, we took the spectrum of the signal using a nitrogen cooled CCD camera attached to monochromator with the optical setup shown in figure3b.

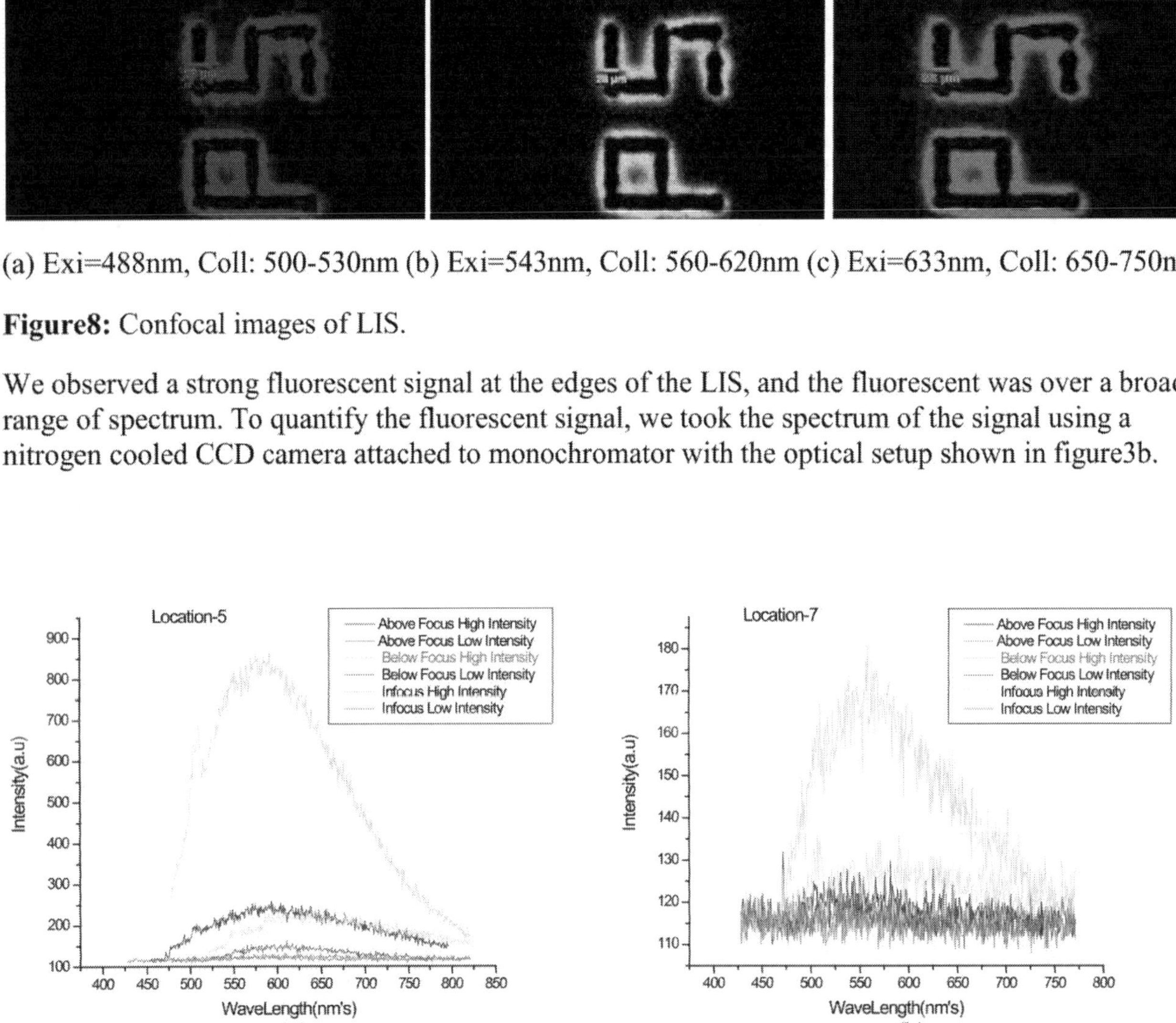

(a) (b)

Figure9: Fluorescent spectrum at the edge of the LIS (a), on the surface of the LIS (b)

The various curves in the Figure9 are from spectrums taken at various levels of focus planes. We got maximum signal when we were on the surface of the film at edge of the LIS. We got many magnitudes less signal compared to the signal at the edge and on the Ag film surface. This spectrum also quantifies the broad range spectrum conclusion we made out of the confocal image data.

Conclusion

- Observed inverse coffee ring effect in concentrated Ag drop cast film.
- Successfully able to draw 8μm size feature on silver nanoparticle film.
- Able to go down 10x times smaller in size compared to typical printing techniques.
- Coated paper shows resistant to laser damage.
- Laser drawn structures were fluorescent at the edges.

References

- Buffat, PH, & Borel, J-P. 1976. Size effect on the melting temperature of gold particles. *Physical review A*, 13(6), 2287–2298.
- Geddes, Chris D, Haishi Cao, Ignacy Gryczynski, Zygmunt Gryczynski, Jiyu Fang, and Joseph R Lakowicz. "Applications of Indocyanine Green to in Vivo Imaging". *J. Phys. Chem. A 107* (2003): 3443-3449.
- Li, Wenbing, Yanyan Guo, and Peng Zhang. "SERS-Active Silver Nanoparticles Prepared by a Simple and Green Method." *The Journal of Physical Chemistry C* 114, no. 14 (2010): 6413-6417.
- Gold, Colloidal, John Turkevich, Cooper Stevenson, and James Hillier. "A study of the nucleation and growth processes". *Discussions of the Faraday Society* 11 (1951): 55-75.
- Rivas, L, S Sanchez-Cortes, and JV Garcia-Ramos. "Growth of silver colloidal particles obtained by citrate reduction to increase the Raman enhancement factor." *Langmuir* 17, no. 3 (2001): 574-577.
- Pillai, ZS. "What factors control the size and shape of silver nanoparticles in the citrate ion reduction method?" *The Journal of Physical Chemistry B* 3, no. 10 (2004): 945-951.

(651f) Controllable Synthesis of Cu_{2-X} E (E=S, Se) Nanocrystals with Localized Surface Plasmon Resonance and Investigation of Conductivity of NC Thin Films

Xin Liu, Xianliang Wang and Mark T. Swihart, Chemical and Biological Engineering, University at Buffalo (SUNY), Buffalo, NY

Inorganic nanocrystals (NCs) are of great interest due to their attractive optical, electronic and optoelectronic properties. Metal nanoparticles (NPs) exhibit localized surface plasmon resonance (LSPR) when photons induce oscillation of the free electrons in conduction band at resonant frequencies. However, in semiconductors, most electrons are confined to the filled valence band and LSPR is not observed. Non-stoichiometric $Cu_{2-x}S$ and $Cu_{2-x}Se$ NCs have a high concentration of free holes in the valance band. Recently, LSPR was observed in these heavily self-doping semiconductor NCs. Here, we demonstrate a facile method for controllable synthesis of $Cu_{2-x}S$ and $Cu_{2-x}Se$ NCs with superior LSPR. Moreover, we studied the electrical conductivity of solution-processed $Cu_{2-x}S$ and $Cu_{2-x}Se$ NCs thin film in ITO-NC film-Al devices. After chemical post-treatment to displace the long-chain surface ligands attached to the particle surfaces, the nanocrystal thin films show surprisingly high electrical conductivity. As a result, they show promise for use in solution-processed field-effect transistors.

(652a) Nanostructured Energetic Materials for Insensitive Munitions

Victor Stepanov, Munitions Engineering and Technology Center, US Army, RDECOM ARDEC, Picatinny, NJ, Hongwei Qiu, Stevens Inst. of Tech., Hoboken, NJ and Ashok Surapaneni, US Army ARDEC, Picatinny Arsenal, NJ

Nanostructured Energetic Materials for Insensitive Munitions

Victor Stepanov[1], Hongwei Qiu[2] and Ashok Surapaneni[1]

[1]US Army, RDECOM-ARDEC, Munitions Engineering Technology Center,

Picatinny Arsenal, NJ 07806, USA

[2]Department of Chemical Engineering and Materials Science, Stevens Institute of Technology

Hoboken, NJ 07030, USA

ABSTRACT

Nanostructured energetic materials are gaining attention due to their favorable sensitivity and performance characteristics. We have previously shown that nanocrystalline RDX-based compositions prepared by rapid expansion of supercritical solutions (RESS) exhibited markedly reduced sensitivity to shock and impact stimuli. In this paper we describe a spray drying technology developed for simple, safe, and effective production of a wide range of nanocomposite energetics. The process combines crystallization and coating into a single step operation. The materials we investigated include cyclotrimethylene trinitramine (RDX) and hexanitrohexaazaisowurtzitane (CL-20). The process typically yields micron sized (1-20 mm) nanostructured granules consisting of discrete nanocrystals uniformly dispersed in a polymeric binder. Here we present the results of structural characterization and initiation sensitivity analysis of these new materials.

Figure 1. RDX/polyvinyl acetate nanocomposite granules prepared by spray drying.

THE GRAPHIC ON THIS PAGE WAS NOT AVAILABLE ON THE E-MEDIA VERSION OF THIS PAPER.

(652b) Spraying Effect On Particle Size and Morphology in Recrystallization of RDX Using Supercritical Carbon Dioxide Anti-Solvent

Byung-Chul Lee, Department of Chemical Engineering and Nano-Bio Technology, Hannam University, Daejeon, South Korea, Youn-Woo Lee, School of Chemical & Biological Engineering, Seoul National University, Seoul, South Korea and Hyoun-Soo Kim, High Explosive Team, Agency for Defense Development, Daejon, South Korea

Supercritical fluid processes have gained great attention as a new and environmentally-benign method of preparing the micro-particles of energetic materials like explosives and propellants. In this work, RDX (cyclotrimethylenetrinitramine) was selected as a target explosive. The micro-particles formation of RDX using supercritical carbon dioxide (CO_2) anti-solvent recrystallization process was performed and the effect of spraying method on the size and morphology of prepared particles was observed. *N,N*-Dimethylformamide was used as an organic solvent for dissolving the RDX. A spraying tube tip nozzle with an inside diameter of 0.0006 inches and a length of 10 ~ 20 mm was installed into the end of a feed line of RDX solution in order to make a pressure difference between the RDX solution phase and the CO_2 phase during spraying the RDX solution into the CO_2 phase. The pressure difference was controlled in the range of 0 to 100 bar by changing the length of the spraying nozzle. In our experiments, other operating conditions were kept constant at 40 °C, 100 bar, and 20 wt. % of RDX concentration in feed solution, 25 ml/min of CO_2 feed flow rate, and 25 ml/min of RDX solution feed flow rate. A big pressure difference generated between CO_2 phase and solvent phase during the spraying of RDX solution into CO_2 phase could produce the RDX particles with better quality. Specifically, the RDX particles of uniform and spherical shape with the average size of less than 5 μm were obtained at the pressure difference of about 100 bar.

(652c) Synthesis and Characterization of Nanostructured Pyrophoric Materials

Lauren Armstrong, RDAR-MEE-M, US Army, RDECOM-ARDEC, Picatinny Arsenal, NJ, Chris Haines, US Army RDECOM-ARDEC, Picatinny, NJ, Zac Doorenbos, Innovative Materials and Processes, LLC, Rapid City, SD, Kendall Mills, US Army, RDECOM-ARDEC and Deepak Kapoor, Materials Division, US Army RDECOM-ARDEC, Picatinny, NJ

Pyrotechnic materials are of high interest to the DoD for a variety of applications, including countermeasure flares, training rounds, and tracers. Recently, significant efforts have been undertaken to exploit the pyrophoricity of metallic nanostructures, which are ideal because of their environmentally benign reaction products. While many such pyrophoric materials are processed via bottom-up processing (e.g. inert gas condensation) such techniques can be relatively expensive. Here, we have taken a top-down approach to synthesizing novel pyrophoric pyrotechnic materials for DoD applications. The top-down approach employs high energy milling of binary, ternary, and quaternary alloys comprised of aluminum, iron, titanium, and magnesium. The high surface area (and resultant nanostructured lamellae) of these milled materials exhibited pyrophoricity. The influence of processing parameters and chemical composition on the combustion properties of these materials will be presented. In addition, the ignitability and sensitivity of these materials will be discussed along with live fire testing results.

(652d) Synthesis and Characterization of Energetic Nanomaterials for Propulsion Applications

William Lewis[1], Christopher Bunker[2], Elena Guliants[1], Shiral Fernando[1], Marcus Smith[2], Barbara Miller[1], Venroy Watson[1] and Xifan Wang[1], (1)University of Dayton Research Institute, Dayton, OH, (2)Propulsion Directorate, Air Force Research Laboratory, Wright-Patterson Air Force Base, OH

The focus of our work for the last several years has been to discover synthesis routes to novel energetic nanostructures suitable for propulsion applications, such as fuel additives, and to characterize the properties and chemical dynamics of these materials. Most recently, we have synthesized organically-capped metal nanoparticles, metal-decorated graphene oxide (GO), GO gels, and GO-wrapped metal nanoparticles. These materials have been characterized using a variety of techniques including DSC, TGA, XRD, SEM & TEM, FTIR, and Raman spectroscopy in order to understand their size, structure, composition and thermal properties. The (often unique) chemical dynamics of these materials have been investigated using time-dependent calorimetry, spectroscopy, and mass spectrometry techniques. Of particular interest is the influence of factors such as size, structure, and passivation method on the resulting chemical reaction pathways and their associated kinetics. We will present our most recent results from these efforts and discuss the outlook for future work.

(652f) Structure – Reactivity Relationship: Milling-Induced Exothermic Nano-Materials

Khachatur V. Manukyan, Ya-Cheng Lin and Alexander Mukasyan, Department of Chemical and Biomolecular Engineering, University of Notre Dame, Notre Dame, IN

An efficient approach that combines short-term (minutes) high-energy dry ball milling and wet grinding to tailor the nanostructure of Ni+Al, Ti+C, Si+C and Ta+C composite reactive particles is reported. Depending on the system and milling conditions two different reaction mechanisms, i.e. sudden ignition and gradual solid-state diffusion for milling-induced reactions are observed. The ignition sensitivity and reactivity of thus prepared materials are investigated as function of milling time. A direct correlation between the milling-induced nanostructures and ignition characteristics (ignition temperature, ignition delay time and kinetics of chemical reactions) are also discussed. Analysis of the microstructures and reaction kinetics suggests that different factors may be responsible for reactivity enhancement. In Ni+Al and Si+C systems such enhancement is primary related to the formation of intimate contact between the reactants. The fine mixing of reactants creates fresh oxygen-free boundaries which reduces the energy barrier for diffusion processes. In metal – carbon systems (e.g. Ti+C, Ta+C) the solid-state dissolution of carbon in the metal crystal lattice is an important factor responsible for reactivity enhancement. Classification of the exothermic systems from the stand point of the dominant reaction mechanism under high energy ball milling conditions is suggested.

(655a) Simulation of the Breakage of Bonded Agglomerates in a Ball Mill

Matthew J. Metzger, Chemical and Biochemical Engineering, Rutgers University, Piscataway, NJ and Benjamin J. Glasser, Chemical and Biochemical Engineering, Rutgers, the State University of New Jersey, Piscataway, NJ

Size reduction is an essential operation across many industries. Despite its prevalence, the efficiency of the conversion of applied force to the creation of new surfaces is extremely poor, none worse than the commonly encountered ball mill. However, tuning of operational parameters may be a means to increase the low efficiency levels, if the proper connection between operating conditions and breakage conditions can be determined. Described here is the implementation of the Bonded Particle Model (BPM) within the Discrete Element Method (DEM) framework to analyze the breakage of bonded agglomerates within a batch ball mill. It is found that agglomerate strength and grinding media diameter affect breakage significantly; whereas grinding media fill level has a minor affect. The majority of breakage occurs near the mill shell, rather than at the point of impact between the media and material and the extent of material breakage has a significant influence on the material flow within the mill.

(655b) A Microhydrodynamic Analysis of Wet Stirred Media Milling for Production of Stable Drug Nanoparticle Suspensions

Afolawemi Afolabi, Chemical, Biological, and Pharmaceutical Engineering, New Jersey Institute of Technology, Newark, NJ and Ecevit Bilgili, Chemical, Biological and Pharmaceutical Engineering, New Jersey Institute of Technology, Newark, NJ

One way to improve the bioavailability of poorly water-soluble drugs is to reduce particle size of drug crystals down to nanoscale via wet stirred media milling. An increase in total surface area per mass loading of the drug and in specific surface area as well as reduced external mass transfer resistance allow a faster dissolution of the poorly-water soluble drug from nanocrystals. To prevent aggregation of nanoparticles, polymers and surfactants are dissolved in water acting as stabilizers via adsorption onto the drug crystals.

In the last two decades, ample experimental data were generated in the area of nanoparticle suspension production via media milling (e.g. [1,2]). Recirculation mode of milling operation, where the nanoparticle suspension in a holding tank continuously circulates through the stirred media mill, has been commonly used in lab, and commercial scales. To explain the impact of some of the processing parameters such as rotor speed, media (bead) loading, bead size, etc., stress intensity and stress number concepts have been widely used [3], which do not account for the complex microhydrodynamics in the mill and the effects of the suspension viscosity. In the present work, the impact of various processing parameters on the process dynamics was studied using Griseofulvin as a model drug. As a major novelty, we analyze and interpret the impact of processing parameters via a microhydrodynamic model [4]. Our analysis suggests that the product of the average bead oscillation velocity and the frequency of single-bead oscillation appears to be a key empirical parameter that governs the breakage dynamics. Physical interpretation of this combined empirical parameter is discussed.

References

1. Bilgili, E., Hamey, R., Scarlett, B., 2006. Nano-milling of Pigment Agglomerates Using a Wet Stirred Media Mill: Elucidation of the Kinetics and Breakage Mechanisms. Chem. Eng. Sci. 61, 149–157.
2. Sommer, M., Stenger, F., Peukert, W., Wagner, N.J., 2006. Agglomeration and Breakage of Nanoparticles in Stirred Media Mills - A Comparison of Different Methods and Models. Chem. Eng. Sci. 61, 135–148.
3. Kwade, A., 1999. Wet Comminution in Stirred Media Mills – Research and its Practical Application. Powder Technol. 105, 14–20.
4. Eskin, D., Zhupanska, O., Hamey, R., Scarlett, B., 2005. Microhydrodynamics of Stirred Media Mills. Powder Technol. 156, 95–102.

(655c) Jamming in Granular Systems

Shrikant Swaminathan[1], Carl Wassgren[1], Jennifer Sinclair Curtis[2], Bruno C. Hancock[3] and William R. Ketterhagen[4], (1)Mechanical Engineering, Purdue University, West Lafayette, IN, (2)University of Florida, Gainesville, FL, (3)Pharmaceutical Development, Pfizer Worldwide Research and Development, Groton, CT, (4)Pfizer Global Research and Development, Pfizer Inc., Groton, CT

Particle jamming and breakage in the gap between a rotating blade and a containing vessel frequently occur in equipment used to process particulate materials. Jamming can be detrimental, for example during mixing, or desired, as in the case of milling. In this study, we present experimental and computational results investigating the frequency of jamming events between a blade tip and a wall as a function of the blade speed, gap width, vessel fill height, and operating time. In addition, several blade tip designs are presented to study their influence on jamming frequency.

- -

(655d) Experimental and Computational Investigation of Breakage of Pharmaceutical Spherical Particles

Shivangi S. Naik and Bodhisattwa Chaudhuri, Pharmaceutics, University of Connecticut, Storrs, CT

Experimental and Computational Investigation of Breakage of Pharmaceutical Spherical Particles

By Shivangi Naik[1,a], Bodhisattwa Chaudhuri[1,b], and Ramesh B. Malla[2,c]

[1]*Department of Pharmaceutical Sciences, University of Connecticut, Storrs, CT 06269-3092*

[2]*Department of Civil & Environmental Engineering, University of Connecticut, Storrs, CT 06269-2037*

Abstract

Particle size reduction of dry granular material by mechanical means, also known as milling or comminution, is undoubtedly a very important unit operation in pharmaceutical, agricultural, food, mineral and paper industries. As comminution is a stochastic and a non-linear process, an attempt was made to understand this complicated process by conducting parametric studies experimentally and computationally using Discrete Element Method (DEM). Studies were performed lactose spheres to understand the effect of hammer speed (rotational), feed rate, and hammer-wall tolerance in a hammer mill. The size and shape of the resulting progeny of particles were analyzed by sieve analysis and microscope/image analysis techniques respectively. The feed rate determines the hold up of material in sizing chamber and hence energy required for size reduction. Greater size reduction was observed at higher speeds and low feed rates owing to the greater centrifugal force experienced by the particles and longer mean free path lengths respectively. Particle shape analysis revealed fragmentation to be the dominant mechanism of size reduction at higher speeds. Increase in impeller wall tolerance resulted in accumulation of powder bed which was found to be significant at low impeller speeds. To develop an understanding of how grain breakage evolves in a granular system under impact and how it affects the dynamics of the system, the breakage force for the material of interest viz. lactose non-pareils were determined. To obtain these values, single particle impact studies were performed using Dynamic Mechanical Analyzer. In this method, the granules were subjected to a quasi-static compression process and the force required to break the granule was recorded. The tests were performed on large sample size to obtain statistical significance. The flow and fragmentation of non-pareils were simulated using DEM, an explicit numerical technique scheme, which calculates interaction forces between grains for each grain-grain contact, and the resulting motion of each grain. The fragmentation of each spherical particle was simulated using the Grady's Algorithm. The diameter of the resultant progeny particle was evaluated base on the fracture toughness of material, propagation velocity of longitudinal elastic waves in the material, and the induced strain rate.

(655e) Microwave Assisted Breakage Through Quartz Inversion in Low Grade Iron Ore

Gurulaxmi Srikakulapu, Anshuman Parida, Pradip Banerjee, Mark Denys and Venugopalan T, Research and Development, Tata Steel Limited, Jamshedpur, India

Indian low grade iron ores contains quartz as gangue bearing mineral with poor liberation characterstics. Banded iron formations such as Banded Hematite Jasper has alternate layers of Hematite, Quartz considered to be hard ore. The presence of quartz in low grade iron ore causes the breakage at mineral - gangue interface selectively during its inversion by microwave treatment. The present study revealed the crack propogation mechanism is due to quartz transition from α to β phase (volumetric expansion). Detail studies showed that crack propogation can reduce energy consumption during grinding and impove liberation of ore. Hardness and compressive strength of ore are reduced from 925 to 825 HV (hematite), 1660 to 1120 HV (quartz) and 5 to 2 kN/cm^2 respectively. The effect of inversion and simultaneous recrystallization to form new born hematite grains is showed during flotation of iron ore. Fe(T) is upgraded from 50% to 59% and quartz from 25% to 12% respectively for flotation concentrate.

(655f) High Shear Rotor-Stator Wet Milling for Drug Substance: Expanding Capability and Improving Scalability

Allison Harter, Luke Schenck, Ivan Lee, Luis Sierra, Vincent Capodanno and Aaron Cote, Chemical Process Development and Commercialization, Merck & Co., Inc., Rahway, NJ

Rotor-stator wet mills are used in the pharmaceutical industry as a means to control final particle size distribution. Wet milling can be used to create seed for crystallization and also allows for normalization of the particle size to allow batches with consistent API physical properties to be delivered for downstream drug product operations. Wet milling is robust, relatively easy to use, and broadly applicable. Wet milling has significant cost advantages over dry milling as it overcomes challenges around yield losses and eliminates the need for expensive dry solids containment. Wet milling also eliminates API physical property instability resulting from crystal lattice disorder that can be caused by dry milling.

Historical rotor-stator wet mill technologies are generally capable of achieving particles sizes down to ~25 microns. Newer higher shear wet mills allow for a reduction of particle size down to ~10 microns. In addition to the improved particle size reduction, recent new wet mill designs better maintain geometric consistency across the product line, which provides an opportunity for enhanced scalability. Traditional scale-up for wet mills usually involves maintaining the tip speed (i.e. shear rate) of the rotor across scales and generally allows for effective scale-up of the terminal particle size attained. However, traditional scale-up models have limited predictive capability in regards to the particle breakage kinetics leading up to the terminal particle size. Studies presented here confirm the importance of maintaining tip speed across scales but identified additional parameters that are important to include in scale-up models addressing particle breakage kinetics. Additional aspects of hydrodynamics, shear rates, and equipment properties were assessed as part of these scale-up model optimization efforts. Specific processing parameters evaluated included flow rate, solids concentration, fluid viscosity, starting particle size distribution, and pressure drop.

Mill head geometrical design is also evaluated within the scope of studies presented here. Custom mill heads were designed to verify optimized scale-up models. These custom mill heads provided further insight into the importance of mill efficiency and slot events. This, in turn, allowed for more accurate scale-up of not only the terminal particle size but also the particle breakage kinetics. The optimized models reduce the reliance on in process controls for determining the end point of milling.

(655g) Impact of Multi-Particle Interactions On the Attainment of A Self-Similar Particle Size Distribution During Dense-Phase Dry Milling

Ecevit Bilgili, Maxx Capece and Rajesh N. Dave, Chemical, Biological and Pharmaceutical Engineering, New Jersey Institute of Technology, Newark, NJ

In dense-phase comminution systems, particle concentration is high and particles with a multitude of sizes undergo mechanical *multi-particle interactions*through collisions or enduring contacts. Austin and Bagga (1981) determined that presence of fines has a cushioning effect during a dry ball milling process causing the specific rate of breakage, particularly of the coarsest particles, to decrease. Gutsche and Fuerstenau (1999) also observed the retardation in the breakage of coarse particles caused by the addition of fines in particle bed compression experiments. They explained that fine particles distribute the force-flux over the surface of coarse particles and stabilize the particle preventing breakage. They also showed that the stabilization of a coarse particle is a function of the size and the mass fraction of the fines. In other words, multi-particle mechanical interactions among particles of different sizes at the particle ensemble scale lead to a population-dependent breakage probability. Other studies (Austin et al., 1981; Meloy and Williams, 1992; Rajamani and Guo, 1992) led to similar findings for both dry and wet milling processes. Traditional time-continuous linear population balance models (TCL-PBMs) cannot explain the impact of the aforementioned multi-particle interactions.

Time-continuous non-linear population balance models (TCNL-PBMs), which have been recently introduced due to the limitations of the TCL-PBMs, describe the impact of the multi-particle interactions and thus are capable of predicting many types of complex non-first-order breakage kinetics during size reduction operations (Bilgili and Scarlett, 2005). Bilgili et al. (2006) decomposed the specific breakage rate function S_i into a first-order specific breakage rate function k_i and a population dependent functional $F_i[\ ]$. The functional $F_i[\ ]$ describes different types of non-linear (non-first-order) breakage kinetics and explicitly accounts for multi-particle interactions. TCNL-PBM recovers the TCL-PBM in the limit of weak multi-particle interactions, i.e., $F_i[\ ]\rightarrow 1$, and the linear time-variant PBMs in the limit of uniform kinetics, i.e., $F_i[\ ]\rightarrow f(t)$, where $f(t)$ is a time-dependent function (Bilgili et al., 2006). Numerical simulations using the TCNL-PBM predicted the experimentally observed complex and non-intuitive breakage behavior such as cushioning effect, crossing-effect, and acceleration effect in batch and continuous milling operations.

In this study, we investigate the evolution of the particle size distribution toward a self-similar distribution during dry milling via computer simulations. Three models were considered in the simulations: a TCL-PBM, Model B (TCNL-PBM with non-uniform kinetics), and Model D (TCNL-PBM with uniform kinetics). A power-law function for the first-order specific breakage rate and difference-similar (normalized) form of the breakage distribution function were assumed for all three models. Evolution of the particle size distribution during batch dry milling of quartz in a tumbling ball mill (Austin et al., 1990) was fitted to determine the parameters adopting the approach in Capece et al. (2011). Goodness-of-fit and statistical significance of the parameters estimated were evaluated to discriminate the models. The experimental data exhibited strong slowing-down effect due to the cushioning action of the finer particles on the coarser ones. Both Model B and Model D, being non-linear, fitted the data much better than the TCL-PBM. Having statistically significant parameters, the non-linear models exhibited a similar fit. It was not possible to differentiate Model D (uniform kinetics) from Model B (non-uniform kinetics) based on goodness-of-fit and statistical significance.

In the second part of this study, we performed simulations with different initial particle size distributions (Gaussian distributions with different mean and standard deviation and bimodal distributions) a) to investigate their effects on the attainment of a self-similar size distribution and b) to assess if Model B and Model D can be discriminated. The model parameters were kept the same as those obtained from the data fitting, as mentioned above. The TCL-PBM solution attained a self-similar particle size distribution very fast, whereas Models B and D solutions tended towards such size distribution at prolonged milling times. In other words, the non-linear effects (slowing-down phenomenon due to the multi-particle interactions) led to a delay of the attainment of a self-similar size distribution. In fact, the delay was more pronounced in the case of Model B than in Model D. Surprisingly, both Model B and Model D simulations show similar predictions for some initial size distributions, thus making the discrimination of these two non-linear models extremely challenging. Our simulations suggest that these non-linear models can only be discriminated by fitting the evolution of the particle size distribution for multiple initial (feed) particle size distributions simultaneously in the parameter estimation stage.

References:

Austin, L.G., Bagga, P., 1981. An analysis of fine dry grinding in ball mills. Powder Technology 28, 83–90.

Austin, L.G., Shah, J., Wang, J., Gallagher, E., Luckie, P.T., 1981. An analysis of ball-and-race milling. Part I. The Hardgrove mill. Powder Technology 29, 263–275.

Austin, L.G., Yekeler, M., Dumm, T., Hogg, R., 1990. Kinetics and shape factors of ultrafine dry grinding in a laboratory tumbling ball mill. Particle & Particle Systems Characterization 7, 242–247.

Bilgili, E., Scarlett, B., 2005. Population balance modeling of non-linear effects in milling processes. Powder Technology 153, 59–71.

Bilgili, E., Yepes, J., Scarlett, B., 2006. Formulation of a non-linear framework for population balance modeling of batch grinding: beyond first-order kinetics. Chemical Engineering Science 61, 33–44.

Capece, M., Bilgili, E., Dave, R., 2011. Identification of the breakage rate and distribution parameters in a non-linear population balance model for batch milling. Powder Technology 208, 195–204.

Gutsche, O., Fuerstenau, D.W., 1999. Fracture kinetics of particle bed comminution—ramifications for fines production and mill optimization. Powder Technology 105, 113–118.

Meloy, T.P., Williams, M.C., 1992. Problems in population balance modeling of wet grinding. Powder Technology 71, 273–279.

Rajamani, R.K., Guo, D., 1992. Acceleration and deceleration of breakage rates in wet ball mills. International Journal of Mineral Processing 34, 103–118.

(692a) Molecular Simulation Studies of Quantitative Correlation Between the Stability of Drug Nanocrystals Suspension System and Certain Properties of Both the Drug API and the Stabilizer

Wusheng Zhu, Frank Romanski and M. Silvina Tomassone, Chemical and Biochemical Engineering, Rutgers University, Piscataway, NJ

Molecular Simulation Studies of Quantitative Correlation between the Stability of Drug Nanocrystals Suspension System and Certain Properties of Both the Drug API and the Stabilizer

Wusheng Zhu, Frank Romanski, and M. Silvina Tomassone

Department of Chemical and Biochemical Engineering, Rutgers University, Piscataway, New Jersey 08854

Molecular simulations have been applied in the suspension system of poorly water-soluble (BCS Class II) drug nanocrystals in presence of a polymer or surfactant stabilizer. Since the binding energy between the drug crystal surfaces and the stabilizer is believed to link the inhibitory effect of the stabilizer that reflects the relative stability of the drug suspension system, comparison of the simulated drug-stabilizer binding energy, which is evaluated as a statistical average over different cleavage drug crystal faces, is utilized to gauge the relative stabilization capability of the suspension system. To efficiently screen an effective stabilizer for a given drug API that will result in good suspension stability, our research work is to find out the correlation between the drug-stabilizer binding energy and certain properties of the API and the stabilizer. If such a correlation is mapped out, an effective stabilizer can be easily chosen for a given API achieving good suspension stability without doing numerous real screening tests. In previous studies, we have built up quantitative correlations for two cases: i) correlation between the drug-stabilizer binding energy and certain properties such as supersaturation ratio and the molar water solubility of the drug API in presence of the same stabilizer; ii) correlation between the drug-stabilizer binding energy and certain properties such as hydrophobicity and molecular area of the stabilizer absorption on the same drug API nanocrystals. The current ongoing work suggests that a hierarchical high dimensional representation model can be employed to build up the full correlation between the binding energy and the properties of both the drug API and the stabilizer. The correlations obtained for the same stabilizer or the same drug API act as the first order approximation in the high dimensional representation model. The second order correlation can be fitted by simulated binding energies between various stabilizers adsorption on various drug APIs.

(692b) The Properties of Various Coating Polymers and Their Effects On Surface Frictional Behavior As Coatings On Pharmaceutical Tablets

Padma Narayan[1], Karen Balwinski[2], Debora Holbrook[2], Madhusudhan Kodam[1], Meaghan Blake[3] and Karl Jacob[1], (1)Engineering Sciences- Solids Processing, The Dow Chemical Company, Midland, MI, (2)Dow Wolff Cellulosics, The Dow Chemical Company, Midland, MI, (3)Analytical Sciences, The Dow Chemical Company, Midland, MI

Pharmaceutical tablets are typically coated for cosmetic and functional purposes, and the properties of coatings are important to ensure good performance and marketability. For immediate release applications, coatings provide a good barrier for taste masking, ease of swallowing tablets, and aesthetics. Smooth coatings are usually desirable for appearance and 'feel', hence the type of polymer and processing method used can influence the texture of the coating surface. Friction between tablets can also affect their performance in downstream manufacturing steps, such as during conveying, gravity filling, and packaging. Tablets with low values of frictional resistance may have improved throughput and minimal defect formation during conveying operations, especially when extremely large batch sizes (hundreds of kilograms) are processed. Assessment of in-process roughness and tackiness can also help during a coating operation, to prevent tablets from sticking and inducing further defects on the surface.

This study investigates the inter-relationship between physical properties of various coating materials and formulations, such as puncture and adhesion forces, surface roughness, and tablet-tablet sliding friction. A Schulze ring shear tester and Aero-Flow avalanche tester, both typically used for measurement of powders, were utilized for testing the frictional behavior of tablets. It was found that coatings with higher adhesion and low puncture forces (more hard and elastic materials) had lower values of friction than softer and plastically deforming materials. The coating surface roughness parameters did not show correlations with tablet sliding frictional behavior; however had a strong relationship with its pull-off (adhesion) force to the tablet surface. Based on this experimental study, it can be concluded that parameters relating to the elasticity and strength of the coatings influence their frictional behavior, agreeing with previous studies conducted on other material surfaces (Bellemare, et al., 2008 and Hancock, et al., 2010). Future work would include measuring the values of Young's modulus, yield strength, and strain hardening exponent of the films as well as understanding the effects of coating process parameters on the morphology and frictional characteristics of the tablet surfaces.

(692c) Surface Modification of Pollen Shape Carriers for Improved Dry Powder Inhalation Efficiency

Hung Loong Giam, Thi Quynh Ngoc Nguyen and **Raymond Lau**, School of Chemical and Biomedical Engineering, Nanyang Technological University, Singapore, Singapore

Pollen-shape particles were found to have great potential as drug carriers for dry powder inhalation. However, the relatively strong interparticle force on the carrier surface limits the detachment of drug particles after the drug-carrier mixture is released from inhalers. In this study, surface etching technique was employed to modify the carrier particle surface. Four surface etching temperature were examined. A low surface etching temperature were not able to modify the particle surface while a high surface etching temperature destroyed the pollen-shape morphology. Two of the surface etching conditions were then utilized and the physical properties of the resulting particles were characterized. Budesonide was used as the model drug and Rotahaler® is used as the model inhaler. The drug delivery efficiency using the two type of carrier particles are also characterized by Andersen Cascade Impactor (ACI) using inhalation rates of 30 liter/min and 60 liter/min. While there is a reduction in the emitted dose (ED), the fine particle fraction (FPF) of the drug was increased when pollen-shape carriers that were processed by surface etching were used.

ENCAPSULATION OF LACCASE IN CHITOSAN MICRO-CARRIERS BY TWO-FLUID AND THREE-FLUID NOZZLE

Ondrej Kaspar, Frantisek Stepanek, Institute of Chemical Technology Prague, Czech Republic

Introduction

Spray drying is a widely used particle formation process by which aqueous or organic solutions, emulsions and suspensions can be converted into a dry powder with a relatively narrow particle size distribution. Thanks to short contact times, spray drying is well suited even for heat-sensitive materials (enzymes, drugs) [1]. One of the most commonly used compounds for microencapsulation purposes is chitosan thanks to its availability and excellent properties, such as being nontoxic, bioadhesive, and biodegradable [2]. Spray dried microspheres made of pure chitosan cannot be kept suspended in aqueous media for a long time because of their swelling and dissolution. Therefore, non cross-linked chitosan microspheres prepared by spray-drying technique are unsuitable for the purpose of enzyme immobilization. Chitosan has been cross-linked by polyvalent tripolyphosphate (TPP) anions.

The effect of composition (in particular, the TPP/chitosan ratio) and cross-linking method (*ex-situ* vs. *in-situ*) on the properties of chitosan microspheres formed by spray drying by 2-fluid/3-fluid nozzle has been recently described [3]. The particle size distribution, particle morphology, zeta potential and swelling behavior have been systematically investigated. A novel cross-linking method, based on in-situ contact of TPP and chitosan solutions directly at the nozzle orifice (3-fluid nozzle), has been developed and compared with a more conventional approach of spray-drying a pre-mixed TPP/chitosan colloid solution using a 2-fluid nozzle (Fig. 1).

The main objective of this work is encapsulation of enzyme laccase by *ex-situ* and *in-situ* methodology. Laccase (EC 1.10.3.2) is an enzyme belonging to the family of copper-containing oxidases, which are widely distributed in fungi or higher plants [4]. It can oxidize ortho- and para-diphenols, polyphenols, anilines and a number of inorganic ions with the concomitant reduction of molecular oxygen to water. Particles of immobilized laccase could be used in a lot of industrial applications (decolorisation of waste water, biobleaching, antioxidant detectors) [5-7].

Materials and methods

Materials

Low molecular weight chitosan (75-85 % deacetylated), 2,2'-azino-bis(3-ethylbenzthiazoline-6-sulfonic acid) diammonium salt (ABTS) and STPP - sodium tripolyphosphate (>98 % purity) were purchased from Sigma-Aldrich (Germany), acetic acid from Penta (>99 % purity), sodium acetate anhydrous (>99%) was purchased from Fluka, sodium phosphate dibasic dodecahydrate (>99 %) and potassium phosphate monobasic (>99.5 %) from Fluka. Laccase obtained from wood rotten fungi *Trametes hirsute* was a kind gift from TU Graz.

Preparation of chitosan microparticles by spray drying

Chitosan solution (0.5% w/v) has been prepared by dissolving the required amount of chitosan in 25 ml of aqueous solution of acetic acid (2% v/v) at room temperature. After complete dissolution of the polymer required amount of laccase was added. In a case of 2-fluid nozzle cross-linking has been achieved by adding different volumes (0-2 ml) of a 1% (w/v) TPP solution drop-wise with a syringe (type 25Gx1") to the chitosan solution under constant stirring and then further stirred at 650 rpm for 15 min to ensure complete cross-linking of chitosan - "*in-situ* cross-linking". After addition of TPP the solution changes from transparent to opalescent. Opalescence occurs due to the formation of chitosan-TPP nanoparticles. The colloidal suspension of cross-linked chitosan nanoparticles was then spray-dried by a 2-fluid or to form microspheres with uniformly dispersed laccase. The volume of added laccase was 25 up to 200uL. In a case of 3-fluid nozzle was chitosan solution containing laccase and TPP solution pumped simultaneously into nozzle and spray-dried - "*in-situ* cross-linking". The laboratory spray-drier Buchi B-290 has been equipped with a high performance cyclone in open mode configuration. The values of operation parameters that ensured stable and robust process for the entire range of compositions for 2-fluid and 3-fluid nozzle: aspirator rate 90%, liquid flow rate 6.4 ml/min (skin-forming phase) and 3ml (inner phase - for 3 fluid nozzle), and atomizing gas flow rate 414 normlitres/hr (2-fluid nozzle) and 670 normlitres/hr (3-fluid nozzle). Inlet temperature was varied in a range 100 - 190°C. The outlet temperature was typically 40°C (for inlet temperature 100°C) to 85°C (inlet temperature 190°C).

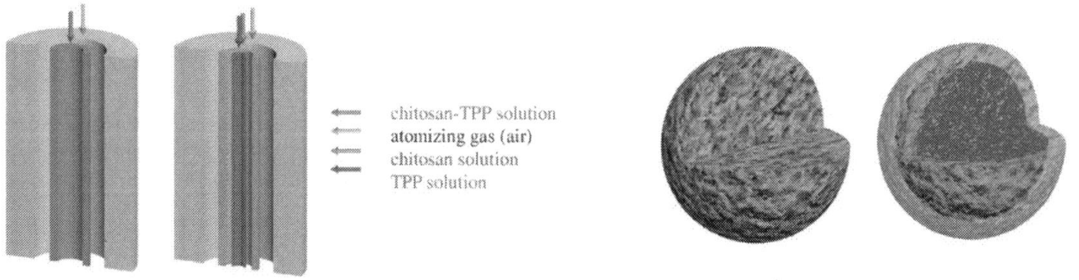

a) b)

Fig. 1 a) Scheme of a 2-fluid and a 3-fluid nozzle b) Schematic droplet structures produced by 2-fluid (left) and 3-fluid (right) nozzle.

Particle size and morphology characterization

Surface morphology of microparticles was examined by Scanning Electron Microscope (SEM) Jeol JCM-5700. Before measurement the samples of prepared microparticles were sputter-coated (Emitech K550X) by a thin layer of gold (5 nm). SEM micrographs were taken at an acceleration voltage of 20 kV. The internal structure of particles produced by the 3-fluid nozzle was determined by a Laser Scanning Confocal Microscope (Olympus Fluoview FV1000 confocal system with Olympus IX81 inverted microscope). The particle size distribution was measured by the static light scattering method (Horiba Partica LA-950 S2) in the wet mode. Chitosan particles were dispersed in absolute ethanol (in which they are not soluble) and placed to an ultrasound bath for 5 minutes. The suspension was then added to the measurement cuvette filled with ethanol and measured five times.

Release and enzymatic studies

Specific enzymatic activity of laccase stock solution was determined by a reaction with ABTS in a 0.1M acetate buffer at pH 5.0. Absorbance of the product (ABTS radical) was measured at 420 nm (characteristic absorbance peak for ABTS radical, $\varepsilon_{420nm} = 36000\ M^{-1}\ cm^{-1}$) at 25°C and 50°C. The value of laccase specific activity was determined by a calibration measurement (0.51 ± 0.02 nmol.min^{-}

^{1}.uL^{-1}). To measure enzymatic activity of laccase immobilized in chitosan powder, a given amount of the sample was placed inside an envelope made of filtrate paper which keeps chitosan microparticles but is fully permeable for ABTS substrate. The envelope was held in the cuvette by a plastic holder with an opening for the beam as shown in Fig. 2. This arrangement was chosen to prevent unwanted disruptions of the signal inside the cuvette by freely floating chitosan microparticles. The measuring cell was filled with 2.8 ml of acetate buffer in each experiment.

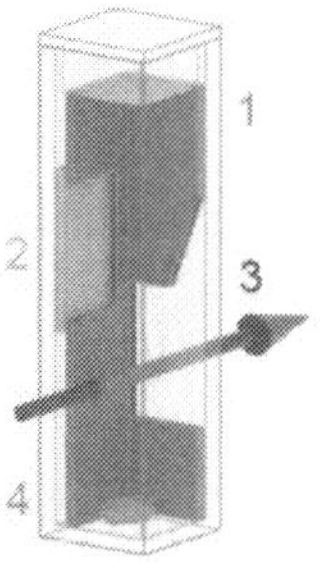

Fig. 2: Spectrophotometric cuvette with a custom attachment for the measurement of release kinetics (1 – plastic holder; 2 – sample; 3 – light beam; 4 – stirrer).

Results

This section is structured as follows. The dependence of inlet parameters (inlet temperature and TPP/chitosan mass ratio) on the laccase activity for two and three-fluid nozzle is discussed. The results will be presented in the form of parametric studies aiming at identifying general trends that can be used for particle design.

Influence of drying temperature (2 and 3-fluid nozzle)

Influence of drying temperature (100 to 190°C) on laccase activity and sample properties has been investigated. At higher temperatures (160 to 190°C) laccase has been significantly deactivated (air outlet temperature > 70°C). Suitable inlet temperature had been chosen 130°C (dry, nonstick product). Lower temperature leads to sticky and wet product. The results of the experiment are showed on Table 1.

Table 1. Influence of drying temperature on laccase activity.

Inlet temperature [°C]	Outlet temperature [°C]	Deactivation with respect to original enzyme [%]	Powder character
190	85	70	overheated
160	75	24	overheated
130	55	16	dry
100	40	14	sticky

Morphology and structure of particles

SEM images of chitosan particles produced by 2-fluid and 3-fluid nozzle are showed at Fig. 3a, b and c. Microparticles of pure chitosan are typically brain-like structured of spherical shape. Cross-linked (same TPP/chitosan ratio) particles produced by 2-fluid nozzle are of irregular collapsed shape and particles produced by 3-fluid nozzle are spherical with smoother surface due to different cross-linking process. These micro-carriers are much more stable [3] and apparently have better flow properties to compare with commonly prepared chitosan-TPP particles of irregular shape. Theoretical core-shell structure has been proven by laser scanning confocal microscopy (Fig. 3d).

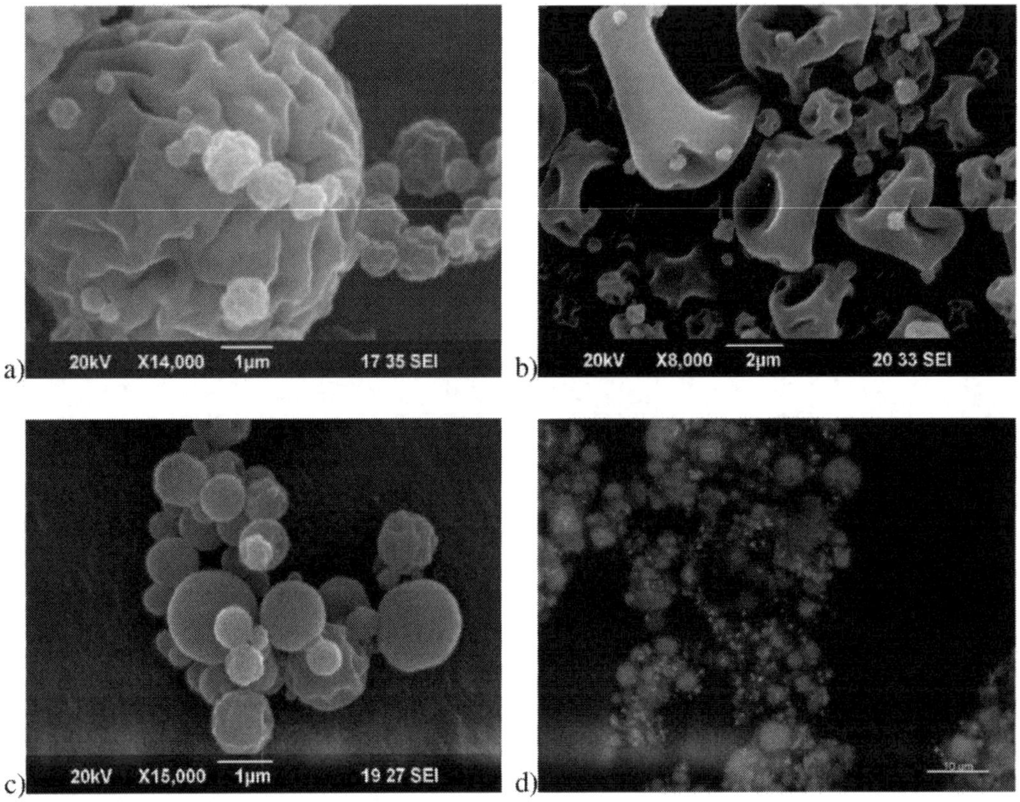

Fig. 3: a) SEM images of uncross-linked b) *ex-situ* cross-linked and c) *in-situ* cross-linked chitosan particles d) LSCM image of core shell structure (curcumin - core, FITC - shell).

Influence of laccase amount, reaction temperature and TPP content (2-fluid nozzle)

Firstly the activity of chitosan powder with different immobilized laccase content at 25 and 50°C has been investigated (Fig. 4a). It can be seen that the reaction temperature has a significant influence on enzymatic activity. In fact, it has been reported in the literature the optimum temperature for laccase *T. hirsute* is around 50°C [4], which is consistent with our findings. Higher TPP content (TPP/chitosan ratio) leads to denser cross-linked chitosan matrix without tendency to swell rapidly in aqueous medium. Increase in TPP/chitosan ratio generally leads to slower substrate diffusion through cross-linked shell to immobilized enzyme (Fig. 4b). Encapsulation efficiency for stable cross-linked samples (TPP/chitosan > 0.08) has been higher than 98%.

a) 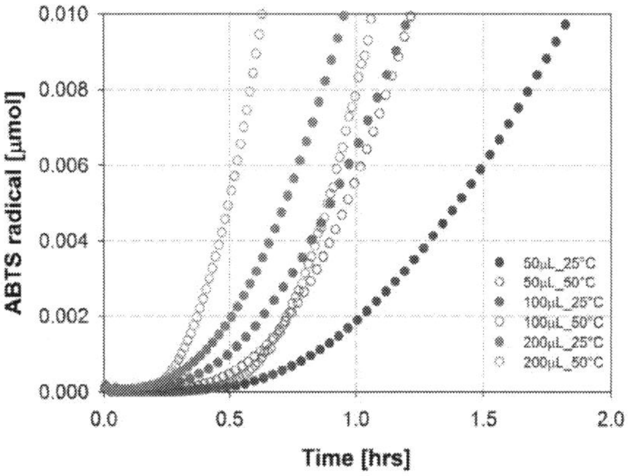

b) 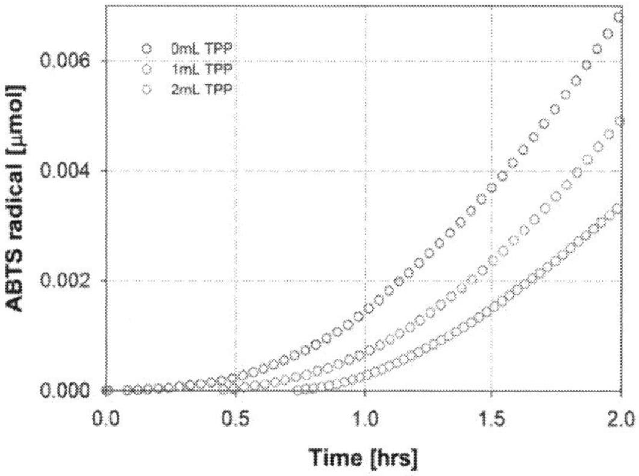

Fig. 4: a) Influence of laccase amount, temperature and b) TPP content on enzymatic activity TPP/chitosan ratio a) 0.08 b) 0 to 0.16.

Influence of TPP/chitosan ratio (varied by chitosan and TPP concentration) for 3-fluid nozzle

Similarly to 2-fluid nozzle higher TPP concentration leads to denser cross-linked matrix and slower substrate diffusion in and out of particle (Fig. 4a). Higher chitosan concentration in outer phase caused lower TPP/chitosan ratio and stronger swelling activity in aqueous medium (Fig. 4b). Flow rates of inner (TPP) and outer (chitosan and laccase) fluid has been 3 and 6.4 mL/min. This set-up has been chosen to achieve higher cross-linked fraction without nozzle clogging and deposition of waste material on inner surface of drying chamber. Encapsulation efficiency for stable cross-linked samples (TPP/chitosan ratio > 0.05) has been 91 to 95%.

a) 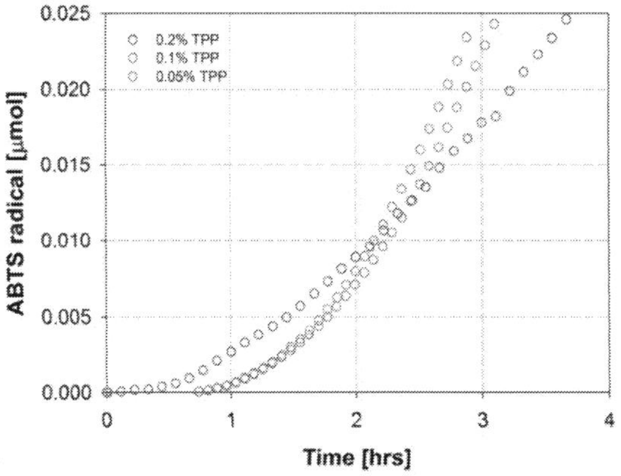

b) 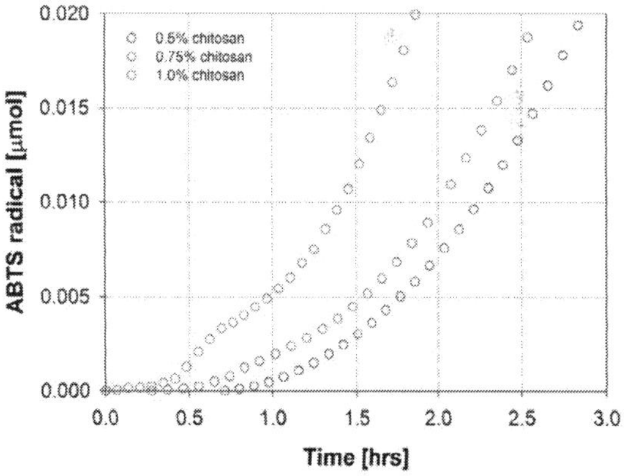

Fig. 5: a) Influence of TPP/chitosan ratio on enzymatic activity TPP/chitosan ratio a) 0.05 to 0.19 b) 0.05 to 0.09.

Comparison of particle size distribution and specific enzymatic activity for 2 and 3-fluid nozzle

Particles produced by 3-luid nozzle have been smaller (mean size approx. 2.7 μm) than particles produced by 2-fluid nozzle (3.6 μm) of same composition due to higher flow of atomizing gas (wider nozzle orifice) to avoid deposition of spray-dried material on inner surface of the drying chamber. Typical particle size distributions for both types of nozzles are showed in Fig. 5a. Comparison of enzymatic activity for particles with same composition is showed in Fig. 5b. Enzymatic activity for 3-fluid nozzle product has been higher to compare with 2-fluid nozzle. Possible explanation is difference between cross-linking methods. In a case of 2-fluid nozzle is an enzyme immobilized into uniformly cross-linked chitosan matrix. In the second case chitosan droplets are cross-linked from inside (core) to outside (shell) during very rapid spray drying process (1-2 seconds) and density of cross-linked chitosan strains of shell and core is different. Diffusion of substrate through hydrated shell is faster and particles are much more stable due positively charged chitosan chains on the surface without tendency to form stable agglomerates. On the other hand encapsulation efficiency for 3-fluid nozzle (91 to 95%) is slightly lower than for commonly used 2-fluid nozzle (>98%).

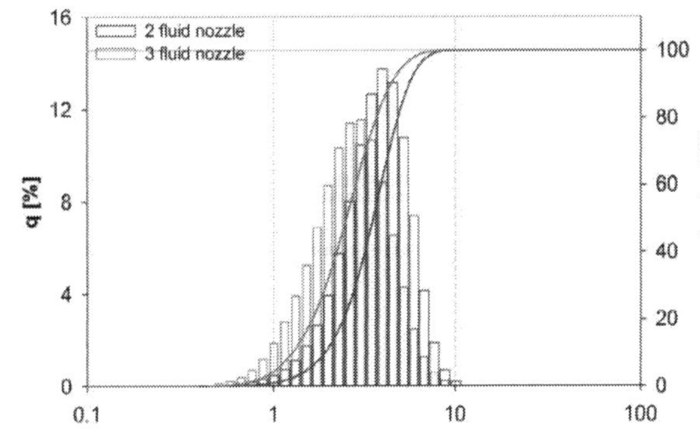

a)

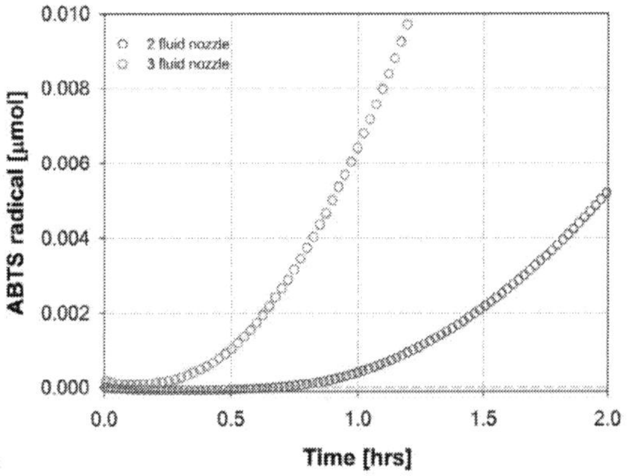

b)

Fig. 6: Particle size distribution and enzymatic activity for samples produced by both types of nozzle (same composition – TPP/chitosan ratio 0.08).

Conclusion

Laccase has been successfully encapsulated by both types of nozzles without significant loss of enzymatic activity – 16% loss for spray-drying at 130°C respect to original. Encapsulation efficiency for 2 and 3-fluid nozzle has been higher than 98% and 91% respectively. Encapsulation efficiency of 3-fluid nozzle could be improved by optimalization of flow ratio of inner and outer fluids to minimize uncross-linked fraction. Particles produced by 3-fluid nozzle are more stable than particle produced by 2-fluid nozzle (higher zeta potential) and could be used in many industrial applications (e.g., decolorization of dyes, biobleaching, food processing, biosensors).

Acknowledgments

This work has been supported by the European Research Council (200580-Chobotix), by the Specific University Research (MSMT 21/2012) and by the Specific Science Cooperation Program AKTION (MEB061108).

References

1. Vehring, R., Foss, W.R.Lechuga-Ballesteros, D., Particle formation in spray drying. Journal of Aerosol Science, 2007. 38(7): pp. 728-746.

2. Agnihotri, S.A., Mallikarjuna, N.N.Aminabhavi, T.M., Recent advances on chitosan-based micro- and nanoparticles in drug delivery. Journal of Controlled Release, 2004. 100(1): pp. 5-28.

3. Kaspar, O., Jakubec, M., Stepanek, F., Characterization of spray dried chitosan–TPP microparticles formed by two- and three-fluid nozzles. Powder Technology, 2012 (in-press).

4. Arora, D.S.Sharma, R.K., Ligninolytic Fungal Laccases and Their Biotechnological Applications. Applied Biochemistry and Biotechnology, 2010. 160(6): pp. 1760-1788.

5. D'Annibale, A., Stazi, S.R., Vinciguerra, V., Di Mattia, E.Sermanni, G.G., Characterization of immobilized laccase from Lentinula edodes and its use in olive-mill wastewater treatment. Process Biochemistry, 1999. 34(6-7): pp. 697-706.

6. Peralta-Zamora, P., Pereira, C.M., Tiburtius, E.R.L., Moraes, S.G., Rosa, M.A., Minussi, R.C.Duran, N., Decolorization of reactive dyes by immobilized laccase. Applied Catalysis B-Environmental, 2003. 42(2): pp. 131-144.

7. Fernandes, S.C., de Oliveira, I.R.W.Z., Fatibello, O., Spinelli, A.Vieira, I.C., Biosensor based on laccase immobilized on microspheres of chitosan crosslinked with tripolyphosphate. Sensors and Actuators B-Chemical, 2008. 133(1): pp. 202-207.

(692e) Engineered Biodegradable Janus Particles for Drug Delivery

Jennifer Winkler, Chemical Engineering, Rutgers University, Piscataway, NJ, Frank Romanski, BASF, Iselin, NJ and Maria. S Tomassone, Department of Chemical and Biochemical Engineering, Rutgers University, Piscataway, NJ

Recent advances in the field of nanotechnology have generated interest in the creation of "Janus" particles, whose dual surface chemistries and unique properties set them apart from spherical isotropic particles. Janus particles present two distinct faces with different shapes, compositions, chemistries, and polarities. Polymeric and hybrid polymer-lipid Janus particles have applications as compartmentalized drug carriers, biological sensors, surfactants, or display materials for electronics. However, biodegradable Janus particles have yet to make their debut in the pharmaceutical industry because current synthesis methods rely on microfluidic devices capable of producing just one single particle at a time. Furthermore, lipid-based materials do not meet the low viscosity requirements of microfluidic devices. Here, we describe a modified emulsion-solvent evaporation template for the large-scale production of biodegradable Janus micro- and nanoparticles from GRAS materials. Biodegradable anisotropic particles are produced through emulsification of the oil phase containing the two GRAS components and the aqueous phase containing stabilizers, followed b homogenization and solvent evaporation. During the slow 6-hour evaporation process, the two immiscible components segregate to an entropically favorable state of separation. However, the immiscible polymer and/or lipid components are forced to co-precipitate within the same oil droplet "microreactor" during evaporation, leading to the formation of an interpenetrating network connecting the two components. Polymeric Janus particles have been formed from 50/50 combinations of GRAS polymers such as PLGA, polycaprolactone, polyvinylpyrrolidone, and poly(4-vinylpyrrolidone). Polymer-lipid "ice cream cone" particles have been formed from 75/25 combinations of GRAS polymers listed above and GRAS lipids such as precirol and compritol. Drug loading is achieved by co-dissolving the API with the polymer and/or lipid components in the dispersed phase prior to emulsification. In some cases, drug loading tremendously alters the phase separation and self-assembly of polymeric Janus particles. Spherical particles with surface pox marks were produced when 5% w/w griseofulvin, omeprazole, or acetaminophen was co-dissolved with the 50/50 polymer mixture in the dispersed phase. However, a Janus-like morphology is still obtained when 5% w/w curcumin or quercetin is introduced into the system. It is hypothesized that certain APIs form interactions with both polymers, thus acting as block copolymer compatibilizers that reduce the interfacial tension between two immiscible polymers. No APIs were found to disrupt the phase separation behavior of the polymer-lipid hybrid particles, presumably because the polymer and lipid are highly dissimilar in structure and chemistry.

(696a) Computer Simulation of a Full Chemical Looping Combustion System

James M. Parker, CPFD Software, LLC, Albuquerque, NM

A full three-dimensional model of a chemical looping combustion (CLC) system using coal as a fuel source has been created. The relevant chemical reactions and thermal effects are modeled inside the full CLC loop, which consists of an air reactor, fuel reactor, riser, and cyclone. The particle-fluid dynamics within the full loop have been modeled with an Eulerian-Lagrangian approach which allows the disparate fluidization regimes inside the CLC loop to be modeled simultaneously. Multiple heterogeneous and homogenous reactions are considered at the particle level including the oxidation-reduction reactions of the ilmenite carrier and gasification of the coal-based carbon. Additionally, the temperature-dependent release of volatile components is modeled in each coal particle. The model is fully thermal and the effects of reaction energy and water vaporization are coupled to the particle-fluid dynamics and reaction chemistry. Results detailing the fluidization and transport behavior, temperature profiles, and reaction conversion in the CLC loop are presented.

(696b) Oxygen Uncoupling Enhanced Iron Oxides for Solid Fuel Gasification – Oxygen Carrier Development, Reactive Testing, and Process Analyses

Feng He[1], Nathan Galinsky[2], Yan Huang[3] and Fanxing Li[2], (1)Department of Chemical and Bimolecular Engineering , North Carolina State University, Raleigh, NC, (2)Chemical and Biomolecular Engineering, North Carolina State University, Raleigh, NC, (3)Department of Chemical and Bimolecular Engineering, North Carolina State University, Raleigh, NC

Oxygen Uncoupling Enhanced Iron Oxides for Solid Fuel Gasification – Oxygen Carrier Development, Reactive Testing, and Process Analyses

*Feng He, Nathan Galinsky, Yan Huang, Fanxing Li**

Department of Chemical and Bimolecular Engineering

North Carolina State University

Raleigh, NC 27695

* Email: fli5@ncsu.edu

Carbon dioxide emitted from fossil energy conversion is estimated to account for ~ 20% of the greenhouse effect. In order to mitigate the anthropogenic CO_2 emissions, effective carbon capture, utilization and storage (CCUS) approaches need to be developed and adopted for fossil fuel conversions. Chemical looping processes, which include chemical looping combustion (CLC) and chemical looping gasification (CLG), utilize novel, indirect strategy for fossil fuel conversion and CO_2 capture. The CLC process is developed to produce heat/electricity. CLG, on the other hand, converts fuels into separate streams of sequestrable CO_2, hydrogen, and heat/electricity.

Although Fe-based metal oxide is considered as the only suitable oxygen carrier for the CLG process, the reaction between solid fuels such as coal and iron oxide tends to be slow. In order to improve the reaction rate of the aforementioned reaction, we propose to incorporate a secondary oxygen carrier such as CuO to the iron oxide based oxygen carrier for chemical looping gasification of solid fuels.

To validate the feasibility of this approach, oxygen carriers with varying combinations of copper and iron oxides are synthesized. Reactions between the oxygen carriers and coal/coal char are performed in both a Thermal-Gravimetric Analyzer (TGA) and a bench-top fluidized bed reactor. Equilibrium based ASPEN Plus® process simulation is also carried out to predict the reactor and process performances. The simulation results indicate that copper oxide addition simplifies the process heat integration scheme, and hence a high energy conversion efficiency for the proposed bimetallic CLG concept can be expected. The potential environmental impacts of the chemical loping process are also evaluated in the present study.

THREE DIMENSIONAL SIMULATION OF CATALYTIC CRACKING REACTIONS IN AN INDUSTRIAL SCALE RISER USING A 11-LUMP KINETIC

Ariane C. Barbosa; Gabriela C. Lopes; Leonardo M. Rosa; Milton Mori
Unicamp, Campinas - SP, Brazil

Waldir P. Martignoni
Petrobras, Rio de Janeiro - RJ, Brazil

Introduction

Fluid Catalytic Cracking (FCC) is one of the most important processes in the oil refinery. It converts heavy hydrocarbon petroleum fractions into more valuable products such as gasoline, middle distillates, and light olefins. A FCC unit is composed mainly by a riser reactor, cyclones and a catalyst regenerator.

Although most cracking in the FCC process occurs via catalytic reactions, thermal cracking reactions are also present in these reactors. According to Bollas et al. [1], 2 - 5 wt% of the products formed during the FCC process correspond to dry gas, which is attributed to "less favored" types of cracking reactions, since these components are not in agreement with the mechanism of catalytic reactions.

To increase knowledge about the presence of the thermal cracking reactions in FCC process, unlike most studies about this subject, the thermal and catalytic reactions are studied together in this work. To represent the kinetics of cracking, the 10-lumps model proposed by Jacob el al. [2] was modified through the addition of a new lump named "dry gas" to represent the results of the thermal reactions. In this 10-lumps kinetic model, the feed is divided into four volatile groups of components (heavy paraffinic molecules, heavy naphthenic molecules, heavy aromatic substituent groups, carbon atoms among aromatic rings and light paraffinic molecules), thus it can be used to represent different charge feedstocks. This is an important aspect because the feed composition can be very different depending on the petroleum origin, gasoline quality and yield. Another important aspect on the FCC process simulation concerns the asymmetry of the flow inside the riser reactor. Lopes et al. [3] emphasized the use of a three-dimensional flow model to predict the behavior of the fluid catalytic cracking in an industrial reactor. They used a 4-lumps kinetic model to represent the cracking reactions and showed the presence of non-uniform patterns that can affect the performance of the reactor. Faced with the need to perform three-dimensional simulations to predict the non-uniform distribution of phases in multiphase systems, the present study extends this contribution, using a more detailed cracking kinetic model and a three-dimensional model to predict the dynamic behavior inside an industrial riser reactor.

Modeling and Simulation

The geometry of the simulated riser was based on the study of Lopes et al. [3]. The reactor has 34.2m of height and 0.8m of diameter, as shown in Figure 1. The numerical mesh created was composed of approximately 700,000 hexahedral control volumes.

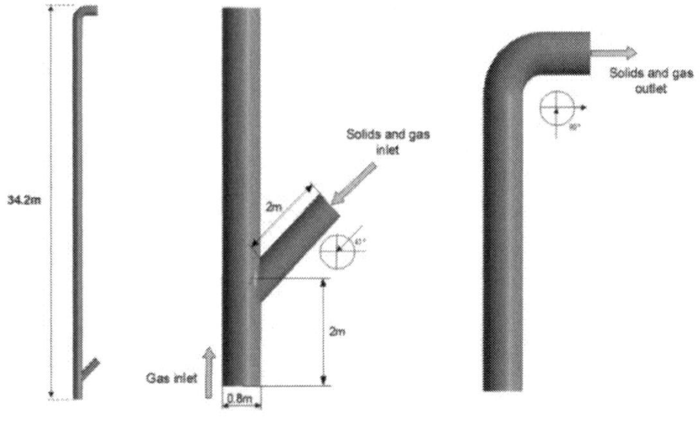

Figure 1. Geometry of the riser [3]

Table 1: Operation Conditions

Feed oil mass flux (kg m^{-2} s^{-1})	40
Catalyst inlet temperature(K)	900
Feed temperature (K)	500
Steam (%)	3
Catalyst-to-oil ratio (kg/kg)	7

The operating conditions applied are the same used by Ali et al. [4]. They show industrial data obtained in a riser reactor with the same height and diameter as applied in the simulations. These data were used to validate the results obtained in the present work. The gas phase is injected into the riser at a temperature of 500K and a mass flux of the 40kg m^{-2} s^{-1}, while solid phase is fed at a temperature of 900K and a mass flux of the 280kg $m^{-2}s^{-1}$, as shown in Table 1. Since the vaporization of the heavy oil is fast, it is adopted a hypothesis that it is injected into the reactor in the gas phase.

It is know that the thermal cracking reactions are favored by high temperatures, occurring mainly in regions of inefficient mixture between the feed oil and the hot catalyst particles. In order to minimize the effect of thermal reactions, five different configurations were proposed to feed the catalyst particles: (1) the catalyst is fed from the base of the reactor with heavy oil (hypothetical case with perfect mixing); (2) the catalyst is injected by a lateral entrance with circular section; (3) the catalyst is injected by a lateral entrance with square section; (4) the catalyst is injected by a lateral entrance with circular section and, additionally, the catalyst particles have an angular component of velocity at that face. (5) a more realistic geometric configuration was also proposed, in which the heavy oil is injected into the reactor by twelve ducts of 0.5in located above the catalyst entrance. Water vapor was injected into the base of the reactor in order to help particles transport.

Cases 1-4 consider the feedstock (heavy oil + steam) is fed through the base of the reactor. The software used in these cases is CFX 12.0 (Ansys), while in Case 5, which used Fluent version 14.0 from Ansys. The mesh used in Case 5 consists of approximately one million hexahedral.

These five cases are outlined in Figure 2.

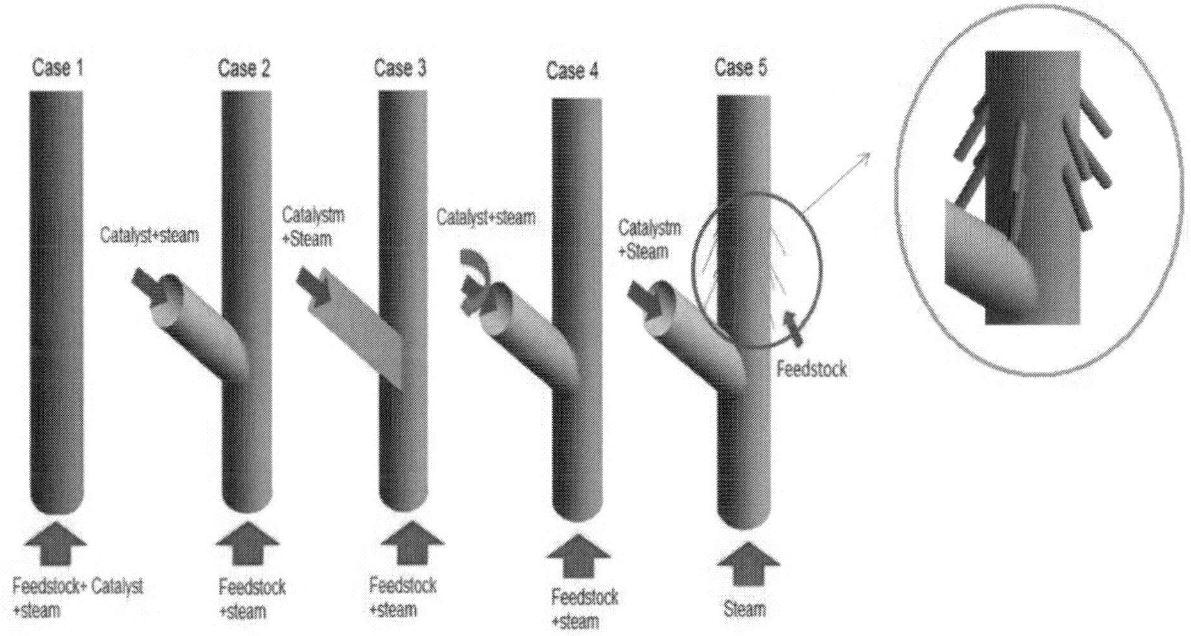

Figure 2. Configurations proposed to feed the catalyst particles

An Eulerian-Eulerian approach was applied to simulate the dynamic behavior of the gas and solids two-phase flow. The drag between these phases was modeled using the Gidaspow drag model, while the heat transfer between them was predicted using the Ranz-Marshall correlation for the Nusselt number. The turbulence in the gas phase was modeled using standard k-ε model. The total simulated time was 15 seconds, of which the last 5 seconds were considered for the calculation of time-averaged values. The time step for the solution was 10^{-3} seconds, with a maximum of ten iterations per time step and a convergence criterion of 10^{-4} RMS (root mean square).

Cracking reactions

In order to simulate the thermal cracking reactions that occur together with catalytic reactions, a new group (DG) was added in the 10-lump kinetic model proposed by Jacob et al. [2] (shown in Figure 3a). This new group represents the product of thermal cracking, which the representative reactions of this adapted model are shown in Figure 13b.

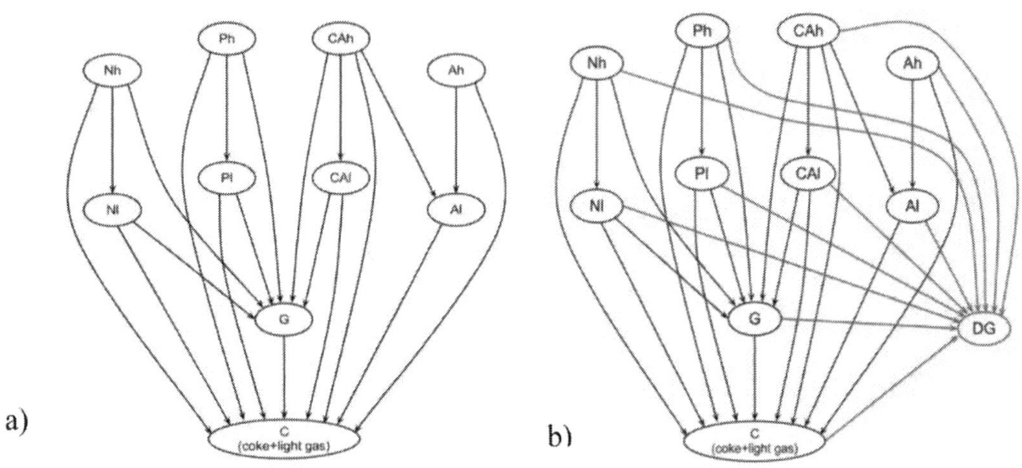

Figure 3. a) 10-lump kinetic model proposed by Jacob et al. [2] b)Adapted kinetic model (11-lumps)

The density, specific heat, viscosity, thermal conductivity and heat of formation of the reactive species, used in this study were collected from the literature [3, 5]. The values for their molecular weight were taken from the study of Pitault et al. [6]. Since the 10-lumps model contains one single group to represent both the coke and the light gases (the C-group), its properties were defined as the weighted average of these species present in it. From others works, it is verified that 30% of coke and 70% of light gas composes the C-groups. This proportion is considered in the present work, in order to determine molar mass of the C-group. The molecular weight used for a dry gas was given by Peixoto e Medeiros [7] and the catalyst properties used in this study are the same applied by Lopes et al. [3].

All catalytic cracking reactions are first-order (n=1) and thermal cracking reaction kinetic are second-order (n=2). The reaction rates are given by:

The reaction rate for catalytic cracking reactions [5] is given by

$$R_{i,rc} = \frac{10.24}{10.24 + exp(4.29 * 0.3 * C_c)} \left(\frac{1}{1 + C_{Ah} K_h}\right) \left(\frac{\epsilon_s}{1 - \epsilon_s}\right) M_{w,i} \sum_{j \neq 1}^{9} v K_{i,rc \to j} * C_i$$

The thermal cracking reactions can be written as

$$R_{i,rt} = M_{w,i} \sum_{j \neq 1}^{10} v K_{i,rt \to j} * C_i^2$$

where ε_s is the solids holdup, ρ_s, the solid density, K_h is the adsorption constant and equals to 5.68 x 10^{-5}, C_{ah} is the mass percentage of the heavy aromatics, $M_{w,i}$ is the molar mass of the component I, v is the stoichiometric, C_i is defined as the molar concentration of the reactants, and K_i is the kinetic constant of the cracking reactions of specie i, described by the Arrhenius equation. The kinetic constant for catalytic cracking were taken from the study of Arbel et al. [8] and for thermal cracking were obtained in the study by Bollas et al. [1].

The 10-lump kinetic model proposed by Jacob et al. [2] is a flexible model, since it allows the simulation of the cracking of heavy oils with different composition. Four different compositions were tested with the configuration used in the simulation of Case 2. They are defined in Table 2 where the charge S is equally composed by the four heavy lumps, P1 and P2 represent the charge rich in paraffin, and A1 stands for the charge rich in aromatics.

Table 2: Compositions of charge stocks (% wt)

Molecules	P1	P2	A1	S
Paraffinics	53	40	17	22
Naphthenics	33	36	26	22
Aromatics	9	7	22	22
Aromatics substituents groups	55	15	33	22

Results and discussion

The results of the simulations are presented below. The catalyst particles distribution obtained in the simulation of the five different cases (Figure 2) are shown in Figure 4. It can be observed that the Case 1 presents a perfect mixture of both gas and solid phases, while in the other cases the solid particles are distributed non-uniformly inside the reactor.

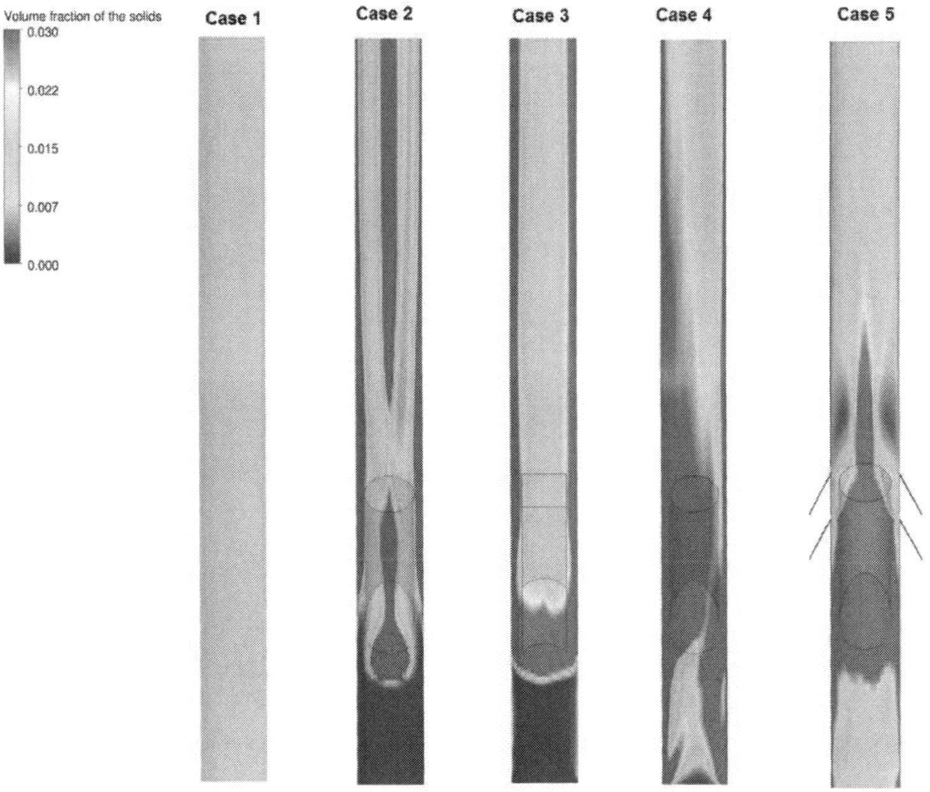

Figure 4. Field of the solids holdup for cases 1-5

The solids distribution in the riser has a significant impact on the gas temperature patterns, as can be seen in Figure 5. As expected, Case 1 presents a homogeneous temperature. However, in the other cases, high temperature gradients are observed. The lateral entrance with square section in Case 3 propitiate a more uniform distribution of solids, in the axial direction to the catalyst entrance, when compared with Case 2, resulting in a more uniform temperature distribution. The simulation of Case 4, in which the solid particles have an angular component of velocity at the inlet, results in solids and temperature distributions even more uniform.

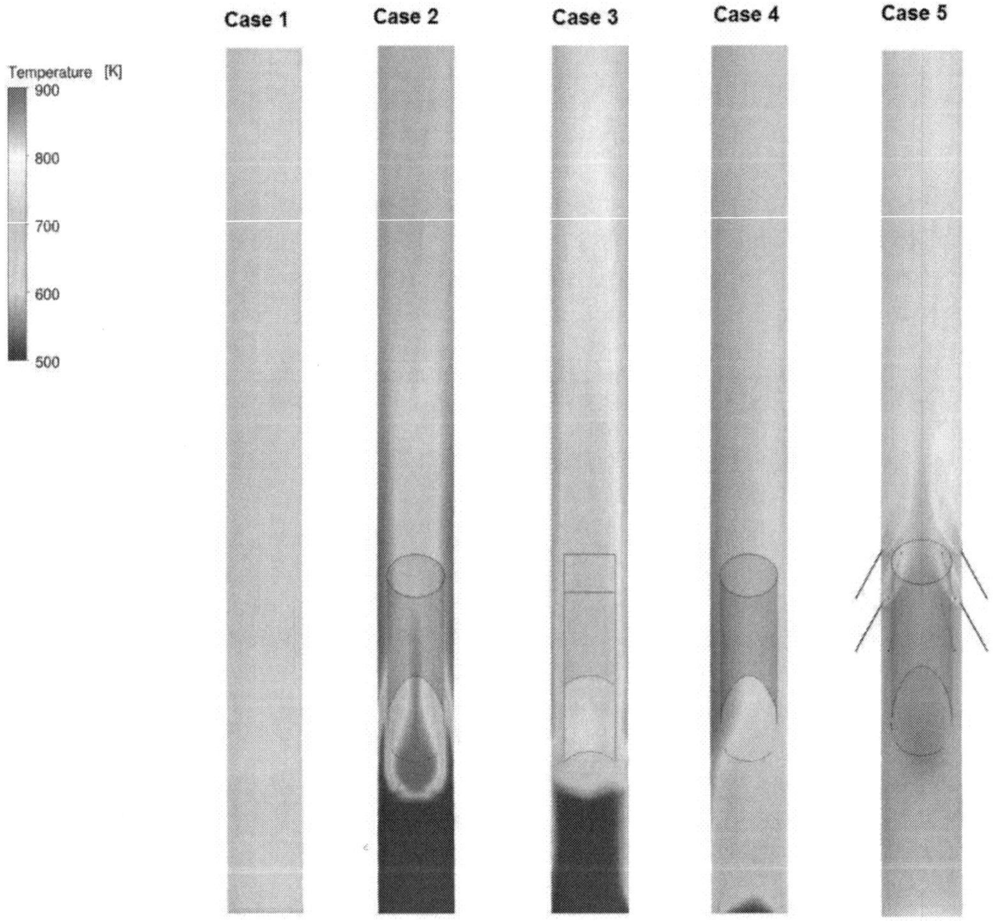

Figure 5. Fields of the gas temperature for Cases 1-5.

The yields of the dry gas obtained at the risers outlets are shown in the Table 3. After Case 1, Case 5 shows the lowest yield of dry gas. In this configuration, heavy oil is injected into the reactor above the catalyst feed entrance as shown in Figure 2, thereby when the reactants come across the catalyst particles, these are more uniformly distributed. Furthermore, before the contact between the catalysts and the feedstock, these particles initially exchange heat transfer to the fluidization vapor that enters from the base of the riser, decreasing their temperature. Consequently with lower temperature the thermal cracking reactions are less favored for the configuration used in Case 5.

The results of these simulations show that the effect of the thermal cracking reactions cannot be avoided in the FCC process. This fact can be verified by looking the results of the simulation of Case1, in which the perfect mixture of the phases show the value of 1.84% of dry gas.

Table 3. Yield of the dry gas for the five cases

Simulation	Yield DG (% wt)
Case 1	1.84
Case 2	2.86
Case 3	2.86
Case 4	2.96
Case 5	2.50

Results obtained in the simulation of Case 5 indicate a better mixture between the phases. It is known that this kind of configuration is common in industrial FCC units. Different feedstock compositions were used to test the ability of the 11-lumps model to predict the cracking of feedstock rich in paraffinic and aromatic compounds.

The gasoline yields at the riser outlet are shown in Figure 6. These results are compared with a plant data reported by Ali et al. [4] and with a similar simulation using 4-lumps model [3].

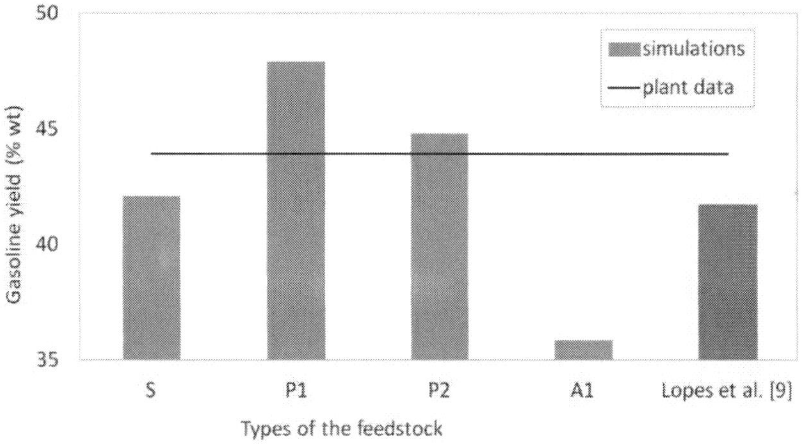

Figure 4. Gasoline yields at the riser outlet

As expected, depending on the composition of the heavy oil, different quantities of gasoline are produced. Feedstock rich in paraffinic compounds resulted in produced higher gasoline yield, as compared with other cases. In contrast, the feed with high concentration of aromatic resulted in the lowest gasoline yield.

When these results were compared against the 4-lumps model, which has just one group to represent the feedstock, the gasoline yield is closer to that obtained with the injection of feedstock type "S". Differently from the 11-lumps model, the 4-lumps model requires an adjustment of its kinetic parameters to predict different types of feed.

Conclusions

An industrial FCC riser reactor was simulated in this work. In order to analyze the conditions that propitiate the thermal cracking reactions inside this equipment, different geometric configurations were proposed to the feedstock and the catalyst particles inlets in the riser. In addition, a 10-lumps kinetic model was modified by the addition of a new lump representing the products of the thermal cracking reactions. It can be concluded that the thermal cracking reactions cannot be avoided, even considering the perfect mixture of the phases. However, the simulation results showed that it is possible to reduce the thermal cracking effects by using an efficient mixture between the gas and the solid phases inside the reactor.

References

1. Bollas, George M.; Angelos A. Lappas; Dimitris K. Iatridis; Iacovos A. Vasalos (2007), " Five-lump Kinetic Model with Selective Catalyst Deactivation for the Predition of the Product Selectivity in the Fluid Catalytic Cracking Process," *Catalysis Today,* v.127, p.31-43.

2. Jacob, Solomon M.; Benjamin Gross; Vern M. Weekman (1976), "A Lumping and Reaction Scheme for Catalytic Cracking," *A.I.Ch.E.Journal*, v.22, p.701-703.

3. Lopes, Gabriela C.; Leonardo M. Rosa; Milton Mori; José R. Nunhez; Waldir P. Martignoni (2011), "Three-Dimensional Modeling of Fluid Catalytic Cracking Industrial Riser Flow and Reactions," *Computers and Chemical Engineering*, v.35, p.2159-2168.

4. Ali, Hany; Sohrab Rohani; Jean P. Corriou (1997), "Modeling and control of a riser type fluid catalytic cracking (FCC) unit," *Chemical Engineering Research and Design*, v.75, p.401-412.

5. Nayak, Subramanya V.; Saket L. Joshi; Vivek V. Ranade (2005), " Modeling of Vaporization and Cracking of Liquid Oil Injected in a Gas-Solid Riser," *Chemical Engineering Science*, v.60, p.6049-6066.

6. Pitault, Isabelle.; David Nevicato; Michel Forissier; Jean-René Bernard (1994), "Kinetic Model Based on a Molecular Description for Catalytic Cracking of Vaccum Gas Oil," *Chemical Engineering Science*, v.49, p.4249-4262.

7. Peixoto, Fernando C.; José L. Medeiros (2001), "Reactions in Multiindexed Continuous Mixtures: Catalytic Cracking of Petroleum Fractions," *AIChe journal*, v.47, p.935-947.

8. Arbel, Arnon; Zupeng Huang; Irven h. Rinard; Reuel Shinnar.; Ajit V. Sapre (1995), " Dynamic and Control of Fluidized Catalytic Crackers, 1. Modeling of the Current Generation of FCC´s." *Ind. Eng. Chem* v.34, p.1228-1243.

9. Lopes, Gabriela C.; Leonardo M. Rosa; Milton Mori; José R. Nunhez; Waldir P. Martignoni (2011), "The Importance of Using Thee-Phase 3-D Model in the Simulation of Industrial FCC Risers." *Chemical Engineering Transactions*, v.24, p.1417-1422.

(696d) Development of Rotary Kiln Furnace for Iron Reduction and Slag Separation

Hui Li[1], Wen-bin Yang[2], Wei Zhao[3], Xiaodan Fan[4], Qiang Zhi[3], Xingyan Zhang[3] and Yong Min[1], (1)Xi'an University of Architecture &Technology, Institute of Powder Engineering, Xi'an, China, (2)Material Institute, Xi'an University of Architecture and Technology,, Xi'an, China, (3)Xi'an University of Architecture and Technology Xi'an 710055, Shaanxi, (4)Xi'an University of Architecture and Technology, Xi'an 710055, Shaanxi, P.R. China

Development of Rotary Kiln Furnace for Iron Reduction and Slag Separation

Hui Li, Wenbin Yang, Wei Zhao, Xiaodan Fang, Qiang Zhi, Xingyan Zhang, Yong Min
Xi'an University of Architecture and Technology Xi'an 710055, Shaanxi

Abstract:

There is a big challenge in the steel industry to use hematite ore due to its special "oolitic-like" composite particle structure. Both magnetic selection and floating selection were investigated with a very little of success. In this paper, we would like to introduce a novel process, which had directly reduced the hematite ore into iron and separated from the slag. Experimental rotary kiln furnace was made and used to investigate the iron reduction and slag separation processes. A pilot scale rotary kiln furnace was also build to scale up the process with industrial grade hematite ore as the raw materials. The initial result indicated that total iron content is greater than 88%, which reached the product requirements of the Chinese national standards. The comprehensive studies of material formulation, process conditions, as well as characterization will be introduced during the presentation.

Keywords: rotary kiln furnace, iron reduction, slag separation, iron slag separation

(696e) Gas-Solid Flow Hydrodynamics Characteristics in a Circulating Fluidized Bed Riser with External Heat Exchanger

Guoliang Song, Xiaozhe Mou, YunKai Sun, Haigang Wang and Qinggang Lv, Institute of Engineering Thermophysics, Chinese Academy of Sciences, Beijing, China

To investigate the gas-solids flow hydrodynamics characteristics in a cold circulating fluidized bed (CFB) loop with external heat exchanger (EHE), electrical capacitance tomography (ECT) and pressure sensor are used to measure the solids concentration and split-flow rates in the EHE with a new type of ash-controlled valve. The CFB facility includes the riser, cyclone separator, external heat exchanger chamber with heating surface and chamber without heating surface. Two ECT sensors with eight electrodes were mounted in the pneumatic pipe between the controlled valve and the EHE. The experiment results indicate that the split-flow rates increase linearly with the increase of the opening of the controlled valve. The solid flow rate from the EHE to the CFB riser is increased with the increase of the fluidization air flow rate in the chamber with heating surface. The increase of the air flow rate in the chamber without heating surface will enhance the transportation of the solids from the chamber without heating surface to the chamber with heating surface. The test results for solids flow and regulating characteristics provide useful information for CFB boiler with EHE.

(696f) Investigation the Gas-Solids Distribution in a Large-Scale Circulating Fluidised Bed with Electrical Capacitance Tomography

Haigang Wang Sr., Chinese Academy of Sciences, Institute of Engineering Thermophysics, Beijing, China, Guorong Zhao, Chinese Academy of Sciences, Institute of Engineering Thermophyscis, Beijing, China, Guizhi Qiu, Chiense Academy of Sciences, Institute of Engineering Thermophyscis, Beijing, China, Yunkai Sun Sr., Chiense Academy of Sciences, Institute of Engineering Thermophysics, Beijing, China, Qinggang Lu, Chinese Academy of Sciences, Institute of Engineering Thermophysics, Beiing, China and Wuqiang Yang, University of Manchester, School of Electrical and Electronic Engineering, Manchester, United Kingdom

The gas-solids hydrodynamic behaviour is one of the key issues to scale up a circulating fluidised bed (CFB) from small scale to large scale with multi-cyclone separators system. To investigate the solids distribution in a large scale CFB with six cyclone separators, electrical capacitance tomography is used to measure the solids concentration in the inlet and the standpipe of the cyclone. A rectangular ECT sensor has been designed and used to measure the solids distribution in a cross-section in the inlet of cyclone. Six circular ECT sensors with eight electrodes were mounted in the standpipe of the cyclone to measure the solid distribution among the six cyclone separator system. For image reconstruction, sensitivity maps have been calculated using a simulation package, Maxwell from Ansoft. Dynamic test with different fluidisation velocity and initial material load are given and analyzed. The preliminary test results show that the solids distribution of the cold model with axis-symmetrically or centre-symmetrically arrangement of cyclones can be measured in different bed inventories and superficial velocities. For the above two type of arrangement, the solid concentration in the six cyclones are different. In terms of uniform of gas and solid mass flow distribution, it is better in centre-symmetrically arrangement than axis-symmetrically arrangement.

(696g) Circulating Fluid Bed Riser Computational Fluid Dynamics (CFD) Model Validation with Electrical Capacitance Volume Tomography (ECVT)

Dirk Van Essendelft[1], **Justin Weber**[1] and Tingwen Li[2], (1)NETL, Morgantown, WV, (2)URS Corp. National Energy Technology Laboratory, Morgantown, WV

Model verification and validation (V&V) is critical for the application of that model to predict the performance of a device or system. Without the V&V, the model prediction cannot be trusted. Applying V&V to multiphase flow applications is a daunting challenge and continues to be a focus of the US Department of Energy's National Energy Technology Laboratory (NETL). Recently, a relatively new technique, namely electrical capacitance volume tomography (ECVT), has been implemented on the 30cm diameter riser of the cold flow circulating fluid bed at NETL in Morgantown, WV. This technique has the ability to accurately measure three dimensional distributions of solid fractions in the riser without disturbing the flow structures. The measured values can be directly compared to various computational fluid dynamic (CFD) models as a - validation tool. Three multiphase CFD models were compared against the ECVT data including NETL's MFiX, ANSYS FLUENT, and Computational Particle Fluid Dynamics' (CPFD) BARRACUDA. Experimental pressure measurements were also taken and compared to the simulation predictions. The resulting work demonstrates that identifying the best model parameters can accurately match the experimental data. However, deciding on the correct model parameters before running the models presents a significant challenge.

(705a) Tunable Assembly of Metallic and Magnetic Nanoparticle Clusters in DMSO Solvent Systems

Jennifer N. Duggan[1], Pranav S. Vengsarkar[2] and Christopher B. Roberts[2], (1)Chemical Engineering, Auburn University, Auburn University, AL, (2)Department of Chemical Engineering, Auburn University, Auburn, AL

A variety of different techniques have been developed in the last several years to synthesize various metal and magnetic nanoparticles of controlled size, but these popular methods are often time-intensive and require the use of harsh and expensive solvents and reagents. One step to further simplify and improve nanoparticle synthesis would involve development of a method that uses a single molecule "functional solvent" for both solvation and nanoparticle stabilization. This study explores the production of cobalt and gold nanoparticles using simple and sustainable synthesis techniques which employ the use of a dimethyl sulfoxide (DMSO) functional solvent. Previous work in our lab has demonstrated the successful use of DMSO as a functional solvent to produce monodisperse samples of palladium nanoparticles. It was found that both the sulfur and oxygen functional groups on the DMSO molecules interact with the nanoparticle surfaces via a resonance hybrid structure which very effectively stabilize the nanoparticles. Similarly, this synthesis technique has been extended to produce other nanoparticle systems in our lab. For example, DMSO has recently been found to successfully function as both the solvent and the stabilizing agent to produce cobalt nanoparticles of precise size that can be tunably assembled into nano-scale clusters of controlled sizes by variations in synthesis temperature, thereby affording unique opportunities to affect the clusters' physicochemical properties, such as magnetism. Moreover, the solubility and dispersibility of these cobalt nanoparticle clusters can then be further manipulated by introducing polar co-solvents (e.g. water, ethanol) to the nanocluster dispersions after synthesis to redisperse nanoparticles and clusters of different sizes. It has been found that the redispersibility of these cobalt nanoparticle clusters into individual single-crystalline nanoparticles is simply the result of the preferential solvation that may occur between the DMSO molecules already present in the solution and the newly introduced co-solvent. Subsequently, it has been found that coupling DMSO with other molecules, such as fatty acids, produces nanoparticle dispersions that are stable for a longer duration than those produced without a fatty acid stabilizing agent. The use of UV-vis spectroscopy to understand the reduction of both the bulk-scale cobalt and gold metals to the nano-scale and analyze any unreduced metallic intermediates that may persist during nanoparticle formation will be discussed. Furthermore, the size of these cobalt and gold nano-scale particles were investigated using TEM, and the interaction between the nanoparticle surface and the DMSO solvent/fatty acid stabilizing agent was studied using FT-IR.

Structure and properties of surface modified magnetic hollow silica spheres

Kovacik Pavel, Kremlackova Zuzana, Soltys Marek, Stepanek Frantisek

ICT Prague, Department of Chemical Engineering, Laboratory of Chemical Robotics

kovacikp@vscht.cz

1 INTRODUCTION

In recent years there has been an increasing interest in iron oxide/silica composites. Due to their special properties such as large surface area and low material density these microparticles are very attractive as catalysts [1], microwave absorption materials [2] or in drug delivery systems [3]. Major methods for the preparation of iron oxide/silica composite microparticles are: sol-gel process [4]; layer-by-layer techniques [5]; coprecipitation [6] and surface reaction [7].

Coating applied to the surface of materials can improve its properties as it allows control of the interaction of a material with its environment [8]. Nowadays, the application range of coatings extends much beyond the simple decoration and protection aspects, and functional coatings have become an enabling technology in a vast variety of different high-tech areas. Fields in which such high-tech coatings are applied range from computer chips and hard disk manufacturing [8] to the use of special coatings in biomedical and aviation applications [9].

The aim of this work is to study the possibility of active control of substance's diffusion across the silica shell of prepared spheres. Depending on the method and its parameters we prepared hollow spheres with a mean size ranged from tens of nanometers to tens of micrometers.. The resultant spheres have a porous silica (SiO_2) shell and a hollow core which includes nanoparticles of iron oxides (Fe_2O_3 / Fe_3O_4). We tested two approaches for controlling the diffusion across the silica shell: i) so-called polymer brushes systems and ii) covering the surface of spheres by a phase change material (Fig. 1). We chose thermoresponsive polymer poly(N-isopropylacrylamide) (PNIPAM) as polymer brushes and palm oil as the phase-change material. Both of them were bound on the surface and inside the pores of prepared hollow microspheres and the release kinetics were studied.

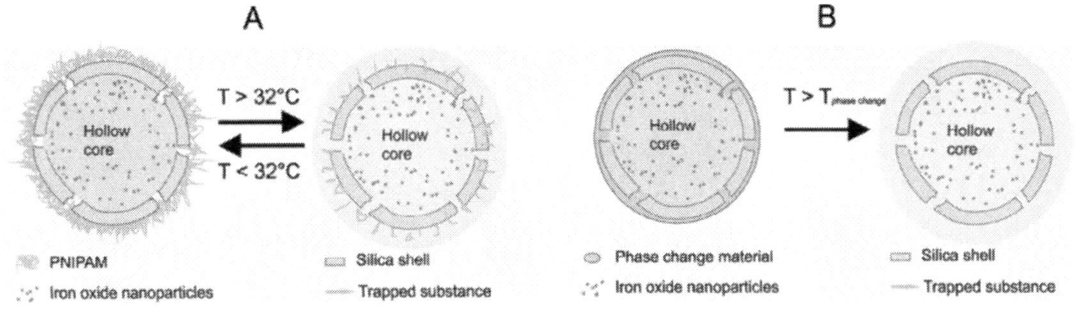

Fig. 1: The diffusion across a silica shell can be controlled by using polymer brushes (A) or phase change material (B).

2 Experimental

2.1 *Materials*

Tetraethyl orthosilicate (TEOS, 98%), n-octylamine (99%), oleic acid (65-88%), acetone (99%), ammonium hydroxide solution (25%), 3-aminopropyl-triethoxysilane (APES), iron(III) chloride hexahydrate, iron(II) chloride tetrahydrate, n-heptane, palm oil, N-isopropylacrylamide (NIPAM), ammonium persulfate (APS) and 3-(trimethoxysilyl)propyl methacrylate (MPS) were purchased from Sigma-Aldrich. Nitric acid (p.a., 65%), ethanol and N,N'-methylenebisacrylamide (BIS) were purchased from Fluka. All chemicals were used as received. Deionized water (Aqual 25, 0.07 μS/cm) was used for all reactions and treatment processes.

2.2 *Synthesis of Iron Oxide Nanoparticles*

The iron oxide nanoparticles used for this work were prepared by the co-precipitation method [10] and stabilized with oleic acid to make them dispersible in non-aqueous media. In a typical synthesis, 0.25 g of $FeCl_3.6H_2O$ and 0.125 g of $FeCl_2.4H_2O$ were dissolved in 40 ml of deionised water in a three-neck flask equipped with a reflux condenser and heated to 80°C. Then 0.2 ml of oleic acid dissolved in 5 ml of acetone were added. Iron oxide precipitation was induced by adding 5 ml of NH_4OH; the mixture was stirred at 500 rpm for the next 10 min, followed by five additions of 0.2 ml of oleic acid at 5 minute intervals. The mixture was stirred for another 30 min and then cooled down to ambient temperature. Dry iron oxide nanoparticles dispersible in non-aqueous media were obtained by freeze-drying.

2.3 *Synthesis of Hollow Silica Microspheres*

Hollow composite iron oxide/silica microparticles with a porous shell were prepared via the soft templating route [11]. Tetraethyl orthosilicate (TEOS) is used as a silica source and octylamine (OA) as a liquid template (Figure 2). The overall synthesis procedure is shown schematically in Fig. 2. Typical synthesis and whole fabrication procedure is fully described in our publication [11].

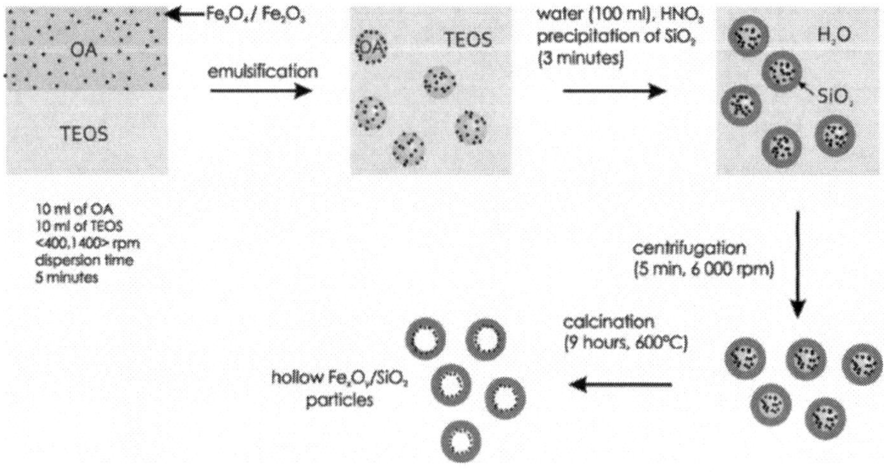

Fig. 2: Scheme of the preparation of composite hollow silica microspheres.

2.4 *SYNTHESIS OF HOLLOW SILICA NANOSPHERES*

The composite hollow-core mesoporous silica nanospheres were prepared via a soft templating route using tetraethyl orthosilicate (TEOS) as a silica precursor, APES as co-structure agent and oleic acid as a liquid template [12]. The overall synthesis procedure is shown schematically in Fig. 3.

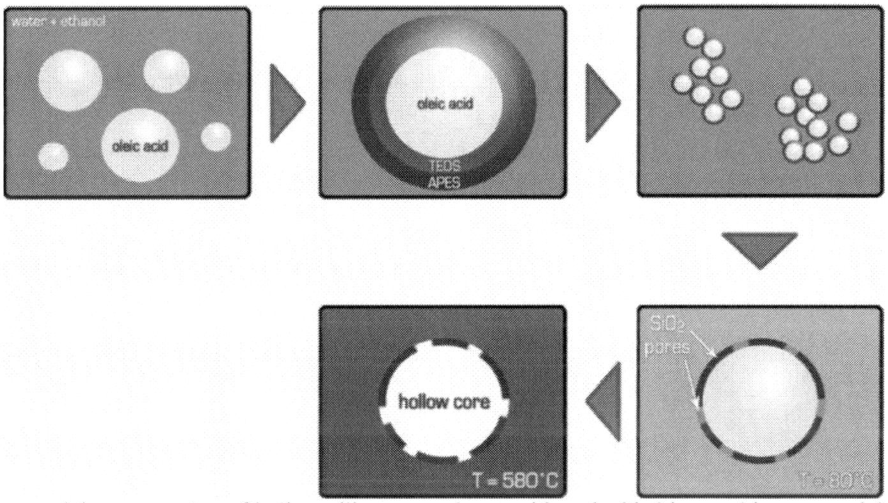

Figure 3: Scheme of the preparation of hollow silica nanospheres with embedded iron oxide nanoparticles via the soft templating route.

In a typical synthesis, the required amount of iron oxide nanoparticles (14 mg, 5% wt) was added to oleic acid (1 mmol, 0.282 g) and sonicated for 5 min to achieve full dispersion. In the next step the oleic acid containing the iron oxide nanoparticles was dispersed in a mixture solution of deionised water (1600 mmol, 28.8g) and ethanol (50 mmol, 2.3 g) at room temperature. To the solution was added a mixture of TEOS (6.7 mmol, 1.4 g) and of APES (1 mmol, 0.221 g) with stirring (1400 rpm). After 30 min, the stirring was stopped, and the reaction mixture was kept at room temperature for 2 h and then aged at 80°C for 1 day. The precipitate was filtered, dried at 60°C and calcined at 580°C for 6 h to remove the templates.

2.5 *GRAFTING OF POLYMER BRUSHES*

The thermoresponsive polymer PNIPAM was chosen for its favourable properties and lower critical solution temperature (LCST, 32 °C) [13]. When PNIPAM is below LCST the amide groups interact strongly with water molecules through hydrogen bonding and the polymer chains swell. In case the polymer is grafted in the pores of silica microspheres it will block the diffusion of trapped material across a silica shell. Above LCST, the hydrogen bonds between water and the amide groups are disrupted, water acts as a poor solvent and is expelled from PNIPAM, and as a result the polymer network collapses and the chains of polymer shrink dramatically. The temperature change causes volume response and the idea is that it will open the way for diffusion across the silica shell. PNIPAM was grafted on the initiator-modified particles as follows: 0.1 g of prepared composite hollow silica microspheres were sonicated for 5 min in a solution of water (16 ml), ethanol (4 ml) and MPS (10 mg). After that 0.2 g of NIPAM and 0.03 g of BIS (crosslinker) were added and the reaction solution was kept under stirring (500 rpm) and nitrogen atmosphere for 30 min. Then the reaction solution was placed in a 70°C oil bath and 0.005 g of APS was added to start the polymerization. After 1 hour the final PNIPAM/iron oxide/silica particles were collected by centrifugation and washed several times with ethanol/water solution (50 % wt).

2.6 *IMPREGNATION BY PHASE CHANGE MATERIAL*

Because the resultant composite microspheres with incorporated iron oxide nanoparticles are able to rise their temperature by about 20°C in few minutes (Fig. 4) we looked for a phase-change material which is solid at room temperature and has its melting point around 40°C. This temperature we can be reached on the surface of microspheres by using radiofrequency field producing by our laboratory magnetic coil (PowerCube 32/400). For this reason and because we had experiences with glass surface modification with palm oil [14] we chose palm oil (melting point ~35°C) for the first experiments as a suitable material. The palm oil was coated on the surface of particles in a very simple way as follows: A solution of palm oil (20 mg) and n-heptane (5 ml) was prepared. 20 mg of prepared composite hollow silica microspheres were dispersed in 500 □l of this solution and then the n-heptane was evaporated by normal pressure and 50°C. Based on previous BET analysis the specific pore volume of the silica shell was 0.427 cm^3/g, hence the theoretical quantity of palm oil (density 0.885 g/cm^3) necessary to completely fill the pore space of 20 mg of silica particles would be 7.5 mg. An excess of the palm oil was therefore applied.

2.7 *CHARACTERISATION*

Scanning Electron Microscopy (SEM) images were taken on a Jeol JCM-5700 scanning electron microscope. The samples were sputter-coated with a 5 nm thin film of gold prior to imaging. Some particles were mechanically broken by using a mortar with a pestle in order to reveal their core-shell structure. Confocal Laser Scanning Microscopy (CLSM) images were taken on the Olympus Fluoview FV1000 confocal laser microscope. For silica visualization the fluorescent dye Fluorescein isothiocyanate (FITC) was used, which was added to the TEOS solution before starting precipitation. The particle size distribution was evaluated by the laser diffraction particle size analyzer (Horiba Partica LA 950/V2). The temperature rise with time of prepared composite microparticles was measured by a 400 kHz, 2.5 kW generator of external alternating magnetic field (PowerCube 32/400). The permeability of the composite iron oxide/silica shell was characterized indirectly by measuring the uptake/release kinetics using UV/VIS spectrophotometry (Specord 205 BU).

3 RESULTS AND DISCUSSION

3.1 *SPHERE SIZE AND STRUCTURE*

The mass of iron oxide nanoparticles was systematically varied to give resultant composite silica spheres with a mass fraction of 0, 1.0, 2.0 and 3.0 %. The fabricated spheres possess a regular spherical shape (Fig. 4) with a relatively narrow size distribution and no dependence on the mass fraction of iron oxide. The presence of hollow cores in the particles was confirmed by SEM of mechanically broken particles (Fig. 3C). In Fig. 3D it is also possible to identify clusters of iron oxide deposited on the internal surface of the silica shell.

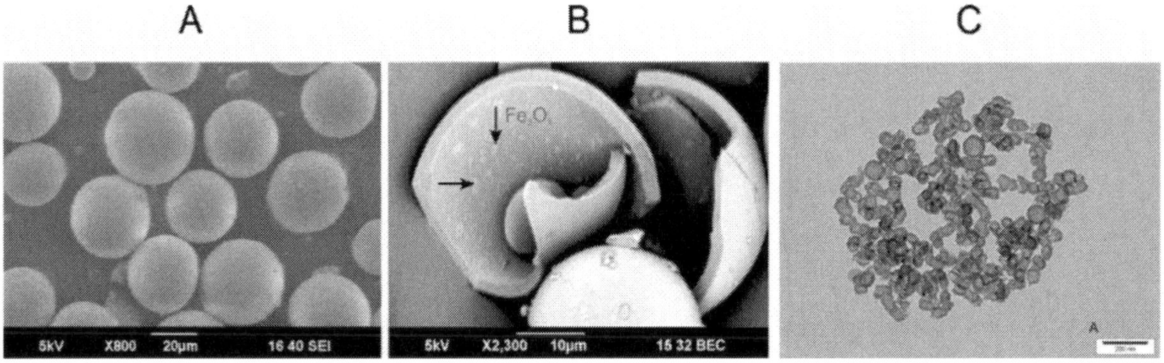

Fig. 4: A) SEM image of composite silica microspheres (1 % iron oxide). B) Core-shell structure of prepared composite silica microspheres (% iron oxide). It is possible to indetify clusters of iron oxide deposited on the internal surface of the silica shell. C) TEM image of composite hollow silica nanospheres (1 % mass fraction).

For detailed discussion about particle size distribution and about the way how to control the mean size of the resultant microspheres please see our publication [11].

3.2 *RADIOFREQUENCY HEATING*

Prepared composite microspheres were inserted into an external alternating magnetic field and the temperature rise with time was recorded. The results prove that the radiofrequency heating is possible to use as a tool for rising the temperature on the surface of prepared microspheres and cross the LCST temperature for pnipam (32°C) and the melting point for palm oil (35°C) (Fig. 5).

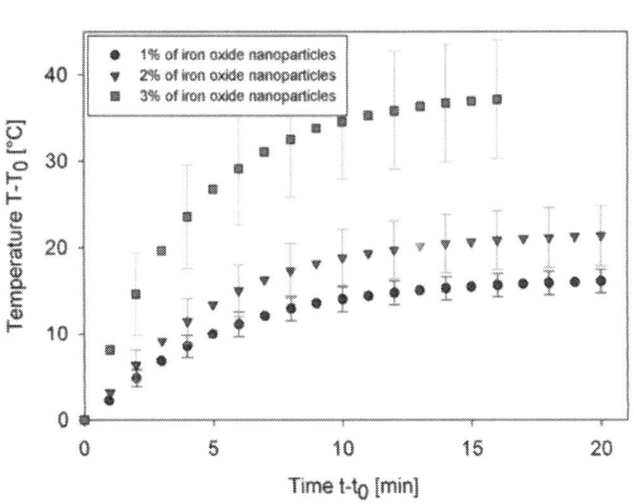

Fig. 5: Effect of iron oxide concentration on the heating curves of composite hollow silica microspheres in a 400 kHZ, 1.25 kW alternating magnetic field.

Being able to reach the desired temperature by a remote mechanism was the requirement for starting to study the release kinetics of trapped material inside the hollow core of prepared composite microspheres.

3.3 *SURFACE MODIFICATION AND RELEASE KINETICS FROM SURFACE MODIFIED MICROSPHERES*

The selective modification of the surface of the prepared microspheres by PNIPAM is shown in the Figure 6A, surface modification with palm oil in the Figure 6B

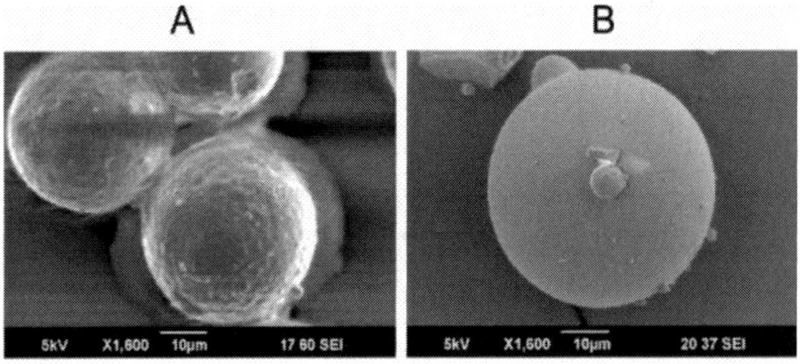

Fig. 6: (A) PNIPAM-modified hollow silica microspheres followed the metodology described in section 2.5., (B) Palm oil-modified hollow silica microspheres.

Vitamin B12 was loaded into hollow cores of microspheres and the release rates were measured at two temperatures (25 and 40°C). The loading method is well described and fully discussed in our recent publication [11]. In case PNIPAM modified particles, the particles were first modified and then loaded. For the case of palm oil modified particles the process was reverse, i.e. the particles were loaded with vitamin B12 first, and only then modified.

We made first release experiments for temperature 25°C and in both cases the release kinetics is slowed down compared with the reference which was non-modified composite silica microspheres (Fig. 7). The released amount is 85% lower for PNIPAM-modified microspheres than for non-modified microspheres and 70% for palm oil-modified microspheres. The release kinetics for 40°C is under investigation and the data will be presented in our future work.

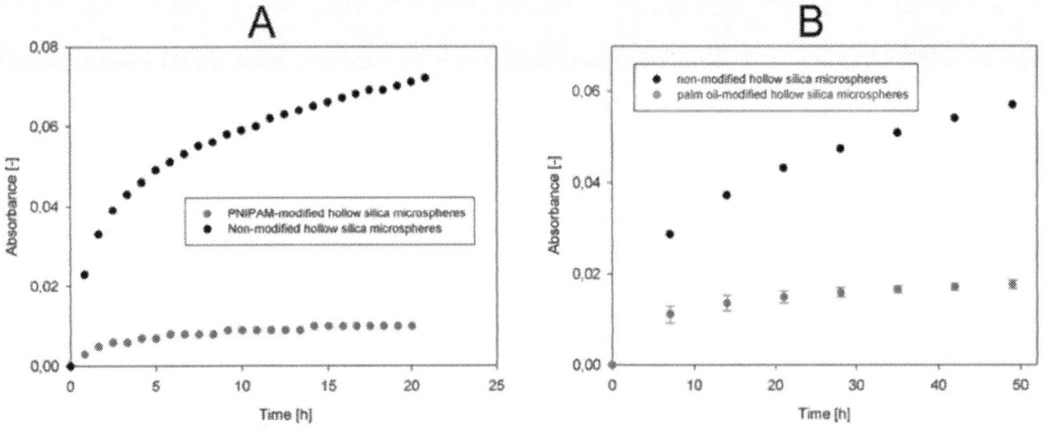

Fig. 7: A) Release kinetics from PNIPAM-modified hollow silica microspheres 25°C. B) Release kinetics from palm oil-modified hollow silica microspheres 25°C.

4 CONCLUSIONS

Hollow composite iron oxide/silica microparticles and naoparticles with a porous shell were prepared. Their properties were characterized via different methods. The presence of a hollow core was confirmed by SEM and TEM microscopy. Particle size distribution was measured by a laser scattering method and was found to be independent of the presence of iron oxide nanoparticles. The composite microparticles are able to heat up in the presence of a radiofrequency magnetic field with a rate from 3 to 9 °C/min. This property in combination with surface modification can be used as a remote mechanism for influencing the release kinetics of trapped material inside the hollow core of microspheres. First results obtained for microspheres modified by thermoresponsive polymer PNIPAM and hydrophobic phase-change material palm oil indicated that the idea is right. However, further work must be done in this area and that is the subject of on-going studies.

5 References

[1] Sadasivan S., Sukhorukov G.B., "Fabrication of hollow multifunctional spheres containing MCM-41 nanoparticles and magnetite nanoparticles using layer-by-layer method", *J. Colloid Interface Sci.* 304, 437–441(2006).

[2] Fuertes A.B., Valdés-Solís T., Sevilla M., "Fabrication of monodisperse mesoporous carbon capsules decorated with ferrite nanoparticles", *J. Phys. Chem.* C 112 , 3648–3654 (2008).

[3] Zhou J., Wu W., Caruntu D., Yu M.H., Martin A., Chen J.F., O'Connor C.J., Zhou W.L., "Synthesis of porous magnetic hollow silica nanospheres for nanomedicine application", *J. Phys. Chem. C* 111, 17473–17477 (2007).

[4] Chen I., Wang C., Chen C.Y., "Synthesis of mono-disperse hollow nickel spheres with a double shell structure using poly(methyl methacrylate) as a template", *Scr. Mater.* 58 , 37–40 (2008).

[5] Caruso F., Spasova M., Susha A., Giersig M., Caruso R.A., "Magnetic nanocomposite particles and hollow spheres constructed by a sequential layering approach", *Chem. Mater.* 13 , 109–1016 (2001).

[6] Wu W., Caruntua D., Martin A., Yu M.H., Connor C.J.O., Zhou W.L., Chen J.F., "Synthesis of magnetic hollow silica using polystyrene bead as a template", *J. Magn. Magn. Mater.* 311 , 578–582 (2007).

[7] Lin C.R., Wang C.C., Chen I.H., "Synthesis and magnetic behavior of silica-coated cobalt ferrite hollow spheres", *J. Appl. Phys.* 99 , 08N707 (2006).

[8] Advincula R.C., Brittain W.J., Kaster K.C., Rühe J., "Polymer brushes, Wiley-VCH, Weinheim", 2004, ISBN 3-527-31033-9.

[9] Zhulina E.B., Singh C., Balasz A.C, "Forming patterned films with tethered diblock copolymers", *Macromolecules,* 26, 6338-6348 (1996).

[10] Maity D., Agrawal D.C., "Synthesis of iron oxide nanoparticles under oxidizing environment and their stabilization in aqueous and non-aqueous media", *J. Magn. Magn. Mater.* 308, 46-55 (2007).

[11] Kovačík P., Kremláčková Z., Štěpánek F., "Investigation of radiofrequency induced release kinetics from magnetic hollow silica microspheres", *Micropor. Mesopor. Mater.*, in press (2012).

[12] L. Han et al., Anionic surfactants templating route for synthesizing silica hollow spheres withdifferent shell porosity, Solid State Sciences 13 (2011) 721-728.

[13] Čejková J, Hanuš J., Štěpánek F., "Investigation of internal microstructure and thermo-responsive properties of composite PNIPAM/silica microcapsules", *Journal of Colloid and Interface Science* 346, 352–360 (2010).

[14] Tokárová V., Pittermannová A., Čech J., Ulbrich P., Štěpánek F., "Thermo-responsive adhesion properties of composite hydrogel microcapsules", *Soft Matter* 8, 1087-1095 (2012).

(705c) Distinguishing Between Amine and Carboxylate Bonds to Iron Oxide Nanoparticles

Myles A. Ikenberry, Chemical Engineering, Kansas State University, Manhattan, KS and Keith L. Hohn, Chemical Engineering, Kansa State University, Manhattan, KS

Iron oxide nanoparticles are the subject of vast research areas, spanning many different disciplines including biology and catalysis. In such research, molecules are attached to the particle surfaces in order to impart certain properties and enable further functionalizations. Common moieties which are important for iron oxide functionalization include carboxylates and amines.

In addition to their application in the functionalization of synthesized particles, amines and carboxylates are frequently present during particle syntheses. However, there has been no conclusive analytic framework by which it could be determined whether or not amine containing ligands such as oleylamine had become attached to the particle surface. Also, because many ligand molecules contain both an amine and carboxylate moiety, previous research has been unable to distinguish between various orientations of such molecules on the iron oxide nanoparticle surface.

A combination of spectroscopic techniques is presented which will enable future research on iron oxide magnetic nanoparticles to determine whether the ligands are attached to iron oxide nanoparticles through amine or carboxylate moieties.

(705d) Magnetic High Density Polyethylene Nanocomposites Reinfored with in-Situ Synthesized Fe@FeO Core-Shell Nanoparticles

Qingliang He[1], Jiahua Zhu[1], Suying Wei[2] and Zhanhu Guo[1], (1)Dan F. Smith Department of Chemical Engineering, Lamar University, Beaumont, TX, (2)Chemistry and Biochemistry, Lamar University, Beaumont, TX

Magnetic high density polyethylene (HDPE) nanocomposites with different loadings of Fe@FeO core-shell nanoparticles were fabricated through a direct thermal decomposition of iron precursor-$Fe(CO)_5$ in refluxing HDPE-solvent complex. Attenuated total reflection-Fourier transform infrared spectroscopy (ATR-FTIR) was utilized to explore the functional groups of these nanocomposites, and the dispersion of the Fe@FeO core-shell nanoparticles was investigated by transmission electron microscopy (TEM). Mössbauer spectrum analysis was used to determine the exact species of these magnetic nanoparticles. The X-ray diffraction (XRD) analysis and differential scanning calorimetry (DSC) characterization were used to further explore the crystalline structure of HDPE with/without these nanoparticles. Magnetic property investigation revealed a soft ferromagnetic behavior for these HDPE PNCs at room temperature and the coercivity was decreased with increasing the particle loading. Thermal gravimetric analysis (TGA) demonstrated that the thermal stability of these HDPE PNCs was enhanced in the presence of the NPs. The melt rheological properties and dielectric properties of these HDPE nanocomposites were also evaluated and will be presented as well.

(705e) Two-Step Gas Phase Synthesis of Janus-Like and Core-Shell Fe_xO_y-TiO_2 Composite Nanoparticles

Juha Harra[1], Juha-Pekka Nikkanen[2], Mikko Aromaa[1], Heikki Suhonen[1], Mari Honkanen[2], Saara Heinonen[2], Erkki Levänen[2] and Jyrki M. Mäkelä[1], (1)Aerosol Physics Laboratory, Tampere University of Technology, Tampere, Finland, (2)Department of Materials Science, Tampere University of Technology, Tampere, Finland

Composite nanoparticles are of a great interest because the properties of several different materials can be combined. In this study, Fe_xO_y-TiO_2 nanoparticles were synthesized in two steps. Iron oxide and titanium dioxide were chosen because they are known to have magnetic and photocatalytic properties, respectively. Fe_xO_y nanoparticles were synthesized using the Liquid Flame Spray technique by feeding a solution of ferrocene dissolved in xylene into a high-temperature H_2-O_2 flame. The flame-made nanoparticles were collected with an electrostatic precipitator and mixed with titanium(IV) isopropoxide (TTIP) liquid. The mixture was sprayed into a tube furnace where TTIP was thermally decomposed to form TiO_2. Both Fe_xO_y and Fe_xO_y-TiO_2 nanopowders were characterized with TEM and XRD. The TEM images show that by altering the process parameters, the structure of the composite nanoparticles can be tailored between Janus-like and core-shell particles consisting of an agglomerated Fe_xO_y particle attached to or inside a spherical TiO_2 particle, respectively. Mixing nanopowder into precursor liquid that is thermally decomposable is a versatile and easy method to synthesize composite nanopowders which are formed of particles containing two or more different materials and properties, as demonstrated in this study.

(707a) Blade Design Effect in Powder Phenomena and Residence Time Distribution Inside the Feed Frame

Daniel Mateo, Chemical Engineering, University of Puerto Rico Mayaguez Campus, San Juan, PR, Fernando J. Muzzio, Department of Chemical and Biochemical Engineering, Rutgers, The State University of New Jersey, Piscataway, NJ, Carlos Velázquez, Department of Chemical Engineering, University of Puerto Rico at Mayaguez, Mayaguez, PR and Rafael Mendez, Chemical Engineering, University of Puerto Rico, Mayaguez, PR

Blade Design Effect in Powder Phenomena and Residence Time Distribution Inside the Feed Frame

Daniel Mateo[a], Fernando J. Muzzio[b], Carlos Velázquez[a] and Rafael Méndez[a].

[a] Department of Chemical Engineering, University of Puerto Rico at Mayaguez, PO Box 9000 Mayaguez PR 00681, Puerto Rico

[b]Department of Chemical and Biochemical Engineering, Rutgers University, 98 Brett Road, Piscataway, NJ 08854, United States

Tablet compression has great significance in the pharmaceutical industry since most of the drugs are in the tablet dosage form. The tablet press feed frame is used to fill powder into the empty dies. Die filling is one of the key steps to control final properties of tablets. The Discrete Element Method (DEM) is a computational method for simulating the dynamics of particle processes. Using DEM, a standard feed frame taken from a Manesty Betapress was simulated, which represents the tableting process without the compression step. This work examines how the blade design affects the performance of a feed frame system mainly to evaluate the effect on powder segregation and particle attrition. Three different blade designs were analyzed, rectangular, cylindrical and a 45° angle blades. The three simulations were done at the same feed frame operating conditions, blade speed 24 rpm and turret speed 29 rpm. Die weight variability, feed frame hold up and exit contributions to the die weights were analyzed but no significance difference were observed between each simulation. Results shows that die segregation can be reduced with the cylindrical or angle blades. Segregation reduction is due to normal forces that particles received. The results show a linear relationship between compressive forces versus particle volume. Bigger particles receive a greater compressive force and can be related to particle attrition. In the cylindrical and angle blade simulations the force applied to bigger particles is lower than in the rectangular blade simulation, therefore less particle attrition. Thus it demonstrated that blade deign is a significant factor in size segregation reduction and control of particle attrition. The residence time distribution (RTD) was also studied in order to evaluate the feed frame performance. RTD can be related to total shear applied, particle attrition and to the possibility of powder stratification in the die for granular material sensitive to total shear applied.

(707b) Study the Effect of the Low Shear Continuous Tumble Mixer Parameters On Powder Phenomena, Residence Time, and Blend Uniformity

Miguel Florian, Chemical Engineering, Engineering Research Center for Organic Particulate System, University of Puerto Rico Mayaguez, Mayaguez, PR, Carlos Velázquez, Department of Chemical Engineering, University of Puerto Rico at Mayaguez, Mayaguez, PR and Rafael Mendez, Chemical Engineering, University of Puerto Rico, Mayaguez, PR

Study the Effect of the Low Shear Continuous Tumble Mixer Parameters on Powder Phenomena, Residence Time, and Blend Uniformity

Miguel Florián, Carlos Velázquez, and Rafael Méndez.

Department of Chemical Engineering, University of Puerto Rico at Mayaguez, PO Box 9000 Mayaguez PR 00681, Puerto Rico

Industries such as the pharmaceutical, food, cosmetic, catalysis, ceramic, and detergent have as a common denominator the granular material mixing step. This operation is performed primarily in batch mode, with some limitations, such as sampling methodology, sample homogeneity determination, and unknown process scale-up behavior. All these limitations hinder the possibility of achieving optimal production, reducing product quality, performance and often lead to batch failures. The pharmaceutical industry started recently to consider continuous manufacturing to be in the capacity to control instabilities on line, to make processes more efficient, optimizing the production and reducing the principal problems of the batch process. One of the more important operations is the powder mixing, which could be the principal process in the drug manufacturing. In this work, a new continuous low shear mixing device consisting of two, short, acrylic, hollow concentric cylinders has been developed using as a model the batch drum mixers to study granular mixing. The complete system consists of two feeders that deliver the active pharmaceutical ingredient and the excipients. The principal goal is understand the effect of the three different flow rate, three API concentrations (2.5, 20.5 and 20%), three speed mixer rotations (30, 50 and 70 RPM) and feeding angle on the relative standard deviation (RSD) of the mixing achieved. The experimental part is complemented with DEM simulations of the continuous mixing process to study the cohesion effect, velocity profile, particle trajectory and the powder phenomena includes the pattern flows, avalanche deformation and flow through tumble orifices to relate these with the properties of the materials characterized previously. A set of experiments with the same operational parameters and particle characteristics similar to the simulations was run to validate these process obtaining similar results with the experimental part. Simulations results show a good agreement with the experimental based on the mixing RSD, the velocity profile and avalanche behavior. The principal results obtained demonstrated that the tumble mixer is capable of achieving a good mixing index based on the RSD of the outlet concentration, the mixing homogeneity was affected by the operation conditions and powder properties such as the particle size distribution and the cohesion, hold up and the average residence time of the mixer.

NUMERICAL STUDY OF SOLID PARTICLE AXIAL MIXING IN A FIXED CYLINDRICAL DRUM WITH ROTATING PADDLES

Z. Zeren[1,2], H. Neau[1,2], P. Fede[1,2], O. Simonin[1,2] and B. Descales[3], W. Stephen[4]

[1] Université de Toulouse; INPT, UPS; IMFT; France

[2] CNRS; Institut de Mécanique des Fluides de Toulouse (IMFT), France

[3] Ineos Technologies, Lavéra, France

[4] Ineos Technologies, Naperville, USA

Abstract

Axial mixture characterization is a widespread problem in granular particle blending processes such as in an horizontal drum mixer. The homogeneous mixture of particles is obtained by blending the particles via rotating paddles in a fixed cylindrical drum. This problem, common to many technological devices, is crucial in the manufacture of a broad variety of industrial products, such as polypropylene. The granular flow behavior in these systems is still poorly understood and the numerical study of such configurations receives increasing academic and industrial attention. In this paper, a study is conducted to investigate the effects of different aspects of the reactor design on the axial transport of monodisperse, uniform density and spherical polypropylene particles. Results show that principally the shape of the paddles is the important design consideration to enhance the axial transport of particles.

Introduction

Granular mixing of particles is a widespread procedure in many industries such as the plastics production, pharmaceutics, concrete and food manufacture. The manufacture of many products in these areas strongly depends on the homogeneous mixing of particles such as propylene where the polypropylene particles are produced through the polymerization reactions in fixed cylindrical drums with rotating paddles. The quality of mixing directly determines the quality of the final product (8). To enhance axial granular flow and to obtain a good level of mixing, the shapes of the paddles and operating conditions (angular velocity, solid mass) are important parameters. However, the granular flow inside the drum is not easily accessible for the experimental apparatus and the effects of these parameters on the flow are not easy to measure (5). Computational Fluid Dynamics (CFD) tools are then useful in order to get information about the detailed mechanics of the flow.

This study is a collaboration between the Institut de Mécanique des Fluides de Toulouse (IMFT), and Ineos Technologies. Important part of the investigation is the characterization of the axial mixing of the solid particles. To this purpose, an axial propagation velocity and a diffusion coefficient using this velocity are computed in order to characterize the mixing of particles. The results are in qualitative coherence with the experimental data conducted at the facilities of the industrial partner.

Mathematical Modeling

The simulations have been performed using the Euler-Euler code NEPTUNE_CFD. NEPTUNE_CFD is a multiphase flow solver developed in the framework of the NEPTUNE project, financially supported by CEA, EDF, IRSN and AREVA-NP. In the Euler-Euler approach, mean equations are solved for both phases and are coupled through the interphase transfer terms (mass, momentum and energy). The turbulence of the gas phase is treated by the two-equation "k-ε" eddy-viscosity model with additional terms taking into account the effect of the presence of the solid phase. Particle phase equations are derived in the general frame of the kinetic theory of granular media supplemented by the interstitial fluid effect (2, 3). Wen-Yu drag model limited by the Ergun's equation for the dense flows is used for the drag force acting on the particles (1). Apart from the drag force for the solid phase, the Archimedean force and the gravity are also taken into account in the momentum transport equations of both phases. The kinetic theory-based approach for the solid phase is theoretically inadequate for the closely packed dense flows where the particle-particle contacts are of long durations such as observed inside the horizontal drum. Indeed, the long duration contacts lead to the frictional force effects, which cannot be taken into account by hard sphere collision models. Following (4), a semi-empirical model is used in this study in order to compute the frictional stress tensor for the solid phase in the momentum equation. The model has been further developed (7), by taking into account the quasi-static state fluctuations of the strain-rates as discussed in (6). The model coefficients are chosen as in the proposition of (7).

Numerical Simulations Overview

Four simulations have been performed to investigate the effect of different operating conditions. The simulations are summarized on Table 1. Solid mass is the initial mass of particles inside the domain. There are two paddle types tested, Mixer A and Mixer B. Mixer A is the initial simple paddle shape and Mixer B is with a substantially different paddle shape.

Table 1. Simulations carried out

Simulations	Angular velocity	Solid Mass	Mixer Type
Sim1	40 rpm	31.7 kg	Mixer A
Sim2	40 rpm	45.2 kg	Mixer A
Sim3	15 rpm	45.2 kg	Mixer B
Sim4	40 rpm	31.7 kg	Mixer B

Mesh

Figure 1 presents 3D perspective view of the mesh generated. The length of the cylinder is one order of magnitude larger than its diameter. Around 40 paddles are present inside the domain that enhances the mixing of the particles. The mesh contains in total 2 500 000 hexahedral cells. The cells have uniform length in axial direction and they are more or less of a cubical shape.

Initial and Boundary Conditions

Initially, the particles are at rest and there is no gas motion, neither. The paddles start to rotate and to mix the solid phase. The entire domain is enclosed with walls so there is no inlet or outlet. Gas and solid phases have frictional interactions with the walls (no-slip condition).

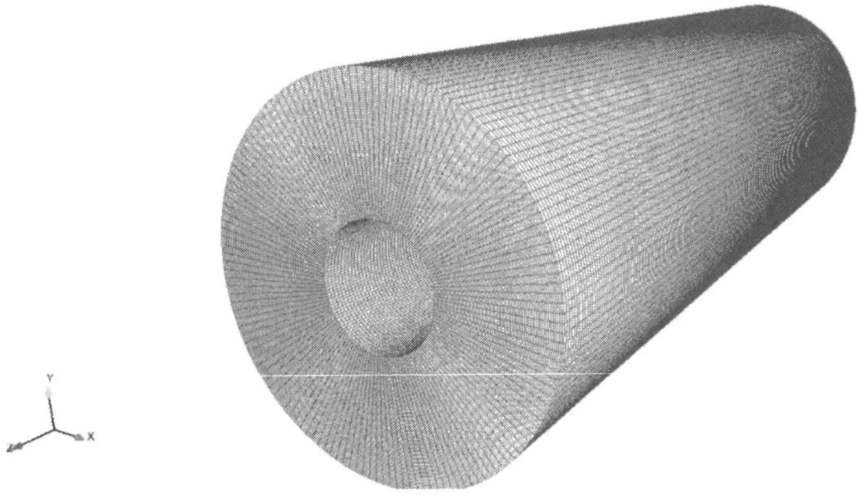

Figure 1. 3D view of the mesh

Material Properties

Solid phase is composed of monodisperse particles, the diameter is 700 μm and the material density is 900 kg/m^3. These characteristics are of polypropylene particles used in the real process. Gas phase is the air with a density of 1.2 kg/m^3 and the dynamic viscosity is 1.8x10^{-5} kg/m/s.

Performing the Simulations and Computer Performances

Simulations were performed in order to characterize the axial propagation of the colored particles and to estimate a dispersion coefficient. The calculations have been done starting after several complete rotations of the paddles. All the simulations have been performed on the supercomputer CALMIP (Toulouse) on a parallel platform. 60 s of physical time cost almost 45 days with 256 processors. The time step of the simulations is not constant. On the mean, it is around 5.10^{-5} s.

Results and Discussion

In order to characterize the axial propagation of particles, colored particles were used. At time t=0 s, the particles in the first axial 10 cm of the domain were colored and their collective motion is predicted using the scalar transport equation. The temporal evolution of these particles' motion gives an idea about the velocity of the axial propagation U_c. This velocity is calculated by using the convection equation of the colored particles written as:

$$\frac{\partial Y_c}{\partial t} = -U_c \frac{\partial Y_c}{\partial z} \tag{1}$$

where Y_c is the local colored particle concentration.

For the computation of the gradients in the equation (1), temporal evolution of the concentration profiles are extracted on several axial lines along the length of the domain after each complete rotation. The figures presented in the sections below correspond to only one line, as the results corresponding to different lines were more or less the same. The line is so chosen that the results would not be affected by the presence of the walls such as of the drum and of the paddles. It is to be noted that using the equation (1) to calculate the convection velocity U_c assumes that the radial distribution of particles is homogeneous.

A diffusion coefficient D_t can then be calculated as:

$$D_t = -\frac{U_c}{\left(\frac{dY_c / dz}{Y_c}\right)} \tag{2}$$

It should be noted that this definition of the diffusion coefficient is of an empiric nature. More rigorous estimation of the diffusion coefficient requires a well-based definition and a well-defined methodology. Nevertheless as we will see in the following sections, using the formula (2) gives reliable results in terms of the flow qualitative behavior.

Effect of the solid mass

Figure 2 shows the influence of the solid mass. Each line corresponds to one complete rotation of the paddles starting from the one on the top (Y_c=1000). Arrows show the direction of the temporal evolution of the concentration profiles. Qualitatively, the profiles are smooth up to 10 cm and between the 30-40 cm. Near the front of the colored particle region (around 20 cm), remarkable oscillations are present. In any case, the profiles are the same after 6-7 rotations. The estimated propagation velocity is around 5 mm/s for both simulations. To conclude on the effect of the solid mass, it does not have a large influence on the axial propagation velocity of the particles.

Effect of the angular velocity

Figure 3 shows the effect of the velocity of rotation. On the left figure, the angular velocity is 15 rpm and on the right, it is 40 rpm. Qualitatively, it is clear from the figure on the left that reducing the velocity results in smoother axial profiles of the particle concentration. As seen on the plots, after 10 rotations, the particle front arrives at 30 cm for both cases. It can be concluded that the velocity of rotation also does not have a large effect on the axial transport of the particles. However, it can be pointed out that the slow motion of the paddles result more flat concentration profiles and temporal evolution of the profiles.

Here, it should be noted that the solid mass of these two simulations are different. However, in the previous sections, it has been shown that the solid mass does not have a large effect on the axial transport.

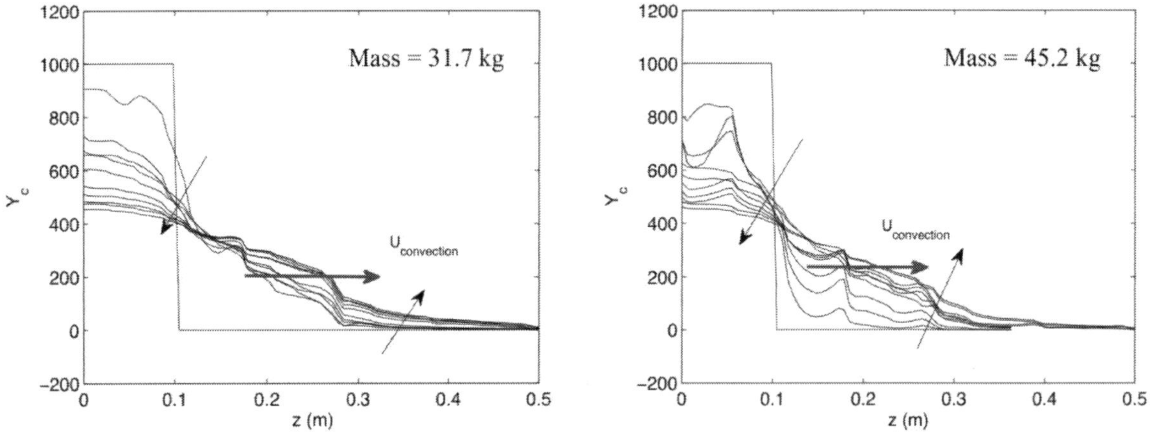

Figure 2. Temporal evolution of colored particle concentration profiles on a line along the axis of the domain: Effect of the solid mass (Mixer A, 40 rpm)

Effect of the paddle shape

Figure 4 shows the effect of the paddle shape. Two different paddle shapes have been tested. On the left, the results with the initial paddle shape and on the right, the results with substantially different paddle shape are presented. Clearly on the figures, particle front propagates faster for the first paddle shape than the second one. In addition, the profiles are much flatter for the second paddle shape than those for the first paddle. The initial paddle shape enhances the axial propagation largely.

The estimated diffusion coefficients D_t for the initial and second paddle shapes are, respectively, $2.0x10^{-3}$ and $0.5x10^{-3}$ m^2/s. We can conclude that the shape of the paddles can largely enhance the axial transport of particles. These results are in qualitative coherence with the experimental results, however, more quantitative comparisons are to be made in the recent future.

Conclusion and Perspectives

3D simulations of an horizontal fixed drum with rotating paddles have been carried out with the Euler-Euler code NEPTUNE_CFD. Different operating conditions have been investigated such as the solid mass, velocity of rotation of paddles and paddles' geometrical shape on the axial transport of colored particles. The simulations show that the solid mass and the angular velocity are much less important than the shape of the paddles in terms of the enhancement of the axial transport. However, substantially different paddle shape can increase the transport up to 4 times. Quantitative comparisons with the experimental data provided by the industrial partner will take place in the future in order to verify the validity of the results. However, the proper characterization of the axial particle transport remains an open problem and more efficient methods should be thought of. Results' sensibility to the frictional model coefficients, particle-wall boundary conditions and the polydispersion are in our future prospects. In the context of the collaboration IMFT-INEOS, more realistic representation of the polypropylene production process is in progress where the polymerization reactions are to be taken into account.

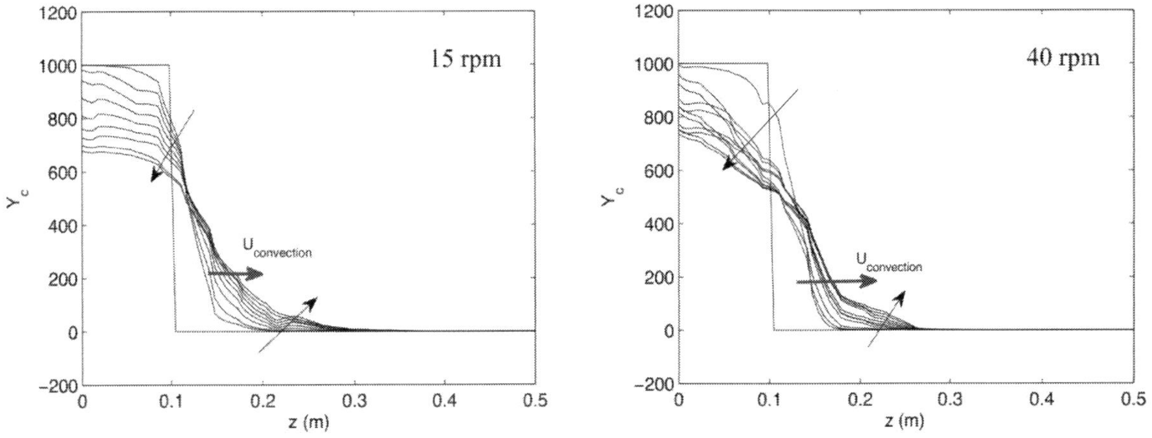

Figure 3. Temporal evolution of colored particle concentration profiles on a line along the axis of the domain: Effect of the angular velocity (Mixer B, solid masses of 45.2 kg and 31.7 kg, respectively)

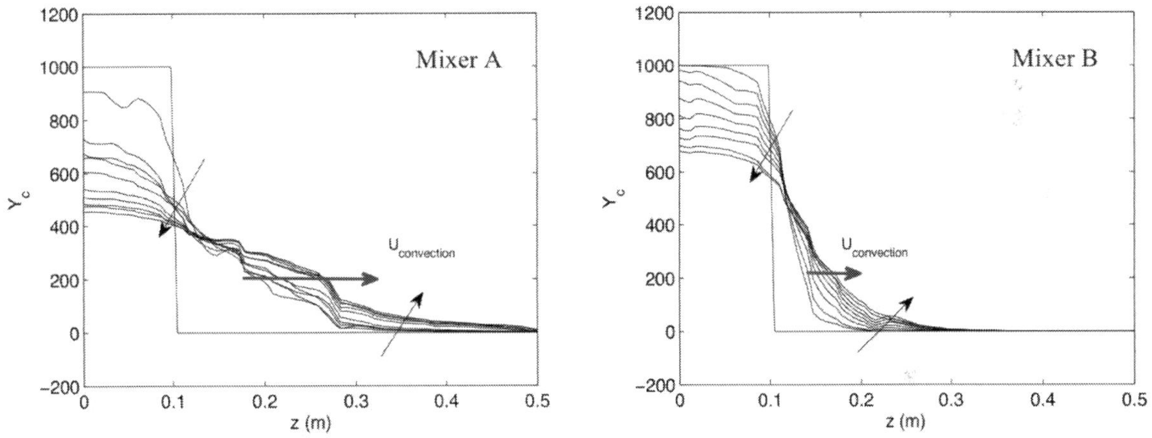

Figure 4. Temporal evolution of scalar concentration profiles on a line along the axis of the domain: Effect of the paddle shape (40 rpm, 31.7 kg)

Acknowledgements

This work was granted access to the HPC resources of CALMIP under the allocation 2012-P0111.

References

1. Gobin, A., Neau, H., Simonin, O., Llinas, J. R., Reiling, V. et Selo, J. L. (2003), "Fluid dynamic numerical simulation of a gas phase polymerization reactor", *International Journal for Numerical Methods in Fluids*, 43, pp. 1199-1220.

2. Balzer G., Boëlle A., Simonin O. (1995), "Eulerian Gas-Solid Flow Modelling of Dense Fluidized Bed", *FLUIDIZATION VIII, Proc. International Symposium of the Engineering Foundation*, p409-418.

3. Jenkins, J. T. & Richman, M. W. (1985), "Grad's 13-moment system for a dense gas of inelastic spheres", *Arch. Rat. Mech. Anal.*, 87, p355–377.

4. Johnson, P.C. et Jackson, R. (1987), "Frictional–collisional constitutive relations for granular materials, with application to plane shearing", *J. Fluid Mech.*, 176, p67–93.

5. Khan, Z. S., Van Bussel, F., Schaber, M., Seemann, R., Scheel, M. and Di Michiel, M. (2011), "High-speed measurement of axial grain transport in a rotating drum", *New Journal of Physics*, 13, 105005

6. Savage, S. B. (1998), "Analyses of slow high-concentration flows of granular materials", *J. Fluid Mech.*, 377, p1-26.

7. Srivastava, A. et Sundaresan, S. (2003), "Analysis of a frictional–kinetic model for gas–particle flow", *Powder Technology*, 129, p72-85.

8. Wightman, C. and Muzzio, F. J. (1996), "Mixing of granular material in a drum mixer undergoing rotational and rocking motions", *Powder Technology*, 98, p113-124.

(707d) Percolation Segregation in a Sheared Bed of Particles

Kevin J. Brennan and Carl R. Wassgren, Mechanical Engineering, Purdue University, West Lafayette, IN

Segregation negatively impacts many powder processing operations, but few tools exist to quantitatively evaluate and predict this phenomenon. In order to assess segregation due to percolation quantitatively, a discrete element method (DEM) model has been developed to simulate a shear cell containing a bimodal size distribution of particles. The steady state flux at which fine particles percolate downward through the bed is recorded for a wide range of shear rates, normal loads, fines concentrations, and particle size ratios. The simulation results are compared to the analytical random, fluctuating sieve model developed by Savage and Lun (1988).

- -

(707e) Characterization of in-Situ near-Infrared Imaging to Aid Pharmaceutical Powder Blend Formulation Development

Juan G. Osorio[1], Gina Stuessy[2], Gabor J. Kemeny[2] and Fernando J. Muzzio[1], (1)Chemical and Biochemical Engineering, Rutgers, The State University of New Jersey, Piscataway, NJ, (2)Middleton Research, Middleton, WI

Pharmaceutical powder blending is a complex process widely studied in recent years. The characterization of blends as a bulk (macro-mixing characterization) has been the main focus of past studies. The present study introduces a new in-situ near-infrared imaging technique (imMix™) designed to characterize micro-mixing (particle to particle interaction and arrangement) in pharmaceutical powder blends. This technique uses in-line / non-contact monitoring of the blending process eliminating the bias introduced by any powder sampling technique commonly used. A Science-Based Calibration (SBC) method, which uses pure component spectral data to create a calibration model, was used to create concentration maps of the blends studied here. The advantage of SBC over PLS or PCA calibration methods is that it does not require a large amount of samples to create a calibration curve. The imMix system showed to be useful in monitoring the distribution and aggregate sizes of the model API and excipients used for this study. It was also able to measure changes in the constituents in a 1-liter test blender as a function of blending time and other experimental parameters. The use of this technique can help in the development of pharmaceutical powder blend formulations.

(707f) Segregation and Mixing of Granular Mixtures in Heap Formation

Yi Fan[1], Paul B. Umbanhowar[1], Julio M. Ottino[2] and Richard Lueptow[1], (1)Mechanical Engineering, Northwestern University, Evanston, IL, (2)Chemical and Biological Engineering, Northwestern University, Evanston, IL

Segregation and mixing of granular mixtures during heap formation have important consequences in industrial and agricultural engineering. In this study, binary mixtures of large and small spheres are poured into quasi-two-dimensional silos; the latter allow straightforward visualization of the segregation dynamics. This work investigates a larger number and wider range of control parameters than previous studies, including particle size ratio, flow rate, flowing layer length, silo thickness, absolute particle size, and heap rise velocity. Depending on these parameters, three different final configurations are observed: stratification, segregation and mixing. The boundary between stratified and non-stratified states is primarily controlled by the two-dimensional flow rate, and the critical flow rate for this transition depends weakly on particle size ratio and flowing layer length. In contrast, the transition from segregation to mixing is controlled by the rise velocity of the heap, and this critical rise velocity depends strongly on the particle size ratio.

(707g) Mixing of Silica Aerogel (Cabot Nanogel) Particles of Different Sizes

Robert Pfeffer and Ding Wang, Chemical Engineering, Arizona State University, Tempe, AZ

Particulate hydrophobic silica aerogel (Cabot Nanogel) is extremely porous, has the lowest density and lowest thermal conductivity of any solid, and is a much better insulator than air. Therefore it is advantageous to mix coarse and fine Nanogel particles to produce a uniform mixture that has a packing density that is higher than that of either of the particles by themselves. Here we present two different methods for mixing coarse and fine Nanogel particles without segregation and simultaneously increasing the packing density. The first method combines a negative pressure (vacuum) with an external force field, such as sound waves or vibration to fill the voids between the coarse particles with fine particles. The second method adds a small amount of liquid to the particles in a slowly rotating tumbler mixer. However, if water is used, the hydrophobicity of the Nanogel surface will repel the water and if an organic solvent is used, the solvent will enter the pores of the Nanogel. To overcome this problem, we used an ethanol- water solution that has a lower surface tension than water, and a contact angle close to 90 degrees. The mixtures produced by both methods are uniform without segregation and have packing densities very close to the predicted theoretical maximum for binary particles.

(707h) Assessing Powder Segregation Potential by near Infrared (NIR) Spectroscopy and Correlating Bench Scale Segregation Tendency to Tabletting Performance

Xiaorong He[1], Xi Han[2], Nadia Ladyzhynsky[1] and Rich Deanne[1], (1)R&D Pharmaceutical Development, Boehringer Ingelheim, Ridgefield, CT, (2)Otto H. York Department of Chemical, Biological and Pharmaceutical Engineering, New Jersey Institute of Technology, Newark, NJ

Segregation is one of the common problems associated with low drug load formulations, particularly for those made by a direct compression process. Segregation of active pharmaceutical ingredient (API) from excipients can lead to variation in content uniformity, which in turn may result in failure to meet product specification. Segregation not only stems from variation of material properties such as differences of API and excipient particle size, shape and cohesion, but also is highly specific to process handling and specific equipment used. The objective of this study was to examine the effects of particle size and cohesion on the segregation tendency of pharmaceutical powders measured by bench scale sifting and fluidization segregation testers. The test blends consisted of 5% aspirin, 64.5% lactose, 30% microcrystalline cellulose (MCC) and 0.5% magnesium stearate with various aspirin particle size and cohesion value. Aspirin cohesion was varied by coating aspirin particles with Cab-O-Sil M5P. The blends were manufactured by a high speed rotary press using a specially designed hopper, which was intended to simulate segregation tendency at large scale tablet manufacturing. The segregation tendency of the blends was also assessed using bench scale sifting and fluidization segregation testers. NIR methods were developed to analyze aspirin content uniformity of blends, segregation samples and tablets. The sifting segregation tester and testing protocol (ASTM D 6940-04) were modified to provide better correlation between the test results and tablet content uniformity. It was found that coating aspirin with Cab-O-Sil M5P can significantly reduce aspirin's cohesion but had little impact on its segregation tendency. The effect of particle size on segregation tendency was quite intriguing. It is commonly believed that large particle size difference between the drug and excipients is a primary factor contributing to segregation. However, our study indicated that one has to consider not only just the particle size difference but also the interaction between aspirin and excipients. Contrary to the common belief, the batch with larger particle size difference had lower segregation potential, which was probably attributed to favorable interaction between aspirin and MCC. In addition to identify critical physical properties of the blends that influence segregation tendency, the study also highlighted the usefulness of NIR for quantitative analysis as well as the importance of using performance based approach to establish correlation between bench scale powder testing and larger scale manufacturing performance.

(707i) Understanding and Preventing Complex Segregation Behavior

Kerry D. Johanson, Material Flow Solutions Inc., Gainesville, FL

Differences in almost any particle scale or bulk property can cause separation of particles. Industry and academia have focused on attempting to understand these various causes by the study of bimodal mixtures which are designed to facilitate one type of separation over the other. The basic premise is that if one can separate the cause and fully describe the behavior, then one can extend that understanding to more complex situations (i.e. multiple components or complex geometries). While this is instructive and is an accepted engineering approach to understanding behavior, it fails with real world mixtures where blends consist of three to six unique ingredients. Complex interactions occur when more than two materials are present in a mixture and simple bimodal segregation behavior is often not additive. Thus, one cannot successfully describe the overall separation behavior as a linear combination or superposition of two simple component interactions. Particle or component separation can occur within a subset of ingredients because of their unique interactions. This paper will discuss multi-component segregation data collected on mixtures with three or more ingredients. This data will be compared with advancements in modeling segregation using a convective-dispersion approach that is driven by mechanistic fluxes. This approach will highlight and explain why a symbiotic segregation relationship may occur with a set of ingredients, and why some ingredients seem to enhance the segregation potential of other ingredients. This paper will present data and theory which describe the interaction of sifting, angle of repose, and air entrainment segregation mechanisms and the role cohesion plays with each. In some cases the addition of cohesion reduces the segregation potential. This is usually in cases where segregation is dominated by one segregation mechanism. However, in other systems, segregation increases when cohesion is added due to a change in mechanism and complex interactions with multiple components. The theory will show how to model such complex systems to predict segregation potential. It will also provide a systematic methodology to design products to mitigate segregation from a product point of view. This approach will provide a basis to help formulators design a particle system to mitigate or reduce segregation tendencies.

(711a) Lead-Free Nanosolders and Their Application for Nanowire Assembly and Joining

Fan Gao and Zhiyong Gu, Chemical Engineering, University of Massachusetts Lowell, Lowell, MA

Due to the requirements for high-performance interconnects in advanced packaging, the development of nanotechnology has to meet the dramatic scaling down of the electronics packaging size from micron-scale to nanoscale. Solder-based connections have great potential and provide ample opportunities to form robust joints between individual nanocomponents and integrate them into a nanosystem with desired functionality and geometry for device applications. Prior study has shown that nanosolders can be used to assemble and join various metallic and metal oxide nanowires. In this study, the approach of nanosoldering is shown to be useful for the construction of functional nanostructures from nanomaterials by utilizing conventional reflow process with proper flux selection and utilization. The interaction between lead-free tin-based nanosolders onto metallic nanowires and reactive substrates such as copper are studied. The melting and wetting behavior of nanosolders on a reactive surface provide primary model of nanoscale joining formed on 2-dimensional planar. Such 1-D cylindrical nanowire structure provides the critical information to process the joining techniques in micro/nano electronics packaging to devices in a much smaller scale.

(711b) Electromechanically Driven Nonlinear Dynamics of Voids in Metallic Thin Films Under Anisotropic Mechanical Stress

Dwaipayan Dasgupta, Georgios I. Sfyris and Dimitrios Maroudas, Chemical Engineering, University of Massachusetts Amherst, Amhest, MA

Void migration and morphological evolution, driven by the combined action of mechanical stresses and electric fields, are the leading mechanisms of failure in metallic thin films, which are used widely in microelectronics. This electromechanically driven void dynamics is extremely rich and complex and it depends strongly on the void size, the strength of the applied electric field, mechanical loading conditions, as well as diffusional anisotropy parameters for diffusion on the void surface. In this presentation, we report results for the effects on the current-driven nonlinear void dynamics of biaxial mechanical loading of the film, in the form of a general applied stress tensor.

We have studied the electromechanically driven surface morphological evolution of voids in single crystalline thin films of face-centered cubic metals for <100>-oriented film planes, characterized by four-fold symmetry of surface diffusional anisotropy. The thin film is subjected to an electric field and a mechanical load expressed by a transversely isotropic biaxial stress tensor. We have investigated all the possible loading modes, i.e., purely tensile loading, purely compressive loading, as well as a more general loading where the applied stress components can be either tensile or compressive, resulting in a mixed (tensile + compressive) loading mode in addition to the loading anisotropy. We have conducted self-consistent dynamical simulations of the driven void surface morphological response according to a well validated, two-dimensional, and fully nonlinear model. The model describes driven mass transport on the void surface and accounts for curvature-driven and stress-driven surface diffusion, electromigration driven by the electric field, and surface diffusional anisotropy. In the simulations, a Galerkin boundary-integral method for the solution of the electrostatic and elastostatic boundary-value problems is combined with a front tracking method for monitoring the shape evolution of the void surface.

We have analyzed the effects of the various mechanical loading modes on the surface morphological evolution of the void for electric-field strengths that are above and below a critical level, which corresponds to a Hopf-bifurcation transition from steady states to time-periodic states in the unstressed film; these asymptotic states of voids are solitons that migrate along the electric-field direction at constant speed. We have spanned the entire range of mechanical loading anisotropy, from uniaxial loading to isotropic biaxial loading. We have mapped the void stability domain boundaries and found that under isotropic or transversely isotropic biaxial loading that is either purely tensile or purely compressive, both steady and time-periodic states can be stabilized in the driven void morphological response; transitions between these states occur at supercritical Hopf bifurcation points. Upon increasing the stress level beyond a certain limit, the void tip extends across the film width causing film failure. Under the same loading anisotropy and stress level, purely compressive and purely tensile stresses result in very similar void morphological responses in terms of oscillation amplitude at the time-periodic states, ranges of stress that result in stable steady and time-periodic states, and level of stress that the film can sustain prior to failure. However, under mixed tensile and compressive biaxial loading, we found that the stable steady states in the void morphological response become completely suppressed and only time-periodic states can be stabilized for stress levels below the critical stress for film failure. Moreover, under such mixed loading, the levels of stress that the film can sustain prior to its failure are much lower than those it can sustain under purely tensile or purely compressive loading. We conclude that such mixed mechanical loading modes make metallic films with voids more vulnerable to electromechanical failure.

(711c) Atomic Layer Deposition Enabled Synthesis of Nanostructured Composite $BiFeO_3/CoFe_2O_4$ Thin Films for Multiferroic Applications

Calvin D. Pham, Department of Chemical and Biomolecular Engineering, University of California Los Angeles, Los Angeles, CA and Jane P. Chang, Department of Chemical and Biomolecular Engineering, University of California, Los Angeles, Los Angeles, CA

Multiferroic materials, that can either exist as single-phase materials or multi-phase composites, exhibit two or more forms of ferroic order such as (anti)ferroelectricity, (anti)ferromagnetism, or ferroelasticity and have been proposed for use in future non-volatile memory technology. Atomic layer deposition (ALD) is proposed as a scalable approach to synthesize multiferroic thin films and to enable the synthesis of multiferroic composites which utilize conformal deposition onto 3-D nanostructures. Challenges that must be overcome in the ALD of multiferroic materials is the amorphous nature of as-deposited films and the difficulty in attaining the desired crystallinity and structure that would enable multiferroic properties to emerge from these materials.

In this work, multiferroic $BiFeO_3$ was deposited by ALD as a single-phase multiferroic thin film as well as the ferroelectric component in a composite multiferroic using a ferrimagnetic $CoFe_2O_4$ mesoporous template that was synthesized using an evaporation induced di-block copolymer self-assembly technique. The ALD process used the metallorganic precursors $Bi(tmhd)_3$ (tmhd = 2,2,6,6-tetramethylheptane-3,5 dione) and $Fe(tmhd)_3$ alongside oxygen atoms produced from a coaxial waveguide microwave powered atomic beam source. A variety of ALD process conditions were studied, such as the effects of process temperature, precursor pulsing time, and precursor pulsing ratio on film composition, growth rate, and crystallization. The ALD films were able to be grown with a composition ratio Bi:Fe close to unity and with a controlled nanostructure and growth rate of ~0.7 Å/cycle. In order to achieve the desired crystalline material after rapid thermal processing, the composition and nanostructure of the as-deposited films must first be controlled via the ALD process to fit within narrow windows.

To compare the performance of the multiferroic ALD films to more well established synthesis methods, measurements of magnetic and ferro/piezoelectric properties were accomplished using SQUID magnetometry and piezoresponse force microscopy, respectively. Magnetic measurements showed that the out-of-plane remnant magnetization of a composite film at room temperature was approximately 66.4 emu/cm^3 while the coercive field was approximately 1950 Oe which was comparable to epitaxial films grown by other methods such as PLD. The magnetoelectric coupling effects in the composite films were studied to assess the effectiveness of the nanostructured material approach.

(711d) Atomic Layer Deposition of AlN Thin Films As Gate Dielectrics for Wide Bandgap Semiconductors

Ya-Chuan Perng and Jane P. Chang, Department of Chemical and Biomolecular Engineering, University of California, Los Angeles, Los Angeles, CA

Aluminum nitride (AlN) is a potential dielectric layer for wide bandgap semiconductor based power electronic devices, such as those demanded in radio frequency, high-speed and high-temperature communication, because of its wide bandgap and high dielectric constant. In particular, for 4H-SiC, AlN is also a promising interfacial layer due to their similar atomic arrangement, small lattice mismatch (1.3%) and comparable thermal expansion coefficients. Although various deposition techniques have been investigated to synthesize AlN thin films with atomic controllability over a large substrate remains a challenge. Atomic layer deposition (ALD) was thus used in this work to grow AlN thin films.

AlN deposition was performed in an ultra-high vacuum chamber with base pressure of 10^{-7} Torr using trimethylaluminum (TMA) and ammonia (NH_3) as precursors. It was discovered that ALD of AlN is possible only when the minute amount of moisture in NH_3, which competed with and inhibited the nitride growth, was completely eliminated. The ALD window was found to be 500-570 ^{o}C with a growth rate of 1.5 Å/cycle. The deposited film composition was evaluated via *in-situ* x-ray photoelectron spectroscopy (XPS) with Al/N determined to be 1.2. *In-situ* reflective high-energy electron diffraction (RHEED) measurements showed as-deposited AlN films were crystalline, which was confirmed by x-ray diffraction (XRD). AlN/4H-SiC MIS capacitors were fabricated to examine the electrical properties with the dielectric constant of AlN determined to be 8.3 and a leakage current density of 10^{-3} A/cm^2 at 4.3 MV/cm. The ALD AlN passivated AlGaN/GaN hetero-structure demonstrated an increase in the carrier density and a 3% decrease in mobility compared to those of non-passivated hetero-structure. With amorphous Al_2O_3 surface passivation, the mobility decreases by 22% with increased carrier density, showing that the crystalline AlN providing a superior property on passivating the hetero-structures.

Title : Observations, Formation Mechanisms and Reduction Methods of 3-Dimensional Cross-Slip Dislocations in Silicon Single Crystal

Paper number : ID#270201:

Author : Do Won Song, Kyo Kim†

LG Siltron Inc, Gyeonggi-do, Korea
† University of Seoul, Seoul, Korea

As part of Materials Engineering and Science Division
Session: Nanoelectronic Materials and Devices II
scheduled for Thursday, November 1, 2012: 1:50 PM - 2:07 PM, Westmoreland East (Westin)

1. Introduction

1) Requirements of Si Wafer

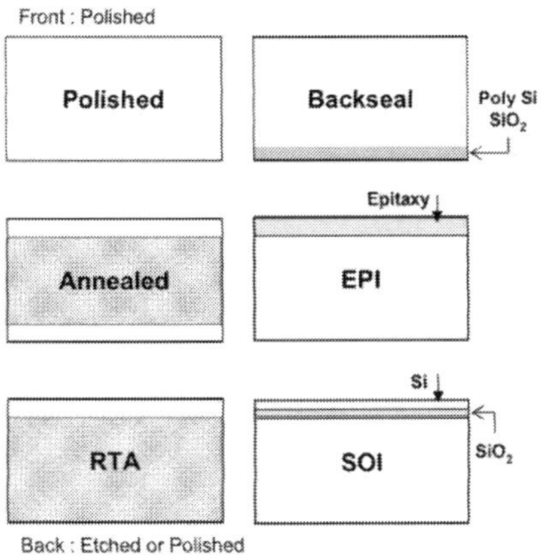

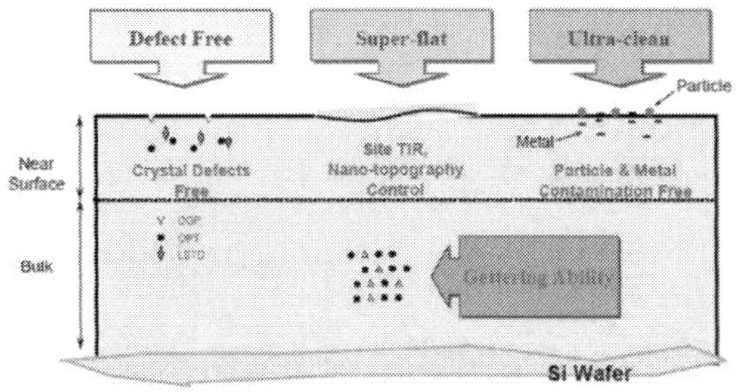

EPI : Epitaxial, RTA : Rapid Thermal Anneal, SOI : Silicon On Insulation, COP : Crystal Originated Particle, OPT : Oxygen Precipitates, LSTD : Laser Scattering Tomography Defect

1 / 9

SILTRON

1. Introduction

2) Defects & Analyzing Method in Silicon Crystal

Various Crystal Defects

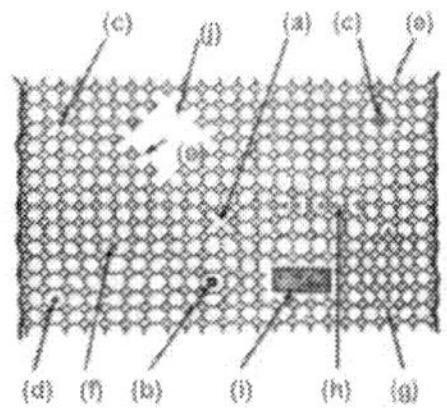

Point defect
- intrinsic
  - **vacancy** (a), **self interstitial atom** (b)
- extrinsic
  - **substitutional impurity atom** (c)
  - **interstitial impurity atom** (d)
  - **dangling bond** (e)

Line defect
- **dislocation : edge-** (f), **screw-**

Plane defect
- **stacking fault**
  - **interstitial-** (g), **vacancy-type** (h)
- **twin, grain boundary**

Volume defect : **precipitate** (i), **void** (j)

Defects of Si Wafer & Analyzing Method

Species	Vacancy Related Defects			Interstitial Related Defects		
Dimension	Volume Defects	Plane Defects	Point Defects	Point Defects	Plane Defects	Volume Defects
Etching + Visual Light	**COP** FPD					**LDP** EPD
Heat Treat + Etching + Visual Light		**O-band** (OiSF)			**T-zone**	
Heat Treat + Electrical Measurement	V-Cluster	O-Band	**VDP**	**IDP**	T-zone	I-Cluster
Eye	Bubble (Air Pocket)					

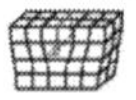
edge-dislocation

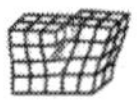
screw-dislocation

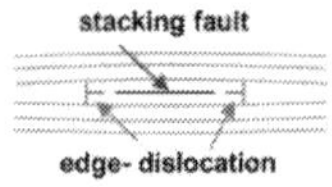

OISF : wet oxidation at 1100℃ for 60min
(Wright 5min, 10,000X, by SEM)

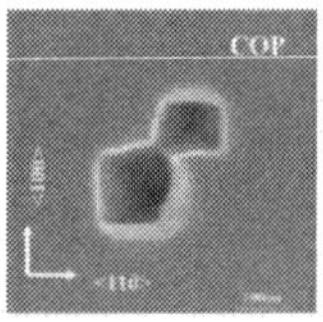

FPD : Flow Pattern Defect, LDP : Large Dislocation Pit, EPD : Etch Pit Defects, OiSF : Oxidation induced Stacking Fault, VDP : Vacancy Dominated Point Defect, IDP : Interstitial Dominated Point Defect

2. Observations of Line Defects

1) Defects in Wafer

Particle Counter

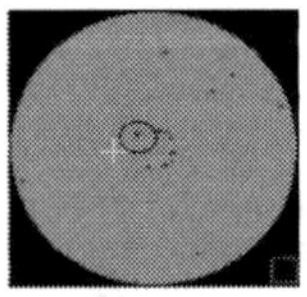

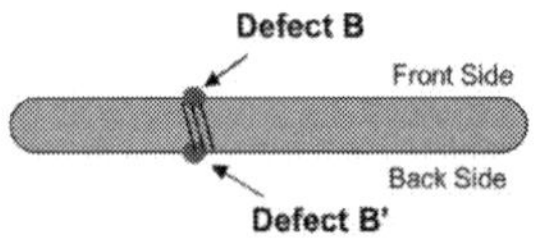

CSD Position (r,z)

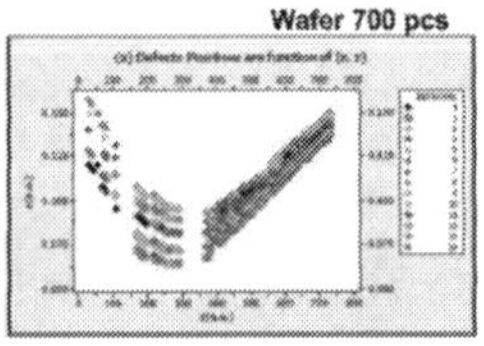

CSD Position (r,θ)

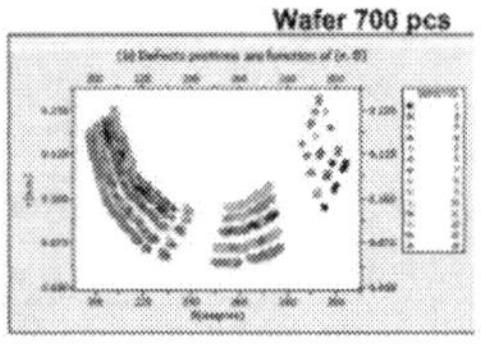

EPD Test **Wright etching 5 min**

Defect B　　Defect B'

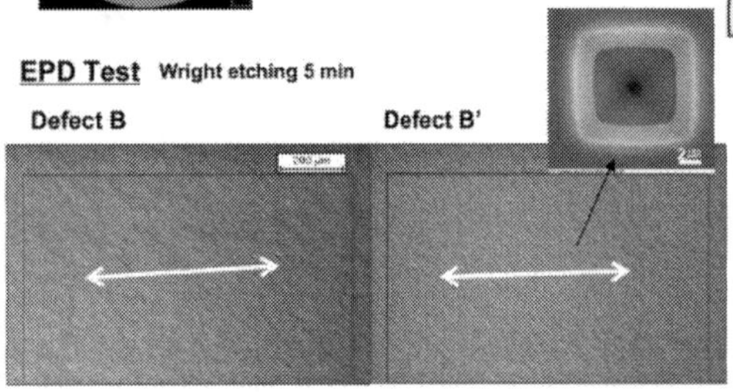

Line Defects in Ingot

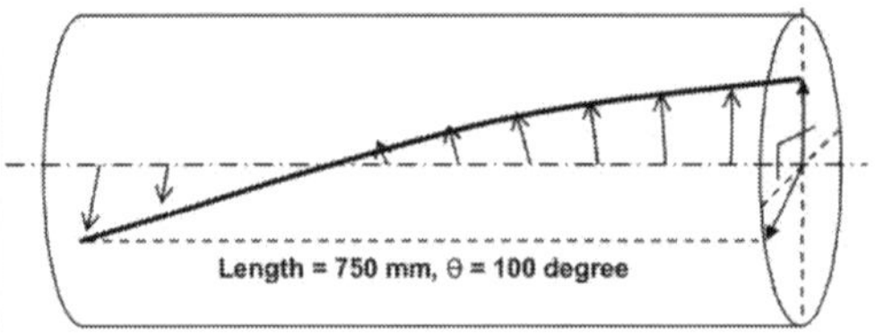

EPD : Etch Pit Defect, CSD : Cross-Slip Dislocation

2. Observations of Cross-Slip Dislocation

2) HR-TEM Images

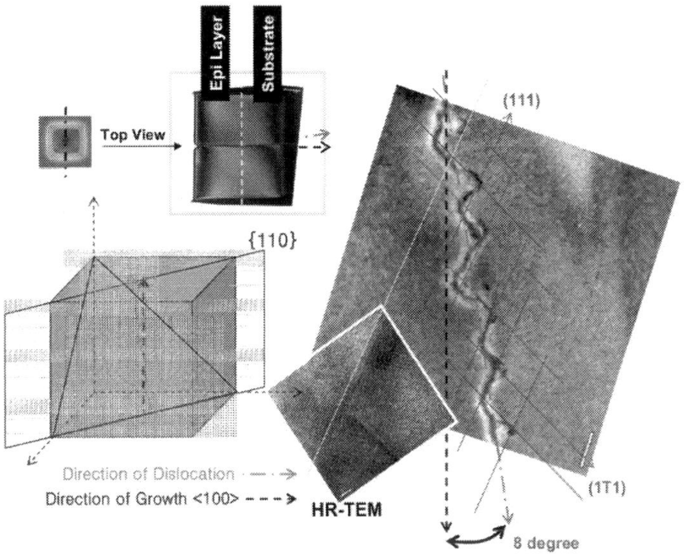

TEM Sample : PT DEPO about 800~900 Å (metal sputter), Etched by Usr3 Beam

3. Effects on Gate-Oxide Insulation

BV Fail Rate(%)

Gate Oxide Thickness : 80A

Electric field	<4	<8	<10	<12
Fail Mode	A	ΣAB	ΣA~C	ΣA~C+
Reference	0.0	0.0	0.0	0.6
CSD 1	0.0	0.6	1.2	1.5
CSD 2	0.0	0.0	0.6	0.9

(MV/cm)

BV Fail Full Map

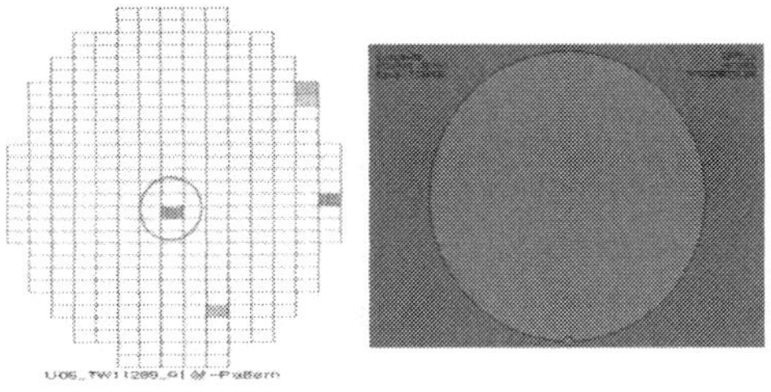

Full MAP

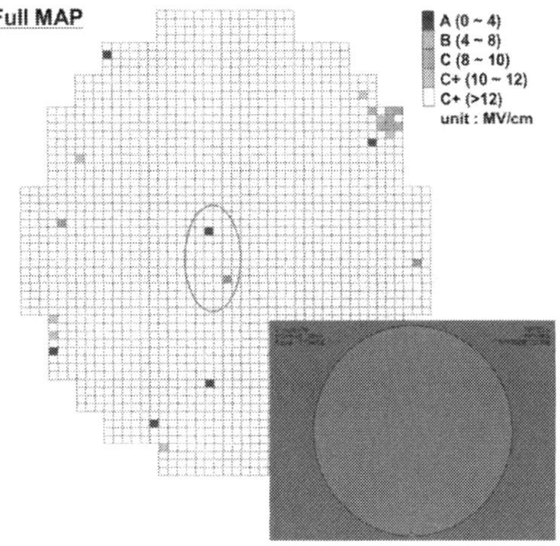

4. Formation Mechanisms

1) Screw Dislocation Growth Model

☑ **Gibbs-Thomson Equation on the Interface**
(Added Constitutional, Capillary and Kinetic Undercooling)

HMCZ Melt Flow → **Screw Dislocation**

$$T^{I} = T_m\left(1 + 2H\frac{\gamma}{L_{\upsilon}}\right) + mC - \mu^{-1}V_n$$

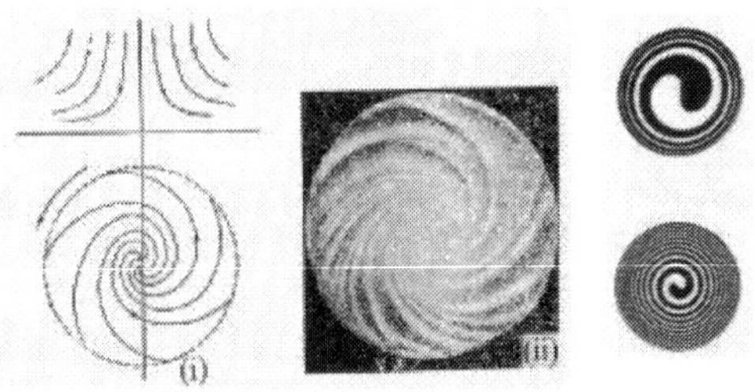

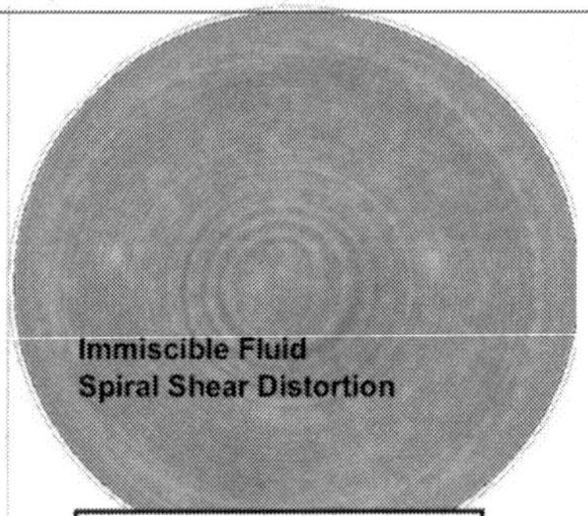

Magic Mirror Image

※ W.A. Tiller "The Science of Crystallization" (1991)

4. Formation Mechanisms

2) Climb & Movement of a Cross-Slip

☑ **Climb & Movement of a Cross-Slip; Sessile Frank Partial**
(Jogged Screw-Dislocation)

Interface Height → **Thermal Stress** → **Jogged Screw Dislocation** → **Cross-Slip**

Scanning Infra-Red Depolarization Image

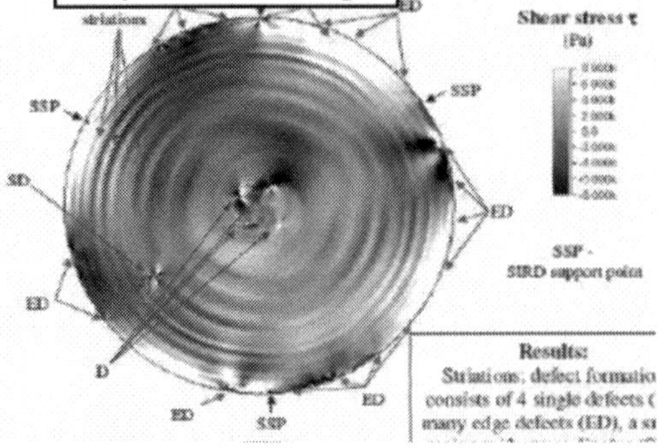

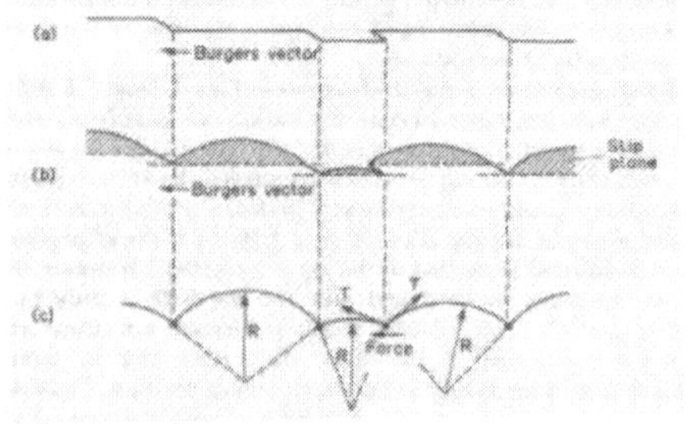

※ D. Hull & D.J. Bacon "Introduction to Dislocation" (1984)

4. Formation Mechanisms

3) Conclusions

Asymmetric Melt Flow → Spiral Shear Distortion → Mechanical & Thermal Stress → Screw Dislocation → Jogged Screw Dislocation → Cross-Slip

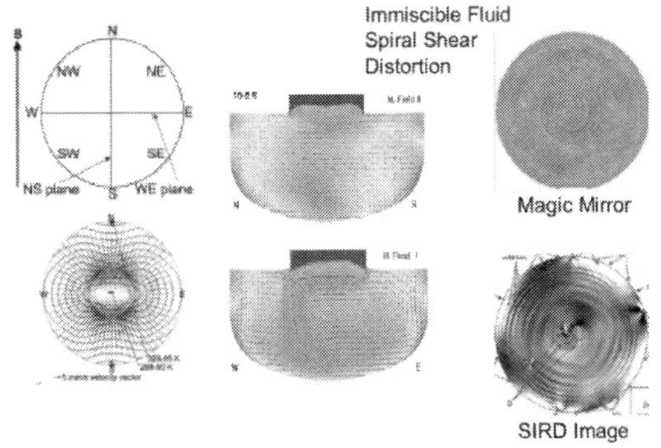

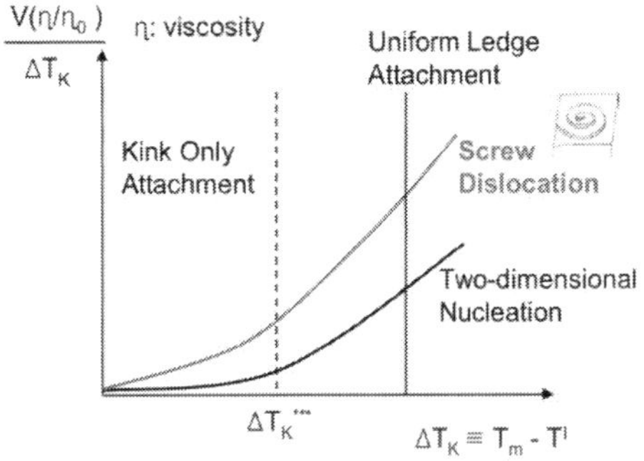

※ International Journal of Heat and Mass Transfer 48 (2005)

※ W.A. Tiller "The Science of Crystallization" (1991)

SIRD : Scanning Infra-Red Depolarization

5. Reduction Method

Decrease Curvature → Increase Gradient → Decrease Spiral Shear Distortion

$$T^I = T_m\left(1 + 2H\frac{\gamma}{L_\upsilon}\right) + mC - \mu^{-1}V_n$$

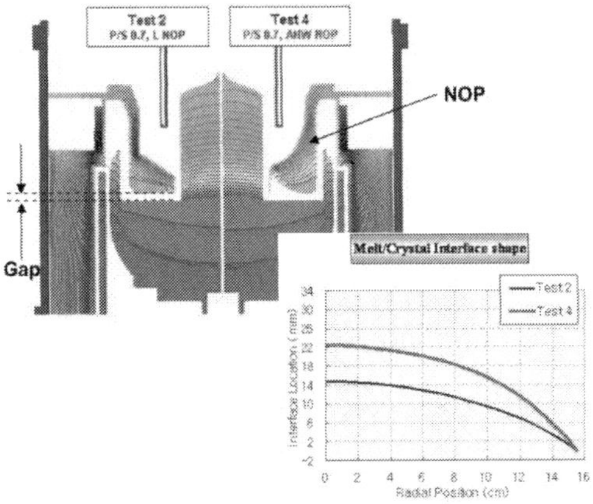

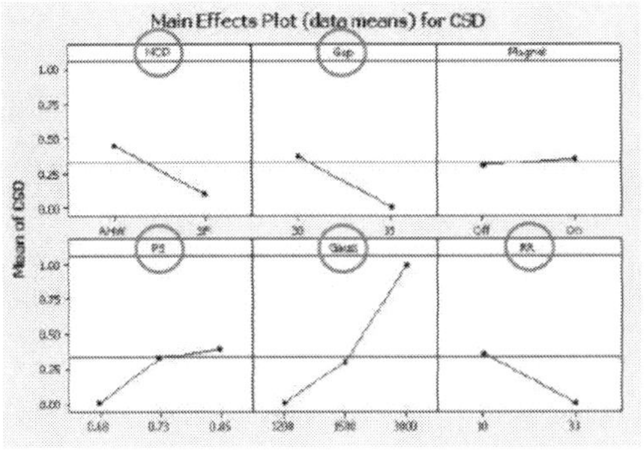

PS : Pulling speed of crystal, RR : Rotation ration between crystal and crucible, NOP : Apparatus of heat shield, Gap : distance between NOP and melt

(711f) Numerical Methods for Solving Mode-Related Problems in Photonic Devices

Meng-Mu Shih, Electrical and Computer Engineering, University of Florida, Gainesville, FL

This work demonstrates a numerical process of solving the mode-related problems in photonic devices with periodic structures. The periodic structures can make the photonic devices behave quite differently than without. In quantum photonics, this periodicity can cause the perturbation effect and show a stop band in the frequency and wave-vector diagram. This effect also can make the propagation constant of the photonic device become a complex number. A visualized method to approach the solution effectively and efficiently is therefore of the central interest.

This work will use the light-emitting devices with hybrid materials of semiconductor and metal as the computational example. This photonic device can emit the light at telecommunication wavelength. This process will cover several sub-processes such as material selection and optical properties, material and optical matching, mathematical issues such as boundary value, linear algebra, and numerical methods. The numerical method has to consider the trade-offs between the precision of the solution and the time required to get the solution. The above results obtained by the photonic method will be compared with the results by the optical method. The results by both methods are close. This photonic method can be applied to photonic devices with different materials and structures for wide applications.

(711g) Analyzing the Dynamics of the Horizontal Ribbon Growth Process for Solar Silicon

Parthiv Daggolu, Andrew Yeckel and Jeffrey J. Derby, Chemical Engineering and Materials Science, University of Minnesota, Minneapolis, MN

Horizontal ribbon growth (HRG) promises the growth of crystalline silicon at rates that are orders of magnitude greater than vertical ribbon growth technologies. If successful, this process would enable the production of higher-quality, near-single-crystalline silicon wafers at fraction of the cost of current production techniques. This fascinating process was first conceived by Shockley in late 1950's for silicon growth and was practiced by Bleil on the late 1960's for germanium growth. Large-scale development efforts were subsequently carried out by Kudo in Japan in the late 1970's and by the Energy Materials Corporation in the US in the early 1980's. However, after encouraging early results, experimental advances and process development efforts stalled, and this technique was abandoned in favor of growth methods that were easier to develop.

Unlike vertical meniscus-defined crystal growth processes, such as edge-defined film-fed growth (EFG), which are inherently stable, there are many failure modes that must be avoided in the HRG process. We argue that its successful operation will rely on a thorough understanding of system design and control–issues that are perhaps only feasibly addressed via computational modeling of the system. Towards these ends, we present a comprehensive thermal-capillary model based on finite-element methods to study the coupled phenomena of heat transfer and interfacial phenomena (solidification and capillarity) in the HRG process. Bifurcation analysis coupled with transient computations using this model reveals process limitations consistent with known failure modes and suggests operating windows that may allow for stable process operation.

(711h) Melt Growth of CZT with Convex Interfaces VIA A Bell-Curve Gradient Freeze Profile

Nan Zhang, Andrew Yeckel and Jeffrey J. Derby, Chemical Engineering and Materials Science, University of Minnesota, Minneapolis, MN

The availability of large, single crystals of cadmium zinc telluride (CZT) with uniform properties is key to improving the performance of gamma radiation detectors. The quality and cost of CZT substrate is primarily determined by crystal growth outcomes. However, the growth of these crystals is still not well understood, and most modifications to crystal growth have been driven by intuition obtained from growth experience with traditional semiconductor systems that are very differently from CZT. We aim to advance the practice of crystal growth by applying physics-based, computational models to understand the most important factors affecting these processes.

In this presentation, a novel, bell-curve furnace temperature profile is presented and predicted to achieve macroscopically convex solid-liquid interface shapes during melt growth of CZT in an electrodynamic gradient freeze (EDG) furnace. A strategy is also presented to dynamically adapt this furnace profile so that uniform, convex interface shapes are maintained through an entire CZT growth run. This approach represents a significant advance over traditional gradient-freeze profiles, which always result in concave interface shapes, and furnace heat transfer modifications, such as redesigned ampule support pedestals, that can achieve convex interfaces, but only over only a small portion of the growth run.

Realizing a convex solidification interface via this adaptive bell-curve furnace profile is postulated to result in better crystallinity and higher yields than conventional CZT growth techniques. Importantly, this strategy may be applied to any Bridgman configuration that utilizes multiple, controllable heating zones.

(712a) Examining Correlations of Electrostatic Discharge Sensitivity and Electrical Conductivity of Composite Energetic Materials

Eric Collins, Texas Tech University, Lubbock, TX, Chelsea Weir, Mechanical Engineering, Texas Tech University, Lubbock, TX and Michelle Pantoya, Mechanical Engineering Department, Texas Tech University, Lubbock, TX

Composite energetic material response to electrical stimuli was investigated and a correlation between the composite's electrical conductivity and ignition sensitivity was examined. The composites consisted of micrometer particle aluminum combined with another metal, metal oxide, or fluoropolymer. Of the nine tested mixtures, aluminum and copper oxide was the only mixture to ignite by electrostatic discharge (ESD) with minimum ignition energy of 25 mJ and an electrical conductivity of 3.45 nS. In an ESD event, energy is delivered through the spark and through joule heating of the particles. For this reason, thermal ignition was also examined using a hot wire. This study showed a similar trend in minimum ignition energy for ignition triggered by a discharged spark compared with ignition triggered by a thermal hot wire source. However, a discharged spark concentrates energy locally supplying greater power for ignition. This study begins to explore critical parameters influencing electrostatic discharge ignition, specifically electrical resistivity and conductivity of particulate mixtures.

NANOSCALE ALUMINUM, SILICON AND ALUMINUM/SILICON MIXTURE FORMATION FOR USE IN PYROTECHNIC APPLICATIONS

Mark Miller[1], Zac Doorenbos[2], Matt Puszynski[2], Lauren Armstrong[3], Deepak Kapoor[3] and Jan Puszynski[1,2]

[1]Chemical and Biological Engineering Department, South Dakota School of Mines and Technology, Rapid City, SD 57701
[2]Innovative Materials and Processes, LLC, Rapid City, SD 57702
[3]Armament Research, Development and Engineering Center, Picatinny Arsenal, NJ 07806

ABSTRACT

Over the past two decades, work in the area of energetic, pyrotechnic and explosive materials has focused on the use of nanoscale materials to tune energetic outputs and combustion characteristics [1-3]. This work presents a method for the formation of nanoscale aluminum, silicon, and aluminum/silicon composite particulates. The effect of attrition milling parameters (e.g. mill configuration, ratio of media to material, time, tip speed, and type of solvent) on the specific surface area and a reactive metal content of different fuel materials is presented. The combustion characteristics of the different fuel materials were tested as a part of nanothermite composite materials and compared to traditional nanothermite materials derived from commercially available aluminum and silicon nanopowders.

INTRODUCTION

It has been shown in the literature that the addition of metal nanoparticles to energetics can increase the burn rates and enhance detonation velocity [4]. The combustion enthalpy (-7.4 kcal/g) and high specific surface area (>20 m^2/g) of nano-scale aluminum has made this a material of choice as an additive or fuel for energetic material formulations. Silicon is another nanoscale metal that is of interest as an additive. Silicon has a similar energy density compared to aluminum. Nanoscale silicon led to similar results as an additive in energetic material as nanoscale aluminum, but it is less susceptible to aging and has a higher ignition temperature. These characteristics may provide safer and more reliable energetic composites derived from this fuel.

A major limiting factor for the uses of these nanoscale additives in energetic, pyrotechnic and explosive materials is their high cost. The ability to produce nanoscale aluminum and silicon powders at a significantly lower cost can be accomplished using a wet attrition milling process. This milling process allows for the specific surface area of the nanoscale materials to be tuned in the range between 5-26 m^2/g and 21-106 m^2/g for aluminum and silicon, respectively. This same technology can be used for the formation of composite mixtures of aluminum and silicon with specific surface areas ranging between 35-50 m^2/g. The stoichiometric ratio of aluminum and silicon in these fuel composites might be adjusted to the desired application. This paper presents the effect of attrition milling parameters on key physical characteristics of aluminum, silicon and aluminum/silicon nanocomposite particulates.

EXPERIMENTAL

Materials

The following materials were used for the formation of nanoscale aluminum, silicon and aluminum/silicon mixture:

- 0.4 mm zirconium grinding media (Union Process)
- Aluminum flake (Alfa Aesar: 11 micron)
- Silicon metal powder (AEE: 1-5 micron)
- Decane (Sigma-Aldrich: anhydrous >99%)
- Heptane (Fisher Scientific: reagent grade)
- Toluene (Fisher Scientific: reagent grade)
- Isopropyl alcohol (IPA) (Fisher Scientific: anhydrous)

Wet Attrition Milling

Wet attrition milling was conducted by first weighing out a desired amount of zirconia grinding media. The amount of a material to be milled was calculated based on the desired weight ratio of grinding media to a material (15:1-50:1). The material was added to the bottom of the milling container, followed by the addition of grinding media. The desired amount of solvent to cover all grinding media and the powder was then added to the milling container and thoroughly premixed by hand. The milling container was then assembled and placed onto the attrition mill. The picture of the wet attrition mill model 01-HDDM purchased from the Union Process company and located at South Dakota School of Mines and Technology (SDSM&T) is shown in Figure 1. Cooling water was turned on to allow flow through the cooling jacket of the milling container. The material was then milled at the selected operating conditions (mill speed and time). The rheology of the slurry within the milling container was observed and small amounts of solvent were added as needed.

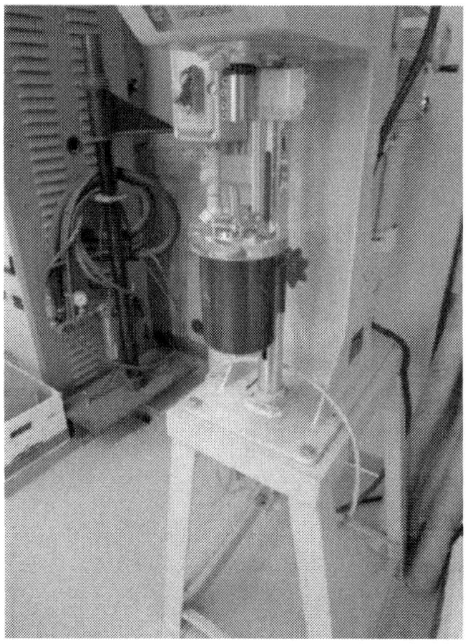

Figure 1: *A photograph showing the wet attrition mill model 01-HDDM purchased from the Union Process company and located at SDSM&T.*

Recovery of Wet Milled Material

In order to recover the milled material, the slurry containing the milling media and material was poured onto a 300 micron (50 mesh) sieve tray with a collection tray for milled powder and the solvent underneath. The grinding media was washed with the fresh solvent until it was clean. Next, a roto-vap was used to remove the solvent. Pior to the exposure of the milled material to ambient air, oxygen passivation process was conducted. The temperature of the material in the roto-vap was reduced to ambient conditions and a small amount of air was bled under vacuum for an extended period of time. This process is dependent on the quantity of the material to be passivated and has to be carefully optimized to prevent its self-heating and possible ignition. The temperature of the passivated nanoparticulate material should be monitored and the intake of oxygen immediately stopped when the recorded temperature rise is more than 5°C. Recovered material from the roto-vap was dried in a vacuum oven for 2 hours at 100°C to insure that all solvent was removed. Next, the material was sieved through 300 micron (50 mesh) sieve tray and collected into an air tight container.

Characterization of Milled Material

Specific surface area (SSA) of the milled materials was determined using BET (model GEMINI III). The reactive content of aluminum nanoflakes was measured using a volumetric reactive aluminum testing setup developed at SDSM&T.

RESULTS

The Effect of Grinding Media Concentration and Attrition Milling Time

Increasing the weight ratio of grinding media to the milled material has resulted in increased specific surface area (SSA) of milled powders over the same period of time. Figure 2 presents the SSA of silicon milled for 2, 4, and 6 hours of milling in isopropyl alcohol (IPA) at three different grinding media concentrations. The results presented in Figure 3 indicate that the SSA of attrition milled aluminum follows a similar trend as silicon (Figure 2)

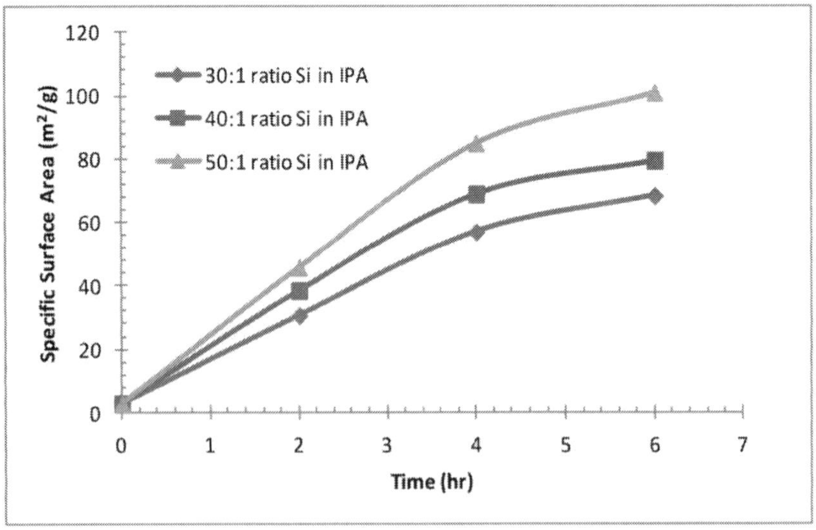

Figure 2: *The effect of grinding media concentration on SSA of attrition milled silicon in IPA as a function of time.*

with increased milling time. When decane was used as a solvent, a slight decrease in SSA was observed compared to IPA solvent. This could be attributed to a partial reactivity of aluminum with this alcohol.

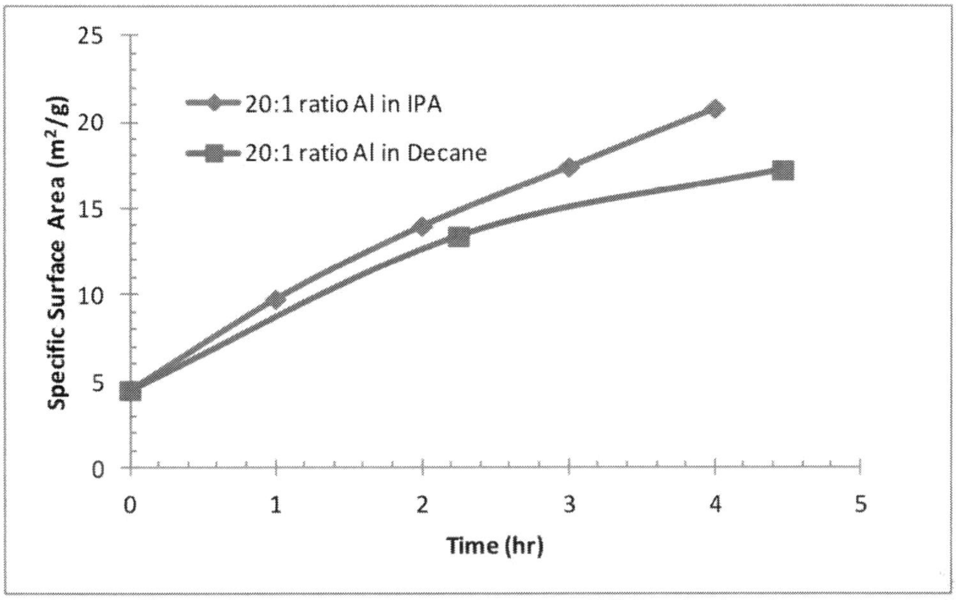

Figure 3: *The effect of attrition milling time on SSA of aluminum for two different solvents (decane and IPA) with a constant grinding media to milled powder ratio.*

Effect of Solvents

Chemical properties of a solvent used in an attritor have a significant effect on powder dispersion and rheology and therefore on milling characteristics. This is especially evident for the milling of silicon, as indicated in Table 1.

Table 1: *SSA results for milled nanoscale silicon using three different solvents with milling parameters.*

Experiment #	Material	Grinding media to material (g:g)	Time (hrs)	Solvent	SSA (m^2/g)
1	AEE 1-5 micron Si	30:1	4	Heptane	22.4
2	AEE 1-5 micron Si	30:1	4	Toluene	33.1
3	AEE 1-5 micron Si	30:1	4	IPA	43.6

The effect of a solvent type was not as significant for the formation of the nanoscale aluminum as shown in Table 2. The reactive content of aluminum nanoflakes was in the range of 84-90 wt% for powders with SSA between 15-17 m^2/g.

Table 2: *SSA results for milled nanoscale flake aluminum using three different solvent with constant milling parameters.*

Experiment #	Material	Grinding media to material (g:g)	Time (hrs)	Solvent	SSA (m^2/g)
1	Flake Al	20:1	3.5	Heptane	17.6
2	Flake Al	20:1	3.5	Decane	16.6
3	Flake Al	20:1	3.5	IPA	15.4

Morphology of Attrition Milled Aluminum, Silicon and Al/Si Nanocomposites

SEM images of milled aluminum, silicon, and aluminum/silicon composites were distinctly different from each other. A SEM photograph of the milled aluminum with a SSA of 26 m^2/g formed flakes, which is shown in Figure 4.

Figure 4: *SEM image of milled aluminum flake with SSA of 26 m^2/g.*

Figure 5: *SEM image of milled silicon powder with SSA of 79.4 m^2/g.*

Figure 6: *SEM image of the milled Al/Si composite with SSA of 48.5 m^2/g.*

The milled nanoflake aluminum looked very similar to the starting micron-size aluminum, but had a much higher SSA of 26 m^2/g compared to the micron-size aluminum having SSA of 4.5 m^2/g. The milled silicon formed agglomerates consisting of very fine nanoparticles. SEM image of the milled silicon with SSA of 79.4 m^2/g is shown in Figure 5. The milled aluminum/silicon composite appeared to be a combination of both previously described SEM image morphologies. SEM image of the milled aluminum/silicon composite with a SSA of 48.5 m^2/g is shown in Figure 6.

Dynamic Combustion Pressures of Nanothermites Derived from Attrition Milled Aluminum and Silicon/Aluminum Nanoparticulates

The different milled materials were tested in nanothermite composite materials in a 2.5 mL steel vessel after being flushed with argon. The nanothermite composite materials were then compared with nanothermite materials derived from commercial fuel nanopowders using 80 nm spherical aluminum from NovaCentrix. The dynamic pressure curves are shown in Figure 7. It should be noted that attrition milled silicon and commercially available nanoscale silicon did not ignite under specific ignition conditions set in this study [5].

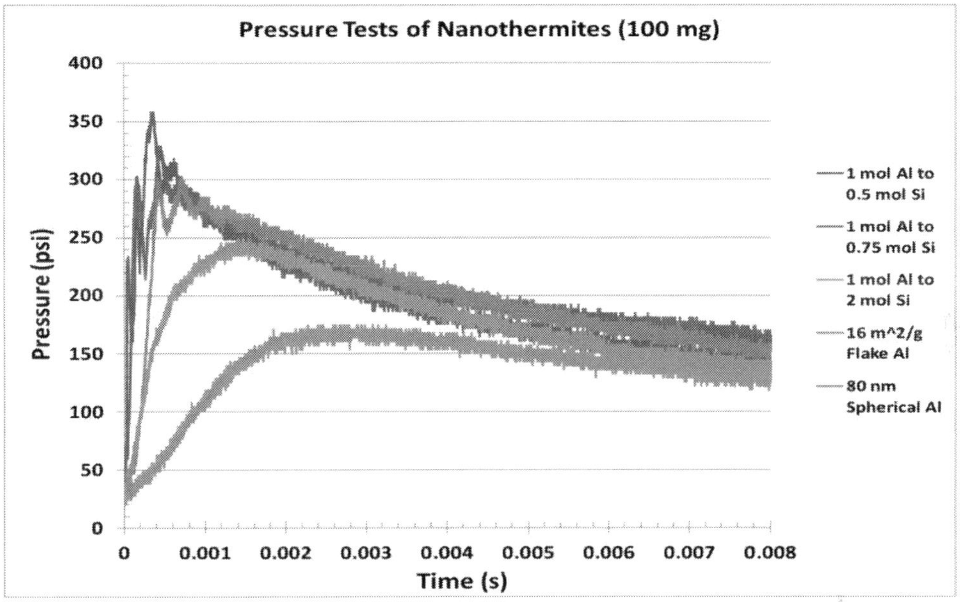

Figure 7: *Dynamic pressure curves of nanothermite mixtures as an oxidizer and nanoscale fuels consisting of bismuth trioxide under current study normalized to 100 mg. Tests were performed in a 2.5 mL steel vessel after being flushed with argon [5].*

CONCLUSIONS

It was demonstrated in this study that the use of an attrition milling technique can lead to the formation of nanoscale aluminum with a comparable SSA and reactive Al content to 80 nm spherical aluminum (NovaCentrix). The ratio of grinding media to material has a significant effect on the time needed to form fuel materials with a higher SSA. As the ratio of grinding media to milled material increases, shorter milling times are needed to obtain the nanoparticulates with the same SSA. Solvent properties have shown a large effect on silicon attrition milling. The effect of a solvent type on aluminum was much less noticeable, but the reason of using decane compared to IPA or heptane is its much lower vapor pressure. Decane is a much less volatile solvent allowing the viscosity during the milling to be more consistent. Also by using a hydrocarbon instead of an alcohol prevents the possible reaction of aluminum or silicon with the OH groups. Attrition milling processing allows for a predictable way of producing and tuning nanoscale materials for their use in pyrotechnic applications.

ACKNOWLEDGEMENTS

The authors acknowledge the support from the ARDEC at Picatinny Arsenal, NJ (Contract No. W15QKN-06-D-0006) and the Army Research Laboratory at Aberdeen Proving Grounds (Contract No. W911NF-08-2-0022).

REFERENCES

1. Son S.F. Yetter R.A. and Yang Y., J. Propulsion and Power, 23(4), 643-644, (2007)
2. Puszynski J.A., Fulian C.J., and Swiatkiewicz J.J., J. Propulsion and Power, 23(4), 698-706, (2007).
3. Piercey D.G., and Klapotke T.M., Central European Journal of Energetic Materials, 7(2), 115-129, (2010)
4. Brousseau P., Dorsett H.E., and Cliff M.W.; 12th International Detonation Symposium, Aug. 11-16, 2002.
5. Doorenbos Z., Walters I., Kapoor D., Balas-Hummers W., and Puszynski J.A., Journal of Energetic Materials and Chemical Propulsion, (submitted 2012)

(712c) On Mechanism of Thermal Initiation in Fully-Dense 2Al.3CuO Nanocomposite Powder

Rayon Williams, Department of Chemical, Biological and Pharmaceutical Engineering, New Jersey Institute of Technology, Newark, NJ, Mirko Schoenitz, NJIT, Newark, NJ, Alexandre Ermoline, Chemical, Pharmaceutical, and Biological Engineering, New Jersey Institute of Technology, Newark, NJ and Edward L. Dreizin, Otto H. York Department of Chemical, Biological, and Pharmaceutical Engineering, New Jersey Institute of Technology, Newark, NJ

Several composite 2Al·3CuO thermite powders with CuO inclusions of different dimensions embedded in Al matrix were prepared by Arrested Reactive Milling. Different milling times were used to adjust the dimensions of CuO inclusions for different powder samples. Thermal initiation in these powders was studied using thermo-analytical techniques (DSC/TGA), ignition on an electrically heated filament, and ignition by electro-static discharge. In ignition experiments, pressure generated by igniting powders was monitored in addition to the produced optical signatures. Each experimental technique involves a different range of heating rates, varied from 10^{-2} to 10^{7} K/s. The results were compared to prediction of a recently developed model describing kinetics of heterogeneous Al/CuO reaction. The model was expanded to account for oxygen release by the CuO, which is partially reduced by subsolidus reaction and decomposes upon further heating. The effect of the oxygen release on thermal diffusivity of the composite powder was also considered while interpreting the ignition experiments. This paper will present experimental data, comparisons with the model predictions and discuss implications of the present measurements for development of a comprehensive model describing thermal initiation in Al-CuO nanocomposite materials for a broad range of heating rates.

(712d) Particle Design of HMX Using Supercritical Anti-Solvent Processes

Bumjoon Seo, School of Chemical and Biological Engineering and Institute of Chemical Processes, Seoul National University, Seoul, South Korea, Youn-Woo Lee, School of Chemical & Biological Engineering, Seoul National University, Seoul, South Korea, Byung-Chul Lee, Department of Chemical Engineering and Nano-Bio Technology, Hannam University, Daejeon, South Korea and Hyoun-Soo Kim, High Explosive Team, Agency for Defense Development, Daejon, South Korea

HMX (Cyclotrimethylenetrinitramine) is a widely used explosive for military and civil use. Like all the other explosives, the performance and stability of HMX depend on its particle properties, especially size and morphology. Firstly, among the four polymorphs of HMX, β–HMX is desired because it shows greatest performance and stability due to its highest density and stability. On the other hand, the effect of particle size on the performance and stability of HMX is still controversial. However, preparation of nano-energetic materials itself is of great interest because it is a state of the art technology. Supercritical anti-solvent process can be used for the recrystallization of HMX because nano- or microparticles production is possible using this process. Two widely used supercritical anti-solvent processes, ASES (Aerosol Solvent Extraction System) process and GAS (Gas Anti-Solvent) process, were used in this work. It was found that the morphology and crystal phase of HMX is tunable by using different processes and solvents. β-HMX was obtained by GAS process when acetone, dimethyl sulfoxide , cyclohexanone, and γ-butyrolactone were used as the solvents. Also, it was found that additives can be used to change the crystal phase of HMX. Crystal phase changed from α to γ when PVA (polyvinyl alcohol) was co-precipitated using ASES process where dimethyl sulfoxide is used as the solvent. Under all the experimental conditions for GAS process, size reduction from the original mean particle size of 100μm to nearly 10μm was observed. When ASES process was used, although β-HMX was not obtained, nano- or microcrystals of other forms were obtained. The effects of operating parameters such as CO_2 addition rate, temperature, initial solution concentration, and stirring rate on the particle size were investigated for the case of recrystallization into β-HMX using GAS process and acetone as the solvent.

(712e) Nanostructured Aluminum-Fluorocarbon Particles for Energetic Applications

Travis R. Sippel[1], Steven F. Son[2] and Lori J. Groven[1], (1)Mechanical Engineering, Purdue University, West Lafayette, IN, (2)School of Mechanical Engineering, Purdue University, West Lafayette, IN

Metal-fluorocarbon energetics have been considered in a number of energetic applications. Aluminum-fluorocarbon mixtures are of particular interest due to the high exothermicity of aluminum fluorination. Like many heterogenous energetics, producing materials having desirable reaction rates and ignitability often requires the reduction of diffusion distances, which can be accomplished by mechanical activation (MA). In this ongoing work, we report the modification of aluminum by inclusion of fluorocarbons by MA of micrometer aluminum and fluorocarbon mixtures. The morphological characteristics of composites made using high and low energy ball milling will be discussed. In addition, we characterize ignition and combustion of the composite particles using differential scanning calorimetry, oxygen calorimetry, simple flame tests, camera flash ignition, and other high heating rate experiments. Thermal characterization shows MA Al-fluorocarbon composites exhibit drastically modified heating and ignition behavior that are similar to those of nanoparticle physical mixtures for some MA composites, ignition via camera flash is possible.

(712f) Combustion of Nanoscale Silicon with Polytetrafluoroethylene

Brandon C. Terry[1], Lori J. Groven[2], Paul Redner[3], Youdong Lin[4], Alexander Mukasyan[5], Deepak Kapoor[6] and Steven F. Son[2], (1)School of Aeronautics and Astronautics, Purdue University, West Lafayette, IN, (2)School of Mechanical Engineering, Purdue University, West Lafayette, IN, (3)Advanced Materials Branch, US Army, RDECOM-ARDEC, Picatinny Arsenal, NJ, (4)Chemical and Biomolecular Engineering, University of Notre Dame, Notre Dame, IN, (5)Chemical Engineering, University of Notre Dame, Notre Dame, IN, (6)Materials Division, US Army RDECOM-ARDEC, Picatinny, NJ

Nanoscale silicon (nSi) based composite energetic materials have the capability to improve fuzes, initiation systems, and high radiant intensity applications. There are also advantages, including higher reaction temperatures and a thinner passivation layer, of using silicon as a replacement for nano-aluminum in many reactive compositions. While various oxidizers have been observed to show promising combustion behavior with nSi, the high heat of fluorination for Si/polytetrafluoroethylene (PTFE) combustion suggests that this system is of particular interest.

In this study, four nSi powders with varied specific surface area, oxygen content, and morphology were investigated and their combustion parameters compared. Specifically, three commercially available nSi powders and highly porous nano-silicon powder synthesized by a combustion synthesis method were studied. Reactive mixtures consisted of nSi, nano-featured PTFE, and FC-2175. Combustion experiments in ambient air were conducted to compare flame temperatures, spectral emissions, and deflagration rates. Visual and infrared spectrometers were used to analyze spectral emissions and determine grey body combustion temperatures. It was demonstrated that all investigated mixtures combust at 1 atm and depending on the nSi, the maximum combustion temperature varies from 2000 to 2600 K and burning rate varies from 0.6 to 1.7 mm/s.

(746a) DEM-CFD Simulation of the Drying of Heterogenous Particles in a Wall-Heated Rotated Drum

Deliang Shi, Collaborative Sciences R&D, SC Johnson, Racine, WI and Joseph McCarthy, Chemical and Petroleum Engineering, University of Pittsburgh, Pittsburgh, PA

The drying process in gas-solid flows involves complex mass (liquid transfer between particles, liquid evaporation to the gas phase and mass transfer in the gas phase), momentum (solid contacts, gas flow, inter-phase momentum transfer) and heat (heat transfer on solid contacts, convective heat transfer in gas phase, inter-phase heat transfer) transfer. These coupled transport mechanisms make analytical approaches to this problem daunting, nevertheless, by incorporating mass transfer (liquid evaporation from particle surface to gas phase and mass transfer in the gas phase) modeling into an our existing gas-solid DEM-based heat transfer code, the drying process is successfully simulated. Here we examine the prototype case of a wall-heated rotary kiln dryer with non-porous particles. Our results show a constant rate drying period as is typically observed experimentally. Results are reported for both mono-disperse and bi-disperse cases as well as for materials with high and low thermal diffusivities. Dimensionless parameters are used to rationalize our varying results.

(746b) Drying of Biomass Pellets: Experimental Study and Comparison of the Performance of a Conventional Fluidized Bed and a Rotating Fluidized Bed in a Static Geometry

Philippe Eliaers and Juray De Wilde, Materials and Process Engineering, UCL, Louvain-la-Neuve, Belgium

The use of dry woody biomass for combustion or gasification improves the process efficiency and reduces net air emissions. Dry biomass is also required for pelletization which often precedes these two processes. Finally, dry biomass facilitates transport and storage because of its reduced biological activity its reduced density.

In this work, the intensification of the drying of biomass particles using rotating fluidized beds in a static geometry [1] was experimentally studied. Experiments in both a 10 cm diameter, 1.5 m tall conventional, that is, gravitational fluidized bed and in a 43 cm diameter, 5 cm length rotating fluidized bed in a static geometry (RFB-SG) were carried out in a range of operating conditions and resulting solids outlet humidity. A schematic representation of the RFB-SG and the rotating bed of biomass particles are shown in Figure 1.

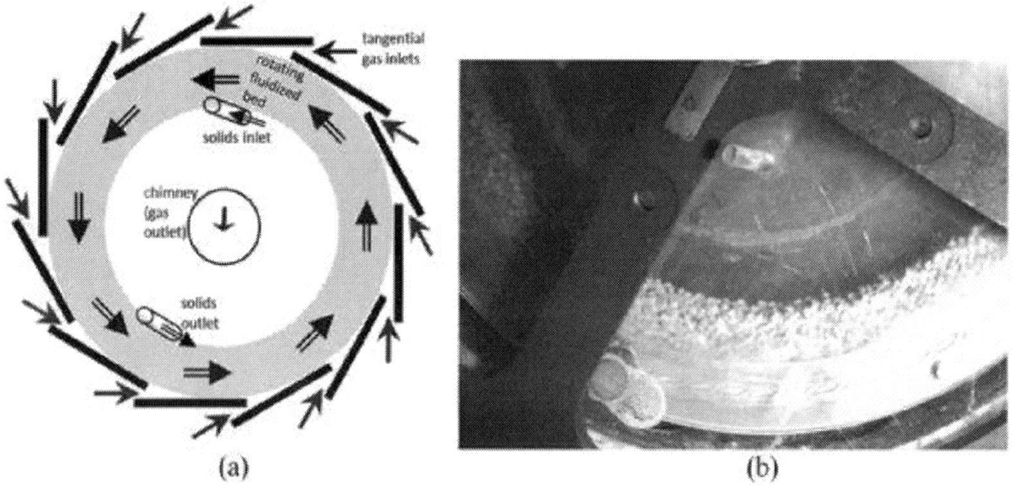

Figure 1. (a) Schematic representation of a Rotating Fluidized Bed in a Static Geometry;

(b) Rotating bed of biomass particles during drying.

The comparison between the specific drying rates (the quantity of water transferred per second and per m^3 drying chamber) in the two types of reactor (Figure 2) shows the process intensification that can be achieved. In the range of operating conditions studied, the process intensification factor was between 10 and 13.

The process intensification results from 3 factors: (i) the higher gas-solid slip velocities resulting in improved mass and heat transfer, (ii) the higher particle bed density and (iii) the higher particle bed uniformity. The relative importance of the three contributions is analyzed. The efficiency of the air utilization was also measured and is discussed. Finally, possible improvements of the RFB-SG design to reduce the air consumption are presented.

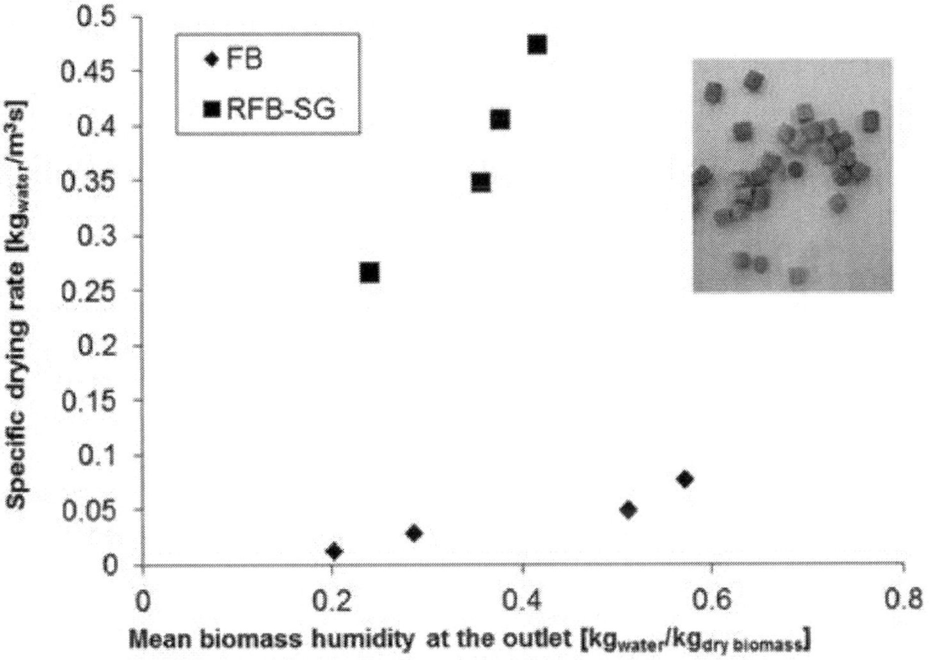

Figure 2. Comparison of the specific drying rates obtained in the conventional fluidized bed and in the rotating fluidized bed in a static geometry for varying mean biomass outlet humidity. Beech wood particles are shown in the picture.

[1] J. De Wilde and A.de Broqueville: Rotating fluidized beds in a static geometry: Experimental proof of concept. AIChE Journal, 53, p. 793–810, 2007.

(746c) Application of Global Sensitivity Analysis As Preparatory Step for Reduction of a Drying Model of Pharmaceutical Granules

Séverine T.F.C. Mortier[1,2], Ingmar Nopens[1], Thomas De Beer[2] and Krist V. Gernaey[3], (1)Department of Mathematical Modelling, Statistics and Bioinformatics, Ghent University, Ghent, Belgium, (2)Department of Pharmaceutical Analysis, Ghent University, Ghent, Belgium, (3)Department of Chemical and Biochemical Engineering, Technical University of Denmark (DTU), Kgs. Lyngby, Denmark

A shift from batch towards continuous manufacturing is nowadays gaining interest in the pharmaceutical industry (Leuenberger, 2001). However, this transition requires detailed knowledge of all consecutive unit operations in a continuous manufacturing line in order to design adequate control strategies for guaranteeing product quality at all time. One hereby relies on in-process measurements of critical process and product parameters and real-time adjustment of input variables. Given the complexity of the system, the knowledge development can be facilitated by developing mechanistic models of the multi-phase systems in the process (Mortier et al., 2011). The importance of using mechanistic models and system-based approaches in a Process Analytical Technology (PAT) and a Quality By Design (QbD) context has been acknowledged (Sin et al., 2008; Gernaey et al., 2012). Modelling efforts only started recently in this field. The ultimate objective of this effort is to develop a validated model describing the continuous drying of pharmaceutical wet granules in a six-segmented fluidized bed drying unit, which is part of a full continuous from powder to tablet manufacturing line (ConsiGma™, Collette™, GEA Pharma Systems). A calibrated and validated model describing the drying behaviour of a single pharmaceutical granule in two subsequent phases has already been developed (Mortier et al., 2012). In the first drying phase the evaporation of water from the droplet free surface takes place. When the radius of the wet particle equals the radius of the dry particle, the second drying phase initiates. Water inside the particle then evaporates at a rate which is much lower compared to the first drying phase.

This single granule drying model is subsequently to be embedded in a Population Balance Model (PBM), allowing the description of the evolution of the moisture content for a population of granules with distributed moisture content (internal property of the PBM). The PBM is to contain a continuous negative growth term describing the drying process. The single granule drying model can serve for this purpose, but is too complex to be integrated as such in the PBM (also bearing in mind that the PBM needs potential coupling with a hydrodynamic CFD model in the future). Therefore, a model reduction of the single granule drying model is to be performed. Prior to this step, a global sensitivity analysis (GSA) is applied as a powerful tool to determine the most important parameters and model inputs that need to be retained in the reduced model, hereby accounting for the entire global domain of degrees of freedom. This contribution gives a detailed description of this GSA to illustrate its usefulness.

The Global Sensitivity Analysis (GSA) methodology used was a Monte Carlo (MC) simulation followed by a statistical analysis for both drying phases, since the behaviour of both phases is very different and might be sensitive to other model constituents. A Latin Hypercube sampling technique from uniform distributions was used to generate the degrees of freedom used in the MC simulation (McKay et al., 1979). Five degrees of freedom were preselected (gas temperature, gas flow rate, gas humidity, gas pressure and initial granule temperature). These were chosen on the basis of their sensitivity in a local sensitivity analysis (evidenced earlier in Mortier et al. (2012)) and the fact that they can be manipulated during the operation of the dryer. A preselection also keeps the sample size of the MC analysis reasonable (2000 samples were generated). An uncertainty range for each degree of freedom was determined based on physical limitations of the dryer and physical reality. Each combined set of degrees of freedom was evaluated (MC shot) and the growth term for the first and second drying phase was calculated. Comparing the different sets of degrees of freedom it is obvious that the chosen combination of degrees of freedom for the sensitivity analysis has an influence on both drying time and behaviour. The output of the MC was further processed using a linear regression analysis at different time instants in both drying periods using the so-called Standardized Regression Coefficients (SRC) method (Saltelli et al., 2004; Saltelli et al., 2008).

As the drying model reaches no steady state, the moisture content for both drying phases has to be compared after a certain time has elapsed. The average moisture content for one simulation would be another way of selecting an output to perform the linear regression with the parameters that have been varied. For the first drying phase the output was evaluated after 3 s. An R of 0.97 was obtained, which is above the recommended value of 0.7 indicating that the linear regression model explains a large portion of the observed variance induced by the variance in the degrees of freedom. The SRCs are a robust and reliable measure of sensitivity, even for non-linear models and are used to rank the degrees of freedom according to their sensitivity (Saltelli et al., 2004; Saltelli et al., 2008). They represent the fraction of the variance observed in the MC results that can be explained by the variance imposed by the degree of freedom. The SRC of the gas temperature was found to be 0.93, clearly higher than the SRCs for the other degrees of freedom (ranging between 0-0.29). This indicates that gas temperature is by far the most influential degree of freedom in the first drying phase regardless of the values of the other degrees of freedom. For the second drying phase two time instants were selected: respectively 1 and 11 seconds after the start of the second drying phase. After 1 s the R was 0.75, while after 11 s the R dropped to 0.54 indicating strong non-linear behavior for which the applied GSA method is no longer valid (other GSA techniques should be used here, but this was considered to be outside the scope of this work). For this reason, the output after 1 s was used to rank the degrees of freedom. The gas temperature with an SRC of 0.87 was again found to be the most sensitive degree of freedom.

Since the gas temperature was found to be by far the most sensitive degree of freedom for both drying phases, this input variable was selected to perform the reduction of the full drying model.

K.V. Gernaey, A.E. Cervera-Padrell, and J.M. Woodley, A perspective on Process Systems Engineering in pharmaceutical process development and innovation. Comput. Chem. Eng. (2012) (in press).

H. Leuenberger, New trends in the production of pharmaceutical granules: batch versus continuous processing, Eur. J. Pharm. Biopharm. 52 (3) (2001) 289-296.

M. D. McKay, R. J. Beckman, and W.J. Conover, A comparison of three methods for selecting values of input variables in the analysis of output from a computer code. Technometrics, 21 (2) (1979) 239-245.

S.T.F.C. Mortier, T. De Beer, K.V. Gernaey, J. Vercruysse, M. Fonteyne, J. Remon, C. Vervaet, I. Nopens, Mechanistic modelling of the drying behaviour of single pharmaceutical granules, Eur. J. Pharm. Biopharm. 80 (3) (2012) 682-689.

S.T.F.C. Mortier, T. De Beer, K.V. Gernaey, J. Remon, C. Vervaet, I. Nopens, Mechanistic modelling of fluidized bed drying processes of wet porous granules: a review, Eur. J. Pharm. Biopharm. 79 (2011) 205-225.

A. Saltelli, M. Ratto, T. Andres, F. Campolongo, J. Cariboni, D. Gatelli, M. Saisana, S. Tarantola, Global Sensitivity Analysis The Primer (2008).

A. Saltelli, S. Tarantola, F. Campolongo, M. Ratto, Sensitivity Analysis in Practice: A Guide to Assessing Scientific Models (2004).

G. Sin, P. Odman, N. Petersen, A. Lantz, K.V. Gernaey, Matrix notation for efficient development of first-principles models within PAT applications: integrated modeling of antibiotic production with streptomyces coelicolor, Biotechnol. Bioeng. 101 (1) (2008) 153-171.

(746d) Investigating Contact Drying in An Agitated Filter Dryer: Experiments and Simulations

Ekneet Sahni, Department of Pharmaceutical Sciences, University of Connecticut, Storrs, CT and Bodhisattwa Chaudhuri, Pharmaceutics, University of Connecticut, Storrs, CT

Manufacturing is a critical component in procuring safe medications. However, it involves a series of unit operations each expected to modify material characteristics resulting in desired properties of the final product. Moreover, minimizing the production cost without compromising quality is another challenge. An improved understanding with the aid of first principle models would help in circumventing some of the process development concerns. Filter-dryer is principally installed because of subsequent overriding concerns: product sterility, toxicity, environmental emissions, and product quality.

Glassbeads and Lactose monohydrate were used as model compounds with ethanol as the solvent. The study investigates the effect of critical process parameters (wall temperature, impeller speed, pressure, and bed depth) in an agitated filter-dryer with temperature and solvent concentration profiles as the output attributes. A three dimensional numerical model was developed for demonstrating the contact heat transfer, by incorporating the algorithm for computing capillary and viscous forces exerted by the interstitial fluid between neighboring particles and heat transfer through liquid bridge and by contact conduction. Until recently, most of the DEM-based heat transfer work was either two-dimensional or in static granular beds. A three dimensional numerical model based on the extension of thermal particle dynamics was used to characterize the drying behavior of glassbead and lactose monohydrate in an agitated filter dryer.

The variation of the process parameters in the range investigated revealed that the rate of solvent drying is enhanced with increasing wall temperatures (40 – 80 °C) and decreasing bed depth (28 – 74 mm) due to increased driving force and reduced resistance to heat transfer respectively. The effect of agitation speed (5 – 25 rpm) was variable for both glassbeads and lactose monohydrate suggesting that the influence of operating parameter is dependent on the material properties of the compound under study. This change in the effect of speed was mainly attributed to frictional behavior which was demonstrated from flow dynamics using the velocity profiles. The velocity profiles of lactose showed the heap formation leading to formation of a recirculation zone due to high friction coefficients. Because of low friction and decreased roughness as in the case of glassbeads, recirculation zones were not created. This describes the ability of first principle models to provide comprehensive information and explanations relating the effect of material properties on operating parameter. Although mechanical agitation aids in improving heat transfer, consequences on particle size can be profound. Decrease in the particle size D[4,3] was evident with an increase in impeller speed and decrease in the bed depth due to increased collision frequency and reduction in the fill volume. It was concluded from the particle size distribution studies that attrition dominates the drying process when the loss of solvent drying rate was low and/or the shear rate was high. Under these conditions, the drying time was sufficiently long for particles to encounter countless particle-impeller and particle–particle collisions that can result in small fragments. Particle attrition was much higher towards the end of drying where solvent acting as a lubricant was reduced further. For higher dryer loads, the number of collisions decreased, stronger agglomerates were formed and thus agglomeration becomes more dominant. Thermograms obtained from conventional DSC scans demonstrate the melting of α-lactose monohydrate. This was further confirmed by PXRD. Regardless of small discrepancies, general trends obtained from simulations were similar suggesting that DEM not only offers a much more flexible approach with a direct relation between the material properties and the model parameters but also reduces the number of experiments and helps in better process understanding. Hence, process modeling based product development can reduce the time required to get products to market as well as the health care cost by saving resources.

(746e) Modelling Convective Drying of Iron Ore Pellets

Anna-Lena Ljung, Engineering Sciences and Mathematics, Luleå University of Technology, Lulea, Sweden and Staffan Lundstrom, Engineering Sceince and Mathematics, Lulea University of Technology, Lulea, Sweden

Modelling convective drying of iron ore pellets

A.-L. Ljung[a], T. Staffan Lundström and Vilnis Frishfelds

Fluid Mechanics

Department of Engineering Sciences and Mathematics

Lulea University of Technology

Lulea 971 87, Sweden;

E-mail: anna-lena.ljung@ltu.se

Iron ore pellets are a highly refined product supplied to the steel making industry for use in blast furnaces or direct reduction processes. The use of pellets offers many advantages such as customer adopted products, transportability and mechanical strength yet the production is time and energy consuming. Being such, there is a natural driving force to enhance the pelletization in order to optimize production and improve quality. The aim with this work is to develop numerical models with which the drying zone of an induration furnace can be examined and optimized. Drying of a bed of iron ore pellets is therefore examined numerically on several scales.

A diffusive model taking into account capillary flow of liquid moisture and internal evaporation is developed to examine drying of a single pellet and simulations of a scanned pellet are validated with good agreement. The result clearly shows four stages of drying; i) evaporation of liquid moisture at the pellet surface, ii) surface evaporation coexisting with internal drying as the surface is locally dry, iii) internal evaporation with completely dry surface and iv) internal evaporation at boiling temperatures. A moisture front moving towards the core of the pellet will start to develop at the second drying stage and the results show that the front will have a non-symmetrical form arising from the surrounding fluid flow. With aid of the results from the single pellet model, a discrete two dimensional bed model is developed. The system of pellets is divided by modified Voronoi diagrams and the convective heat transfer of hot fluid flow through the system including dispersion due to random configuration of the pellets is modeled. The results show that the temperature front advances much faster in the gaps between pellets than in the interior of the pellets even if all the heat energy of the air goes in heating of the pellets initially. An uneven distribution in temperature and moisture content between pellets is furthermore displayed in the discrete system. This phenomenon is related to the natural dispersion occurring in random system of pellets. In an established stage of the drying, the moisture volume-fraction increases continuously from the inlet to the outlet as it is collecting water evaporation from the pellets. The latter results in a lower evaporation in the wet area close to the outlet.

(764a) Nano-Al and Nano-Al/CuO Thermite Under Rapid Heating: A Mechanistic Study

Guoqiang Jian, Department of Chemistry and Biochemistry, University of Maryland, College Park, MD, Nicholas Piekiel, Department of Mechanical Engineering, University of Maryland, College Park, MD and Michael Zachariah, Department of Mechanical Engineering and Chemistry, University of Maryland, College Park, MD

The aluminum nanoparticles (Al-NPs) and nano-Al/CuO thermite mixture were investigated in a rapid heating environment by temperature jump time-of-flight mass spectrometry. Upon rapid heating (10^5 K/s-10^6 K/s) of Al-NPs in vacuum, Al species intensity was found to slowly increase until a sudden increase occurs at ~2030 K, suggesting an enhancement of the outward diffusion of Al species which is consistent with the diffusion based mechanism. Presence of hollow particles after Al-NPs heated to 1650 K also reveals the diffusion of Al through the oxide shell. The lack of mass production of Al-containing species below 2030 K indicates the oxide shell prevents the mass ejection of Al-containing species from the aluminum core. Time resolved mass spectra of nano-Al/CuO reactions show the existence of Al, Al_2O, AlO and Al_2O_2 intermediate reaction products, and the Al_2O specie was found to be the main intermediate oxidation product which agrees with the thermochemical calculation results.

(764b) Nano-Structured Aluminum Powders with Modified Protective Surface Layers

Shasha Zhang, Department of Chemical, Biological & Pharmaceutical Engineering, New Jersey Institute of Technology, Harrison, NJ, Mirko Schoenitz, NJIT, Newark, NJ and Edward L. Dreizin, Otto H. York Department of Chemical, Biological, and Pharmaceutical Engineering, New Jersey Institute of Technology, Newark, NJ

Aluminum powder is widely used as a fuel additive in propellants, explosives, and pyrotechnics due to its high volumetric combustion enthalpy and relatively low cost. However, the thermodynamically predicted benefits of aluminum combustion are rarely achieved because of extended ignition delays associated with heterogeneous reactions involving protective alumina layer, which is always present on the aluminum surface. In order to fully exploit aluminum's high reaction energy, this effort focuses on adjusting its combustion dynamics by modifying its surface and structure. The modification is achieved by cryo-milling aluminum with hydrocarbon compounds, which are liquid at room temperature. At the temperature of liquid nitrogen, the hydrocarbons, such as cyclooctane, become solid and form composite with ball-milled aluminum. The natural alumina layers are removed from the particle surfaces; freshly formed surfaces are coated with the softer hydrocarbon. Once the milling is completed, most of the hydrocarbon becomes liquid and is easily separated from the prepared powder. Thus, the prepared powder is nearly pure aluminum. However, its surface remains modified and coated with a protective layer with properties significantly different from those of regular alumina. The particle morphology is typical of mechanically milled metals and includes multiple grain boundaries, which are also expected to be coated with similar, hydrocarbon-modified layers. Prepared powders ignite at temperatures substantially reduced compared to that of pure aluminum. Their oxidation kinetics, as observed from thermo-analytical measurements, is also different from that of pure aluminum. In addition to ignition and oxidation studies, experiments on combustion of the prepared powders using constant volume explosion setup and laser ignition setup will be discussed. Effect of specific hydrocarbon on the properties of the prepared material will be investigated.

(764c) Silicon-Based Nanocomposites As Novel Illuminants (INVITED)

Paul Redner[1], Karl Oyler[2], Christopher Csernica[3], Jesse Sabatini[3], Jay Poret[3], Anthony Shaw[3], Neha Mehta[2], Gartung Cheng[2] and Deepak Kapoor[1], (1)Advanced Materials, US Army, RDECOM-ARDEC, Picatinny Arsenal, NJ, (2)Energetics and Warheads, US Army RDECOM-ARDEC, Picatinny Arsenal, NJ, (3)Pyrotechnics, US Army, RDECOM-ARDEC, Picatinny Arsenal, NJ

Nanoscale and nanostructured energetic materials are enabling technologies for a wide variety of defense applications. Many current and future requirements cannot be achieved by using more conventional materials only as they fall short in terms of performance, strength, sensitivity, or for environmental reasons. By exploiting the "tunable" nature of nanocomposites, one can adjust the energy output, light output, strength of materials, and sensitivity to various stimuli. This is accomplished via altering the constituents, particle or grain size, surface area and surface characteristics of the material.

Porous silicon is a nanostructured material being explored as a novel fuel to replace a variety of materials in illuminant and other energetic applications. A matrix of mixes was generated in which the standard fuel and oxidizers were replaced by silicon and more novel oxidizers. The materials were then characterized based on light output and color, burn time, and ignition temperature. Results will be discussed.

Measuring Thermal Histories Using Fragile Glass Crystallization

Shashank Vummidi Lakshman, New Jersey Institute of Technology, Newark, NJ

Edward L. Dreizin, New Jersey Institute of Technology, Newark, NJ

Mirko Schoenitz, New Jersey Institute of Technology, Newark, NJ

Introduction

New energetic materials are being developed for bio-agent defeat applications [1, 2]. One of the key factors for inactivation of weaponized biological agents is the temperature generated by pyrotechnical events. To characterize the performance of those energetic materials, the temperature profiles of pyrotechnical events needs to be quantified. Direct measurements are difficult since the combustion clouds generated in these events are optically thick hence any emission based pyrometry is limited only to the expansion front of the event. Thermocouples are fragile, sluggish and slow to determine the temperatures.

The current work focuses on the use of thermal witness materials, which are dispersed throughout the combustion environment, collected after and interrogated ex-situ to reveal information about their environment temperature over the time interval of exposure. The thermal witness material proposed here is a powder of bio-active glass 45S5, a fragile glass that undergoes crystallization above 900 K and has a high melting point of 1423 K. Based on the environmental temperature, residence time in the event and the particle size two possible changes occur to the material. The material can crystallize if exposed long enough, and it can melt if its temperature exceeds the melting point. Small particles reach environmental temperatures faster than large particles. Differentiating the effects of crystallization and of melting based on particle size, the temperature and time history of the exposed environment can be determined with certain limitations. Crystallization of glass particles are detected via polarized light microscopy and molten particles typically have a rounded shape compared to crystallized and non-crystallized material.

A transition-regime heat transfer model is used to describe the particle temperature as a function of environmental temperature. For a fixed time and fixed temperature profile, particles of different size attain different temperatures. Experiments are performed based on this principle where glass particles are fed into an acetylene and air flame where the temperatures range from 1400 K – 1800 K and the residence time are in the 10 – 50 ms range.

Experimental

Material

Powder of bioactive glass 45S5 (-53 μm) was obtained from MO-SCI Healthcare, L.L.C. The powder was made by crushing large pieces of 45S5 powder. The size distribution as reported by the manufacturer is 10 – 50 μm. The refractive index reported by the manufacturer is n = 1.55 and the density of 2.7 g/cm^3 [3].

Thermal analysis: crystallization kinetics

Crystallization kinetics were studied by performing Differential Scanning Calorimetry (DSC) using a Netzsch Simultaneous Thermal Analyzer STA409 PC in mixed argon (Matheson Tri-Gas, 99.999%) and air (Airgas, zero grade) environment. DSC experiments were performed using platinum crucibles and lids, with 20-30 mg sample masses. Constant heating rate experiments of 4, 10, 20 and 40 K/min were performed.

Short-time, high temperature exposure experiments

Rapid heating experiments were conducted using a burner operated with acetylene and air mixture. The experimental setup is shown in Figure 1. The burner consist of two concentric tubes, an outer tube with an ID of 5.23 mm and an inner tube with an ID of 2.5 mm. A mixture of acetylene and air was fed in the annulus between the 1st and 2nd tubes. Aerosolized glass particles were fed through the inner tube and collected on carbon coated glass at various distances along the vertical axis of the flame. Glass powder was aerosolized using a custom-made screw feeder. Glass powder was coated on a 0.75 inch diameter screw with a thread count of 16 threads per inch. The loaded screw was fit into a brass holder with an air jet aimed tangentially at the powder, entraining the powder in a controllable stream of air. The aerosol density was controlled by adjusting the RPM of the screw feeder.

The temperature of the flame was controlled by varying the flow rates of acetylene and air. Different particle residence times in the flame were achieved by varying the flow rate of the carrier gas. Temperature measurements were done by inserting a tungsten filament mounted on a stand into the flame. Thermal emission from the filament was captured using a spectrometer EPP2000 (StellarNet Inc.). The fiber optics of the device was collimated and focused at the center of the filament. Temperature of the filament was regressed from the thermal emission spectrum.

Particle velocity was determined by illuminating the particle stream with a 532 nm green laser sheet. The laser was pulsed with a known frequency. Images of the particle streaks were captured using a Nikon D7000 with a Nikkor 40 mm f 2.8 lens. The velocity of the particles was determined by the product of relative distance between each streak and the pulsed frequency. With known axial flame temperature and particle velocity a time *vs.* temperature profile was generated for the particles.

With the known gas temperature, a transition regime heat transfer model was used to describe the heat transfer from the gas to the glass particles. The model consists of a Langmuir layer of thickness δ adjacent to the particle. The thickness of the layer depends upon the size of the particle, temperature and pressure of the gas. The heat transfer within the Langmuir layer is governed by free-molecular model, while outside the layer, heat transfer occurs through the continuum regime. The model assumes that the heat flow from the bulk gas to the Langmuir layer and the heat flow from Langmuir layer to the particle are equal. With that assumption the temperature of the glass particle is determined for a given environmental temperature. The details of the model have been reported previously [4].

In order to differentiate individual crystallized and non-crystallized particles after exposure to the flame environment, a polarized light microscope from Carl Zeiss MicroImaging GmbH was used (Zeiss Axio Lab. A1 microscope). Crystallized particles are birefringent hence light up under crossed polarized light unlike non-crystalline material. This is a useful tool to distinguish crystalline and non-crystalline particles. Particles that melted during exposure assume spherical shape, and can be recognized this way. With the use of an optical liquid with refractive index close to the fresh powder, it is possible to distinguish glass from other particulates that may be present on the collection slide. The optical liquids used where obtained from Cargille Laboratories, NJ and have refractive index n =1.5500 and 1.5600.

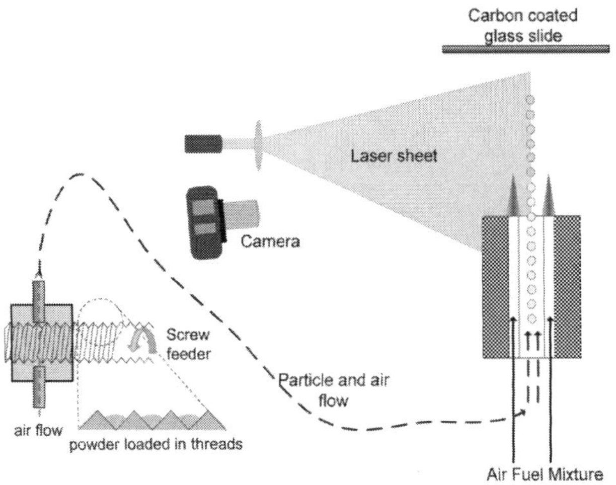

Figure 1: High heating experimental setup

Results

Crystallization kinetics

Figure 2 shows the DSC trace for 45S5 glass powder recorded at a constant heating rate of 40 K/min. The DSC trace shows a glass transition around 825 K, followed by an exothermic peak beginning at 900 K corresponding to the glass crystallization region. These events are well documented in literature [5, 6]. DSC experiments were terminated at 1273 K before the onset of melting (1423 K).

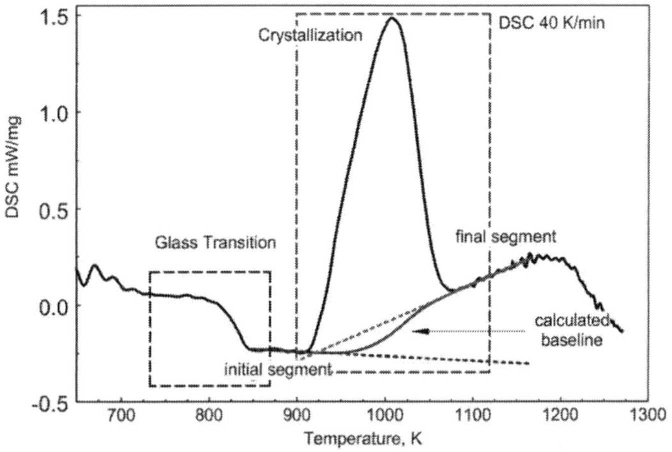

Figure 2: DSC curve of 45S5 powder heated at 40 K/min

Progress of crystallization is determined by integrating the exothermic peak. In order to calculate a baseline, straight lines are fit to the regions before and after the exothermic peak. The baseline is then calculated by iteratively averaging the initial and final straight line segments with the area between DSC signal and baseline as mathematical weights. After the baseline is subtracted from the exothermic peak, the cumulative area under the curve normalized by the total area under the curves defines the progress of crystallization, α. The progress functions for different heating rates are shown in Figure 3. The apparent activation energy as a function of reaction progress was determined by isoconversion analysis [7]. The result of the isoconversion analysis is shown in Figure 4.

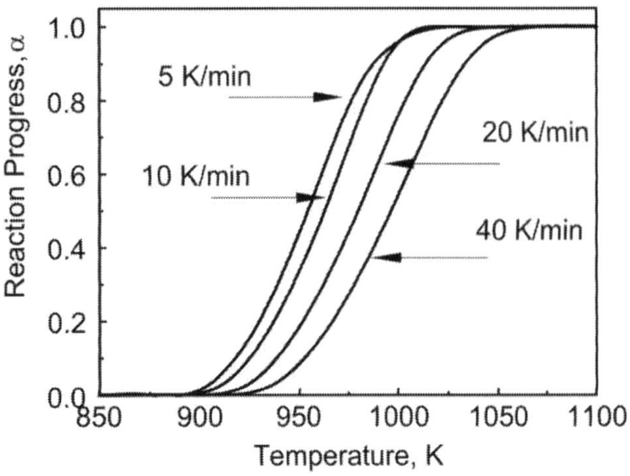

Figure 3: Crystallization progress for different heating rates

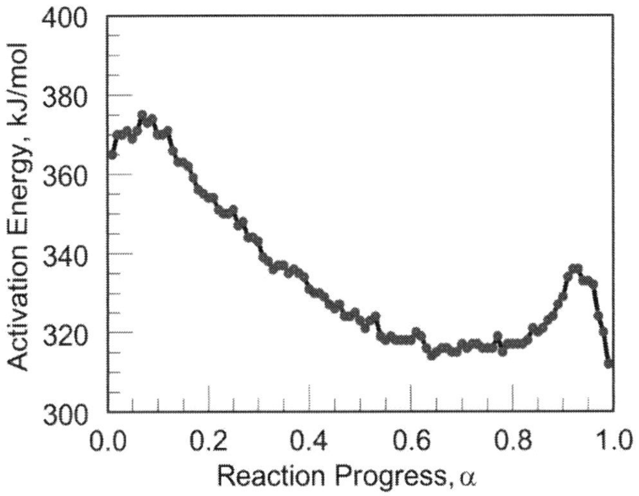

Figure 4: Apparent activation energy determined by isoconversion analysis

The crystallization mechanism is described by the Johnson-Mehl-Avrami mechanism (Equation 1). An activation energy Ea = 334 kJ/mol and n = 0.9 was used to describe the crystallization kinetics. The calculated and the experimental reaction progress curves are shown in Figure 5.

$$\frac{d\alpha}{dt} = A \cdot \exp(\frac{-E_a}{RT}) \cdot n \cdot (1-\alpha) \cdot [-\ln(1-\alpha)]^{(1-\frac{1}{n})} \qquad (1)$$

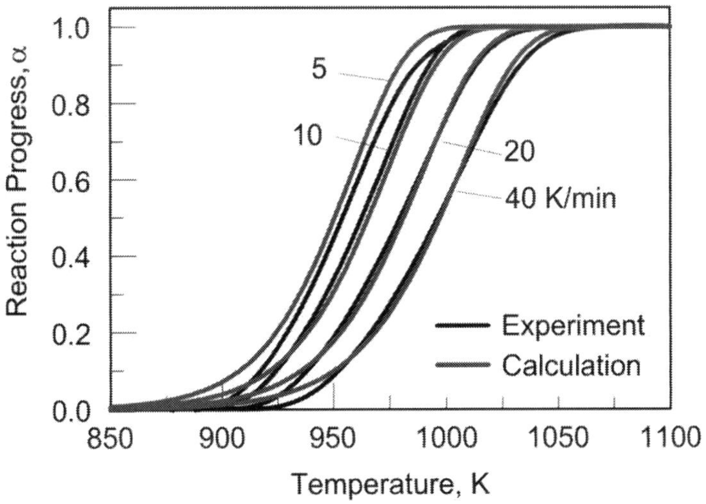

Figure 5: Comparison between observed and calculated crystallization curves

High heating rate experiments

The operating conditions of the burner are shown in Table 1. Temperature measurements were performed at 1mm, 5mm, 10 mm, 15 mm, 20 mm, 30 mm and 40 mm above the burner. The vertical temperature profiles for the two cases in Table 1 are shown in Figure 6. Figure 7 (A) shows a typical glass particle trajectory without the flame as the particle exits the burner. Efforts were made to record the particle trajectory in the flame, but this was difficult to achieve. Hence aluminum particles of comparable size were used to determine the velocity of the particles in the flame (see Figure 7 (B)). The ratio of aluminum particle velocities with and without the flame was approximately 2. This correlation was applied to determine the velocity of the glass particles in the flame. The velocity as a function of height for different flow conditions is calculated.

Figure 8 shows the temperature measured by the tungsten filament as a function of time. Although there is substantial variation of the temperature over time, as a first approximation the environment can be described by a temperature that is constant over a certain period of time.

Table 1: Burner operating conditions

	Acetylene flow rate, scfh	Annulus Air flow rate, scfh	Inner Air flow rate, scfh
Run 1	0.2	5	1
Run 2	0.2	5	2

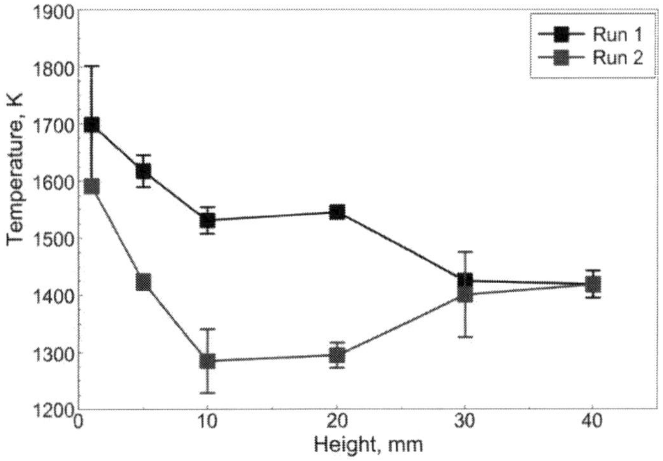

Figure 6: Vertical Temperature Profile

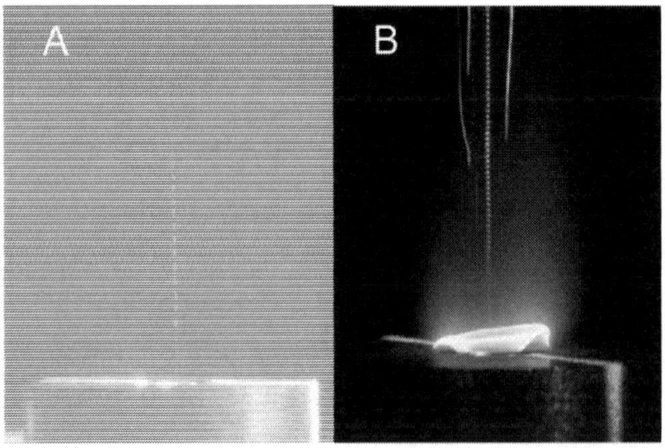

Figure 7: Particle trajectory, A) Glass particles without the flame and B) Aluminum particles within the flame

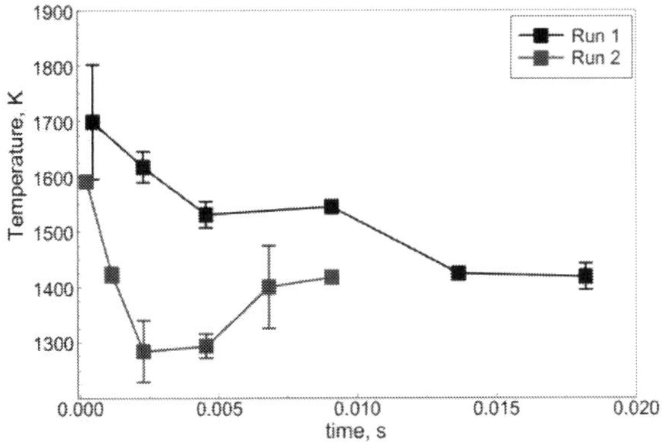

Figure 8: Vertical time-temperature profile of the flame

Since the vertical time-temperature profile and the crystallization kinetics are known, it is possible to predict the nature of the glass powder that is collected through the flame. Figure 9 shows a typical calculated curve for a simple temperature step function, each point in the graph represents where a cold particle is inserted into a hot gas for a given exposure time. The solid lines represent crystallized material at α = 0.5 and the dashed lines represent the environmental conditions where the particles are molten. To recover the time and temperature of the exposed particle we need to know the largest particle that is molten and the largest particle that is crystallized, the crystallization curve and the molten curve for these two particles have a distinct intersection point, this point of intersection describes the environmental temperature and time history of the particle. For example when the 10 μm molten line and the 20 μm crystallized line intersect, then the particle experienced an environment temperature ~1430 K and had a residence time of ~ 8 ms.

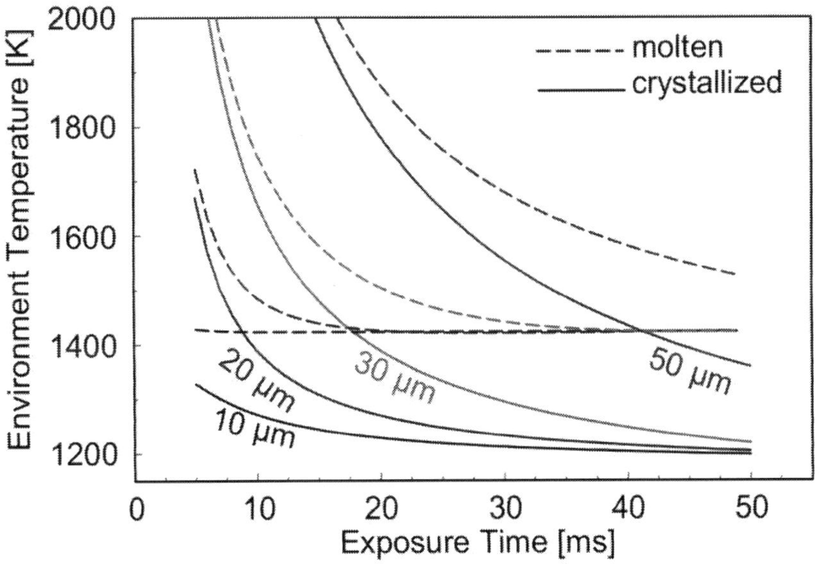

Figure 9: Calculated molten and crystallization curves for different size particles

Figure 10 represents fresh 45S5 powder and crystallized 45S5 powder under cross polarized light. The crystalline material has bright spots which is representative of the crystallites present in the material, whereas the fresh material has no bright spots. The material collected from the burner experiments will be subjected to polarized light and crystallized size fraction can be easily identified. Results from the high heating rate experiments and the predictions from the models will be presented. The feasibility of this approach to determine the temperature-time profiles of explosion/combustion events will be established.

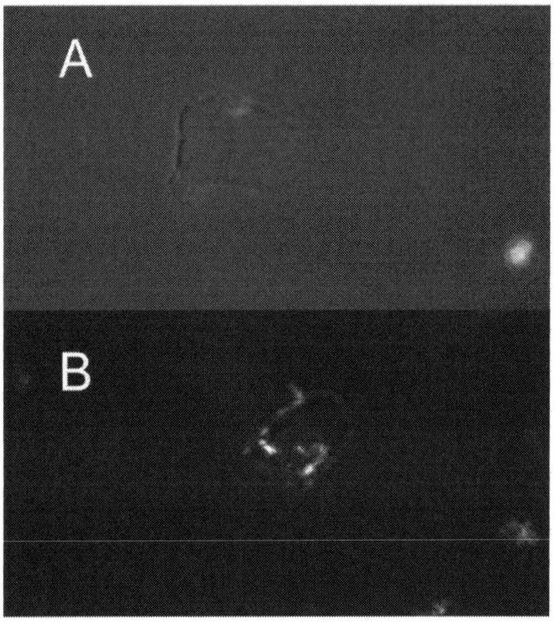

Figure 10: Optical image s of glass particles A) Fresh 45S5 powder and B) Crystallized 45S5 powder, under crossed polarized light

References

1. Grinshpun, S.A., et al., "Method for studying survival of airborne viable microorganisms in combustion environments: Development and evaluation". Aerosol and Air Quality Research, 2010. 10(5): p. 414-424.

2. Grinshpun, S.A., et al., "Inactivation of aerosolized Bacillus atrophaeus (BG) endospores and MS2 viruses by combustion of reactive materials". Environmental Science and Technology, 2012. 46(13): p. 7334-7341.

3. Available from: http://www.mo-sci.com/Mo-Sci_Health_Care/Products/bioactive-glasses/.

4. Mohan, S., M.A. Trunov, and E.L. Dreizin, "Heating and ignition of metal particles in the transition heat transfer regime". Journal of Heat Transfer, 2008. 130(10).

5. Lefebvre, L., et al., "Structural transformations of bioactive glass 45S5 with thermal treatments". Acta Materialia, 2007. 55(10): p. 3305-3313.

6. Clupper, D.C. and L.L. Hench, "Crystallization kinetics of tape cast bioactive glass 45S5". Journal of Non-Crystalline Solids, 2003. 318(1–2): p. 43-48.

7. Vyazovkin, S., "Advanced isoconversional method". Journal of Thermal Analysis and Calorimetry, 1997. 49(3): p. 1493-1499.

(764e) Aluminum-Based Reactive Materials with Customized Combustion Characteristics

Yasmine Aly and Edward L. Dreizin, Otto H. York Department of Chemical, Biological, and Pharmaceutical Engineering, New Jersey Institute of Technology, Newark, NJ

Adding aluminum to propellants, pyrotechnics, and explosives is a common way to boost their energy density. A number of approaches have been investigated that shorten aluminum ignition delay, increase combustion rate, and decrease the tendency of aluminum droplets to agglomerate. Previous work showed that particles of mechanically alloyed Al-Mg and Al-Ti powders burn faster than similarly sized particles of pure aluminum. However, preparation of mechanically alloyed powders with particle sizes matching those of fine aluminum used in energetic formulations was not achieved. This work is focused on preparation of mechanically alloyed and composite Al-Mg and Al-Ti powders in which both internal structures and particle size distributions are adjusted simultaneously. Milling protocol is optimized for each composition to prepare equiaxial, micron-scale particles suitable for laboratory evaluations of their oxidation, ignition, and combustion characteristics. Oxidation of the prepared powders is studied using thermo-analytical measurements. Ignition is characterized experimentally using an electrically heated filament setup. Combustion is studied using a constant volume explosion setup, for the powder cloud combustion, and a laser ignition setup for characterization of combustion rates and temperatures for individual particles. For all materials, ignition and combustion characteristics are compared to those of pure Al and compositions with improved performance are identified. Results will be presented and discussed in this talk.

(764f) Water Processed Nanothermite Inks for Direct Deposition On Low Energy Initiators

Robert R. Nellums[1], Steven F. Son[2], Richard M. Kellett[3], Craig J. Boucher[3] and Lori J. Groven[2], (1)Mechanical Engineering, Purdue University, West Lafayette, IN, (2)School of Mechanical Engineering, Purdue University, West Lafayette, IN, (3)Ensign-Bickford Aerospace and Defense

Pyrotechnic initiators are used in a multitude of applications spanning from airbag inflation systems to energetic detonators. Specifically, semiconductor bridge (SCB) initiators have seen increased use in recent years, particularly in low energy devices. Applications requiring SCB's with dimensions of 80 μm or smaller have sparked interest in the use of nano-sized energetics due to increased interfacial contact area between energetic particles and the bridge. Previous work in this area showed that nanothermites, such as Al-Bi_2O_3, Al-MoO_3, etc., work well for small initiators, but their application typically involves hazardous organic solvents. While water processing has been suggested as an alternative to organic solvents, it has several drawbacks including material degradation, dangerous heating due to reaction of nano-sized aluminum with water, and difficulties in particle dispersion. Currently, ammonium dihydrogen phosphate (ADP) is used to inhibit material degradation and subsequent reaction-induced heating in water processed nanothermites. However, it was found that at the solids loading levels necessary for syringe deposition of nanothermite mixtures, safe processing was not possible with ADP. This work reports a methodology in which the use of a palmatic acid coated nanoaluminum (L-ALEX) and an appropriate dispersant overcomes the above issues. By processing these nanoenergetic mixtures via a LabRAM Resodyn, a high solids loaded ink (30% v/v) results that can be directly applied to the bridge with no visible settling of reactants or cracks. SCB ignition thresholds for these mixtures were found to be comparable to mixtures in which the mixing medium was dimethylformamide (DMF) (35 $mJ*m^{-2}$ vs. 32 $mJ*m^{-2}$). This new processing technique allows for safer processing of sensitive nanoenergetics at high viscosities and has been shown to deliver reliable performance at low energy thresholds for SCB initiators.

AUTHOR INDEX

Abbasi, Alireza 133
Abbasi, Emadoddin 50
Abdulaziz, Yasser 173
Acharya, Sumanta 65
Ackermann, S. 453
Adams, Justin R. 68
Afolabi, Afolawemi 402, 473
Agrawal, Kapil 348
Ahmadi, Goodarz 51
Aigner, Matthias 452
Akseli, Ilgaz 6
Al-Dahhan, Muthanna 108, 173, 199, 385
Alenzi, Adel F. 170
Alivisatos, A. Paul 457
Alizadeh, Ebrahim 130
Allan, Jenna 69
Allcroft, Kody 458
Altmeyer, Andreas 100
Alwan Sr., Ghanim M. 108
Aly, Yasmine 575
Amritkar, Amit 307
Anderson, Brian 408
Angelopoulos, Anastasios 163
Anumolu, Rajasekhar 231
Aradhya, Shreekanta 108
Arastoopour, Hamid 50, 74
Arifin, Darwin 275
Armstrong, Lauren 469, 546
Aromaa, Mikko 11, 210, 517
Aston, Victoria J. 275
Austin III, John S. 3
Azad, Mohammad 392, 402
Azhie, Koosha 70
Bagaria, Hitesh G. 46
Balwinski, Karen 481
Banerjee, Pradip 476
Banerjee, Sudhanya 408
Barbosa, Ariane 493
Barrasso, Dana 53
Barrett, Mark 84
Bartholomew, Calvin H. 122
Barton, P. I. 291
Bayham, Samuel 107
Beaudoin, Stephen P. 37
Becker, Ulrich 308
Beeckman, Jean 383
Beierling, Thorsten 82
Benyahia, Sofiane 148
Berger, Kyle 129
Bermingham, Sean K. 95, 96
Bertrand, Francois 130
Bhakay, Anagha 392
Bhatnagar, Navendu 326
Bi, Xiaotao 185
Bielawski, Christopher W. 46
Bilgili, Ecevit 15, 392, 402, 473, 478
Biserni, Stefano 380
Bishop, Kyle J. M. 10
Blackwood, Daniel O. 288
Blake, Meaghan 481
Blankenhorn, Elizabeth 40
Blattmann, Christoph O. 267
Boddu, Veera 323
Bolda, J. 453
Borghard, William G. 383
Borissova, Antonia 18
Bose, Manaswita 167
Boucher, Craig J. 576
Bousfield, Douglas W. 458
Braun, Markus 308
Breault, Ronald W. 409
Brennan, Kevin J. 207, 527
Brisson, Guillaume 2
Broda, Marcin 211
Brouwer, Gerrit 145
Brown, Robert C. 328, 408
Brozio, Jörg 407
Bu, Dongsheng 7
Buchner, Raymond D. 150, 152
Buckner, Ira Shea 405
Buesser, Beat 158, 254
Bunchatheeravate, Poom 230, 441
Bunker, Christopher 470
Burcham, Christopher L. 95, 96
Burnett, Daniel J. 400
Butch, Christopher J 329
Calzolari, Vittorio 380
Canty, Tod 205
Capece, Maxx 15, 131, 478
Capodanno, Vincent 477
Caram, Hugo 80
Carreto, Victor H. 306
Castellanos, Diana 306
Cha, Bumjoon 319
Chakraborty, Jayanta 125, 157
Chang, Jane P. 535, 536
Chaouki, Jamal 130
Charpentier, Paul A. 69, 70
Chaudhuri, Bodhisattwa 62, 240, 475, 562
Chelst, Isaac 326
Chen, Jie 83
Chen, Yanxin 278, 279, 280
Chen, Yong 209
Cheng, Chong 241
Cheng, Gartung 325, 566
Cheng, Guanzhi 293
Cheng, Jiaming 170
Cheng, Victoria 46
Cheng, Yongpan 147
Cheng, You-Wei 113

AUTHOR INDEX

Cheng, Zhengdong 306
Chern, R. 289
Cho, Hyungtae 319
Chowdhury, Snehaunshu 257
Clark, Samuel M. 310
Cocco, Ray 33, 47, 451
Collins, Eric 545
Corcoran, Amy 324
Costales-Nieves, Carmen 323
Cote, Aaron 477
Crowe, Jakob 329
Crownover, Robin 379
Csernica, Christopher 325, 566
Cuitino, Alberto 383, 387
Culver, James 455
Curtis, Jennifer Sinclair 30, 75, 103, 164, 230, 256, 350, 441, 474
Da Rosa, Leonardo Machado 493
Daggolu, Parthiv 543
Dai, Gance 221
Dalal, Namita 405
Dale, Steven 63
Damavarapu, Reddy 323
Das, Biswa P. 241
Dasgupta, Dwaipayan 534
Davé, Rajesh 15, 38, 131, 392, 394, 402, 478
Davies, Clive E. 2, 172, 386
Davis, Robert H. 48
Davis, Zenda D. 42
Daw, Stuart C. 197
De Beer, Thomas 13, 560
De Haan, André B. 327
De Lasa, Hugo H. 133
De Wilde, Juray 396, 558
Deanne, Rich 531
Deb, Surya 307
Denys, Mark 476
Derby, Jeffrey J. 543, 544
Derksen, Jos 309, 349
Descales, B. 520
Deshmukh, Ankita 167
Diaz, Agustin 306
Dickinson, James E. 114
Dietiker, Jean-Francois 49
Ding, Tongli 436
Djuric, Dejan 100
Dohnal, Jiri 295
Donahue, Carly M. 48
Dooley, Kerry M. 65
Doorenbos, Zac 469, 546
Dreizin, Edward L. 127, 208, 322, 324, 553, 565, 575
Du, Bing 213, 232
Duggan, Jennifer N. 506
Eckert, Charles A. 329
Ege, Paul E. 133
Eggersdorfer, Max 34, 273
Ehrman, Sheryl H. 40
Eichwalder, Gregor 452
Eicke, Matthias Johannes 200
Eilers, Robert 92
Eitzlmayr, Andreas 283, 284
Eldyasti, Ahmed Khiray 411
El-Hedok, Ibrahim A. 66
Eliaers, Philippe 396, 558
Elsner, Martin Peter 200
Emady, Heather N. 61
Engisch Jr., William E. 287
Eniola-Adefeso, Omolola 393
Ermoline, Alexandre 553
Escudero, David R. 184
Evans, James M. B. 291
Fair, Michael 379
Falconer, John 39, 119, 276, 281, 282, 294
Falola, Akinola 18
Fan, Haihong 278, 280
Fan, Jinhe 278
Fan, Junying 203
Fan, L.-S. 52, 72, 107
Fan, Xiaodan 502
Fan, Yi 529
Fassbender, Natalie A. 383
Faulhammer, Eva 380
Fede, Pascal 520
Feke, Donald L. 67
Feng, Jingyu 257
Fernando, Shiral 470
Fisch, Andreas 407
Fletcher, Ray 310
Florian, Miguel 519
Ford, Kenneth J. 80
Fox, Rodney O. 31
Freer, Alexander 455
Freireich, Ben 1, 102
Fuji, Yusuke 441
Fullard, Luke 386
Funke, Adrian 100
Fushimi, Chihiro 146, 147
Galinsky, Nathan 492
Galvin, Janine 148
Galvin, Kevin P. 114, 225
Gan, Yuan 213
Gao, Fan 533
Gao, Jinsen 274
Gao, Xi 113
Gao, Yijie 383
Garg, Rahul 49
Gel, Aytekin 347
Geldenhuis, Petrus 5
Gellman, Andrew J. 8
Gernaey, Krist V. 13, 560

AUTHOR INDEX

Gevgilili, Halil....372
Ghoroi, Chinmay....38, 394
Giam, Hung Loong....482
Gidaspow, Dimitri....27
Gilchrist, James F....80
Girardi, Matthew....126, 252
Givand, J....289
Glasser, Benjamin J....251, 383, 472
Glennon, Brian....84
Glicksman, Howard....40
Gohel, Shitanshu Ratilal....308
Gokhale, Ankush A....305
Good, David....7
Gould, Troy D....119, 276
Grace, John R....185
Grandeury, Arnaud....407
Grawe, Detlef....92
Greif, V....453
Gröhn, Arto J....34, 154, 155, 250
Groven, Lori....361, 555, 556, 576
Gu, Zhiyong....533
Guan, Guoqing....146, 147
Guarnaccio, Lucas....455
Guha, Sudipto....91
Guliants, Elena....470
Guo, Yu....164, 350
Guo, Zhanhu....71, 516
Gupta, Ram B....292, 401
Haapanen, Janne....11, 210
Haines, Chris....469
Halow, Jack....197
Halt, Joe....224
Hamabe, Hisashi....330
Hamey, Rhye....7
Han, Xi....38, 394, 531
Han, Xu....454
Hanberry, Jacob....65
Hancock, Bruno C....103, 164, 256, 288, 350, 474
Harra, Juha....517
Harris, Derek D....161
Harris, Michael....3, 455
Harter, Allison....477
Hashemnia, Kamyar....388
Hays, Roy C....33
He, Feng....492
He, Qingliang....516
He, Xiaorong....6, 531
Heindel, Theodore J....184
Heinonen, Saara....517
Heslinga, Michael J....393
Higashida, Kyohei....118
Higgs III, C. F....165, 170, 360
Hill, Priscilla J....16
Hoerl, Gudrun....284
Hoffmann, Vern K....324
Hohn, Keith L....515
Holbrook, Debora....481
Holloway, William....28, 348
Honkanen, Mari....517
Hou, Guangyang....84
Hou, Qinfu....76
Hrenya, Christine M....48, 81, 129
Hsieh, Daniel....94
Hu, Naiping....163
Hu, Yang....293
Hu, Yaru....280
Hua, Xia....103, 256
Huang, Weixing....213, 232
Huang, Yan....492
Huang, Zhengliang....321
Hubert, Mario....94
Hutter, Sandro....282, 294
Huynh, Phong....398
Ida, Junichi....41
Ierapetritou, Marianthi G....383
Iijima, Motoyuki....12
Ikenberry, Myles A....515
Im, Jisun....44
Irizarry, Roberto....26, 132
Ishizuka, Masanori....117, 146
Islam, Mohammad Ashraful....133
Issangya, Allan S....33
Itaya, Yoshinori....330
Itoi, Tatsuya....353
Izar, Alan....282, 294
Jacob, Karl....1, 481
Jajcevic, Dalibor....351
Jallo, Laila Jai....38
James, Benjamin D....164, 350
Jang, Jungkee....74
Jarboe, Laura....328
Jarmer, Daniel....202
Jensen, Klavs F....120
Ji, Shaowen....305
Ji, Wenjiao....293
Jian, Guoqiang....257, 564
Jiang, Binbo....321
Jiang, Shaoyi....9
Jiang, Yiming....293
Johanson, Kerry D....171, 532
John, Edgar....403
John, Vijay T....149
Johnston, Keith P....46
Johnston, Patrick....328
Jones, James....2
Joshi, Amit....203
Juhnke, Michael....403
Just, Sarah....100
Kagumba, Moses....173
Kalyon, Dilhan....372

AUTHOR INDEX

Kamalssanan, Kaladhar 162
Kamath, Ganesh 326
Kamboj, Lakhinder 45
Kamiya, Hidehiro 12
Kampa, Barbara M. 121
Kan, Chengyou 293
Kandpal, Sanjeev Kumar 458
Kangasniemi, Oskari 210
Kansha, Yasuki 117
Kapoor, Deepak 325, 469, 546, 556, 566
Kapteijn, Freek 121
Karmakar, Srinibas 65
Karri, S. B. Reddy 33, 47, 451
Kaspar, Ondrej 483
Kaus, Mark 123
Kayrak-Talay, Defne 1, 61
Kellett, Richard M. 576
Kemeny, Gabor J. 528
Kenis, Paul J. A. 91
Ketterhagen, William R. 103, 164, 256, 350, 474
Khinast, Johannes G. 100, 126, 283, 284, 351, 380
Khoo, Jiyi 400
Kiel, Darwin 185
Kierzkowska, Agnieszka 211
Kim, Hyo 537
Kim, Hyoun-Soo 468, 554
Kim, Hyung 107
Kim, Suk-Joon 319
Kleinebudde, Peter 100
Klinzing, George 437
Knop, Klaus 100
Knowlton, Ted 33, 47, 451
Kobayashi, Nobusuke 330
Kodam, Madhusudhan 1, 481
Koehl, Maximilian 195
Kofler, Tobias 452
Kong, Song-Charng 408, 410, 435
Koscher, Gerold 283, 284
Kovacik, Pavel 507
Kraxner, M. 452, 453
Kremlackova, Zuzana 507
Kreutzer, Michiel T. 121
Krishnan, Sitaraman 395
Kuai, Niansheng 213, 232
Kuang, Shibo 75
Kumar, Anand 120
Kumar, Pradeep 335
Kumar, Rahul 98
Kushner IV, Joseph 288
Kuusipalo, Jurkka 11
Lacevic, Naida 361
Lacks, Daniel J. 169
Ladyzhynsky, Nadia 6, 531
Lahtonen, Kimmo 210
Lai, Chiajen J. 94
Lakshman, Shashank Vummidi 567
Lamarche, Casey 75
Lan, Xingying 274
Larson, Ronald G. 156
Lash, Melissa H. 162, 277
Lau, Darren Yan Jun 147
Lau, Raymond 482
Law, Yuk Yu 67
Lawrence, Simon 380
Lee, Byung-Chul 468, 554
Lee, Doh C. 456
Lee, Ilsoon 305
Lee, Ivan 477
Lee, Jaeho 46
Lee, Youn-Woo 468, 554
Levänen, Erkki 517
Levilain, Guillaume 200
Lewis, William 470
Li, Fang 27
Li, Fanxing 492
Li, Hui 278, 502
Li, Jianfeng 102
Li, Shu 162, 277
Li, Songgeng 436
Li, Tingwen 347, 505
Li, Xi 113
Li, Yanliang 330
Li, Yukun 241
Li, Zongshan 213
Liang, Xinhua 275
Liévano, Diana M. 390
Liew, Sip C. 3
Lin, Weigang 436
Lin, Ya-Cheng 471
Lin, Youdong 556
Lindner, Jeffrey S. 206
Liotta, Charles L. 329
Litster, James D. 5, 37, 61, 63, 104
Little, Steven R. 162, 277
Liu, Jingjing 18
Liu, Jun 156
Liu, Qin 36
Liu, Sha 123
Liu, Xin 466
Liu, Yang A. 288
Liu, Zhihong 134
Liyanaarachchi, Kathika R. 225
Ljung, Anna-Lena 563
Llusá, Marcos 380
Lopes, Gabriela Cantarelli 493
Lorenz, Heike 92
Lotus, Adria F. 70
Lovette, Michael A. 95, 96
Lu, Guang 99, 195
Lu, Qinggang 504

AUTHOR INDEX

Lubers, Alia M. 119, 276
Luciani, Carla....202
Lueptow, Richard 529
Luhtalampi, Miia 210
Lundstrom, Staffan....563
Lv, Qinggang 503
Ma, Xiaofei 36
Macdonald, Bruce C. 288
Majumder, Aniruddha 17
Mäkelä, Jyrki M. 11, 210, 517
Malla, Ramesh B. 475
Mallapragada, Surya K. 68
Manas-Zloczower, Ica....67
Mannan, M. Sam....306
Manthiram, Karthish 457
Manukyan, Khachatur V. 471
Marashdeh, Qussai 52
Marinack Jr., Martin C. 165, 170, 360
Maroudas, Dimitrios....454, 534
Martignoni, Waldir P. 54, 493
Marus, Gregory....329
Mashuga, Chad 306
Mason, Michael D. 458
Mateo, Daniel 518
Matsusaka, Shuji....441
Matsuyama, Tatsushi....41
McCarthy, Joseph....162, 170, 277, 390, 557
McDaniel, Anthony H. 275
McDonnell, Ryan 3
McKnight, Craig....185
McLaughlin, John B. 51
McMurry, Peter H. 34
Medlin, J. Will....119, 276
Meesters, Gabrie....121, 391
Mehta, Neha 325, 566
Meier, Henry F. 54
Meindersma, G. Wytze....327
Mejia, Andres F. 306
Mendez, Rafael....518, 519
Metzger, Matthew J. 251, 472
Metzger, Philip....129
Meyer, Terry 408
Mezo, Andrew 185
Milioli, Chris 28, 73
Milioli, Fernando 28, 73
Miller, Barbara 470
Miller, Mark....546
Mills, Kendall....469
Min, Yong....278, 279, 280, 502
Misra, Rajiv....10
Mitrano, Peter P. 81
Mohajerani, Amirhossein....388
Mohan, L. Srinivasa....308
Moher, Eric 202
Moon, Il....319
More, Apurva 62
Mori, Milton....493
Morris, Gary....84
Mortier, Séverine T. F. C. 13, 560
Mota-Aguilar, David A. 64
Mou, Xiaozhe....503
Moula, Golam 69
Mudde, Robert F. 145
Mudryy, Ruslan S. 362
Mukasyan, Alexander 471, 556
Muliadi, Ariel....5, 104
Müller, Christoph R. 99, 195, 209, 211
Müller, Uwe 92
Mumin, Abdul 69
Mumtaz, Hassan....95, 96
Muschelknautz, U. 452, 453
Muzzio, Fernando....287, 383, 387, 518, 528
Myerson, Allan S. 83
Myrick, James M. 395
Nagy, Zoltan K. 17
Naik, Punith P. 206
Naik, Shivangi S. 475
Nakhla, George....411
Narayan, Padma....481
Nasr, Hojjat 51
Nastac, Laurentiu....362
Natalie, Kenneth....203
Naud, Darren....361
Neau, Hervé 520
Nedeltchev, Stoyan....108
Negi, Yuvraj Singh....335
Neilson, Bethany....46
Neivandt, David J. 458
Nellums, Robert R. 576
Nguyen, Thi Quynh Ngoc 482
Nicoletti, Phil 409
Nie, Hongqi 322
Nikkanen, Juha-Pekka....517
Niranjan, Manish 125
Nopens, Ingmar 13, 560
Noriler, Dirceu....54
Nowinski, Ann K. 9
Ober, Courtney A. 292, 401
Obregon, Luis....4
O'Brien, Thomas J. 105
O'Donnell, Jennifer M. 66
Oka, Sarang....387
Okubo, Tatsuya....43
Okuyama, Yoshitomo....118
Olsen, Rebecca E. 122
Osorio, Juan G. 528
Ottino, Julio M. 529
Oyler, Karl....325, 566
Ozarkar, Shailesh....308
Pack, Shawn....203

AUTHOR INDEX

Pak, Chaewon....456
Pandey, Sumeet C. 454
Pantoya, Michelle....545
Papavassiliou, Vasilis....150
Parida, Anshuman 476
Parker, James M. 491
Parker, Nathan J....305
Passalacqua, Alberto....31
Paterson, Anthony (Tony) H. J. 2
Pattis, Paul....452
Pawate, Ashtamurthy 91
Peabody, George 40
Pearson, Larry 206
Pease III, Leonard F. 231
Pedro, Aghogho....203
Perfetti, Giacomo 391
Perng, Ya-Chuan 536
Perry, Sarah L....91
Pfeffer, Robert....530
Pham, Calvin D. 535
Pham, Dieu Huong....43
Piekiel, Nicholas 564
Pinkerton, Nathalie M. 407
Pollet, Pamela....329
Polster, Christopher S. 95, 96
Poret, Jay....325, 566
Potoff, Jeffrey J. 326
Pratsinis, Sotiris E....34, 154, 158, 250, 254, 267, 273
Prud'Homme, Robert K. 407
Pruessmann, Klaas....195
Pukadyil, Rohan....285
Puszynski, Jan....325, 546
Puszynski, Matt....546
Qi, Di....152
Qiu, Guizhi 504
Qiu, Hongwei....467
Quiñones, Leonel....4
Radeke, Charles 126, 351
Radl, Stefan 73, 126, 143, 252
Rajniak, P. 289
Ramachandran, Rohit 53
Rangarajan, Deepak 30
Redner, Paul....325, 556, 566
Reese, David 361
Reinalter, Michael....452
Remy, Brenda....203
Reynolds, S. D....289
Ricardo, Guilherme A. N. 54
Riebesehl, Bernd 407
Roberts, Christopher B. 506
Roblegg, Eva....283
Rohani, Parham 123
Rokosz, Joseph 379
Romanski, Frank 480, 490
Rossano, Lucius....204
Rover, Marjorie....328
Saarinen, Jarkko J. 11
Sabatini, Jesse....325, 566
Sadowski, Gabriele....82
Sahni, Ekneet....240, 562
Sakai, Mikio....242, 258, 353
Sankaran, R. Mohan....169
Santhanam, Priya R. 127
Sarkar, Debasis....157
Sarkar, Saurabh....62
Saw, Horng Yuan 2, 172
Sawai, Masakazu....330
Schaefer, Dale W. 163
Scharmach, William J. 150, 152
Scharrer, Georg....100, 351, 380
Schenck, Luke....477
Schiffer, Peter 10
Schoenitz, Mirko....322, 553, 565, 567
Schroeder, Ben....161
Schwerin, William C. 61
Seibert, Kevin D. 202
Seidel-Morgenstern, Andreas....92, 200
Sekimura, Naoto....353
Seo, Bumjoon....554
Sexton, Frederick A....395
Sfyris, Georgios I. 534
Shahnam, Mehrdad....49, 347
Shaikh, Aabid 125
Sharma, Munish 123, 150, 152
Shaw, Anthony....325, 566
Shen, Yangyang....97
Sheskey, Paul 285
Shi, Deliang....557
Shigeto, Yusuke 258
Shih, Meng-Mu....542
Shimojima, Atsushi 43
Shin, Sun Hae Ra....10
Shukla, Nisha....8
Sierra, Luis 477
Simonin, Olivier....520
Singh, Abhishek....157
Singh, G....289
Singh, Jagdeep....44
Singh, Surendra Pal 335
Sinka, I. C. 289
Sippel, Travis R. 555
Smith, Johanna....455
Smith, Laura T. 206
Smith, Marcus 470
Smith, Sean T....161
Smola, Clemens-Michael....283
Snider, Dale....310
Soares, Jason W. 44
Soltys, Marek....507
Son, Steven F....361, 555, 556, 576

AUTHOR INDEX

Son, Sung Real....319
Song, Do Won....537
Song, Guoliang....503
Song, Wenli....436
Sorensen, C. M.....34
Sotiriou, Georgios A....267
Spelt, Jan. K.....388
Srikakulapu, Gurulaxmi....476
Srivastava, Devesh....305
St John, Samuel....163
Stecula, Adrian....6
Steeves, Diane M.....44
Steilberg, Jayne....455
Stepanek, Frantisek....297, 483, 507
Stepanov, Victor....467
Stephen, W....520
Stepien, Milena....11
Štìpánek, F.....295
Stuessy, Gina....528
Su, Ziyang....7
Suami, Akira....330
Subramaniam, Shankar....32
Suda, Toshiyuki....134
Sugawara-Narutaki, Ayae....43
Suhonen, Heikki....517
Sun Sr., Yunkai....504
Sun, Xiaosong....242
Sun, Yunkai....503
Sundaramoorthy, Arul....291
Sundaresan, Sankaran....28, 73, 126, 143, 252, 348
Sung, Pei-Fang....3
Sungkorn, Radompon....349
Sunkara, Bhanukiran....149
Surapaneni, Ashok....467
Suzuki, Sho....41
Swaminathan, Shrikant....5, 474
Sweeney, Jason....203
Swihart, Mark T.....123, 150, 152, 466
Syamlal, Madhava....347
Tafti, Danesh....307
Takhistov, Paul....398
Tan, Jingyi....213, 232
Tanaka, Toshitsugu....118
Tang, Hansong....372
Tang, Yixing....70
Tatarchuk, Bruce J.....42
Teisala, Hannu....11
Temmel, Erik....92
Terry, Brandon C.....556
Themistou, Efrosyni....241
Third, James....99, 195, 209
Thompson, Michael R....285
Thorn, Jonathan O....442
Tiarks, Jordan....408
To, Daniel....38
Toghiani, Rebecca K.....206
Toivakka, Martti....11
Tokarova, Viola....297
Tokumoto, Kazumasa....41
Tomasini, Michael....399
Tomassone, M. S.....97, 399, 480, 490
Toschkoff, Gregor....100
Treffer, Daniel....283, 284
Tringe, Joseph W....39, 281
Tsai, Kuochen....166
Tsianou, Marina....241
Tsuji, Takuya....118
Tsutsumi, Atsushi....117, 146, 147
Tuncer, Alisar....387
Tuominen, Mikko....11
Umbanhowar, Paul B.....529
Valden, Mika....210
Valdesueiro, David....121
Van Essendelft, Dirk....505
Van Hee, P.....391
Van Huynh, Cuong....410
Van Norman, Staci A.....39, 281
Van Ommen, J. Ruud....121, 145
Velázquez, Carlos....64, 518, 519
Vema-Varapu, Chandra....7
Vengsarkar, Pranav S.....506
Venugopalan, T.....476
Vidyapati, V.....32
Vitasari, Caecilia R.....327
Vu, Truc....204
Wafford, Kristin....455
Wagner, Evert....145
Wake, Graeme....386
Wang Sr., Haigang....504
Wang, Aining....52
Wang, Chenchi....94, 204
Wang, Chi-Hwa....147
Wang, Dawei....107
Wang, Ding....530
Wang, Haigang....503
Wang, Jingdai....321
Wang, Li-Jun....113
Wang, Neng....65
Wang, Shuyan....274
Wang, Xianliang....466
Wang, Xifan....470
Wang, Xue Z.....18
Wang, Yingqing....149
Wang, Yiwen....44
Wassgren, Carl....5, 63, 98, 103, 104, 164, 207, 256, 350, 474, 527
Watanabe, Soshi....41
Watson, Venroy....470
Weatherley, Sharleen....285
Webber, Grant B.....225

AUTHOR INDEX

Weber, Justin ... 505
Wechner, S. ... 453
Wegner, Karsten ... 154, 155, 250
Wei, Lixin ... 274
Wei, Suying ... 71, 516
Weimer, Alan W. ... 39, 119, 275, 276, 281, 282, 294
Weir, Chelsea ... 545
Whitten, James E. ... 44
Williams, Rayon ... 553
Willis, Gabriella ... 393
Windhab, Sarah ... 284
Winkler, Jennifer ... 490
Wloch, Sebastian ... 92
Wong, Shin Yee ... 83
Woodfield, Brian F. ... 122
Wormsbecker, Mike ... 185
Worthen, Andrew ... 46
Wu, Cheng ... 113
Wu, Feng ... 280
Xiong, Qingang ... 435
Xue, Zheng ... 46
Yamada, Yoshinori ... 242
Yamamoto, Hideo ... 41
Yang, Wei ... 9
Yang, Wen-Bin ... 502
Yang, Wuqiang ... 504
Yang, Xiaogang ... 221
Yang, Yongrong ... 321
Yao, Yanfei ... 279
Yeckel, Andrew ... 543, 544
Yoon, Kiyoul ... 46
Yoshie, Yuji ... 117, 146
Yu, Aibing ... 75, 76
Yuan, Jingjie ... 213, 232
Yuk, Andrew S. ... 5
Zachariah, Michael ... 36, 257, 564
Zaid, Faraj M. ... 199, 385
Zarate, Nyah V. ... 37
Zeren, Zafer ... 520
Zhan, Jingjing ... 149
Zhang, Lei ... 9
Zhang, Nan ... 544
Zhang, Shasha ... 208, 565
Zhang, Xi ... 71
Zhang, Xingyan ... 502
Zhao, Bo ... 279
Zhao, Guorong ... 504
Zhao, Ling ... 221
Zhao, Wei ... 502
Zhi, Qiang ... 502
Zhong, Kai ... 40
Zhou, Lijin ... 321
Zhou, Yefeng ... 321
Zhou, Zongyan ... 76
Zhu, Jiahua ... 516
Zhu, Jingxu ... 411
Zhu, Wusheng ... 480
Zilner, Rayan ... 323
Zong, Yuan ... 221
Zou, Jiong ... 241

American Institute of Chemical Engineers
3 Park Avenue
New York, New York 10016-5991

CURRAN ASSOCIATES INC
proceedings.com